New Concepts in Polymer Science

Polymers Derived from Isobutylene

Synthesis, properties, application

New Concepts in Polymer Science

Also available in this book series:

Ecological Aspects of Polymer Flame Retardancy
S.M. Lomakin and G.E. Zaikov

Molecular Dynamics of Additives in Polymers
A.L. Kovarski

Structure and Properties of Conducting Polymer Composites
V.E. Gul'

Interaction of Polymers with Bioactive and Corrosive Media
A.L. Iordanskii, T.E. Rudakova and G.E. Zaikov

Immobilization on Polymers
M.I. Shtilman

Radiation Chemistry of Polymers
V.S. Ivanov

Polymeric Composites
R.B. Seymour

Reactive Oligomers
S.G. Entelis, V.V. Evreinov and A.I. Kuzaev

Diffusion of Electrolytes in Polymers
G.E. Zaikov, A.L. Iordanskii and V.S. Markin

Chemical Physics of Polymer Degradation and Stabilization
N.M Emanuel and A.L. Buchachenko

Of related interest:

Journal of Adhesion Science and Technology
Editors: K.L. Mittal and W.J. van Ooij

New Polymeric Materials
Editor-in-Chief: F.E. Karasz

Journal of Biomaterials Science, Polymer Edition
Editors: C.H. Bamford, S.L. Cooper and M. Vert

Composite Interfaces
Editor-in-Chief: H. Ishida

New Concepts in Polymer Science

Polymers Derived from Isobutylene

Synthesis, properties, application

Yu.A. Sangalov[*], *K.S. Minsker*[**] *and G.E. Zaikov*[***]

[*] Institute of Petrochemistry and Catalysis, Bashkirian Academy of Sciences, Ufa, Russia
[**] Bashkirian State University, Ufa, Russia
[***] Institute of Biochemical Physics, Russian Academy of Sciences, Moscow, Russia

///VSP///

UTRECHT • BOSTON • KÖLN • TOKYO - 2001

VSP BV
P.O. Box 346
3700 AH Zeist
The Netherlands

Tel: +31 30 692 5790
Fax: +31 30 693 2081
vsppub@compuserve.com
www.vsppub.com

First published in 2001

ISBN 90-6764-335-1

Printed in The Netherlands by Ridderprint bv, Ridderkerk.

The authors are frankly grateful to the Bayer AG
(Leverkusen, Germany)
for great help in publishing this book.[1]

[1] It is well known that the Bayer AG Company is the most important industrial producer of polyisobutylenes.

CONTENTS

Foreword

Let us begin with a joke[1]. A young bride and her betrothed are talking on the eve of the marriage.

The betrothed: — I love you, and we'll live until death in happiness.

The bride: — Yes, of course, my darling! It will be, and we'll have three children: two daughters and one son.

The betrothed: — Sure! But how do you know that we'll get exactly three children: two daughters and one son?

The bride: — Simply because they are now living with my Mom until we get married.

The authors of this monograph are similarly confident that the book will be useful for the wide range of readers. It is structured according to the "Research, Development, Production" principle. Hence we hope that the book will find its readers not only among scientists and students specialized in chemistry and physics of high-molecular compounds, but also among engineers and technologists specialized in the branch of olefins and polyolefins.

We will be pleased, if readers send to us their critical constructive notes, which will be taken into account in our future work.

Postal address for notes:

Sangalov Yu.A., Minsker K.S., Zaikov G.E.
4 Kosygin str., N.M. Emanuel Institute of Biochemical Physics, Russian Academy of Sciences, Moscow, 117334 Russia;
Fax (7-095) 137-4101;
E-mail: chembio@glasnet.ru.

[1] A Russian proverb says: "Every joke contains some element of truth" or "Every joke is only partially a joke".

Authors' preface

Some fifteen years have passed since the monograph "Isobutylene and its polymers" was published, but the problem in scientific and technological literature all over the world is still of great interest. In this period (1985-2000), many thousands of original articles and patents have been published, which testifies to the importance and vitality of this branch of macromolecular chemistry. Progress in this branch of knowledge is characterized by noticeable accumulation of quantitative experimental data and qualitatively new ideas and scientific directions.

Such a situation is not accidental. Besides the well known application of isobutylene for producing polymers with this as the basic monomer (oligo- and polyisobutylenes, butyl rubbers, isobutylene copolymers with butenes, styrene, etc.), together with alkylphenols, it participates in synthesis of other technically important products: methyl-*tert*-butyl ester, methacrylic acid, methacrylates, aliphatic diamines, a sequence of insecticides, etc. Widening fields of application indicate continuous growth of the isobutylene demand. Just in the USA, the deficit of isobutylene in 1995 was about 8 million tons.

Keeping in mind the industrial production of a rather wide range of isobutylene polymers and composites with isobutylene as starting material, we can talk about the existence of a self-contained field of low-tonnage isobutylene chemistry; in this case it should be taken into account that there are a number of traditional fields that are bound to application of high-molecular polyisobutylene and butyl rubber, which will continue in use with no alternative.

Paradoxical as it may seem, the breadth of investigations and durability of the problem of isobutylene chemistry are produced by the classic, but cation-active, character of the monomer. The monomer is characterized by a simple non-isomerizing structure. These are precisely the factors that have put isobutylene into the list of acknowledged classic objects for fundamental investigations in the field of cationic polymerization of olefin and vinyl monomers and other possible chemical transformations of olefins.

A deserving place in studies of chemistry and technology of isobutylene and its polymers is occupied by molecular engineering as a totality of methods for regulation of architecture of macromolecules (the nature of cyclic groups, chemical composition of macromolecules, length and MMD of chains, etc.) and chemical assembling as directed construction of more complicated macromolecular structures on this basis (regulation of the network, block-, grafted star-like, and other macromolecules, telehelic polyisobutylenes). Notice that the preliminarily planned approved deviation from purely classic kinetic methods of investigation by virtue of the extremely high rates of isobutylene poly- and copolymerization was marked by a new fundamental breakthrough in the field of macrokinetic description of these and other fast chemical processes and, consequently, working out of their unique design on this basis.

The field as a whole connected to different sides of chemistry and technology of isobutylene and its polymers was developed extremely fruitfully and dynamically owing to efforts made by various groups of scientists and engineers, and some enthusiasts. This was the rationale for the necessity rapidly writing a new monograph reflecting the latest achievements in this branch of knowledge. The authors also tried to combine modern practice and requirements in the monographic literature by embracing as comprehensively as possible the new data with a full list of references, while restricting the monograph to a workable length.

The scientific basis of the presentation of the material — the application of the acid–base interaction theory in consideration of the fundamental steps of isobutylene polymerization — is combined with application of quantum–chemical calculations of catalysts, linked active centers, separate elementary stages of the process, and critical analysis of some experimental facts. Besides the remaining main sections, contained in the previously published monograph, the new edition discusses a list of problems, among which are macrokinetic description of isobutylene polymerization, description of the balanced scheme of industrial production of isobutylene polymers, initiation of isobutylene polymerization with the help of immobilized cationic catalysts, and others. Of special importance are ecological aspects of synthesis and application of isobutylene polymers, etc.

The present monograph was written with the active participation of many specialists, among whom, first of all, we must mention our colleagues who are outstanding scientists of high native and international reputation, Academician N.S. Enikolopyan, a full member of the Russian Academy of Sciences, Prof. Al.Al. Berlin, and our followers Prof. S.R. Ivanova (development of the problem of thermocatalytic degradation of isobutylene polymers), Doctor of chemistry R.Z. Biglova (functionalization of isobutylene polymers), Doctor of chemistry V.A. Babkin (quantum–chemical calculations of cationic systems and polymerization processes), Doctor of chemistry Yu.Ya. Nelkenbaum (classification of data on non-traditional methods of isobutylene copolymers production and development of new composites on the basis of isobutylene polymers), and others. To them, as well as to many other specialists of the applied science and engineering and technology workers of industrial enterprises (Prof. V.N. Zaboristov, Prof. K.V. Prokofiev, Doctor of chemistry A.P. Kirillov, and others), who actively develop new scientific ideas relevant to everyday life, the authors are grateful for their help in studies, support and fruitful discussion.

But the readers always have the final say. The authors will be pleased to receive suggestions and critical comment, which will be accepted with frank appreciation.

Sangalov Yu.A., Minsker K.S., Zaikov G.E.

Part I. Scientific and technological aspects of isobutylene production

CHAPTER 1. Properties and production of isobutylene.

1.1. Physical properties.

Under normal conditions, isobutylene (or isobutene, 2-methylpropene, or β-methylpropene, or α,α-dimethylethylene; molecular mass 56.10) is a colorless gas with characteristic unpleasant smell. In 1825, Faraday was the first to isolate and describe this compound from products of high-temperature decomposition of animal grease.

In 1868, pure isobutylene was derived by Butlerov in the reaction of dehydration of *tert*-butyl alcohol with dilute sulfuric acid.

The molecular structure and main physical and chemical constants and technical characteristics of isobutylene [1–5] are given below:

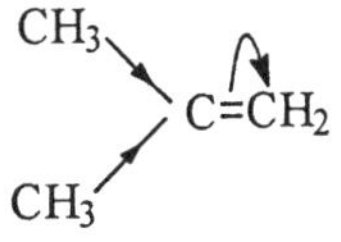

the symmetry group C_{2v}; $r_{C=C}$ = 0.1342 ± 0.0003 nm; $r_{C—C}$ = 0.1508 ± 0.0002 nm; $r_{C—H\ (methyl)}$ = 0.1119 ± 0.0007 nm; $r_{C—H\ (methylene)}$ = 0.1095 ± 0.0020 nm; ∠ C—C=C = 122.2 ± 0.2°; ∠ C—C—C = 116°; ∠ H—C—C = 107.9 ± 0.8°; ∠ C=C—H = 121.3 ± 1.5°; ∠ H—C—H = 106.5°.

Critical constants are: temperature — 417.89 K; pressure — 4.0003 MPa; density — 235 kg/m^3; volume 239 cm^3/mol.

The temperatures, K, (under atmospheric pressure), are: boiling point — 266.3; melting point — 132.8; ignition — 738.2; self-ignition — 313.8.

Refraction index is $N_D^{298.2}$ = 1.3811.

Molar heat capacity under constant pressure is 89.2 J/(mol·K).

Specific heat capacity at 298.2 K is 1.487 kJ/(kg·K).

Density at 298 K is 594.2 kg/m^3.

Heat conductivities, W/(m·K), is 45 at 132.8 K (liquid), 31 at 266.3 K (liquid), 2.8 at 273.2 K (vapor), 6.6 at 373.2 K (vapor).

Characteristic heats, kJ/mol (at 298.2 K), are: formation 16.9, combustion 2702.3, polymerization 53.6, melting (at 132.8 K) 5.9, evaporation: 27.6 at 132.8 K, 22.5 at 266.3 K, 20.4 at 298.2.

Entropy in the standard state at 298.2 K: 293.6 J/(mol·K).

Dielectric permeabilities: 2.208 at 273.2 K, 2.487 at 203.2 K.

Dipole moment (gas state): 0.5.

Viscosities, mPa·s: –2.8 at 132.8 K, –0.2 at 266.3 K.

Temperature dependencies of some physical constants of isobutylene are shown in Table 1.1.

Combustion heat is 45.2 mJ/kg; standard $T_{self\text{-}ignition}$ is 738 K; concentration limits of ignitability of isobutylene mixed with air are 1.8–9.6% (vol.); the maximum combustion rate is 0.375 M/s; the minimum explosive content of oxygen at diluting isobutylene-air mixtures by carbon dioxide is 15.3% (vol.), by nitrogen 12.3% (vol.).

Isobutylene is soluble in ethyl alcohol, ether and hydrocarbons. Solubility in water in T = 283–313 K is 0.17–0.25 l/l (extrapolated to normal conditions). It forms azeotropic mixtures with some compounds, in particular, with waterless ammonia [45% (mass) of NH_3, $T_{boiling}$ = 234.7 K], sulfur dioxide [59% (mass) of SO_2, $T_{boiling}$ = 259.2 K], methylnitrite ($T_{boiling}$ = 257.2 K), ethylene oxide, methyl formiate, etc. It is low-stable in water.

Table 1.1

Temperature dependencies of some physical constants of isobutylene.

Temperature, K	Specific heat capacity, kJ/(kg·K)	Surface tension, MN/m	Temperature, K	Vapor pressure at boiling temperature, kPa	Density, kg/m^3
93.3	1.0459 (sol)		168.1	0.133	
98.8	1.0953 (sol)		176.7	0.667	
105.5	1.1484 (sol)		191.3	1.333	
112.3	1.2133 (sol)		199.8	2.666	
118.9	1.2795 (sol)		209.4	5.333	
139.2	1.9037 (liq)	32.4	215.5	7.999	677.5
145.2	1.9109 (liq)	31.6	224.3	10.101	673.8
152.0	1.9171 (liq)	30.7	226.3	12.211	671.0
159.1	1.9259 (liq)	29.7	228.9	13.33	669.2
166.1	1.9347 (liq)	29.2	229.1	18.35	668.0
172.9	1.9469 (liq)	29.05	232.8	22.42	663.2
179.8	1.9598 (liq)	27.2	236.5	26.665	660.5
186.5	1.9741 (liq)	26.3	238.4	30.21	658.1
210.2	2.0348 (liq)	23.05	243.5	48.12	652.0
216.2	2.0499 (liq)	21.6	249.7	52.41	644.4
227.9	2.0808 (liq)	20.9	251.0	53.33	641.5
239.0	2.1089 (liq)	19.5	254.1	77.14	640.8
242.7	2.1307 (liq)	18.75	263.2	82.32	630.6
248.0	2.1491 (liq)	18.4	266.3	101.325	626.5
253.0	2.1658 (liq)	17.4	269.5	104.12	623.4
263.2		16.1	273.2	131.8	618.9
273.2		14.9	275.5	152.3	617.9
283.2		13.8	280.5	220.1	609.5
293.2		12.6	283.2	285.5	607.1
303.2		11.4	303.2	349.0	
313.2		10.3	298.2	2878	
323.2		9.2			

Inhalation of isobutylene causes a narcotic effect [2]. It is irritating, as well. The intensity of the toxicological effect depends on the monomer concentration and individual characteristics of the organism. The symptoms of poisoning are: indigestion, headache and nausea. Vision disorder and loss of consciousness are observed on later stages. A durable stay in a compartment full of gas may cause death [1, p. 56].

Storing and working with isobutylene require observance of the accident prevention rules, general for all combustible and explosive substances; the flash temperature is 193–200 K.

The maximum permissible concentration MPC = 100 mg/m^3, the emergency class is 4. In air MPC = 0.1 mg/m^3. Up to 10 mg/m^3 it causes no adverse influence on the sanitary regime of water basins. Threshold concentration in water by smell is 0.5 mg/l. The unpleasant smell is already detectable at 0.059–0.11 mg/l.

1.2. Chemical properties.

The presence of the π-bond and two electron-donor CH_3-groups provides high reactivity of the isobutylene molecule. Isobutylene is able to participate in reactions of addition, including telo- and polymerization, and cyclization, as well.

Interacting with bromine it forms saturated products — dibromide and polybromine derivatives of isobutane:

$$(CH_3)_2C{=}CH_2 \xrightarrow{+Br_2} (CH_3)_2CBr{-}C(Br)H_2 \xrightarrow[-HBr]{+Br_2}$$

$$\rightarrow BrCH_2{-}C(CH_3)Br{-}C(Br)H_2$$

In reactions with chlorine it preferentially forms unsaturated products (1-chloro-2-methylpropene-2, 1-chloro-2-methylpropene-1, 1,3-dichloro-2-methylpropene-2):

$$(CH_3)_2C{=}CH_2 \xrightarrow{Cl_2} (CH_3)_2CCl{-}C(Cl)H_2 \xrightarrow[-HCl]{}$$

$$\rightarrow CH_2{=}C(CH_3){-}CH_2Cl \xrightarrow[-HCl]{Cl_2} ClCH_2{-}C(CH_3){=}C(H)Cl$$

$$\rightarrow (CH_3)_2C{=}C(Cl)H$$

This reaction forms the basis of the industrial synthesis of methallylchloride and isobutyric aldehyde:

$$(CH_3)_2CCl{-}CH_2Cl \xrightarrow[-2HCl]{2H_2O} (CH_3)_2COH{-}CH_2OH \xrightarrow[-H_2O]{} (CH_3)_2CH{-}CHO$$

Interaction of isobutylene with hydrogen halides proceeds fast and quantitatively already at low temperatures, especially that with an anhydrous hydrogen halide in the absence of a solvent:

$$(CH_3)_2C{=}CH_2 \xrightarrow{HCl} (CH_3)_3CCl$$

This reaction is of interest for the synthesis of isobutylchloride as a raw methacrylonitrile material to product methacrylonitrile. It is also used to analyze isobutylene.

In the presence of initiators (peroxides, O_2, UV-radiation) HBr forms up to 80% of isobutyl bromide (the Karash reaction):

$$(CH_3)_2C{=}CH_2 + HOCl \rightarrow (CH_3)_2C(OH){-}CH_2Cl.$$

The reaction with hypochlorous acid does not proceed by the Markovnikov rule:

$$(CH_3)_2C = CH_2 + HOCl \rightarrow (CH_3)_2C(OH) - CH_2Cl$$

The mixture of the silver salt of carbonic acid and a halogen behaves in water or acetic acid solution as an acyl bromate or acyl iodate:

$$(NO_2)_2C_6H_3C(=O)OAg + I_2 \longrightarrow (NO_2)_2C_6H_3C(=O)OI + AgI$$

$$(CH_3)_2C{=}CH_2 + 3{,}5\text{-}(NO_2)_2C_6H_3C(=O)OI \longrightarrow (CH_3)_2CICH_2{-}OC(=O)C_6H_3(NO_2)_2\text{-}3{,}5$$

The reaction is used to identify isobutylene.

Along with a dinitro compound, at isobutylene interaction with nitrogen dioxide, nitro-*tert*-butylnitrite is formed, a small part of which oxidizes to a corresponding nitrate:

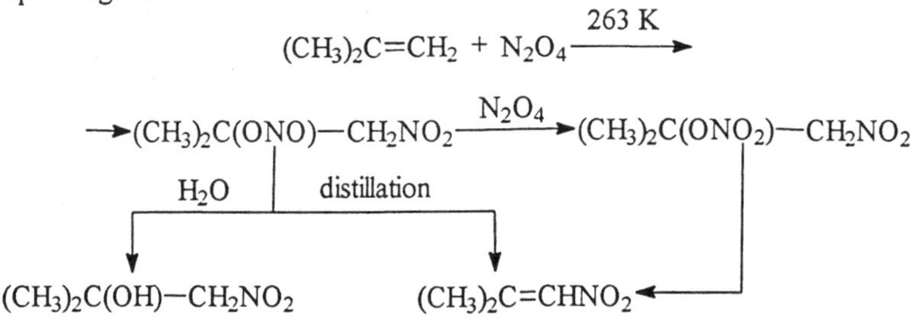

For industrial production of methylmethacrylate, a reaction with participation of isobutylene and N_2O_4 has been developed. In the liquid phase at 273–278 K, an olefin is oxidized to oxyisobutyric acid with the 85% yield (by IB), which in turn is dehydrated and etherified by methanol:

$$(CH_3)_2C{=}CH_2 \xrightarrow{N_2O_4} (CH_3)_2C(ONO_2){-}CH_2NO \longrightarrow (CH_3)_2C(OH){-}COOH$$

Isobutylene reacts easily with nitrogen oxide forming nitrosites:

$$(CH_3)_2C{=}CH_2 + ON{-}ONO \longrightarrow (CH_3)_2C(NO){-}CH_2NO_2$$

At the interaction of isobutylene with nitrosyl chloride in an ether solution, nitroso-*tert*-butyl chloride or α-chlorisobutyraldehyde oxime is formed close up:

$$(CH_3)_2C{=}CH_2 + ONCl \rightarrow (CH_3)_2C(Cl){-}CH_2NO \rightarrow (CH_3)_2C(Cl)CH{=}NOH$$

In the reaction with nitrogen dichloride, chloro-*tert*-butylamine hydrochloride is formed:

$$(CH_3)_2C{=}CH_2 + NCl_2 \xrightarrow[3HCl]{223\ K} (CH_3)_2C(NH_3Cl){-}CH_2Cl + 2Cl_2$$

Isobutylene can be directly nitrated by concentrated HNO_3:

$$(CH_3)_2C{=}CH_2 + HONO_2 \longrightarrow (CH_3)_2C{=}C(H)NO_2 + H_2O$$

The reaction of addition of hydrogen sulfide in the presence of alkaline and some acidic catalysts is used to separate isobutylene from butene-2:

$$(CH_3)_2C{=}CH_2 + H_2S \xrightarrow[H_3PO_4]{340\text{-}355\ K} (CH_3)_3CSH$$

Of great importance is its reaction with sulfuric acid:

$$(CH_3)_2C{=}CH_2 + H_2SO_4 \longrightarrow (CH_3)_3COSO_2OH$$

Tert-butyl alcohol is obtained by hydration of *tert*-butylsulfuric acid.

The reaction with multichlorous sulfur is more complex. At 293 K, the preferred direction is its two-stage addition by the Markovnikov rule via an intermediate episulfonium ion with formation of bis(chlorisobutyl) disulfide [6]:

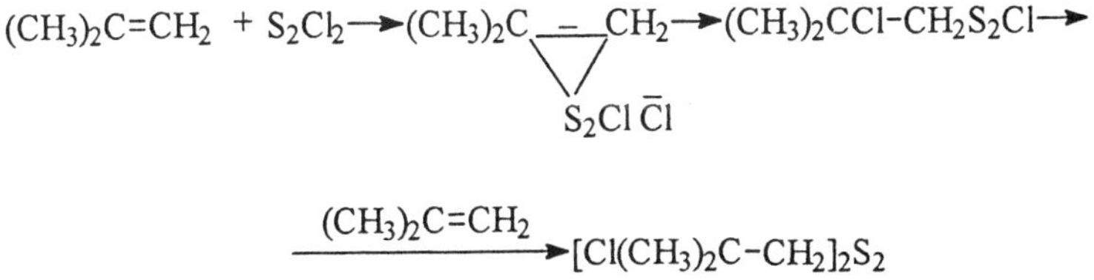

At room temperature in the ether medium, isobutylene reacts with aluminum or lithium-aluminum hydride:

$$(CH_3)_2C{=}CH_2 + AlH_3 \longrightarrow (CH_3)_2CH{-}CH_2AlH_2 \longrightarrow$$

$$\longrightarrow [(CH_3)_2CH{-}CH_2]_2AlH \longrightarrow [(CH_3)_2CH{-}CH_2]_3Al$$

Hydrosilylation and other similar reactions proceed similar to that of hydroalumination.

Of extreme importance is the reaction of direct production of aluminum organic compounds from isobutylene, aluminum and hydrogen:

$$3(CH_3)_2C{=}CH_2 + Al + 3/2H_2 \rightarrow \{(CH_3)_2CH{-}CH_2]_3Al$$

Salts of some metals (copper, silver, mercury, etc.) can rather easily bind to the olefin double bond:

$$(CH_3)_2C{=}CH_2 + HgBr_2 + H_2O \rightarrow (CH_3)_2C(OH){-}CH_2HgBr + HBr$$

In the presence of the Friedel-Crafts catalysts isobutylene easily alkylates aromatic hydrocarbons:

$$C_6H_6 + (CH_3)_2C{=}CH_2 \xrightarrow{[AlCl_3]} C_6H_5C(CH_3)_3$$

Toluene and μ-xylene are alkylated in the same manner, the *tert*-butyric radical being in the *meta*- and *para*-positions to alkyl.

In the presence of initiators or under high temperature and pressure (775 K, 29.5 MPa) and in the absence of a catalyst, isobutylene alkylates paraffin waxes and cycloparaffins. A similar reaction proceeds at cationic alkylation in the presence of acidic catalysts (Al_2O_3, etc.). For example,

$$(CH_3)_2C{=}CH_2 + (CH_3)_3CH \rightarrow \begin{cases} (CH_3)_3{-}CH_2CH(CH_3)CH_3 \\ (CH_3)_2CH{-}CH(CH_3){-}CH(CH_3)CH_3 \end{cases}$$

An interesting modification of reactions of this type is that of alkylation of isobutylene by methanol [7]. At 623–673 K, the reaction of dehydration (dehydrogenation) catalyzed by an aluminum-molybdenum-cobalt catalyst proceeds through an intermediate formation of dimethyl ester and is accompanied by the olefin chain propagation:

$$(CH_3)_2C{=}CH_2 + CH_3OH \rightarrow (CH_3)_2C{=}CH_2 + CH_3OCH_3 \rightarrow$$

$$\rightarrow \begin{cases} \text{iso-}C_5H_{10} + CH_3OH \\ \xrightarrow[-H_2]{} \text{iso-}C_5H_8 + CH_3OH \end{cases}$$

Of importance for properties of final products and obtaining isobutylene of a high purity are its reactions with water and alcohols:

$$(CH_3)_2C{=}CH_2 + H_2O \xrightarrow{H} (CH_3)_3COH$$

$$(CH_3)_2C{=}CH_2 + CH_3OH \xrightarrow{H^+} (CH_3)_3COCH$$

$$(CH_3)_2C{=}CH_2 + C_6H_5OH \xrightarrow[[AlCl_3]]{[HCl]} n\text{-}(CH_3)_3CC_6H_4OH$$

Along with mineral acids, heterogeneous systems, for example, sulfonated copolymers of styrene with divinylbenzene in the H-form or heteropolyacids

($H_3PW_{12}O_{40}$, $H_4SiW_{12}O_{40}$) at 298 K, are efficient at hydration of isobutylene and its reaction with aliphatic alcohols [8, 9]. The use of specific catalysts — polyphenoxyalumoxanes — provides a high selectivity of *ortho*-alkylation of phenol by isobutylene resulting in production of both 2-alkyl- and 2,6-dialkylphenols [10].

Under specific conditions, isobutylene reacts with NH_3 forming *tert*-butylamine:

$$(CH_3)_2C{=}CH_2 + NH_3 \xrightarrow[\text{[Na]}]{\text{520-420 K, 84-94 MPa}} (CH_3)_3CNH_2$$

More effective catalysts of this reaction — alumosilicates and zeolites — have also been proposed [11].

The reaction of oxidative ammonolysis of isobutylene (gas phase, 300–600 K, diluter — water vapor or nitrogen, catalysts — compounds of antimony, lead or molybdenum) provides the basis of a highly efficient method of the production of methacrylonitrile and, at further solvolysis, of methylmethacrylate [12, 13].

Hydrogen cyanide is added to isobutylene forming pivalic acid nitrile:

$$(CH_3)_2C{=}CH_2 + HCN \xrightarrow[\text{[}Al_2O_3\text{]}]{\text{670-725 K}} (CH_3)_3CCN$$

The reaction with acetonitrile in acetic acid medium in the presence of concentrated H_2SO_4 proceeds in a different way:

$$(CH_3)_2C{=}CH_2 + CH_3CN \xrightarrow{H^+,\,HO^-} (CH_3)_3C{-}N{=}\underset{\displaystyle OH}{\underset{|}{C}}CH_3 \rightleftharpoons$$

$$\rightleftharpoons (CH_3)_3CNH{-}\underset{\displaystyle O}{\underset{||}{C}}CH_3$$

Of interest is the reaction of isobutylene with acrylic acid nitrile (548 K, 17 MPa) with formation of C_8-dinitrile, catalytic reduction of which (Ru/Al_2O_3, 393–423 K, 11.3 MPa) results in a high yield of methylnonane diamine — a raw material to produce polyamide fiber [13].

Formaldehyde is added at the double bond of isobutylene forming 4,4-dimethyldioxane-1,3 (the Prince reaction):

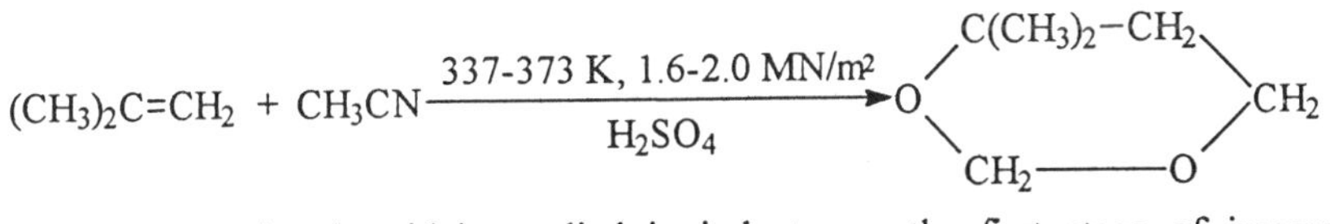

The reaction is widely applied in industry as the first stage of isoprene synthesis. In the second stage, dimethyldioxane decomposes to isoprene, CH_2O and H_2O by heating (633–653 K) in the presence of water vapor and a catalyst (acidic calcium phosphate).

The yield of isoprene can be increased in a single-stage catalytic (V_2O_5 – P_2O_5) reaction of isobutylene with CH_2O in N_2 flow at 513–533 K [14].

Another interesting method for producing isoprene is in the formation of an intermediate non-saturated alcohol:

$$(CH_3)_2C{=}CH_2 + CH_2O \xrightarrow[Al_2O_3,\ \text{water vapor}]{625\ K} CH_2{=}C(CH_3)CH_2CH_2OH \longrightarrow$$

$$\longrightarrow CH_2{=}C(CH_3)CH{=}CH_2 + H_2O$$

Water solutions of formaldehyde saturated by HCl react as a chloromethyl alcohol:

$$(CH_3)_2C{=}CH_2 + CH_2O + HCl \xrightarrow{[CuCl_2]} (CH_3)_2C(Cl){-}CH_2{-}CH_2OH$$

Acetals are added in a similar manner:

$$(CH_3)_2C{=}CH_2 + CH_3OCH_2OCH_3 \xrightarrow[{[BF_3]}]{320\ K,\ \text{pressure}} (CH_3)_2C(OCH_3){-}CH_2{-}CH_2OCH_3$$

Reactions between olefines and carbon oxide — the reaction of carbonylation proceeding in the presence of Fe, Co and Ni carbonyls — are well known in industry. The reaction with isobutylene gives isovalerianic and pivalic (trimethylacetic) acids (the Reppe synthesis):

$$(CH_3)_2C{=}CH_2 + CO + H_2O \longrightarrow \begin{cases} (CH_3)_2C(COOH)CH_3 \\ (CH_3)_2CH{-}CH_2COOH \end{cases}$$

A modification of the above-mentioned reaction is that of oxidative carbonylation of isobutylene, catalyzed by the $PdCl_2$-$CuCl_2$ (HCl, O_2) system under gentle temperature conditions [15]. The scheme of the main reaction proceeding through an intermediate formation of a π-allyl complex is:

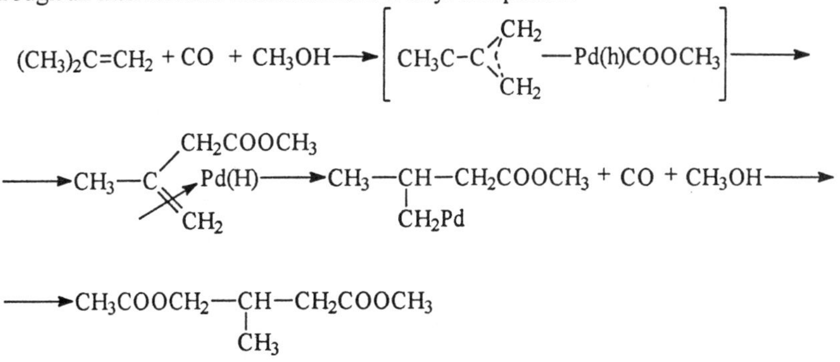

Besides the dimethyl ester of 3-methylglutaric acid, a methyl ester of 3-butenic acid is formed.

In an excess of menthol as the alcohol component, a single-stage synthesis is realized of a valuable drug — validol — the mixture of the initial alcohol and the menthol ester of isovaleric acid [16]:

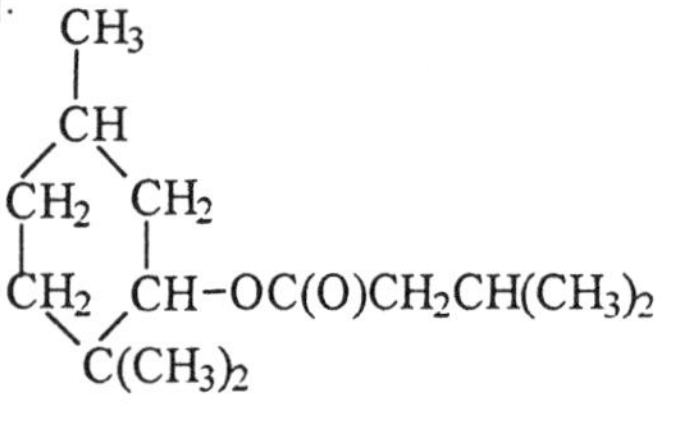

Isobutylene can add H_2 at the double bond:

$$(CH_3)_2C{=}CH_2 + H_2 \xrightarrow[{[V_2O_5/Al_2O_3]}]{675\ K} (CH_3)_2CH{-}CH_3$$

and is easily oxidized; in this case, depending on the reaction conditions and the characteristics of reagents, different products can be obtained. In the presence of $KMnO_4$ in a neutral or alkaline water solution, or OsO_4 with alkaline chlorate, diol-1,2 is formed in the reaction of syn-hydroxylation:

$$(CH_3)_2C{=}CH_2 + 2[HO\bullet] \rightarrow (CH_3)_2C(OH){-}CH_2(OH)$$

A modern version of the reaction (the catalyst causes no corrosion of the equipment) includes the treatment of isobutylene by H_2O and O_2 on an immobilized catalyst — the compounds of Au (oxide, hydroxide, salts) combined with $TeCl_4$ [17].

The reaction is used for analytic detection of the C=C double bond.

Under more rigorous conditions, in acidic solution or on heating, the molecule breaks at the double bond ($KMnO_4$). A similar reaction is observed on H_2O_2 oxidation in an ether solution in the presence of OsO_4:

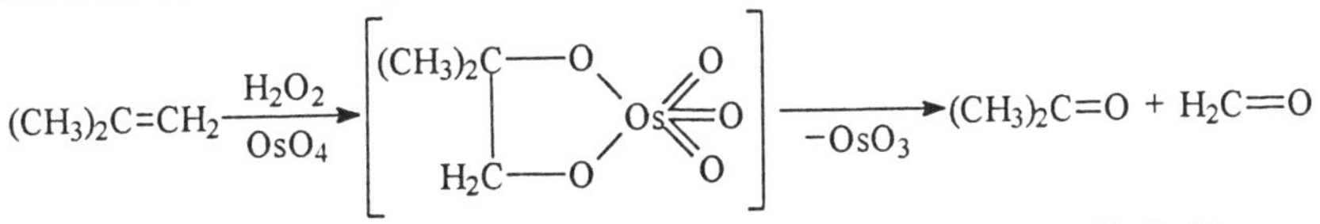

To oxidize the alkene double bond, chromic acid is widely applied either as a solution of CrO_3 in concentrated acetic acid in the presence or the absence of H_2SO_4 or as a water solution of sodium or potassium bichromate with H_2SO_4. The position and structure of the C=C-bond in the molecule are detected with the help of oxidative decomposition. The formation of a ketone indicates the presence of $(CH_3)_2C<$ groups, while formation of carbonic acid implies the presence of $CH_2<$ or $RCH<$ groups:

$$(CH_3)_2C{=}CH_2 + 3(O) \rightarrow (CH_3)_2C{=}O + HCOO$$

Oxidation by chromium trioxide in acetic anhydride solution produces isobutylene oxide. Good results are obtained on oxidation by organic per-acids (peracetic, isopropylbenzene hydroperoxide) by the Prilezhaev reaction:

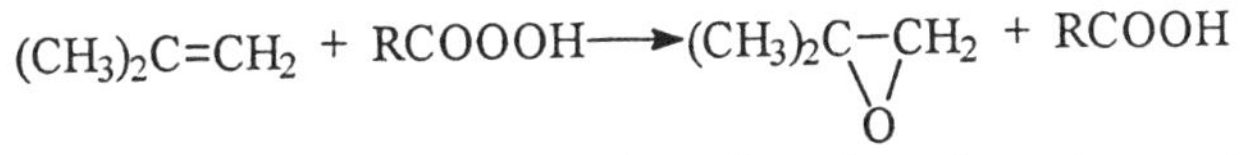

Reduction of α-oxide on a copper catalyst gives the values close to those of isobutyric alcohol yield [18].

Ozone is a characteristic reagent to detect the C=C-bond (the Harris reaction). The reaction appears more selective than in the case of other oxidants. That is why it is used to detect the structure and determine the number of C=C-bonds:

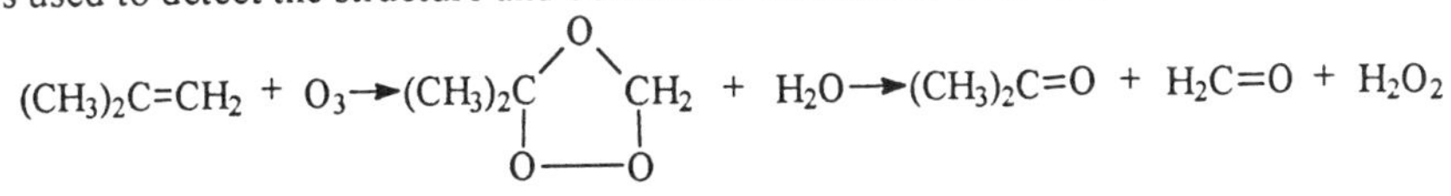

For technological applications, of special interest is the reaction of selective catalytic oxidation of isobutylene to methacrolein (Bi, Co and Mo oxides; Cu and W complexes immobilized on silica gel, 630 K) and then to an acid and esters [12, 13,

19]. According to this scheme, a new process of producing methacrylate esters was developed, most efficient in the case of preliminary hydration of isobutylene to *tert*-butyl alcohol.

A series of 1:1 and 2:1 reactions of cycloaddition are known for isobutylene: the nature of the second component determines the nature of the product and the reaction conditions.

Isobutylene easily participates in the diene synthesis by Diels–Alder, in particular,

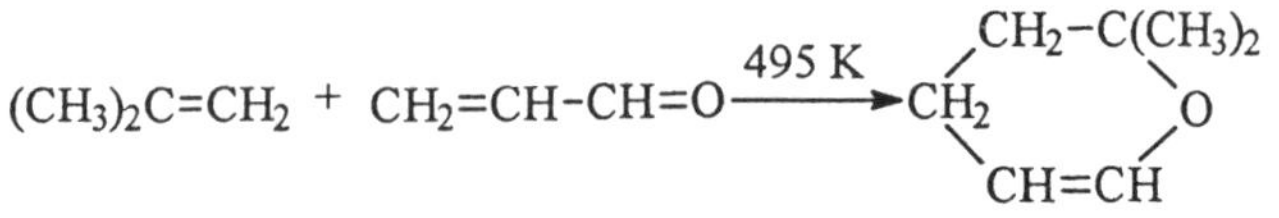

Under increased temperature and pressure, isobutylene reacts with tetrafluoroethylene forming 1,1,2,2-tetrafluor-3,3-dimethylcyclobutane:

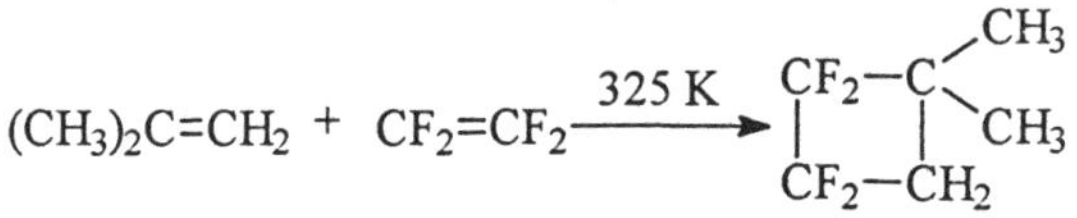

Similar to most olefins, isobutylene participates in the reaction of cyclopropanation. Electrophilic interaction of carbenes with isobutylene produces substituted cyclopropanes:

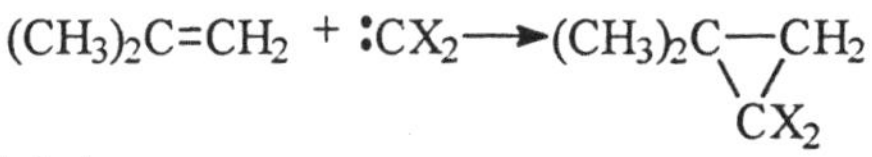

where X = H, alkyl, aryl, halogen.

When activated by the complex formation with dioxane, SO_3 interacts with isobutylene by the mechanism of dimerization combined with sulfonation [20]:

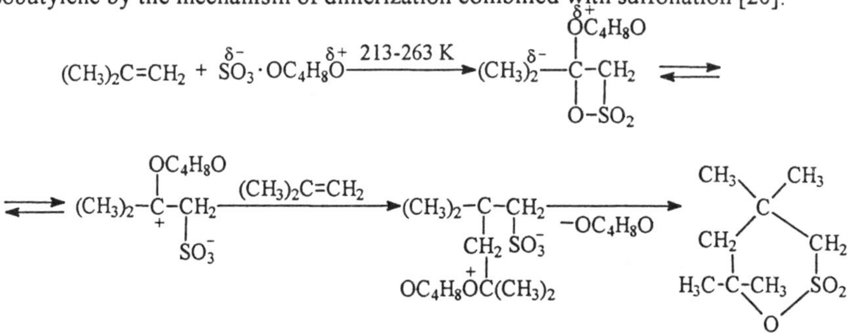

Contrary to the above examples, thermal dimerization of isobutylene (675 K, 3.63 MPa) is accompanied by isomerizational transformations:

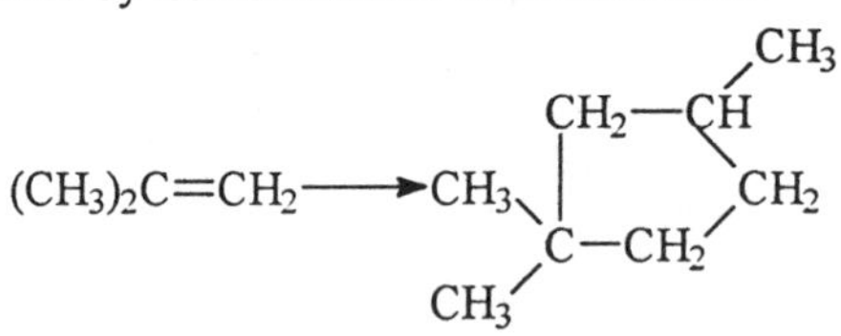

High-temperature (1173–1273 K) transformations of isobutylene are accompanied by the breaking of C—C-bonds [21]:

$$2\,(CH_3)_2C{=}CH_2 \rightarrow CH_3{-}C{\equiv}CH + CH_2{=}C{=}CH_2 + 2CH_4$$

Thermal catalytic cyclization of isobutylene with formation of aromatic compounds is, apparently, coupled with such transformations [22]:

$$3\,(CH_3)_2C{=}CH_2 \xrightarrow[AlH_3/Al_2O_3]{723\text{-}923\ K} C_6H_3(CH_3)_3 + 3CH_4$$

Telomerization should be attributed to the reactions of isobutylene, interesting for synthesis of various functional compounds. Usually, the activity of telogens with C—Hal and C—H bonds in radical telomerization is low. Reactions proceed under rather rigorous conditions (benzene or *tert*-butyl peroxides, 373–473 K, azobis-isobutyronitrile combined with UV radiation) [23]. Some examples of reactions with halogen containing telogens are shown below:

$$(CH_3)_2C = CH_2 + CCl_4 \rightarrow Cl_3C[CH_2 - C(CH_3)_2]_nCl \qquad n=1,2,3$$
$$+ BrCl_3 \rightarrow CCl_3[CH_2 - C(CH_3)_2]_nBr \qquad n=1,2$$
$$+ CHCl_3 \rightarrow \text{low yield}$$
$$+ (CCl_3)_2SO_2 \rightarrow Cl_3CSO_2CCl_2[CH_2 - C(CH_3)_2]_nCl + \qquad n=1,2$$
$$+ SO_2[CCl_2CH_2 - C(CH_3)_2]_2Cl$$

Catalytic telomerization of isobutylene by CCl_4 ($FeCl_2$, benzoin, $(C_2H_5)_2NH{\cdot}HCl$ in *iso*-C_3H_7OH) is a multistage process:

$$(CH_3)_2C = CH_2 \xrightarrow{CCl_4} (CH_3)_2CClCH_2CCl_3 \xrightarrow{\text{catalyst}} (CH_3)_2CClCH_2CCl_2 \rightarrow$$
$$\xrightarrow{(CH_3)_2C=CH_2} [(CH_3)_2CClCH_2]_2CCl_2$$

Despite the radical initiation, the formation of just an adduct and its participation in the reaction as a telogen are explained by the process proceeding in coordinational sphere of a complex catalyst [24].

Parallel (consecutive) proceeding of the multistage telomerization, along with the usual (regular) one, is especially characteristic for telogens with active C—H-bonds. It is accompanied by regrouping and migration of hydrogen, including its transmission to isobutylene with formation of isobutyric radicals, which suffer cross-recombination with radicals of the regular telomerization.

$$3\,(CH_3)_2C{=}CH_2 + ClCH_2COOCH_3 \rightarrow$$
$$\rightarrow H[(CH_3)_2C{-}CH_2]_nCHClCOOCH_3 \qquad n = 1, 2$$
$$\rightarrow H[(CH_3)_2C{-}CH_2]_nCClCOOCH_3 \ (\text{with } CH_2C(CH_3){=}CH_2 \text{ on } C) \qquad n = 2$$
$$\rightarrow CH_2{=}C(CH_3)CH_2[(CH_3)_2C{-}CH_2]_nCHClCOOCH_3 \qquad n = 0, 1$$

$$3\,(CH_3)_2C{=}CH_2 + (CH_3)_2CHCOOCH_3 \rightarrow$$
$$\rightarrow H[(CH_3)_2C{-}CH_2]_nC(CH_3)_2COOCH_3 \qquad n = 1 - 3$$
$$\rightarrow CH_2{=}C(CH_3)CH_2[(CH_3)_2C{-}CH_2]_nC(CH_3)_2COOCH_3 \qquad n = 0, 1, 2$$

At telomerization of isobutylene with acetic aldehyde (radiation from Co^{60}, 288 K) the products of the addition of both radicals —$(CH_3)_2C$—$CHCOCH_3$, $CH_3COCH_2C(CH_3)_2CH_2CH(CH_3)_2$— and ions —$(CH_3)_3C$—$COCH_3$— are formed.

Ionic (cationic) telomerization catalyzed by Lewis acids ($ZnCl_2$, $ZnBr_2$, $SnCl_4$, $TiCl_4$, $HgCl_2$) at 273–298 K proceeds easier than the radical one [25]. Mobile halogens of used telogens activated by Lewis acids — fluor-(*tert*-)haloidalkyls, chlorolefins, simple chloroethers, chloromethylic ethers of carbonic acids, chloroanhydrides — bind to the α-carbon atom of isobutylene. Typical characteristics of cationic processes are observed — stabilization of intermediates for allyl haloidolefins and, in some cases, dihydrochlorination of the reaction products. The most typical examples of isobutylene ionic telomerization with haloid-containing telogens are:

$$(CH_3)_2C=CH_2 + (CH_3)_3CCl \rightarrow (CH_3)_3CCH_2CCl(CH_3)_2 \xrightarrow[-HCl]{} \rightarrow (CH_3)_3CCH=C(CH_3)_2$$

$$+ C_6H_5CHBrCH_3 \rightarrow C_6H_5CH(CH_3)CH=C(CH_3)_2$$

$$(CH_3)_2C=CH_2 + CH_3\text{-}\underset{\displaystyle CH_3}{\underset{|}{C}}=CHCH_2Cl \rightarrow CH_3\text{-}\underset{\displaystyle CH_3}{\underset{|}{C}}=CHCH_2CH_2CCl(CH_3)_2$$

$$+ CH_3\text{-}\underset{\displaystyle CH_3}{\underset{|}{C}}Cl\text{-}CH=CH_2 \rightarrow CH_2=CH\text{-}C(CH_3)_2CH_2CCl(CH_3)_2$$

(low yields due to allyl stabilization of the intermediate state)

$$+ CH_3\text{-}CHCl\text{-}CH=CH_2 \rightarrow CH_3CH=CH\text{-}CH_2CH_2CCl(CH_3)_2$$

$$(CH_3)_2C=CH_2 + CH_3OCH_2Cl \rightarrow CH_3OCH_2CH_2CCl(CH_3)_2$$

$$+ ClCH_2OCH_2Cl \rightarrow [(CH_3)_2CClCH_2CH_2]_2O$$

$$+ ClCH_2CH_2OCH_2Cl \rightarrow ClCH_2CH_2OCH_2CH_2CCl(CH_3)_2$$

$$(CH_3)_2C=CH_2 + CH_3COOCH_2Cl \rightarrow CH_3COOCH_2CH_2CCl(CH_3)_2$$

(in the case of radical telomerization the C—H-bond of the telogen is broken)

$$(CH_3)_2C=CH_2 + CH_3C(O)Cl \rightarrow (CH_3)_2CCl\text{-}CH_2C(O)CH_3 \rightarrow \xrightarrow[-HCl]{} (CH_3)_2C=CHC(O)CH_3$$

A series of telehelic polyisobutylenes with the end $\sim CH_2C(CH_3)_2Cl$ group, including star-like three-ray polymers, was obtained by the reaction of the "living" carbocationic telomerization of isobutylene with BCl_3 in the presence of sterically hindered two-functional co-catalysts (1,3-di(2-methoxy-2-propyl)-5-*tert*-butyl-benzene, *trans*-2,5-diacetoxy-2,5-dimethyl-3-hexene, 1,3,5-*tris*-(2-methoxypropyl)-benzene, *tert*-diacetates, 2,4,4-trimethyl-2-methoxypentane, cumylmethylic ester, dichlorocumyl, etc.) at 200–265 K in chlorine-containing solvents (methylene chloride, dichloroethane, chloroform, etc.) [26–31]. The conversion yield reaches 98–100 mass%. The functionality of synthesized products equals 2 at normal MMD. It was reported in [32] on the production of a new asymmetric telehelic polyisobutylene

$CH_3RC(CH_3)_2CH_2C(CH_3)_2$—$CH_2C(CH_3)_2Cl$, where R is ~$CH$=$CHCH_2CHCH_2$—$CH_2$~. In this case, the polymerization is not "vital" and *in situ* 2-pinanyl chloride is formed as a side product.

Besides the functionality of telomer homologues, determined by nature of the telogen, an additional functionalization is realized due to chemical transformations. For example, the synthesis of ω-haloidalkylcarbonic acids (hydrolysis of CCl_3-groups) and haloid(polyhaloid)olefins of complex structures (haloid hydrogen separation) is easily performed.

Taking into consideration genetic interconnection of the processes of cationic telomerization and polymerization (the former represents the limiting case of the latter), a high activity of isobutylene in electrophilic polymerization without isomeric transformations is evident. It is fundamental for studying the regularities of polymerization and forms the basis of directed synthesis of many polymers. Polymerization of isobutylene is a reaction most important for practical applications and theoretical studies. Besides, due to the presence of end (polyisobutylene) and internal (isobutylene copolymers) double bonds, many of the considered reactions of isobutylene may be realized in a polymer and may be used to modify its properties.

1.3. Methods of isobutylene production.

The processes of isobutylene production can be divided into two groups: those with the maximum yield of isobutylene or C_4 hydrocarbon fraction and those in which they are side products.

The first group includes: 1) high-temperature pyrolysis of liquid or gas raw materials — dry gases of oil processing, casing-head gases of crude oil mining, ethane, propane, butanes, directly distilled and gas gasolines, in some cases raffinates of catalytic reforming (after extraction of aromatic hydrocarbons), gas oil fractions, oil and oil residues; 2) thermal cracking of high-molecular paraffins obtained in oil processing; 3) selective depolymerization of polyolefins (polypropylene, oligo- and polybutylenes, etc.); 4) catalytic dehydrogenation of paraffins (butanes, pentanes, normal paraffins); 5) dehydration of *tert*-butyl alcohol [33].

The second group includes contact catalytic reactions in oil processing in production of fuels accompanied by significant amounts of C_3 and higher olefins.

Pyrolysis and cracking of various oil fractions, directly distilled gasolines, as well as gas hydrocarbons are most important for isobutylene production. The average concentration of isobutylene in the pyrolysis products (18–32% calculated for C_4-hydrocarbons) is higher than in catalytic cracking (10–20%). That is why the quota of pyrolysis in industry is constantly rising (Table 1.2) [12].

Table 1.2.

Structure of isobutylene production from various raw materials (USA), thousand tons (%).

Raw materials	1973	1978	1985	1990
Catalytic cracking	4071 (90)	4039 (87.2)	4400 (80.2)	3357 (71.2)
Pyrolysis	272 (6)	435 (8.8)	726 (13.2)	971 (20.6)
Other sources	177 (4)	195 (4.0)	363 (6.6)	385 (8.2)

Besides isobutylene, on decomposition of hydrocarbon raw materials, hydrogen, methane and various olefins are formed. The processes are accompanied by side reactions of degradation of formed olefins, hydrogenation and dehydrogenation of olefins to paraffins, dienes, acetylene and their derivatives, condensation with formation of high-molecular hydrocarbons, arenes, cyclodienes, etc.

Normal and isomeric paraffins can be used as raw materials to produce C_4 hydrocarbons. Usually, the total yield of butane-butylenes grows with increase of the number of carbon atoms in molecules, as well as at introduction of overheated vapor (Table 1.3). The yield of C_4 hydrocarbons also depends on the pyrolysis regime (Table 1.4) [34].

Table 1.3.

Yields of C_4 hydrocarbons in pyrolysis of various raw materials.

	Ethane	Propane	*n*-Butane	Gaso-line	Light gas oil	Heavy gas oil
	Temperature, K					
	1120–1185	1070–1125	1070–1125	1070–1185	1020–1075	970–1075
Vapor amount added to the raw, % (mass)	60	70–90	90	85–90	—	—
Conversion in one cycle, % (mass)	1.3	2.38	2.75	3.18	3.79	4.1
Yield of products, % (mass)						
$H_2 + CH_4$	15.30	30.00	20.60	18.03	11.96	11.82
C_2H_4	76.98	42.00	36.40	31.40	26.40	24.40
C_3H_6	2.88	16.20	20.50	14.77	14.20	13.90
C_4H_6	1.35	3.15	3.04	3.73	3.89	3.39
ΣC_4H_8	0.61	1.39	12.65	4.14	4.65	4.36
Gasoline	2.88	6.03	4.09	24.20	19.52	17.73
boiler fuel	—	1.21	2.32	3.73	19.38	24.40

The increase of the resources of C_4-olefins is provided by expansion of ethylene production and growth in the number of devices based on heavy raw materials.

In modern oil processing, catalytic reforming and hydrocracking are widely used. The process of coking, in which C_4 hydrocarbons are formed, is also widespread [33, p. 85]. A significant amount of isobutylene is formed at thermal contact cracking (Table 1.5) [35].

The process is very flexible. It may help in forming products from boiler or gas turbine fuel to olefin rich gases.

The reaction of catalytic dehydration of *tert*-butyl alcohol, which is formed in production of propylene oxide, and that of dehydration of isobutane are also sources of isobutylene. In the latter case, of special interest is the usage of hydrogen acceptors, such as Zr_2Fe intermetallide, which allow doubling of the equilibrium yield of isobutylene [36].

Taking into account the growing shortage of isobutylene, let us note new non-traditional methods of its production that are attractive due to the high accessibility of their raw material. Thermal catalytic (ZrO_2, $Co_4(CO)_{12}$, $Co_2Rh(CO)_{12}$, silicalite, 573–

673 K) transformations of methanol, propylene, synthesis gas, its mixtures with propylene make it possible to produce fractions with suitable content of isobutylene, and in some cases, with high selectivity [12, 37–40].

Table 1.4.

Yields of C_4 hydrocarbons in various regimes of pyrolysis of directly distilled gasoline.

	Regimes		
	Ethylene	Propylene	Butylene
Process parameters			
Temperature, K	1055	1025	1000–1005
Vapor amount added to the raw, % (mass)	50	50	30
Time, s	1	0.8–1	1
Yield of products, % (mass)			
iso-C_4H_8	1.63	2.4	2.7
Butene-1	0.8	1.8	2.5
Butene-2	0.31	1.1	1.9
ΣC_4H_8	2.74	5.3	7.1
C_4H_6	2.76	3.1	3.0
C_4H_{10}	0.9	1.0	1.0
ΣC_4	6.4	9.4	11.1
iso-C_4H_8:*n*-C_4H_8 ratio	14.6	0.82	0.61

Table 1.5.

Typical composition of gases [% (mass)] formed in processing of black oil and arlanian oil tar.

Component	Thermal contact cracking		High temperature cracking	
	795 K	815 K	875 K	900 K
H_2	0.5	0.6	0.8	1.0
CH_4	21.2	23.2	18.6	19.1
C_2H_4	13.7	18.3	21.7	19.5
C_2H_6	13.6	15.3	13.4	13.5
C_3H_4	15.6	17.4	21.0	21.8
C_3H_8	12.5	9.2	6.8	6.2
n-C_4H_8	9.7	7.7	11.5	12.1
iso-C_4H_8	6.4	5.2	4.8	5.2
n-C_4H_{10}	5.8	2.5	0.8	0.9
iso-C_4H_{10}	1.0	0.6	0.6	0.7

As fractions of C_4 hydrocarbons are one of the main sources for isobutylene production, methods for their purification are of special interest.

Alkynes can be removed by low temperature hydrogenation in the liquid phase (Table 1.6) [41].

Hydrogenation may be conducted in the presence of metal catalysts (Pt, Pd, Ni on carriers), as well [42].

Separation of the butane–butadiene fraction with simultaneous removal of butadiene by usual refraction appears impossible due to proximity of boiling temperatures of the components and formation of azeotropic mixtures (Table 1.7).

Table 1.6.

Selective hydrogenation of acetylene homologues from C_4 fraction (288–313 K, excessive pressure 255–637 MPa, rate of raw feeding 5.6 kg/m^{3}cat·s).

Components	Usual regime of pyrolysis		Rigid regime of pyrolysis	
	Composition of raw material, % (mass)	Hydrogenation product composition, % (mass)	Composition of raw material, % (mass)	Hydrogenation product composition, % (mass)
Butanes	5.5	5.58	3.14	3.25
Butenes	59.68	61.12	46.41	48.78
Butadiene-1,3	34.59	33.28	49.43	47.95
Ethylacetylene	0.06	0.02	0.10	0.02
Vinylacetylene	0.17		0.92	

Note: Alkynes are hydrogenated by 92–98%; losses of butadiene due to hydrogenation are 3–4%.

Table 1.7.

Boiling points of C_4 hydrocarbon fraction components.

Hydrocarbon	Boiling point, K	Hydrocarbon	Boiling point, K
n-Butane	272.7	Butene-1	266.90
trans-Butene-2	274.04	Isobutylene	266.30
cis-Butene-2	276.89	Isobutane	261.43
Butadiene	268.75		

That is why, after purification of the C_4 fraction from acetylene homologues, butadiene is extracted by chemosorption with the help of copper salts solutions, absorption or extractive rectification using selective solvents (furfural, acetonitrile, dimethylformamide, dimethyl acetamide, N-methylpyrrolidone) [33, p. 150; 43, 44]. Monovalent copper in solutions (273–283 K) forms complex compounds with butadiene, which easily degrade on heating up to 330–355 K. Cooled down the solutions of Cu^{1+} can be used again for binding butadiene.

Residual concentration of unsaturated compounds is 10^{-2}–10^{-3}% mol. and sometimes even lower.

The process of two-stage extractive distillation with acetonitrile and N-methylpyrrolidone is also widely used. If necessary, residual butadiene may be removed by selective hydrogenation (Table 1.8) [33].

Due to broadening of the diversity of the raw materials sources for the C_4 fraction production, they may contain various admixtures of polar compounds: water, aldehydes, alcohols, ethers, carbon oxide and oxysulfide and sulfurous compounds. In this case, preference is given to contact methods of purification — fraction processing under normal conditions or under increased temperature by chemically active compounds, which irreversibly binds admixtures, for example, suspensions of finely ground sodium, potassium, calcium in xylene, aluminum chlorides or alkyl-chlorides as complexes or solutions in polar solvents [43]. The maximum content of mentioned admixtures after purification is ~$2 \cdot 10^{-4}$% (mass).

Table 1.8.

Selective hydrogenation of residual butadiene from the C_4 fraction (298–333 K, 785–1765 MPa, the rate of raw feeding is 2.8–5.6 kg/m^{3}cat·s).

Components	No isomerization		With isomerization	
	Raw material composition, % (mass)	Hydrogenation products composition, % (mass)	Raw material composition, % (mass)	Hydrogenation products composition, % (mass)
Butane	9.2	9.7	8.9	15.0
Isobutylene	45.5	45.5	45.2	45.2
Butene-1	27.6	25.3	27.6	3.9
Butene-2	17.3	19.5	16.5	35.0
Butadiene-1,3	0.4	Absent*	1.8	Absent*

* The concentration of residual butadiene-1,3 is below $200 \cdot 10^{-6}$% (mass).

To decrease the power capacity of purification of C_4 hydrocarbon fraction the process of rectification is substituted by selective absorption. The rapidly developing technique of semi-permeable polymeric membranes also appears promising. For example, to extract butadiene and isoprene from the C_4 fraction, the film membranes from polyethylene, polyisobutylene and other polymers are used [43].

An interesting method to remove butadiene from the C_4 fraction suitable on a laboratory scale was suggested in [45]. Treatment of the fraction by an anionic catalyst — disodium-tetra-α-methylstyrene (223–273 K) — gives ~100% yield of liquid polybutadiene. An additional introduction of monomers (acrylonitrile, isoprene, methylmethacrylate) induces the extraction of isobutylene due to formation of a block-copolymer of the A–B–A type (B is a polybutadiene block).

1.4. Production of isobutylene in industry.

In industry, traditionally isobutylene is produced from the C_4 hydrocarbon fraction by dehydrogenation of isobutane and butane with further isomerization and chemical absorption by sulfuric acid. Recently, high effectiveness has been demonstrated of isobutylene extraction by liquid-phase hydrogenation or interaction with lower alcohols on ion-exchange polymers.

Of importance for production of isobutylene in industry is the method of dehydrogenation of butane fraction (chromium oxide on aluminum oxide, 770–775 K) with further isomerization to isobutylene (H_3PO_4 on silica gel, silicon dioxide or chamotte; Al_2O_3 and other, 575–815 K) and isomerization of butane ($AlCl_3$, $AlBr_3$, BrF_3, HF, HCl, etc.) with further dehydrogenation of isobutane on chromium-aluminum oxide catalysts using a continuous process (810–875 K 980 kPa) [2, 46, 47]. At 50% conversion the yield of isobutylene reaches 42% with 82% selectivity (calculated for isobutane).

To prevent electrophilic polymerization of isobutylene on acidic catalysts, the initial butylene mixture is diluted by water vapor. Coke is formed as a sediment during dehydrogenation, the activity of the catalyst decreasing. Regeneration of the catalyst is performed by air (925 K, 0.117 MPa). After the extraction of butadiene from the C_4 fraction isobutylene is extracted [47, 48].

The sulfuric acid method of isobutylene extraction includes two main stages. The monomer is practically quantitatively absorbed by 60–65% H_2SO_4 (215–315 K), the loss of isobutylene due to a side reaction of oligomerization (mostly to di- and triisobutylenes) being low:

$$(CH_3)_2C=CH_2 + H_2SO_4 \rightarrow (CH_3)_3COSO_2OH$$

The mixture containing *tert*-butylsulfuric acid is removed from absorbers and hydrolyzed by sharp vapor. As a result, *tert*-butyl alcohol is formed, which is dehydrated by 35% H_2SO_4 or Al_2O_3 (635 K) giving isobutylene:

$$(CH_3)_3COSO_2OH + H_2O \rightarrow (CH_3)_3COH + H_2SO_4$$
$$(CH_3)_3COH \rightarrow (CH_3)_2C=CH_2 + H_2O$$

Isobutylene can also be produced by another scheme. The mixture, produced by extraction with 65% H_2SO_4, is diluted by water to 45% concentration of H_2SO_4. At heating (300–320 K) under low pressure isobutylene is eliminated. The monomer is washed in alkali, purified from alcohol and oligomer mixtures at cooling, and after washing by water is liquefied under pressure. Worked out acid is concentrated up to 65% and returned to extraction.

One of the schemes provides for regeneration of isobutylene from sulfuric acid extract without diluting sulfuric acid and, consequently, without its further concentration [49]. The peculiarity of the process lies in the possibility of using C_4 fractions with high content (up to 50%) of butadiene as raw material without their preliminary extraction. The process is of special interest in those cases when butene-1 and butadiene of high purity are produced simultaneously with isobutylene. The degree of isobutylene extraction is 90–95%, and the purity of the rectificate not lower than 99.8% (mass) [47].

Catalytic dehydration of isobutyl alcohol is one of many widely used methods. The initial raw material is usually produced synthetically from propylene, CO and H_2 under high pressure and temperature (hydroformylation in the presence of cobalt-containing catalysts):

$$CH_2{=}CHCH_3 + CO + H_2 \longrightarrow CH_3{-}CH(CH_3)C(=O)H \longrightarrow$$

$$\longrightarrow CH_3{-}CH(CH_3){-}CH_2OH$$

Raw isobutylene is purified from water and alcohol mixtures by cooling, then liquefied under pressure and distilled in a column (about 75 plates, 490.3 kPa).

As in the sulfuric acid method, the process of isobutylene extraction from C_4 fractions with ion-exchange polymers (sulfated styrene copolymers with divinylbenzene of KU-2[1] type, etc.) includes its hydration to *tert*-butyl alcohol which

---

[1] KU-2, KU-2FPP, KIF, etc. are Russian trademarks.

dehydrates further into an olefin [47, 48, 50]. This method is characterized by the absence of highly aggressive corrosive media, the possibility of multiple use of a catalyst and a high (99.95% (mass)) purity of produced isobutylene.

The first stage is performed in two consecutive hydrators filled with an ion-exchange polymer KU-2 (363 K, 1.96 MPa). The C_4 fraction is fed mixed with water. Reaction products are removed from the lower part of the hydrator and separated from non-reacted hydrocarbons by pressure choke. Then they are subjected to extraction and rectification of the recurrent fraction. After rectification, the azeotrope of trimethylcarbinol with water, extracted from the water–alcohol solution, gets free from C_4 hydrocarbons and feeds the dehydrator with KU-2 (383 K, 0.14–0.19 MPa). Gaseous isobutylene is passed through a fractional capacitor and washing column, in which it escapes from the alcohol mixture and isobutylene oligomers, and after compression and condensation is directed to azeotropic desiccation and rectification. At isobutylene extraction on ion-exchange polymers, conversion of olefin is 92% (mass) the extraction from fraction being 90% (mass). Isobutylene rectificate is about 99.95% (mass) pure, the content of admixtures being: isobutane and butylenes below 0.05% (mass), trimethylcarbinol below 0.005% (mass), water below 0.005% (mass). The technological scheme is quite flexible. The decrease of conversion at decreased content of isobutylene in the initial mixture may be prevented by increase of pressure and contact time of the mixture and the catalyst at the first stage. Application of a polar solvent (ethyl cellosolve) provides homogeneity of the fraction in water and significantly increases the efficiency and the time of functioning of the ionite catalyst (up to 4000–5000 h without change) [51].

It appears of interest to conduct the stage of the alcohol dehydration in a reactor, the construction of which allows the usage of the condensation heat of $(CH_3)_3COH$ and H_2O vapors (67 kJ/mol for 85% of alcohol) to transform alcohol to isobutylene, which requires 65.7 kJ/mol [52].

In industry, isobutylene is exctracted from C_4 hydrocarbon fractions via interaction of *tert*-olefin with lower alcohols, in particular, with methanol in the presence of macroporous ion-exchange catalysts such as Amberlist-25, KU-2FPP, CIF[2], $Al_2O_3{\cdot}SiO_2$, zeolites, magnesium silicates or active carbon [53–58], with further decomposition of the *tert*-butyl ester formed. The use of H_2SO_4 and other compounds with acidic properties as catalysts is also possible. However, only ion-exchange catalysts provide easy extraction of the products and high selectivity of the process accompanied by the absence of waste waters:

$$(CH_3)_2C{=}CH_2 + CH_3OH \xrightleftharpoons{\text{Cationite}} (CH_3)_3C^{+}\ {}^{-}OCH_3 \rightleftharpoons (CH_3)_3C{-}OCH_3$$

$$(CH_3)_3C{-}OCH_3 \xrightleftharpoons{\text{Cationite}} (CH_3)_2C{=}CH_2 + CH_3OH$$

The method has special potential due to noticeable growth of the production and consumption of methyl-*tert*-butyl ester[3], which allows broadening of the resources of high-octane non-ethylated gasolines and refuse toxic alkyl-lead and other metallorganic anti-detonators.

---

[2] Russian trademarks.

[3] In 1987 — 2.2 million tons (USA) and 0.87 million tons (Europe) [61].

Ethyl-*tert*-butyl ester, which allows increase of the octane rating of gasoline by 1-2 points above that for methyl-*tert*-butyl ester, is also promising [63].

In interacting with isobutylene, the reactivity of alcohols decreases with the growth of their molecular masses and, within one homologous, with the growth of branching of the alcohol molecule. The reaction is equilibrated (Table 1.9) [59].

Table 1.9.

Equilibrium constants K_{eq} of the olefin reaction with methanol.

Temperature, K	Methylisopropyl ester		Methyl-*tert*-butyl ester	
	K_{eq}	A^*, % (mass)	K_{eq}	A^*, % (mass)
323.2	421	95	279	94
358.2	37.0	84	21.4	79
373.2	13.8	74	9.4	69
398.2	3.1	51	2.5	46

* A is the ester content in the equilibrium mixture with the molar ratio of 1:1.

Table 1.10.

Enthalpy of the synthesis of methyl-*tert*-butyl ester.

Temperature, K	Enthalpy in the ideal gas state, ΔH_f (kJ/mol)			Condensation heat, ΔH_v (kJ/mol)			Reaction enthalpy, ΔH_r (kJ/mol)	
	Isobutylene	Methanol	Methyl-*tert*-butyl ester	Isobutylene	Methanol	Methyl-*tert*-butyl ester	In the gas phase	In the liquid phase
298.2	-16.91	-201.30	-293.08	20.43	37.51	28.30	-74.86	-45.22
323.2	-18.34	-202.22	-295.34	18.80	36.72	27.67	-74.78	-46.93
343.2	-19.51	-202.93	-297.14	17.17	35.50	27.17	-74.76	-49.28
363.2	-20.52	-203.65	-298.90	15.98	33.91	26.29	-74.73	-51.83

The equilibrium degree of the isobutylene transformation into ester grows with the molar ratio methanol:isobutylene and the content of olefin in the initial hydrocarbon fraction.

Data on heats of the synthesis of methyl-*tert*-butyl ester are shown in Table 1.10 [53, 60].

Depending on the conditions of the process, the synthesis of methyl-*tert*-butyl ester may include side reactions, for example, formation of dimethyl ester, dimerization and hydration of isobutylene. *n*-Butylenes and butadiene present in the C_4 fraction just slightly react with CH_3OH (conversion of about 1%) [60]. In practice, 100% selectivity by methyl-*tert*-butyl ester is obtained, if heteropolyacids are used as catalysts [61].

The second stage of isobutylene extraction from the C_4 fraction using ion-exchange polymers is catalytic decomposition of methyl-*tert*-butyl ester to olefin and methanol [62].

The principal reaction is accompanied by side processes of dimethyl ester formation, hydration and dimerization, the extent of which depends both on the reaction conditions and the nature of the catalyst used (Al_2O_3, SiO_2, MgO, $CaSO_4$, $Fe_2(SO_4)_3$, H_3PO_4 on carrier, cationites, etc.).

The reactions discussed belong to the class of heterogeneous catalysis, the leading role played by proton transfer from the solid surface to liquid reagents. That is why the rate (depth) of the reaction is affected by the area of the catalyst surface, sorption ability of reagents on the surface, temperature, and transportation of reagents to grains and active centers of the catalyst, i.e. external and internal diffusion. To select the optimum regime of the process it is important to know the thermodynamic equilibrium constant. The values of the concentration equilibrium constants, cited in literature, depend on the composition of the reaction mass and change over a wide range [64]. On the other hand, the calculated value of the constant (> 1000 at 298 K [64]) is evidently overestimated since the reversibility of CH_3OH interaction with isobutylene has been experimentally established. A strict approach supposes calculation of the influence of the solvent upon the equilibrium of methyl-*tert*-butyl ester synthesis [65]:

$$A + B \overset{K_p}{\Leftrightarrow} C$$

$$A + l \overset{K_1}{\Leftrightarrow} A \cdot l \qquad K_p = K_1 \cdot K_2 \cdot K_3 \cdot K_4$$

$$B + l \overset{K_2}{\Leftrightarrow} B \cdot l \qquad \text{A and B are reagents}$$

$$Al + Bl \overset{K_3}{\Leftrightarrow} C \cdot l + l \qquad \text{C is the reaction product}$$

$$C \cdot l \overset{K_4}{\Leftrightarrow} C + l \qquad \text{l is the solvent}$$

An alcohol may be used as a proton solvent since due to its good sorption by cationite it serves as a proton carrier from solid acid to isobutylene [65]:

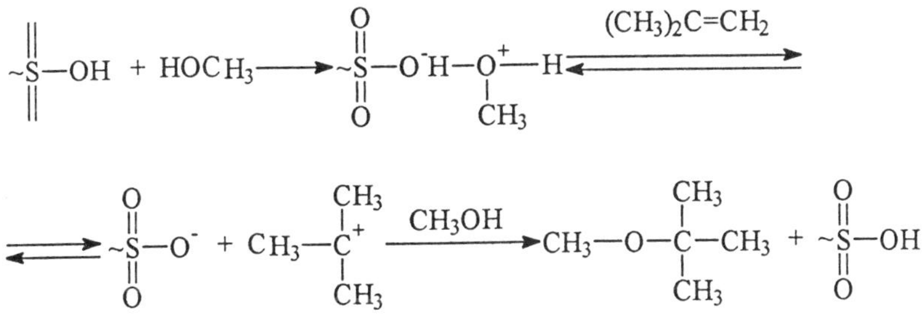

The numerical value of K_{eq} is 15 (333 K) [65]. A close value K_{eq} = 5.3 (323 K) was documented in [64]. The equation suggested, $\lg K_{eq} = 240/T - 6.71$, proves the possibility of obtaining yields of methyl-*tert*-butyl ester exceeding 80% at temperatures above 343 K.

The synthesis (310–450 K) and decomposition (310–475 K) of methyl-*tert*-butyl ester is recommended to be conducted in direct flow reactors with a stationary layer of cationic catalyst in one of two versions [53].

On leaving the first reactor, non-reacted hydrocarbons are removed from the reactionary mixture, and the volume product is channeled directly into the decomposing reactor (version I) or into the second rectification column for separation of excess methanol. The alcohol is removed from the column bottom and is channeled to the reactor of ester synthesis, and from the upper part its azeotrope with ester is channeled to the ester decomposing reactor (version II). Isobutylene formed is purified by washing off and rectification.

In the version II, the decomposing reactor is less loaded and more favorable thermodynamic conditions of the reaction are created. Nevertheless, the power capacity of rectification of the methanol–ester azeotropic mixture does not allow comparison *a priori* of the relative benefits of the versions with or without additional rectification.

The synthesis of methyl-*tert*-butyl ester on sulfocationites is possible in the "boiling layer" of the catalyst [53]; however, the drawback of the method is in erosion of the catalyst powder, change of its dispersion and decrease with time of technological indexes of the process.

A two-stage synthesis of methyl-*tert*-butyl ester has been suggested [53]. It included the conduction of the reaction at the first stage with an excess in the reactor of one of the reagents — methanol — that provided greater transformation of *tert*-olefin, and at the second one with the excess of isobutylene in another reactor. The hydrocarbon fraction with decreased concentration of *tert*-olefin is channeled from the upper parts of the rectification column into the first reactor. In one of the versions, the initial flows of methanol and hydrocarbon fraction are mixed before entering the ester synthesis reactor [53].

Gentler conditions of the synthesis at the first stage (323–363 K, 1–3 MPa) as compared with the second one (333–373 K, 1.47–3.52 MPa), reduce the formation of dimer and oligomer products.

Since the reactions of ester synthesis and decomposition are in equilibrium [59, 60], a continuous outlet of the reaction products is of great importance. As a result, the best isobutylene extraction from hydrocarbon mixtures can be obtained when its interaction with an alcohol and ester decomposition occur in the presence of a heterogeneous catalyst in reaction rectification reactors [53].

The mixture of hydrocarbons containing isobutylene is mixed up with methanol and fed to the middle part of the column reactor. From its upper part the fraction of non-reacted hydrocarbons is removed and, from the bottom one, the reaction product (the ester) is channeled to the next column reactor to be decomposed. *tert*-Olefin is removed from the upper part of the reactor, and methanol from the bottom one (or from a side part), and channeled to the first reactor again.

According to a version of the above discussed method the first stage is run in the reactor with a stationary layer of a catalyst, and the second one, in the rectification column with a packing of sulfocationite catalyst [66]. The construction of the packing provides a high rate of mass transfer under low pressure and effective contact with the catalyst.

The efficiency of C_4 *tert*- and *n*-olefin separation via formation of methyl-*tert*-butyl ester on ionite catalysts in reaction rectification devices may be increased by separate feeding of reagents: hydrocarbon fraction to the bottom, and extragent to the upper part of the catalyst zone of the device [60]. Thus, efficient counterflow is created in the reaction zone due to an excess of corresponding reagents on the borders of the reactionary zone.

The process is characterized by high yields (96–98 mass%) and purity (up to 99.95 mass%) of isobutylene (Table 1.11) [53].

To improve the technology of the complex usage of raw components of the C_4 fraction and to produce valuable products, the basic process of the isobutylene transformation into methyl-*tert*-butyl ester has been modified [67–70] with the help of various ion-exchange resins. Introduction of palladium into macroporous cationic resin (H^+/Pd^0) makes it possible to use effectively catalytic cracking gases as a raw material without previous purification from dienes, alkynes and other hydrocarbons.

In another version, similar catalysts provide cooperative production of the target ester (for isobutylene extraction) and one more product — diisobutylene. At the consecutive contact of the C_4 fraction with the hydrogenation (Pd/Al_2O_3) and etherification (sulfocationite) zones of the catalysts, a high octane rating component for gasoline was produced — methyl-*tert*-butyl ester mixed with saturated and unsaturated cyclic hydrocarbons.

Table 1.11.

Composition of initial and processed pyrolytic C_4 fraction in the process of isobutylene extraction.

Components	Initial fraction, % (mass)	Processed fraction, % (mass)
C_3	0.5	0.85
iso-C_4H_{10}	4.40	7.50
n-C_4H_{10}	11.80	20.20
iso-C_4H_8	42.20	1.70
α-C_4H_8	22.39	38.00
β-*trans*-C_4H_8	10.30	17.60
β-*cis*-C_4H_8	8.0	13.50
C_4H_6	0.41	0.70

The growth of the yield of highly purified isobutylene (via methyl-*tert*-butyl ester) can be reached by preliminary isomerization of butene-1 into isobutylene. In its turn, a high degree of the isobutylene extraction in its reaction with CH_3OH even from fractions very poor in isobutylene makes the latter a valuable raw material for butadiene synthesis (oxidative dehydrogenation).

Another version is no less valuable. It proposes a cooperative polymerization of the components of the removed C_4 fraction in the production of methyl-*tert*-butyl ester in the presence of Al_2O_3 thus obtaining oil-like oligomeric products.

From the point of view of industrial application, the process of C_4 fraction processing by water solution of a metal chloride and hydrochloric acid with further dehydration of *tert*-butyl alcohol is promising due to its high selectivity [71]. The technological scheme of the process consists of four main stages: absorption, degassing, regeneration and rectification.

The absorption is conducted at 273–333 K and pressure up to 980 kPa in the presence of zinc, tin, antimony or bismuth chlorides, and degassing at 333–348 K and atmospheric pressure. From the degassing device the mixture of *tert*-butyl alcohol and *tert*-butyl chloride enters regenerators and distillation–reactionary columns with perforated plates. In the former the alcohol is dehydrated (353–413 K), and in the latter, the reaction of dehydrochlorination is run. After rectification the purity of isobutylene reaches 99.9% (mass), the extent of extraction, 94 mass%. Negligible amounts of isobutylene dimers are formed as side products, as well.

Cheap isobutylene is obtained as a side product in the production of propylene oxide by the method used by the Halkon Company (USA) [72]. *tert*-Butyl hydroperoxide, formed in the liquid phase oxidation of isobutane (O_2, catalyst — alkaline–earth metal salt, 363–423 K, 1.964–4.7 MPa), in the process of selective epoxidation of propylene is transformed into trimethylcarbinol, which in turn is dehydrated into isobutylene with the help of the usual water eliminating catalysts.

For industrial production of isobutylene on a small scale, catalytic dehydration of isobutyl and *tert*-butyl alcohols is used [72]. It provides a high purity of the monomer. This method is realized as the second stage of isobutylene extraction from the C_4 fraction by catalytic hydration (see above).

1.5. Special methods of isobutylene production

A high yield of the isobutylene production is reached in any of the following non-traditional special methods:

1. By pyrolysis of neopentane, methylcyclopentane, as well as arenes containing *tert*-butyl group [1, p.47].

$$(CH_3)_4C \xrightarrow{850\,K} (CH_3)_2C = CH_2 + CH_4$$

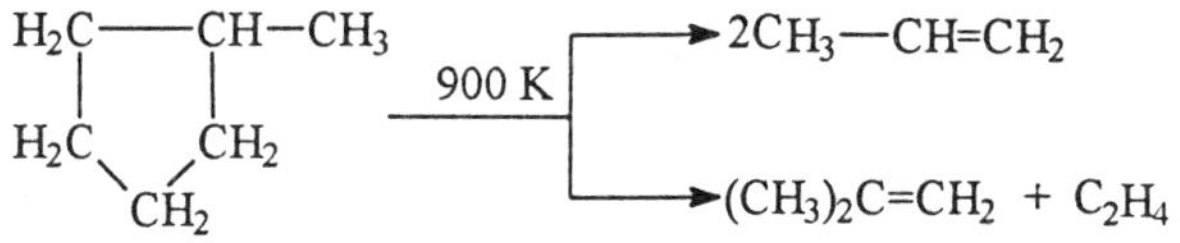

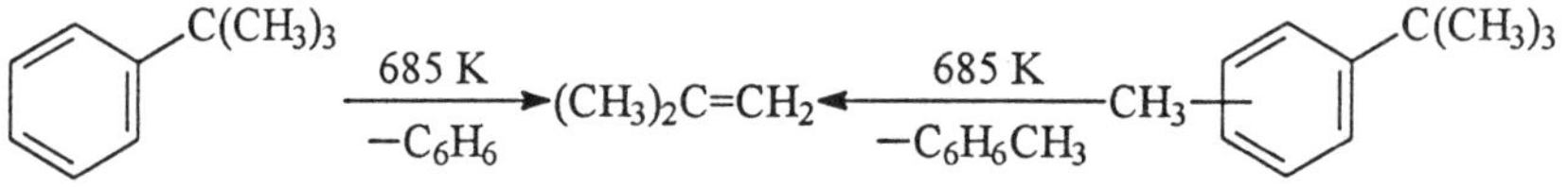

2. By selective thermocatalytic depolymerization of oligo- and polyisobutylenes (zeolites, metal oxides, $NaAlCl_4$, etc., 475–700 K) [46, 73–78].
3. In some reactions of diene synthesis, by interaction of 1,1,2,3-tetramethylcyclohexadiene-2,4 with dimethyl ester of acetylenedicarbonic acid [1, p. 47]:

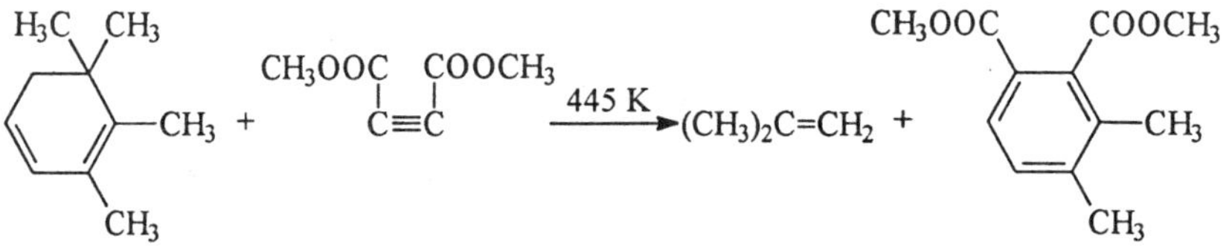

4. From acetone (670–725 K, kaolin; 440–535 K, alumosilicates):

$$2(CH_3)_2C = O \xrightarrow[-H_2O]{} (CH_3)_2C = CH\text{-}C(O)\text{-}CH_3 + H_2O \xrightarrow[-CH_3COOH]{}$$
$$\rightarrow (CH_3)_2C = CH_2$$

5. By the interaction of *tert*-butylmagnesium halogen with carbonyl compounds (pivaloyl chloride, for example). The reaction proceeds abnormally, since the reduction of the >C=O-group occurs instead of the addition of the *tert*-butyl radical to the carbonyl carbon [1, p. 49]:

$$(CH_3)_3C\text{-}C(O)Cl + 2CH_3\text{-}C(CH_3)_2MgCl \rightarrow$$
$$\rightarrow 2(CH_3)_2C = CH_2 + (CH_3)_3C\text{-}CH_2OMgCl + MgCl_2$$

6. As a product of the reaction of phenylmagnesium bromides with 2-chlorine-2-methylpentanone-4 (the latter is easily produced from mesityl oxide by HCl addition):

$$(CH_3)_2C(Cl){—}CH_2{—}C(O){—}CH_3 + C_6H_5MgBr \rightarrow$$
$$\rightarrow (CH_3)_2C{=}CH_2 + C_6H_5C(O)CH_3 + MgBrCl$$

7. In the group of monoolefins with four carbon atoms one may perform isomerization, similar to that of *n*-butane into isobutane. For example, under special conditions, butene-1 may transform not only into both stereoisomeric forms of butene-2, but also into isobutylene.

Isomerization of butenes into isobutylene proceeds with acidic catalysts:

$$\left.\begin{array}{l} CH_3{—}CH_2{—}CH{=}CH_2 \\ CH_3{—}CH{=}CH{—}CH_3 \end{array}\right\} \xrightarrow{H^+} [CH_3{—}CH_2{—}\overset{+}{C}H{—}CH_3] \xrightarrow[{—H^+}]{} $$

$$\longrightarrow (CH_3)_2C{=}CH_2$$

At 573 K, the concentration of isobutylene in the reaction mixture may reach that of both *n*-butenes.

In the laboratory, isobutylene can be easily obtained by dehydration of isobutyl or *tert*-butyl alcohol in the presence of Al_2O_3 (520–575 K). Dehydration of *tert*-butyl alcohol is also performed at heating water solution of the alcohol and oxalic acid (up to 350–355 K) or in the presence of cationites (KU-23[4], 363 K).

The gas produced is channeled through a spiral glass rectifier immersed into a bath of ice and salt, to separate alcohol and water. Isobutylene is condensed by cooling with the mixture of solid carbon dioxide and ether. Liquid isobutylene is distilled at low temperature.

The method of thermal (383 K) decomposition of triisobutylaluminum or its realkylation by a heavier olefin under the same conditions may also be recommended to produce the small amounts of pure isobutylene [79].

1.6. Analysis of isobutylene.

Most often the purity of isobutylene is determined with the help of H_2SO_4 or HCl, some special reactions, and also by gas chromatography, IR- and UV-spectroscopy [1, p. 54].

Qualitative determination of isobutylene is based on its transformation into *tert*-butyl alcohol with further identification of the alcohol, for example, as 3,5-dinitrophenylbenzoate (T_{melt} = 414.5–415.5 K). The reaction of quantitative addition of dry HCl or HBr to isobutylene with further identification of *tert*-butyl halogenide may be also used.

---

[4] Russian trademark.

To identify isobutylene one may use its reaction with 2,4-dinitrobenzenesulfonyl chloride, in which chloro-*tert*-butyl (2,4-dinitrophenyl) sulfide (T_{melt} = 359–360 K) is formed, or with 3,5-dinitrobenzoliodate, with the formation of 3,5-dinitrobenzoate of iodine-*tert*-butyl alcohol (T_{melt} = 355.4–356.2 K). Isobutylene can also be determined in reaction with phenol (catalyst H_2SO_4) as 4-*tert*-butylphenol.

Quantitatively with enough accuracy isobutylene is determined by the reaction with anhydrous HCl. The reaction proceeds instantaneously, but the accuracy of determination is affected by the presence of pentenes and other heavier homologues of ethylene with a tertiary carbon atom. The gas containing isobutylene is washed with NaOH solution, dried above anhydrous $CaCl_2$ and then liquefied. The calculated quantity of liquid isobutylene is treated by excess of gaseous HCl or HBr, and the content of isobutylene is determined by the decrease of volume.

A colorimetric method of isobutylene analysis is based on the color reaction between mercury acetate and diphenylcarbazide. The range of determination is 2.5–20 μg in 5.3 ml of the solution.

Isobutylene forms water-insoluble complexes with mercury salts. Mercury oxide is dissolved in a minimum amount of 70% nitric acid with heating; then the liquid is diluted up to the necessary volume and carefully neutralized by NaOH (until the solution becomes turbid). The titer is stated in comparative titration by potassium rhodanide of known concentration with a Fe^{3+} salt as an indicator. A portion of gas at fixed temperature and pressure is absorbed by an accurately measured volume of a reagent. The reaction mixture is transferred to a glass container and heated up to 373.2 K. Extracted sediment is filtered after cooling down the solution to room temperature and washed off by bidistilled water until the washing water no longer produces sediment at addition of ammonium sulfite. The sediment on the filter is dissolved in 10 ml of 70% HNO_3, Hg^{1+} is oxidized by a small amount of $KMnO_4$ to Hg^{2+}, and excess of the oxidant neutralized by oxalic acid. After dilution by water up to 100–150 ml and addition of 2 ml of iron ammonium sulfate, the liquid is titrated by 0.1 N potassium rhodanide to determine the amount of mercury and, consequently, the amount of isobutylene. Note: 1 ml of 0.1 from a normal solution of potassium rhodanide corresponds to 0.1601 ml of isobutylene at 273.2 K and 98.07 kPa. The method is characterized by high accuracy and speed and may be applied even in the cases when isobutylene contains lower and higher homologues of ethylene and diolefins.

Isobutylene is also detected via titration by the bromide–bromate method. Interacting in acidic medium, both salts produce bromine, which binds across the double bond:

$$KBrO_3 + 5KBr + 3H_2SO_4 \rightarrow 3H_2O + 3K_2SO_4 + 3Br_2$$
$$(CH_3)_2C{=}CH_2 + Br_2 \rightarrow (CH_3)_2CBr{-}CH_2Br$$

A measured amount of gas is shaken in a closed vessel with excess of the reagent in the presence of dilute sulfuric acid. When the reaction finishes, an accurately measured volume of potassium iodide solution in a definite concentration is added to the mixture. The amount of free iodine, equivalent to unreacted bromine, is detected by titration with sodium hyposulfite solution. By the amount of iodine formed the number of double bonds and, consequently, isobutylene content in the gas sample are calculated.

The content of isobutylene may be quantitatively analyzed by the O_3 absorption (170–575 K) (with the help of a home-made ADS-4 device — Inst. Chem.

Phys., AS USSR). The reaction rate is limited by the rate at which reagents are fed to the reaction zone. Kinetic parameters are: at 293 K in the gas phase $k = (3.6 \div 14 \pm 2)$ $m^3/(mol \cdot s)$, $E_a = 11.72 \pm 1.67$ kJ/mol; $A = 1.4 \cdot 10^3$ $m^3/(mol \cdot s)$; at 293 K in the liquid phase (CCl_4) $k = (0.2 \div 0.97) \cdot 10^2$ $m^3/(mol \cdot s)$ [80, p. 69].

Physical methods of analysis are quicker and more accurate than chemical ones. In this case the super-legible rectification as an accessory operation appears useful.

In some cases, the difference in refraction indexes of C_4 hydrocarbons allows us to apply refractometry to detect isobutylene.

The degree of the isobutylene purity may be stated cryoscopically.

Good results are obtained with IR-spectroscopy of C_4-olefins under elevated pressure (with detection accuracy about 0.1%), mass-spectroscopy (isobutylene preliminarily transformed into *tert*-butyl chloride), their combination, the Raman–Mandelschtamm spectroscopy and analytical (preparative) gas chromatography.

To control continuously the production of isobutylene, ozonometry, IR- and ESR-spectroscopy, mass-spectroscopy and, especially, gas chromatography are used [81–83].

One of the methods of chromatographic analysis is based on using a detector of combustion heat [81, p. 136]. A column (length 8 m, diameter 5.5 m) is filled with the 0.25–0.5 mm fraction of diatomic bricks, impregnated with triethyleneglycol butyrate (4% of the bricks' mass) and 40% solution of $AgNO_3$ in ethyleneglycol (25% of the bricks' mass). The conditions of the analysis are: the temperature of the column 295 ± 3 K, the rate of the gas-carrier 1 $\mu m^3/s$; the probe amount to analyze 2 μm^3. Relative retention times are shown below:

Ethane	0.19	Isobutylene	1
Propane	0.22	β-Isoamylene	1.30
Isobutane	0.26	*cis*-β-Butylene	1.73
Butane	0.30	α-Butylene	1.87
Isopentane	0.45	α-Isoamylene	2.22
Pentane	0.54	γ-Isoamylene	2.29
Trans-β-Butylene	0.70	Butadiene	3.90
β-Isoamylene	1.03	Isobutylene dimers	5.7 and 6.3

When isobutylene contains admixtures with extremely different boiling points no direct dependence between the peak area and the component concentration is observed. That is why, to determine the content of any component, the calculated peak areas should be multiplied by the corresponding correction coefficient (shown below):

Isopentane	1.12	Propylene	0.79
Pentane	1.05	*cis*-β-Butylene	0.82
Trans-β-Butylene	0.78	α-Butylene	0.84
β-Isoamylene	1.03	Isobutylene dimers	1.7

To detect *n*- and *iso*-butylenes the same stationary phase is used with the following conditions of analysis: the column 3 m long, 30% of the liquid phase to the diatomic bricks, 295 ± 3 K, the rate of the gas-carrier leaving the column 0.65–0.75 $\mu m^3/s$; the volume of the probe 0.7 μm^3.

The content of isobutylene in gaseous products may also be determined with 6% vaseline oil on TZQ (tripolite of Zikeevsky quary) as the immovable phase, the column of 6 m long and 3 m in diameter; the gas-carrier rate (helium) 1 $\mu m^3/s$.

Characteristic absorption bands in the IR-spectra of isobutylene are (cm^{-1}): $CH_2{=}\nu_{as}$ 3080, δ 1750–1800 (overtone), δ 890; $>C{=}CH_2$ ν 1655; $>C(CH_3)_2$ ν_{as} 2960, ν_s 2870, δ 1385–1380 and 1370–1365.

ESR-spectrum of isobutylene is characterized by: δ_{CH_2} = 4.62 m.d.; δ_{CH_3} = 1.68 m.d. ($^4J^{'H-'H}$ = 1.25 Hz).

References to Chapter 1

1. Vatsulik, P., Chemistry of monomers, (Translated from Czech), 1960, v. 2, 738 p., Moscow, Izdatinlit. (Rus)
2. Encyclopedia of polymers, 1972, v. 1, pp. 804–807, Moscow, Sov. Encyclopedia. (Rus)
3. Chemical Encyclopedia, 1988, v. 1, p. 639, Moscow, Sov. Encyclopedia. (Rus).
4. Tokue, J., Furukawa, T., and Kuchitsu, K., J. Mol. Struct., 1974, v. 23(1), pp. 33–52.
5. Yaws, C.Z., Chem. Eng., 1976, v. 83(2), pp. 107–115.
6. Filinov, Yu.P., Bogatchuk, Yu.Ya., Litovchenko, N.P., Pakholkov, V.G., Gashko, G.P., and Zryakin, I.N., Ukr. Khim. Zh., 1984, v. 50(10), pp. 1062–1065. (Rus)
7. Akatiev, V.V., Gokhberg, P.Ya., and Sharf, V.Z., Neftekhimia, 1987, v. 27(1), pp. 83–86. (Rus)
8. Meuladijk, J., Joosten, G.E.H., Stamhuis, E.J., and Reenackers, A.A., J. Mol. Catal., 1989, v. 54(2), pp. 183–197.
9. Kozhevnikov, I.V., Khankhasaeva, S.Ts., and Kulikov, S.M., Kinet. i Catal., 1989, v. 30(1), pp. 50–54. (Rus)
10. Kurashev, M.V. and Korneev, N.N., Neftekhimia, 1988, v. 28(2), pp. 176–182. (Rus)
11. Deeva, M. And Ford, M.E., J. Org. Chem., 1988, v. 53(19), pp. 4594–4596.
12. Melnikova, L.M., Khim. Prom. Za Rubezhom, 1987, No. 1, pp. 28–45. (Rus)
13. Khaimova, T.G. and Mkhitarova, D.F., Neftepererab. I Neftekhimia, 1985, No. 2, pp. 29–31, Moscow. (Rus)
14. Mamori, A., J. Catal., 1987, v. 106(1), pp. 280–286.
15. Gubaidullin, L.Yu., El'man, A.R., and Belaya, L.A., Homogeneous Catalysis in Synthesis of Functional Derivatives of Olefins and Dienes, 1990, 32 p., Inf. Release, Moscow, NIITEKhIM. (Rus)
16. El'man, A.R., Matveev, V.A., Slivinski, E.V., and Loktev, S.M., Catalysis and Catalytic Processes in Chemical Pharmaindustry, 1989, pp. 169–171, Thesis, 2nd Conf. of All-Union Sci. and Chem.-Technol. Inst. of Med. and Microbiol. Ind., Moscow. (Rus)
17. Kumadzaeva, T. and Kitamura, M., Hydrohylation of Olefin Compounds, 1987, Jap. Claim No. 61-118334, R.Zh. Khim, 13N, 8p. (Rus)
18. Farberov, M.I., Bondarenko, A.V., Tsilurik, T.V., Srednev, S.S., and Shutova, I.L., Zh. Prikl. Khim., 1984, v. 57(9), pp. 2071–2076. (Rus)
19. Belousova, V.M., Kashuba, E.V., and Zyashenko, L.V., React. Kinet. and Catal. Lett., 1987, v. 33(1), pp. 99–104; 1986, v. 32(1), pp. 33–38.
20. Roberts, D.W. and Williams, D.L., Sulfone Chemistry, 1987, v. 43(6), pp. 1027–1062.
21. Alimova, E.V., Shapovalova, L.P., Tmenov, D.N., and Dolia, L.P., Khim. Tekhnol., 1984, No. 4, pp. 29–32, Kiev. (Rus)
22. Slaugh, L.H., Conversion of Isobutylene to Aromatics or Polyisobutylene, Pat. USA No. 4293723, NZhKhIM, 17L 248P, 1982.
23. Freidlina, R.Kh., Velichko, F.K., Zlotski, S.S., Rakhmankulov, D.L., and Terentiev, A.B., Radical Telomerization, 1988, 288 p., Moscow, Khimia. (Rus)
24. Zvezdin, V.L., Domrachev, G.A., and Sangalov, Yu.A., Doklady Akademii Nauk SSSR, 1971, v. 198(1), pp. 102–104. (Rus)
25. Petrov, A.A. and Genusov, M.L., Ionic Telomerization, 1968, 296 p., Leningrad, Khimia. (Rus)

26. Wang, B. and Mishra, M.K., Polym. Bull., 1987, v. 17(3), pp. 205–211, 213–219.
27. Mishra, M.K., Wang, B., and Kennedy, J.P., Polym. Bull., 1987, v. 17(4), pp. 307–314.
28. Faust, R., Nagy, A., and Kennedy, J.P., J. Macromol. Sci., 1987, v. 24(6), pp. 595–609.
29. Mishra, M.K. and Kennedy, J.P., J. Macromol. Sci., 1987, v. 24(8), pp. 933–948.
30. Quanying, W., Bin, C., and Yixin, Z., Polym. Bull., 1988, v. 20(3), pp. 221–228 and 229–236.
31. Zsuga, M., Faust, R., and Kennedy, J.P., Polym. Bull., 1989, v. 21(3), pp. 273–280.
32. Nemes, S. and Kennedy, J.P., Polym. Bull., 1989, v. 21(3), pp. 293–300.
33. Chernyi, I.R., Production of Monomers and Raws for Oil-Chemical Synthesis, 1973, 264 p., Moscow, Khimia. (Rus)
34. Paushkin, Ya.M., In: New Oil-Chemical Products and Processes, 1971, 81 p., Moscow, CNIITECHNeftekhim. (Rus)
35. Botnikov, Ya. Kh., In: Problems of Processing of High-sulfuric Oils, 1966, p. 161, Moscow, CNIITECHNeftekhim. (Rus)
36. Lunin, V.V. and Chetina, O.V., Neftekhimia, 1990, v. 30(2), pp. 202–206. (Rus)
37. Machashi, T., Maruya, K., Domen, K., Aika, K., and Onishi, T., Chem. Lett., 1984, No. 5, pp. 747–748.
38. Maruya, K., Machashi, T., Haraoka, T., Narui, S., Osakawa, Y., Domen, K., and Onishi, T., Bull. Chem. Soc. Jap., 1988, v. 61(3), pp. 667–671.
39. Pat. No. 73959, 1986 (Finland); RzhKhim 16N 1P, 1988. (Rus)
40. Pat. No. 2562066, 1985 (Finland); RzhKhim 14N 7P, 1986. (Rus)
41. Zdonic, S.B., Green, E.J., and Halle, L.P., Oil and Gas, 1966, v. 64(48), pp. 62–66.
42. Lakhman, L.I., Tumanova, V.A., Karelski, V.V., Radchenko, E.D., Salamova, T.M., and Berenblum, A.S., Heftepererabotka i neftekhimia, 1978, No. 10, pp. 56–67. (Rus)
43. Prokofiev, K.V. and Sangalov, Yu.A., Khim. I Tekhnol. Topliv i Masel, 1986, No. 3, pp. 2–4. (Rus)
44. Emmerich, G., Techn. Mitt. Krupp., 1986, No. 1, pp. 25–28.
45. Pat. No. 200326/3, 1983 (GDR); RzhKhim 24 p. 496 P, 1983. (Rus)
46. Guterbock, G., Polyisobutylene and Isobutylene Copolymers, 1962, 363 p., Leningrad, Gostoptekhizdat. (Rus)
47. Kirpichnikov, P.A., Beresnev, V.V., and Popova, L.M., Album of Technological Schemes of Main Productions in Synthetic Cautchouc Industry, 1976, 112 p., Leningrad, Khimia. (Rus)
48. Chaplits, D.N., Industrial Methods of Isobutylene Extraction, 1971, 42 p., Moscow, CNIITECHNeftekhim. (Rus)
49. Dalin, M.A., In: Modern State and Perspectives of C_4–C_{20} Olefins' Production, 1972, 74 p., Moscow, CNIITECHNeftekhim. (Rus)
50. Chem. Eng., 1983, v. 90(25), p. 60.
51. Ionites in Chemical Technology, Nikolsky, B.P. and Romankova, P.G. (Eds.), 1982, 416 p., Leningrad, Khimia. (Rus)
52. Chaplits, D.N., Samokhvalova, L.D., Sobolev, V.M., and Turiaev, I.Ya., Khim. Prom., 1966, No. 10, pp. 734–736. (Rus)
53. Chaplits, D.N., Extraction of Isobutylene and Isoamylenes from Hydrocarbon Fractions, 1981, 35 p., Moscow, CNIITECHNeftekhim. (Rus)
54. Hydrocarbon Processes, 1989, No. 11, p. 96.

55. Vibihal, J., Chem Prom., 1980, v. 30(11), pp. 585–589.
56. Fattore, V., Mauri, M., Massi, O., and Paret, G., Hydrocarbon Processes, 1981, v. 60(8), pp. 101–106.
57. Beran, S., Bolcha, J., Nedomova, K., and Wichterolova, B., React. Kinet. and Catal. Lett., 1984, v. 26(3–4), pp. 437–439.
58. Rustamov, M.I., Guseinova, A.D., Asker-Zade, S.M., and Yunusov, S.G., Khim. Prom., 1986, No. 6, pp. 327–328. (Rus)
59. Setinek, K., Prokor, Z., Kraus, M., and Zitny, Z., Chem. Prom., 1977, v. 27(9), pp. 451–455.
60. Chaplits, D.N., In: Study and Development of the Production Technology for Monomers and Synthetic Caoutchoucs, 1979, pp. 119–227, Moscow, CNIITECHNeftekhim. (Rus)
61. Chernyshkova, F.A., Neftekhimia, 1991, v. 31(5), pp. 579–591. (Rus)
62. Choo, K.J., Golden, D.M., and Benson, S.W., Intern. Chem. Kinet., 1974, v. 6(5), pp. 631–641.
63. Anderson, E., Chem. and Eng. Mews, 1988, v. 66(43), pp. 11–12.
64. Safronov, V.V., Sharonov, K.G., Rozhnov, A.M., and Alenin, V.I., Khim. Prom., 1988, No. 10, pp. 590–592. (Rus)
65. Muja, J., Goidea, D., Sandulescu, D., Marculescu, N., Andreescu, G., and Anghel, G., Rev. Chem., 1982, v. 33(10), pp. 903–911.
66. Richards, D., Chem. Eng. (USA), 1991, v. 98(7), p. 44.
67. Pittman, C.U., Polym. News, 1987, v. 12(10), p. 307.
68. Jiang, G., China Synth. Rubber Ind., 1988, No. 3, pp. 210–213.
69. Pat. No. 3036481 (FRG); RzhKhim 9N 40P, 1983. (Rus)
70. Pat. No. 261297 (France); RzhKhim 21N 20P, 1989. (Rus)
71. Pat. No. 235404 (ChSSR); RzhKhim 17 N 5P, 1987. (Rus)
72. Schultz, D.L., Rubb. World, 1980, v. 181(4), pp. 43–44.
73. Lapidus, A.L., Tishkina, T.F., Lipovich, V.G., Isakov, Ya.I., and Minachev, Kh.M., Neftekhimia, 1976, v. 16(5), pp. 723–728. (Rus)
74. Muganlinski, F.F., Kasianov, V.V., Usenko, M.I., Gusseinov, M.M., Kakhramov, V.B., and Lapin V.B., Neftekhimia, 1978, v. 18(1), pp. 62–67. (Rus)
75. Sangalov, Yu.A., Prochukhan, Yu.A., and Minsker, K.S., Vysokomol. Soed., 1978, v. B20(3), pp. 686–688. (Rus)
76. Sangalov, Yu.A., Minsker, K.S., and Berlin Al.Al., Vysokomol. Soed., 1983, v. A25(7), pp. 1451–1457. (Rus)
77. Ivanova S.R., Gumerova, E.F., Berlin, Al.Al., Minsker, K.S., and Zaikov, G.E., Uspekhi Khimii, 1991, v. 60(2), pp. 430–447. (Rus)
78. Ivanova, S.R., Gumerova, E.F., Minsker, K.S., Zaikov, G.E., and Berlin, Al.Al., Prog. Polym. Sci., 1990, v. 15(7), pp. 193–215.
79. Tolstikov, G.A. and Yuriev, V.P., Organoaluminum Synthesis, 1979, 290 p., Moscow, Nauka. (Rus)
80. Razumovski, S.D. and Zaikov, G.E., Ozone and Its Reactions with Organic Compounds, 1974, 322 p., Moscow, Nauka. (Rus)
81. Baranova, V.G., Analytic Control of Production of Main Monomers for Synthetic Caoutchoucs, 1967, 212 p., Leningrad, Khimia. (Rus)
82. Turkeltaub, N.M. and Zhukhovitski, A.A., Zav. Lab., 1957, v. 23(9), pp. 1120–1124. (Rus)
83. Sokolov, V.A. and Kuzmina, A.P., Zav. Lab., 1957, v. 23(9), pp. 1034–1037. (Rus)

Chapter 2. Isobutylene polymerization.

2.1. General ideas of polymerization. Catalysts.

According to the cationic mechanism isobutylene is easily polymerized in the presence of electrophilic catalysts forming a wide range of products, from lowest oligomers (n = 2–3) to polymers with molecular mass of several millions.

Cationic polymerization is an ionic process, in which the growing active particle carries a positive charge:

$$K^{\delta+}A^{\delta-} + nCH_2 = C(CH_3)_2 \rightarrow$$

$$\rightarrow K[-CH_2 - C(CH_3)_2]_{n-1} - CH_2 - \overset{\delta+}{C}(CH_3)_2 A^{\delta-}$$

The affinity to a proton ΔH^0_{298} for isobutylene in the reaction $X_{(g)} + H^+_{(g)} \rightarrow XH^+_{(g)}$ is estimated [1, p. 127] as 695 kJ/mol for *iso*-$C_4H_9^+$ and 820.6 kJ/mol for *tert*-$C_4H_9^+$.

The reaction is activated by substances that are electron acceptors, i.e. Lewis acids: 1) proton acids (H_2SO_4, H_3PO_4, $HClO_4$, HCl, CCl_3COOH, etc.); 2) aprotic acids — the Friedel-Crafts reagents of the general formula MeX_n (Me is a metal, usually, of groups III–IV of the Periodic Chart, X is a halogen), in particular, BF_3, $AlCl_3$, $SnCl_4$, $TiCl_4$, $ZnCl_2$, etc.; 3) metal alkyl-halides, usually, of groups II–III of the Periodic System (R_2AlBr, $RAlCl_2$, etc.); 4) salts of stable carbonium ions (triphenylmethyl — $(C_6H_5)_3C^{\delta+}A^{\delta-}$, tropylium — $C_7H_7^{\delta+}A^{\delta-}$, where A is $SbCl_6^-, PF_6^-, AlCl_4^-$, etc.), an oxonium compound, for example, $R_3O^{\delta+}A^{\delta-}$, where A is $BF_4^-, SbCl_6^-$, etc., R is CH_3, C_2H_5, etc.

The process is also activated by oxides and salts of some metals (Al_2O_3, $Ni(ClO_4)_2$, $MgCl_2$, $Al(CF_3SO_3)_3$, etc.), natural (Floridin, bentonite, attapulgite) and synthetic alumosilicates, high power radiation. The number of catalytic systems and methods of polymerization activation, as well as their activity and selectivity, are rather manifold with consideration of possible combinations of compounds of various classes. Almost without exception it appears that the most active compounds cause formation of polymers with the highest molecular mass. Activity of proton acids depends on their strength. Soft acids do not activate cationic polymerization. Usually, products of low molecular mass are formed in isobutylene polymerization in the presence of proton acids. Activity of proton acids in cationic polymerization may vary with regard to the reaction conditions (temperature, pressure, medium polarity). Nevertheless, for isobutylene polymerization under particular conditions (195 K), activity of catalysts of this group decreases in the sequence: $BF_3 > AlCl_3 > AlBr_3 > TiCl_4 > TiBr_4 > SnCl_4 > BCl_3 > BBr_3$.

To activate the cationic polymerization by aprotic acids small amounts of another ionogenic substance — co-catalyst (promoter) — is usually used, in particular, water, hydrogen halide acids (HX), alcohols, alkylenoxides, α,β-ester halides, etc. However, up to the present time there is no general opinion on the necessity of using cocatalysts for activation of isobutylene polymerization.

In 1933 a mechanism of initiation was suggested, according to which direct interaction of an aprotic acid with a monomer proceeds [2]:

$$AlCl_3 + CH_2 = CRR' \rightarrow Cl_3\overset{\delta-}{Al}\text{-}CH_2\text{-}\overset{\delta+}{C}RR'$$

Variants of initiation by direct addition of a Lewis acid to a double bond of an olefin are reactions of electron transfer:

$$H_2C = C(C_6H_5)_2 + SbCl_5 \rightarrow [\dot{C}H_2C(C_6H_5)_2]^{\delta+}SbCl_5^{\delta-}$$

$$2[\dot{C}H_2C(C_6H_5)_2]^{\delta+}SbCl_5^{\delta-} \rightarrow \dot{S}bCl_5^{-}(C_6H_5)_2\overset{\delta+}{C}\text{-}CH_2CH_2\text{-}\overset{\delta+}{C}(C_6H_5)_2\dot{S}bCl_5^{-}$$

and elimination (after addition) of an intermediate H-acid with formation of a complex counterion:

$$Cl_4\overset{\delta-}{Ti}\text{-}CH_2\text{-}\overset{\delta+}{C}RR' \rightarrow TiCl_3\text{-}CH = CRR' + HCl$$

$$Cl_4Ti + HCl + CH_2 = CRR' \rightarrow CH_3\text{-}\overset{\delta+}{C}RR'TiCl_5^{\delta-}.$$

The effect of a proton donor in the hidden form is:

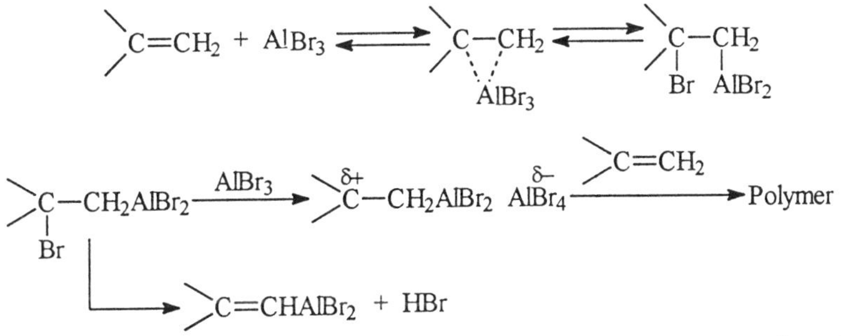

In accordance with [6], polymerization may also be activated by autoionization by Friedel-Crafts halides:

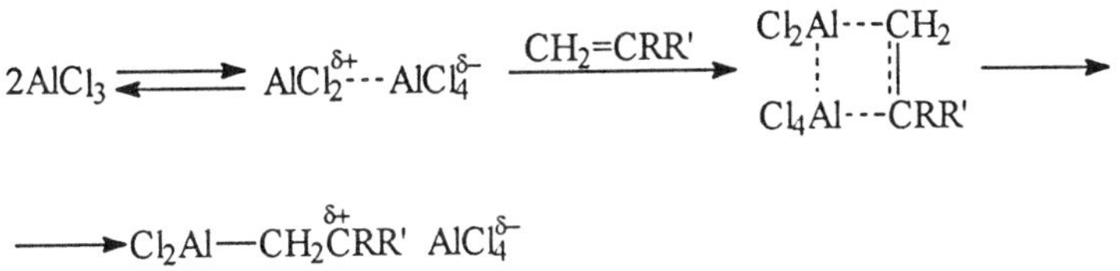

Proceeding through self-ionization by the scheme

$$Br_2Al^{\delta+}AlBr_4^{\delta-} + CH_2 = C(CH_3)_2 \rightarrow Br_2AlCH_2\overset{\delta+}{C}(CH_3)_2AlBr_4^{\delta-}$$

is confirmed in [7, 8], where fast and quantitative polymerization of isobutylene in a system almost free from humidity has been described. It is supposed that mixed Friedel–Crafts halides of the $AlBr_3{\cdot}TiCl_4$ type act similarly [7–9].

If H atoms exist in the β-position to C=C-bond in olefins, initiation by capture of hydride ion by an aprotic acid is possible [10]:

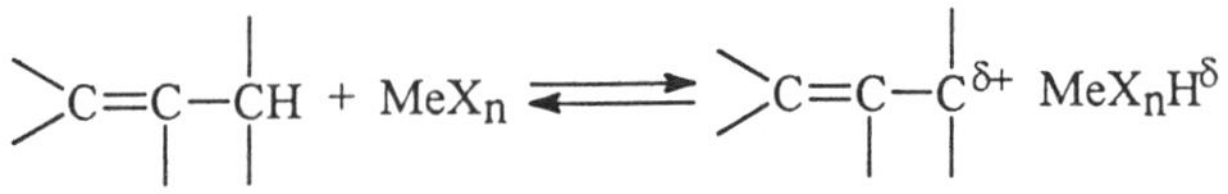

However, analysis of schemes of cationic polymerization self-initiation by Lewis acids, including those described in [4], indicates high specificity of these reactions. That is why, at best, these reactions are relevant for particular schemes and process conditions only. For example, direct addition of an acid requires its high excess, self-ionization — the presence of a polar solvent, and formation of a hydride ion — obligatory existence of labile (allyl, for example) hydrogen in a monomer.

Results of many recent studies have often disproved the possibility of excitation of olefin polymerization processes by Lewis acids only [11, p. 291], apparently, excluding such systems as $AlCl_3$, $AlBr_3$, $C_2H_5AlCl_2$ [12, 13]. It should be noted that experimental difficulties related to purification and drying of a monomer, solvent, catalysts, walls of reactive vessels, etc. provide no unambiguity of a self-excitation mechanism of the cationic polymerization by individual aprotic acids.

Obviously, in the general case, rebuilding of π-complexes proceeds in the presence of a protogenic additive which produces a modified reaction mechanism:

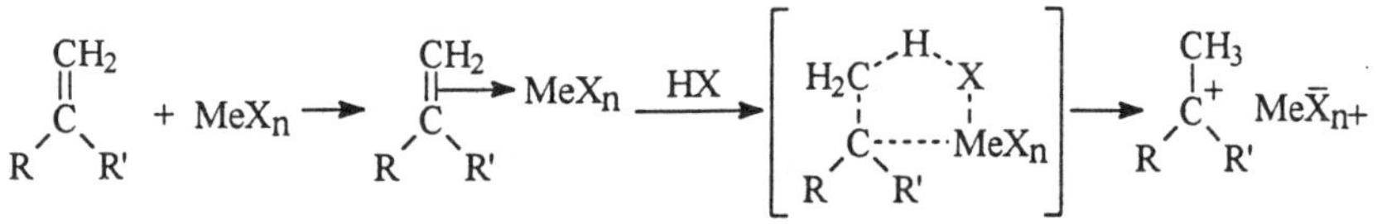

An alternative for proton or aprotic excitation of polymerization in the presence of Lewis acids appears only at extremely low concentrations of possible activating bases. If a protogenic additive is present in the system before the beginning of the catalysis stage (initiation) in amounts comparable with the amount of the Lewis acid, then protonation of an olefin and formation of corresponding complex counterion will always happen. Under similar conditions the donor–acceptor interaction of Lewis and Brönsted acids proceeds with formation of complex compounds, which contrary to initial proton donors represent harder but non-ionized acids. Ionization of complexes and initiation of polymerization connected with it in media with low dielectric permeability, which are olefins and hydrocarbon solvents, proceeds only at participation of monomers or accepting bases.

The effect of acidic properties increase in systems at the complex formation (Table 2.1) is accompanied by a significant growth of their catalytic activity.

Table 2.1.

Values of the acidity function of some proton and complex acids.

Proton acid	$-H_0$	Reference	Proton acid	$-H_0$	Reference
HF	9.7	[14]	$HF(HCl)Br_3(AlCl_3)$	16.6	[18]
H_2SO_4	11.9	[15]	$HSO_3F{\cdot}SbF_5$	17.5	[19]
HSO_3Cl	12.8	[16]	$9HF{\cdot}SbF_5$	20	[20]
HSO_3F	13.9	[17]	$HF{\cdot}SbF_5$	Over 20	[20]

In this manner it can be stated now that the process of cationic polymerization of olefins does not usually begin or proceeds only slowly on mixing pure and dry monomer and a catalyst — aprotic acid. In many systems, the most careful purification and drying off of reagents cause only a small, often unrepeatable, decrease of the initial rate of the process. It may be that in some cases one of components serves as a co-catalyst in the system, for example, the solvent, one of co-monomers (during co-polymerization) or a product of side reactions of the catalyst.

In most cases, complex acids are initiators of the cationic polymerization of isobutylene, i.e. the reagents serve as proton donors: $HClO_4$, $H^{\delta+}(AlCl_3X)^{\delta-}$, $H^{\delta+}(BF_3OH)^{\delta-}$; $H^{\delta+}(R_2AlClOH)^{\delta-}$ (where X is Cl, OH, etc.).

Another large group comprises compounds with a preliminarily formed cation–active particle of the type $R^{\delta+}(BF_3OR)^{\delta-}$, $K^{\delta+}(AlCl_4)^{\delta-}$, and so on. Such systems are usually formed directly at the initiation stage in the presence of a monomer. Their value as initiators of polymerization is defined by the acidity strength, stability and other factors.

Other initiating particles should also be mentioned. For example, a special case of initiation is the introduction of isobutylene by "active and adjusted to isobutylene" *tert*-carbon atom–oxygen bond. The case appears at formation of alkyl(aryl)acetate complexes with BCl_3 [21]. In the case of polymerization proceeding in the phosgene medium (catalyst $AlCl_3$ or $TiCl_4$) [22], carbonium ions $C^+(O)Cl$ with admixtures of carboxonium or chloronium structures are apparently active. The behavior of complex organic acid haloid–hydride–Lewis acid systems is unusual. They display proton activity under stoichiometric ratio of components and generate RC^+O acylium ions under the component ratio of 1:2 [23, 24]. It has been proved that only the dimeric complex ion $Al_2Br_7^-$ provides the necessary concentration of acylium ions. That is why corresponding systems serve for introduction of end carboxyl groups into polyisobutylene macromolecule.

A self-contained group of initiators is composed from systems with a positively charged atom of a halogen. Beside syn-catalytic systems $Cl_2 + AlR_3$ [25], other catalysts are suggested, in which chloronium ions and carbonium ions stabilized by chlorine are active. For example, activation of isobutylene polymerization by 2,5-dichloro-2,5-dimethylhexane combined with BCl_3 is ascribed to the $\overline{C(CH_3)_2CH_2CH_2C(CH_3)_2C}\overset{+}{l}BCl_4$ ion [26]. In the $AlCl_3$–CO_2Cl_2 system, initiation is bound to carbonium ion $CH_2\overset{+}{=}Cl$ [27]. Of interest is that at protonation vinyl chloride give chlorocarbonium structures, inactive in polyvinylchloride polymerization but able to initiate isobutylene polymerization [28, 29].

Consideration of known catalysts and various schemes of initiation predetermines expediency of the general approach to initiation of electrophilic polymerization processes in the context of the acid–base interaction theory. Initiating properties of electrophilic catalysts are defined by the measure of acidic properties, because cationic polymerization represents an original sequence of events proceeding due to the principle of acid neutralization (active center) by a base (monomer) [30, 31]. Such interactions may be classified as a peculiar case of acid–base interactions in non-water media, particular nature of acid (Lewis, Brönsted or their combination) and base (monomer, electron–donor solvent) defining the processes' specificity.

From the point of view of quantum chemistry, any acid–base interaction represents a superposition of wave functions ψ_0 and ψ_1, which characterize a system with no charge transfer (electrostatic, dipole and other interactions) and after electron transfer from donor to acceptor, respectively:

$$\psi_{AD} = a\psi_0(AD) + b\psi_1(A^-D^+).$$

Under such interaction the change of energy is the sum of an electrostatic component, bound only to charge characteristics of donor and acceptor centers, and a covalent term, which is a function of overlapping orbitals of appropriate symmetry [32]:

$$\Delta E = -\frac{q_D \cdot q_A}{R_{DA} \cdot \varepsilon} + 2\Sigma_1 \Sigma_2 \frac{\left(c_D^m \cdot c_A^n \beta_{DA}\right)^2}{E_m^* - E_n^*},$$

where q_D, q_A are full charges on donor and acceptor centers, respectively; R_{DA} is the distance between the donor and acceptor; ε is the dielectric constant of the solvent; Σ_1, Σ_2 are occupied and free orbitals of donor and acceptor, respectively; c_D^m, c_A^n are coefficients of the donor orbital m on donor and acceptor orbital n on the acceptor atom; β_{DA} is the resonance integral between the donor and acceptor at the distance R_{DA}; E_m^*, E_n^* are energies of the donor orbital m in the acid field and acceptor orbital n in the base field, respectively.

In the generalized form, convenient for the analysis, the contribution of both types of interaction is calculated in semiquantitative HSAB theory (hard and soft acids and bases), according to which reactions of acids and bases should be considered with due regard for the strength factor (hardness) and polarizability (softness) [33].

$$\lg k = S_a \cdot S_b + \sigma_a \cdot \sigma_b,$$

where k is the kinetic constant; S_a, S_b and σ_a, σ_b are forces and factors of polarizability of acids and bases, respectively.

It is significant that the nature of the cationic particles in electrophilic polymerization of olefins ($H^{\delta+}C^{\delta+}$) allows analysis of the elementary stages with due regard for both strength and hardness–softness factors. The more so, implying that the hardiness factor is a synonym of strength, and the softness factor, of weakness, the suggested approach also demonstrates their non-coincidence and independence. For example, at initiation of polymerization by H–acids, hard acidic protons somehow level differences in basicity of olefin monomers and, vice versa, soft acidic centers differ by increased selectivity in relation to them. Hard acidic (hard) centers ($H^{\delta+}$) of "superhard" or "magic" acids according to the HSAB concept in preference react with hard (weakly polarized) bases, in particular, with simple carbon–carbon bonds, causing them to break [34].

Polymerization:

Chain termination:

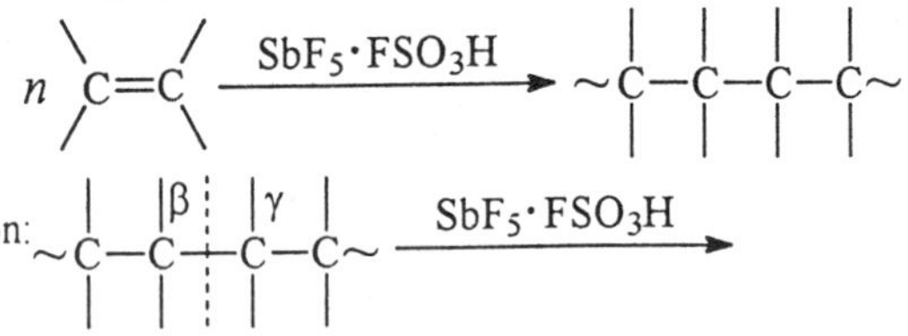

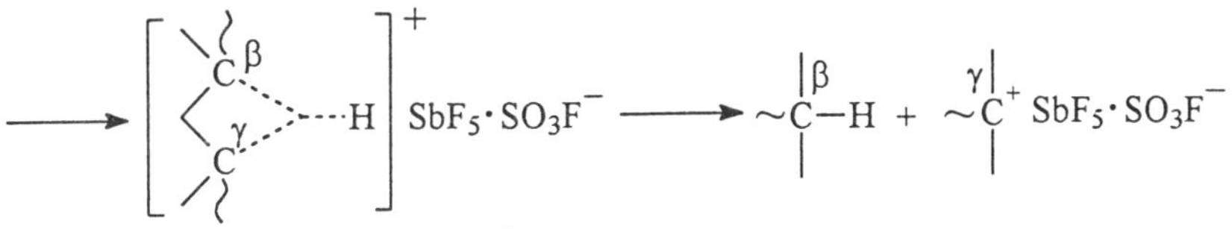

In the case of soft (weak) acids ($H^{\delta+}$) interactions with soft bases able to be polarized (for example, in the case of olefin monomers — with alkyl– or aryl–substituted ones of ethylene) are more probable. Processes in such systems proceed at the expense of increase of acid–base properties of components in the transition state as a result of polarization. In the limiting case of such interactions the acid–base mechanism transforms into the consistent (concert) one:

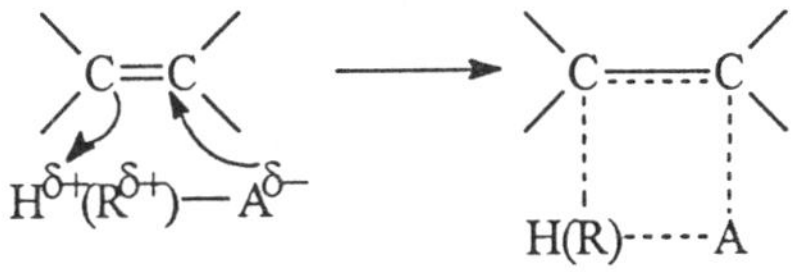

Obviously, most known cationic catalyst—olefin monomer systems are intermediate between the two discussed cases.

The reaction of olefin protonation, simple at first glance, is not always explained by classic notions of carbonium ion formation. The dependency of olefin protonation rate on the Hammett acidity function H_0, i.e. on the nature of the acid anion, and not on the concentration of hydrogen ions, may bear witness to the role of "non–classic" ions as predecessors of the classic structure [1]

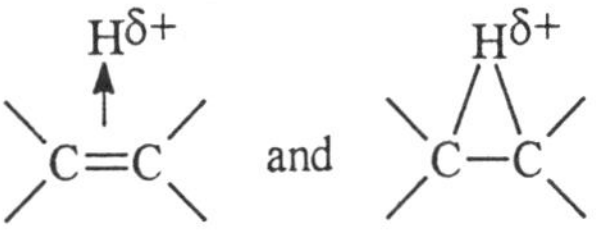

If low nucleophilicity of isobutylene double bonds and the charge on the β-carbon atom (–0.09) are taken into account, an interaction of this kind seems more profitable than direct attack on the C_β atom.

The influence of ionic (electrostatic) and covalent (orbital controlled) types of interactions may be traced at initiation of olefin polymerization by carbonium ions, which is also considered as the polymeric chain propagation. In line with the HSAB concept "amorphism", i.e. the ability to react with both "hard" and "soft" bases, is typical for carbocations [33].

As indicated in Table 2.2 displaying some characteristics of single–charge ion sequence, the electrostatic parameter (e^2/r) of carbocations is significantly lower than that of alkaline metal ions.

The ability to form covalent bonds ($I - Q$), vice versa, is more typical for carbonium ions, because H^+, Li^+, Na^+ ions are similar to inert gases in their electron structure, and carbocations contain unpaired electrons in their external 2p-envelope. Taking into account that processes of cationic polymerization of olefins are performed in non-aqueous media with low or moderate dielectric permeability, participation of carbocations in ionic interactions with hard bases depends on their reactions with p- and π-electrons of soft bases (olefins), dual behavior of carbonium ions as a hard or soft acid in relation to particular conditions of polymerization being responsible for the result of various types of competition of possible reactions in the system.

Table 2.2.

Characteristics of single–charged ions.

Cation	Radius *r*, nm	Ionization potential *I*, kJ/mol	Hydration heat *Q*, kJ/mol	e^2/r *	$I - Q$*
H^+	$1 \cdot 10^{-4}$	1306.3	1101.1	$1 \cdot 10^5$	49
Li^+	7.8	519.2	523.35	1.28	−1
Na^+	9.8	494.05	418.69	1.02	18
CH_3^+	17.5	942	255.39	0.57	164
CH_5^+	20.5	795.5	226.09	0.49	136
$(CH_3)_2CH^+$	23.8	720.1	192.59	0.38	126
$(CH_3)_3C^+$	28.0	661.5	167.47	0.36	118

* These are suggested as electrostatic (e^2/r) and covalent ($I - Q$) characteristics of ions [35].

In line with regularities of the acid–base interaction, the activity of complex "Lewis acid–cocatalyst" systems in electrophilic polymerization of olefins depends on their strength (Table 2.3) [36].

Table 2.3.

Diagram of electrophilic activity of "Lewis acid–electron donor additive" system*.

$BF_3 \cdot R_2O$	$AlCl_3 \cdot R_2O$	$RAlCl_2 \cdot R_2O$	$R_2AlCl \cdot R_2O$	$R_3Al \cdot R_2O$
$BF_3 \cdot ROH$	$AlCl_3 \cdot ROH$	$RAlCl_2 \cdot ROH$	$R_2AlCl \cdot ROH$	$R_3Al \cdot ROH$
$BF_3 \cdot H_2O$	$AlCl_3 \cdot H_2O$	$RAlCl_2 \cdot H_2O$	$R_2AlCl \cdot H_2O$	$R_3Al \cdot H_2O$
$BF_3 \cdot Hhal$	$AlCl_3 \cdot Hhal$	$RAlCl_2 \cdot HHal$	$R_2AlCl \cdot Hhal$	$R_3Al \cdot HHal$

* It is derived from the data in periodicals and patents [18]; the range of activity of the systems lies below lines in each column; Lewis acid radicals may be the same or different.

As the strength of Lewis acids decreases from left to the right (Table 2.3), the selection of activators from unitypical series assisting AC generation is narrowed. The hardest acid (BF_3) causes ionization of all activating bases [37] and in this sense levels their individuality. Obviously individuality of activating bases (the ability to be ionized) is also leveled in the opposite case — if a soft Lewis acid is used in due regard to its low acceptor properties. Moreover, high mobility of organic groups at the Al atom is typical for R_3Al and partly for R_2AlCl. In the case of intermediate strength acids (Table 2.3, R_nAlCl_{3-n} with $n < 3$) a differentiating effect of the Lewis acid is displayed in relation to co–catalytic additives, diving a different capability for their ionization and, consequently, initiation of electrophilic process is observed.

To initiate polymerization a great number of exciters[1] — Lewis and Brönsted acids — are used, which are easily classified in respect to determination of the role of Lewis and Brönsted acids complex formation and the effect of proton acceptors in this process. The symmetrization system may be based on the most typical signs of

[1] A definite drawback is the absence of a general terminology and strict definitions of the terms "catalyst", "initiator", as well as "cocatalyst" and "co-initiator" in relation to cationic processes. In many cases terms of initiation and catalysis in respect to action of substances causing cationic polymerization are equally tolerable. The term "exciter" is also suitable.

complex systems — chemical and physical nature of components, their stoichiometry, catalytic activity, phase state of catalysts, etc. — but obviously subdivision of the systems due to their acidic–catalytic properties is the most suitable:

comparatively soft solid acids, which include hydrated surfaces of oxides and salts of metals, as well as natural compounds (crystal hydrates of metal salts included);

complexes of hard Lewis acids–Friedel–Crafts metal halides (based on B, Al, Ti, Sn, etc.) with various cationogenes;

"superacidic" systems — the complexes of hard Lewis and Brönsted acids.

Such classification of catalysts is generally coincident with the change of their electrophilic activity (grows from the first to the fourth group). Changing of the phase state (first two groups — heterogeneous exciters, the third and fourth groups — mostly homogeneous ones) may not be taken into consideration, because from the point of view of the acid–base catalysis the difference between homogeneous and heterogeneous catalysts are not significant [38].

Soft solid acids are related to the class of the most stable exciters of cationic polymerization in the range of 320–575 K, which is in respect of the absence of labile groups in their structure. Activity of acids in processes of olefin oligomerization is not high [39] and is highly dependent on the catalyst composition and conditions of their preparation [40]. Identification of acidic centers allows relation of soft solid acids to the class of complex catalysts. For example, coordination — unsaturated atoms of Al – the Lewis acidic *Z*-centers — are present on dehydrated surface of Al_2O_3, which in the presence of water (residual or injected) are able to transform into Brönsted acids:

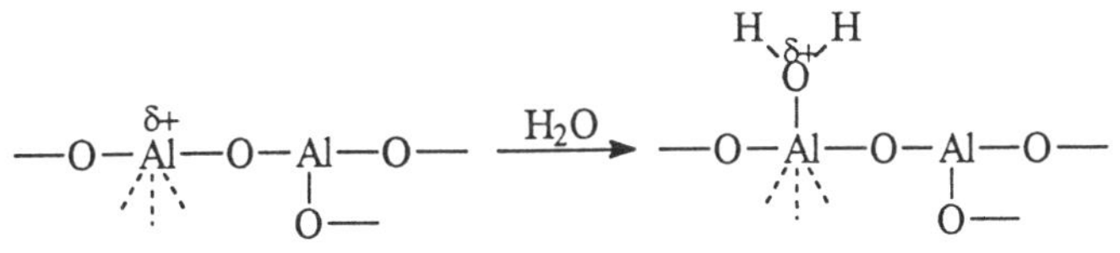

Z-center B-center

Of importance is that OH-groups on the surface of fluorinated Al_2O_3 are not carriers of Brönsted activity [41], and only coordination-linked H_2O is the only source of proton centers differing by increased acidic strength ($H_0 \approx -16.3$).

A significant feature of soft solid acids is the presence of several types of ACs, which usually implies little selectivity of exciters of electrophilic processes. In particular, at increased temperatures acidic centers of Al_2O_3 of moderate strength are responsible for the fission of double bonds in olefins, and the more acidic ones for skeleton isomerization [40].

In due regard to the above-considered scheme, active *B*–centers in alumosilicates are also formed:

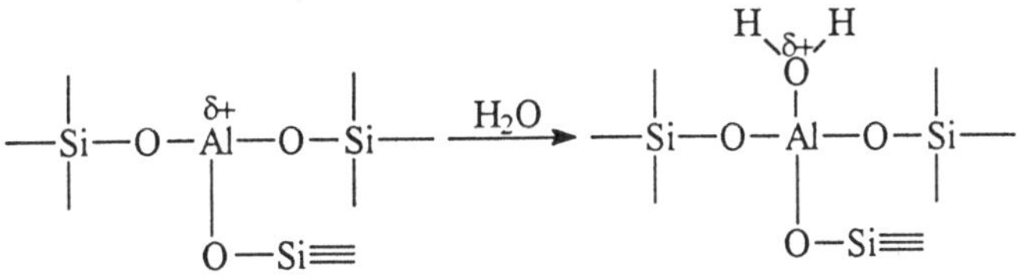

Z-center B-center

Hard acidic centers of alumosilicates are ascribed to oxonic hydrogen ions, formed at water dissociation during adding to structure of the solid acid:

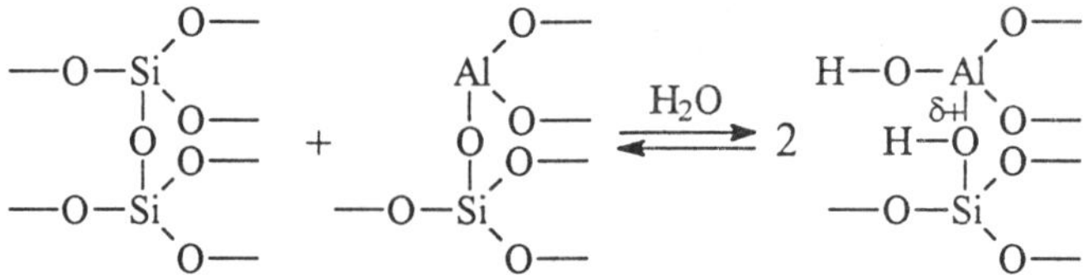

A change in the environment of the Al atom is displayed in a significant change of acidic properties and activity during polymerization of olefins. For example, Floridin in the amount of 0.35 mmol/g, containing acidic centers possessing $H_0 < -8.2$, polymerizes isobutylene at 198 K into a product with a molecular mass of about 800 [42, p. 15]. Isobutylene is easily oligomerized in the presence of synthetic alumosilicate at 313 K forming di-, tri- and tetramers [11, p. 118].

It is typical that the activity of alumosilicate in polymerization of olefins abruptly decreases after the exchange of surface H-atoms for Na ions [43] and correlates with its Brönsted acidity [44], determined by the exchange of NH_4^+ ions, and not by total acidity by butylamine. Oxidative and acidic–catalytic properties characterize synthetic zeolites containing as cation elements of groups I and II of the Periodic Chart. This is the class of hydrated carcass alumosilicates, Lewis centers of which are transformed into Brönsted *B*–centers by adding H_2O [45], high polarizing strength of the exchange cation (increase of the cation charge ratio to its radius, $-e/r$) promoting easier protonation of surface groups — OH-groups and adsorbed H_2O molecules [46, 47].

Activity and selectivity of zeolites in transformation of butenes (323 K) may be regulated by changing their composition (by nature of cation, SiO_2/Al_2O_3 ratio, the exchange degree), process conditions, introduction of astoichimetric components [48, 49]. Introduction of CO_2 increases activity and selectivity of zeolites in isobutylene oligomerization into trimers, which is stipulated by an increase of the catalyst activity.

$MgCl_2$ may also be related to the list of solid complex acids (in the presence of low additives of H_2O). Although the catalyst is characterized by relatively low activity in isobutylene polymerization (30% conversion at 263 K, 1–3 h), it allows production of quite high–molecular polymeric products.

The process of isobutylene polymerization excitation is bound to formation of solid Brönsted acids:

$$(MgCl_2)_{n(\text{solid})} \xrightarrow{H_2O} [(MgCl_2)_n OH]^{\delta-}_{\text{solid}} H^{\delta+} \xrightarrow{CH_2=C(CH_3)_2}$$

$$\rightarrow [(MgCl_2)_n OH]^{\delta-}_{\text{solid}} \overset{\delta+}{C}(CH_3)_3$$

Active centers of the Brönsted type are also formed in sulfates of metals (Ag, Cu, Ca, Pb, Zn, Mg, etc.) under coordination with H_2O. In particular, a *B*–center with Ni^{5+} possessing free sp^3d^2-orbitals occurs at transfer of the hydrated form into the anhydrous one. Monohydrates of metal sulfates also display low acidity, stipulated by specific interaction of H_2O molecules with the central metal ion in the octahedral coordination [50]. For isobutylene polymerization, a connection between activity of such catalysts and acidic properties of salts exists, i.e. electron-acceptor ability depending on electronegativity of the metal (on the example of sulfates of Ag, Cu, Ca, Pb, Zn, Mg [40, p. 131]).

Hence in the case of solid acids, application of catalytic properties are also displayed by such soft acids as H_2O (by means of its coordination effect with Lewis solid acids). The specificity of acidic activity of catalysts of this type (low molecular masses and action selectivity) is related to their heterogeneous character and high working temperatures.

Complexes of mineral acids with solid supports (Lewis acids) display catalytic activity in the processes of cationic polymerization of various polymers — isobutylene, first of all.

Individual hydrogen halides do not display any noticeable affinity to polymerization excitation. Usually, interaction of HX with olefins possessing electron–donor substituents (isobutylene, propylene), causes formation of ionic addition products — appropriate alkyl chlorides [51]:

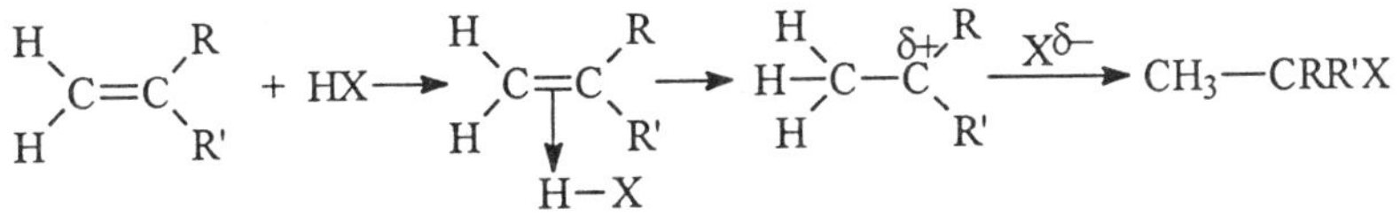

The reaction proceeds through the formation of π–complexes. It is a fast equilibrium process with the formation enthalpy (on the example of HCl interaction with ethylene, propylene, isobutylene, tetramethylethylene) of 10–22 kJ/mol [52], which barely exceeds the values of $\Delta H \leq 9.3$ kJ/mol for analogous complexes with Lewis acids.

HX coordination with metals, salts, etc. adheres properties of catalysts of cation–active monomers' polymerization to mineral acids [53], Brönsted ACs being formed by the standard scheme:

$$HX + Z \rightleftharpoons H^{\delta+}[X \rightarrow Z]^{\delta-}$$

Owing to the coordination effect, the concentration of protons grows, which leads to an increase of the initiation rate and significant slowing down of Cl ions reaction with growing polymeric carbocation.

The nature of the coordinating additive defines the acidic–catalytic properties of HX. On the example of heavy metal sulfate complexes with HCl, the increase of activity in the sequence Zn < Sn < Pb < Cu is shown correlating with the change of an electronegativity parameter of the salt cation $X_i = (1 + 2z)X_0$, where X_0 is the electronegativity of a neutral atom according to Pauling; z is the charge on the metal ion [54–57]. Along with the activity, stability of complexes to HCl elimination increases at oligomerization of isobutylene that correlates with change of the salt structure and acid state at coordination

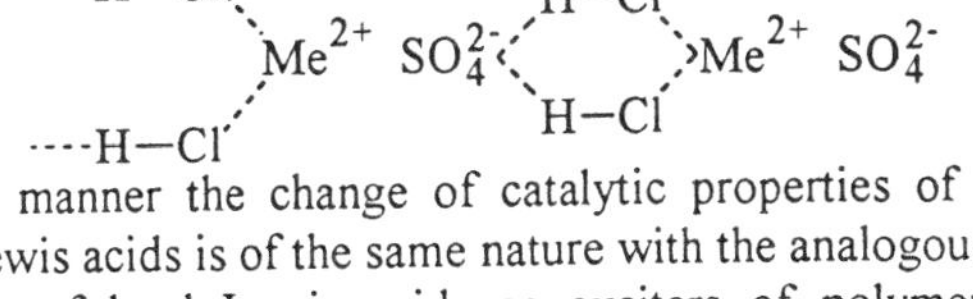

In such a manner the change of catalytic properties of mineral acids HX combined with Lewis acids is of the same nature with the analogous method of H_2O.

Complexes of hard Lewis acids as exciters of polymerization are rather multiple which provides a wide selection of Lewis acids and proton-donor additives.

Activity of complex catalytic systems depends on both the nature of the aprotic and proton acids and conditions of the complex formation, ratio of components, stability of counterion and other factors. Of special role at the interaction

$$B + HCl + MeCl_n \rightarrow BH^{\delta+}MeCl_{n+1}^{\delta-}$$

where B is the base, including a monomer;

is the acidity of the system. The rate of isobutylene polymerization is linearly dependent on the acidity function of initiator H_0 [58].

Efficiency of complex catalysts of HX–MeX$_n$ type in electrophilic processes is defined by the counterion resistance to elimination of ligands at the interaction with carbonium ions [60]. For acids of HBF_4 or $HalCl_4$ type, reactions of the polymer chain limited by the counterion fragment (F^- or Cl^- anions) are not typical, however, this is probably possible for $(CH_3)_2AlCl$ [61]; in this case, in cationic polymerization of olefins formation of chlorinated side products is supposed:

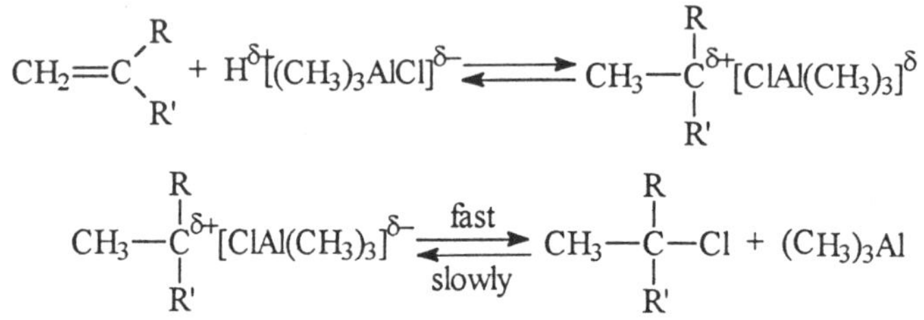

With increase of acidity in the sequence $R_3Al < R_2AlCl < RAlCl_2$ the Al–Cl bond in complex counterions (R_nAlCl_{4-n}, where n = 1 – 3) formed at HCl(RCl) inteaction with appropriate Lewis acid becomes more stable [60]. Alkylation of a carbocation by the carbonic fragment of a complex counterion is the predominant reaction of the polymer chain limiting:

Activity of these systems in formation of high-molecular polyisobutylene grows with stability of counterions in the sequence $(C_2H_5)_3AlCl^- < (CH_3)_3AlCl^- < (C_2H_5)_2AlCl_2^- < (CH_3)_2AlCl_2^-$.

Complexes of hard aprotic acids with H_2O form an important group of cationic polymerization initiators. Strength of the donor–acceptor interaction with proton–donor heteroatom is defined by the nature of the metal–complex former and its electronic configuration.

Of significant importance is the dependence of the catalytic activity of complexes of hard acids with water on stoichiometry. Generally, 1:1–complexes are most effective [62, p. 107]. Additive compounds of alternate composition of the type 2:1 for H_2O/BF_3, (2 to 6):1 for $H_2O/AlCl_3$, $SnCl_4$ hydrates containing from 3 to 8 water molecules, etc., as a rule, are inactive. Absence of the catalytic activity in cationic polymerization of the $SnCl_4 \cdot 3H_2O$ complex is stipulated by formation of oxonic salts $(H_3^+O)_2[SnCl_4(OH)_2]^{2-}$; the same is true for complexes $SbF_5 \cdot 2H_2O$, $BF_3 \cdot 2H_2O$, etc., which form oxonic salts of appropriate acids — $H_3O^{\delta+}SbF_5OH^{\delta-}$; $H_3O^{\delta+}BF_3OH^{\delta-}$, etc.

1:1–Complexes of $BF_3 \cdot H_2O$ easily cause isobutylene polymerization. The maximum polymer yield (close to 100 mass%) is reached at equimolar BCl_3/H_2O ratio in various solvents (183–263 K) [63]. It is suggested that termination of polymeric chains occurs at carbocation chlorination by the $[BCl_3OH]^{\delta-}$ counterion, i.e. with the fission of the relatively weak B–Cl bond.

Higher (stable) $SnCl_4$ hydrates (for example, $SnCl_4{\cdot}3H_2O$, $SnCl_4{\cdot}2H_2O$) are catalytically inactive, but isobutylene polymerization is excited by 1:1–complexes $SnCl_4{\cdot}H_2O$ [64]. It is assumed that 1:1–complexes with participation of a monomer of $SnCl_4{\cdot}M{\cdot}H_2O$ type are more active. In keeping with this, acceleration of isobutylene copolymerization with styrene is observed on adding more active isobutylene. The deterministic role of $SnCl_4{\cdot}H_2O{\cdot}IB$ complexes in understanding regularities of cationic polymerization was also confirmed in refs. [65, 66].

Of great interest are relatively new complex catalysts based on R_nAlHal_{3-n} (n = 1, 2) [18, 36; 67, p. 132]. Mono– and dialkylaluminum halides, which initiate no polymerization under specially pure conditions [68, 73, 74] or display low activity [13], form effective cationic systems combined with H_2O and an alcohol. Halogenaluminum–organic compounds (HAOC) are related to the substances, which are extremely reactive to H-compounds. Generally, these reactions are out of control [75], but controlled interaction of the components is achieved by using diluted (about 100 mol/m^3) solutions of HAOC in inert solvents, under moderate temperature (273–295 K) and keeping HAOC excess by H–compounds during the reaction. Of crucial importance is the method of the proton-donor introduction, which would exclude any possibility of its fluctuations in the solution and, consequently, formation of HAOC decomposition products, for example, via crystallohydrates, crystalloalcoholates, crystallohydrochlorides [70, 76, 77]. Though the presence of aluminum-organic compounds shifts the temperature threshold of H_2O, HCl and CH_3OH elimination from appropriate crystalloadducts, the process is ruled by regularities of topochemical reactions, i.e. it is defined by the solid phase transformation. On the contrary, the rate and character of the H–compound interaction with HAOC in solution depend on the nature of solvent and Lewis acid. In heptane, activity of Al compounds in relation to water, alcohol, HCl decreases in the following sequence: $(C_2H_5)_3Al > (C_2H_5)_2AlCl > C_2H_5AlCl_2$. The interaction is accelerated in benzene, and for water the reverse sequence of activity occurs: $C_2H_5AlCl_2 > (C_2H_5)_2AlCl > (C_2H_5)_3Al$, and for HCl and CH_3OH it is represented by the sequence $(C_2H_5)_3Al > C_2H_5AlCl_2 > (C_2H_5)_2AlCl$.

The deterministic influence on direction of the HAOC reaction with H-compounds, H_2O, for example, is caused by the solvent (Table 2.4).

Interaction of $C_2H_5AlCl_2$ with H_2O is accompanied by Al–C bond break in hard electron–donor solvents and preserving Al–C bond in solvents with low values of donor numbers and comparatively low ε. The intermediate position is occupied by arenes, nitromethane, α–olefins, for which partial hydrolysis of $C_2H_5AlCl_2$ is accompanied by the olefin oligomerization. Hence the decisive role in defining the reaction direction is played by electron–donor properties, but not the solvent polarity.

Of special interest is formation of a HAOC complex with H_2O, at which interaction of components does not either touch Al—C bonds or is not accompanied by the solvent coordination with the reaction product (aliphatic or aromatic solvents). For example, aquacomplexes of HAOC with H_2O are formed along these lines [70, 77] (Table 2.5).

Formation of aquacomplexes in hexane for various HAOC relates to a decrease of Al—C bonds' capability for hydrolysis in the sequence $(C_2H_5)_2AlCl > (C_2H_5)_3Al_2Cl_3 > C_2H_5AlCl_2$ that coincides with the decrease of Al—C bonds' polarity that follows from electronegativity of Al in these compounds (1.37, 1.52, and 1.61, respectively). In consequence of additional activation of Al—Cl and Al—C bonds [76], aquacomplexes of alumoxanes are always formed in arenes (benzene).

Table 2.4.

Influence of the solvent nature on the character of $C_2H_5AlCl_2$ interaction with H_2O.

Solvent	Dielectric constant, ε	Donor number (from [100])	Character of $C_2H_5AlCl_2$ interaction with H_2O	
			With Al–C bond break	No break of Al–C bond
Hexane	1.9	0	–	+
Heptane	1.9	0	–	+
Cyclohexane	2.0	0	–	+
Methyl chloride	6.2	0.1	–	+
Methylene chloride	9.0	0.1	–	+
Higher α–olefins*	2–2.2	—	+	+
Nitromethane	35.9	2.7	+	+
Benzene	2.3	4.9	+	+
Toluene	2.4	—	+	+
Sulfolane	42.0	14.8	+	–
Diethyl ether	4.3	19.2	+	–
Tetrahydrofurane	7.6	20	+	–

* From hexene-1 to decene-1.

Existence of HAOC aquacomplexes is rather strictly proved in [70, 71, 77][2]. In PMR spectra of $C_2H_5AlCl_2 \cdot H_2O$ in C_6D_{12} signals of methyl protons with δ_t 1.13 m.p., J 7.0 Hz (3H) and methylene protons with δ_q 0.52 m.p., J 7.0 Hz (2H) are observed, which are little different from signals of appropriate protons in $C_2H_5AlCl_2$ with respect to position or intensity. UV–spectra of the complex display new bands of charge transfer: an intensive one with λ_{max} 201 nm (lgε 3.09) related to the $\sigma \rightarrow \sigma^*$–transition of electrons, and a weak one with λ_{max} 410 nm (lgε 0.97) conforming with the prohibited $n \rightarrow \pi^*$–transfer of electrons from $2p_y$–orbitals of water oxygen to vacant $3d^0_{xyxzyz}$–orbitals of Al able to form π–bonds. Appearance of the band at 457 cm^{-1} in the IR-spectrum of the complex, related to oscillations of the donor–acceptor (DA) bond Al—O, and noticeable shears of absorption bands of $C_2H_5AlCl_2$ at 490 and 626 cm^{-1} (oscillations of Al—Cl and Al—C bonds, respectively) also testify to water linking in the complex with $C_2H_5AlCl_2$.

Table 2.5.

Types of complexes in the reaction of halogen–containing organoaluminum compounds with H_2O.

Halogenaluminum–organic compound	Solvent	
	Hexane	Benzene
$(C_2H_5)_2AlCl$	$[Cl(C_2H_5)Al]_2O \cdot H_2O$	$[Cl(C_2H_5)Al]_2O \cdot H_2O$
$(C_2H_5)_3Al_2Cl_3$	$[Cl(C_2H_5)Al]_2O \cdot H_2O$ $C_2H_5AlCl_2 \cdot H_2O$	$(C_2Al)_2O \cdot H_2O$ $[Cl(C_2H_5)Al]_2O \cdot H_2O$ $(C_2Al)_2O \cdot H_2O$
$C_2H_5AlCl_2$	$C_2H_5AlCl_2 \cdot H_2O$	$[Cl(C_2H_5)Al]_2O \cdot H_2O$

[2] Probability of HAOC complex formation is mentioned in [79, 80], but no proofs are displayed.

Exceptionally high resistance of the Al—C bond of the complex to hydrolysis under increased temperature (375 K) and anomalous behavior in some reactions, in a ligand–ligand substitution, in particular, should be noted. Water is not displaced from the complex by stronger electron donors (ester). Metallic sodium does not interact with $C_2H_5AlCl_2 \cdot H_2O$ solutions (273 K) which testifies to the absence of free water in the system. Increased electroconductivity of $C_2H_5AlCl_2 \cdot H_2O$ solutions ($\varkappa = 1.5 \cdot 10^{-10}$ $Ohm^{-1} \cdot cm^{-1}$) compared with $C_2H_5AlCl_2$ ($\varkappa = 5 \cdot 10^{-12}$ $Ohm^{-1} \cdot cm^{-1}$) indicates an increase of the structure polarity at the complex formation, however, in accordance with the qualitative reaction with xanthone, $R_nAlCl_{4-n} \cdot H_2O$ complexes ($n < 3$) are not H–acids.

Hence $C_2H_5AlCl_2 \cdot H_2O$ represents a modified Lewis acid, and its structure is more accurately depicted by the formula

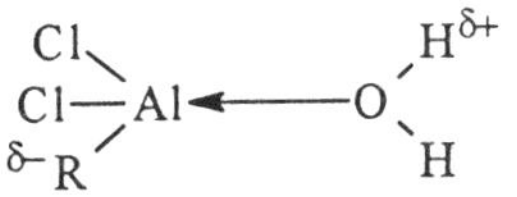

[81], than by the traditional one $H^+(RAlCl_2OH)^-$ [82]. Analogous proofs of $RAlCl_2$ complex formation were obtained for hydrogen sulfide and alcohol [77].

The general property of all studied HAOC complexes is reduction of the Lewis acidity of aluminum at the sacrifice of saturation of free p–orbitals by electrons of a donor additive. Important information on this point was obtained by semi-empirical quantum–mechanical calculations [83–85]. Complexes exist in diluted solutions of indifferent solvents, and they may be considered in the "approximation of an isolated molecule". The structure of $RAlCl_2$ complexes with H_2O, alcohol and HCl was calculated by the SCF MO LCAO method[3] in the CNDO/2 approximation. The complexes were represented as a rectilinear tetrahedron because, in accordance with preliminary estimations, distortion of the configuration at the sacrifice of non-equivalence of substituents at Al atom cause no sufficient change of the electron density and difference in energies of the main state. The most profitable configuration of electrophilic centers was determined by the minimum of the main state energy varying position of the complex components. Changes of the E—H bond strength (where E is a heteroatom of an electron–donor additive) and its capability for protonation are suggested to be strictly connected with the bond degree change, obtained from calculations.

In accordance with the type of H–compound, complexes of various configuration types were found to be stable. In the case of HCl, formation of Brönsted $HalX_4$ acids with a small charge on the hydrogen atom is energetically non-profitable, and the electrophilic center possesses a composition close to the Brönsted acid $H^+(AlX_4)^-$. In the absence of additional polarizing factors, the complex with H_2O is stable and, despite nonequivalence of H atoms (charges of 0.37 and 0.26, degrees of bonds with oxygen of 0.41 and 0.57), detachment of one of them as a proton is not favorable. Consequently, the composition $H^{\delta+}(OHAlRCl_2)^{\delta-}$ can be adhered to the electrophilic center. For the complex with alcohol (R_1OH), the removal of an R^+ cation appears energetically profitable, which leads to occurrence of an electrophilic center $R_1^+(OHAlRCl_2)^-$ in polar and $R_1^{\delta+}(OHAlRCl_2)^{\delta-}$ in non-polar media. For such complexes the stage of initiation, connected with attachment of a proton, must be

[3] The method of self-consistent field in approximate molecular orbitals, approximated by linear combinations of atomic orbitals: additional approximation — complete neglecting of the differential overlap.

absent, but a possibility of direct initiation by a carbocation appears. A difference in the AC nature in systems of $RAlCl_2$ and various activators, which illustrates the peculiar differentiating action of the Lewis acid (see Table 2.3), is confirmed by the experimental data.

Formation of $H^{\delta+}$ is shown for $RAlCl_2(R_2AlCl)$ complexes with HCl in benzene. Complexes with proton attachment, analogous to the known Gustavson complexes, are most probable in these systems.

Carbocationic activity of alcohols (combined with $RAlCl_2$) seems unexpected only at first glance, meaning the known reactions of HAOC alcoholysis. Obviously, interaction of alcohols with $RAlCl_2$ proceeds by the general scheme of carbonium–ion formation from neutral molecules, as is observed, for example, during protonation of alcohols by Brönsted acids [18, 86]. Direct proofs of such a direction of alcohol fragmentation were obtained by mass–spectrometry [87]. Spectra of protonated *tert*–C_4H_9OH and the product of H_2O addition to $C_4H_9^+$ was found to be practically the same. In the case of complexes of alcohols with $R'AlCl_2$ stabilization of OH–groups is the necessary condition of the reaction, which is possible by means of formation of rather stable counterions involving four–coordinated aluminum:

$$ROH + R'AlCl_2 \rightarrow R^{\delta+}[R'AlCl_2OH]^{\delta-}.$$

Experiments on diisobutylene transformation (the model of a cationic monomer with limited ability to grow) using CD_3OH [88] correlate with the above scheme. IR–spectra of products displayed weak absorption bands at 2183 and 2137 cm^{-1}, correlated to $\nu_{C\text{–}D}$. Absorption bands in the range of 3400–3600 cm^{-1} ($\nu_{C\text{–}D}$ and $\nu_{O\text{–}H}$) are absent. The CD_3^+ cation is unstable, and the efficiency of initiation by it is not high enough. More convincing proofs of initiation by a carbocation were obtained for alcohols capable of formation of stable carbocations [88]. Alcohol end groups were identified in the products of diisobutylene oligomerization in the presence of $C_2H_5AlCl_2$–$C_6H_5CH_2OH$: aromatic nucleus absorption at 263 nm (lgε 2.25) in UV–spectra, and signals of phenyl protons with δ_C 7.18 m.p. (5H) in PMR spectra. In $C_2H_5AlCl_2$ activation by *tert*–C_4H_9OH, a fraction of oligomers was extracted (78% yield, molecular mass 290) conforming with $C_4H_9^+$–cation addition to two molecules of diisobutylene. The absence of OH–groups in products and low concentration of C—OR bonds point out a small contribution of reactions of the chain termination at the sacrifice of possible interactions with fragments of a counteranion or with free alcohol:

$$\left.\begin{array}{l} \ldots\text{—}C^{\delta+}C_2H_5AlCl_2OR^{\delta-} \\ \ldots\text{—}C^{\delta+}C_2H_5AlCl_2OH^{\delta-} \end{array}\right\} \longrightarrow \begin{array}{l} \ldots\text{—}COR + C_2H_5AlCl_2 \\ \ldots\text{—}COR + C_2H_5AlCl_2 \end{array}$$

$$\left.\begin{array}{l} \ldots\text{—}C^{\delta+}C_2H_5AlCl_2OR^{\delta-} \\ \ldots\text{—}C^{\delta+}C_2H_5AlCl_2OH^{\delta-} \end{array}\right\} \xrightarrow{ROH} \begin{array}{l} \ldots\text{—}COR + C_2H_6 + Cl_2AlOR \\ \ldots\text{—}COR + C_2H_6 + Cl_2AlOH \end{array}$$

Initiation of the isobutylene polymerization by an alcohol carbocation is also confirmed with the example of the $C_2H_5AlCl_2$–triphenylcarbinol system. In the UV-spectrum (195 K) of polyisobutylene $(\overline{P} = 1230)$ absorption at 231 nm may be related to the end $(C_6H_5)_3C$-group.

The nature of ACs forming at $AlCl_3$ interaction with soft proton-donor compounds (H_2O, ROH) is not clear enough because of the occurrence of side reactions. Alkylation of arenes with the help of $AlCl_3{\cdot}ROH$ [89], where R is the alkylating agent, formally matches with the formation of a carbocation from alcohol. However, a possibility of HCl extraction in the reaction

$$AlCl_3{\cdot}ROH + \text{arene} \rightarrow \text{products} + HCl$$

$$HCl + ROH \rightarrow H_2O + RCl$$

$$RCl + AlCl_3 + \text{arene} \rightarrow \text{products} + HCl$$

allows us to draw no unique conclusion about the nature of the electrophilic center.

The truth of the scheme is also confirmed by the data on formation of Gustavson complexes at alcohol adding to $AlCl_3$ in the presence of arene.

In a like manner, because of HCl extraction, it is impossible to judge uniquely about the existence of aquacomplexes with $AlCl_3$ and formation of an appropriate AC

$$Cl_3Al{\cdot}O\begin{matrix}\diagup H\\ \diagdown H\end{matrix}$$

Obviously, activation of cationic processes in the presence of $AlCl_3$ and H–compounds is more complicated than with $RAlCl_2(R_2AlCl)$ owing to higher acidity.

The typical feature of the behavior of R_nAlCl_{3-n} complexes with weak proton–donors is strict selectivity of their action in electrophilic processes. Among other things $R_nAlCl_{3-n}{\cdot}H_2O$ complexes excite homopolymerization of isobutylene in the presence of α– and β–butylenes [90–94], alkylation of arenes by olefins according to the scheme of one–side consecutive reaction with formation of polyalkylation products [95, 96] and in some cases (toluene–propylene system) strictly positional alkylation [97]. Obviously these properties are directly bound to low acidity of the complexes.

Complexes of hard Lewis and Brönsted acids represent the limiting case of the systems classified by the definite criterion of acidity. They primarily are combinations $HF–SbF_5$, $HSO_3F–SbF_5$, $HF–NbF_5$ with different molar ratio of the components and some others, named magic or superacids [18, 98]. With respect to their strength they are 10^7 times harder than 100% sulfuric acid. Among these are also $HF{\cdot}BF_3$, $HF{\cdot}BF_5$ complexes, stabilized by arenes, and unstable compounds of $H_2SO_4–AlCl_3$ type [99]. Magic acids allow generation of carbonium ions from non–polymerizing olefins [19, 99], study of the decomposition mechanism and comparative resistance of a polymer sequence (polyethylene, polypropylene, polyisobutylene) [100], kinetics of polymerization by stopped jet method (styrene) [101], etc. Generally, magic acids are noteworthy first from the point of view of application of acidic catalysts in new situations for solving special tasks (for example, low–temperature isomerization of hydrocarbons). The frames of their application in synthetic polymeric chemistry and possible perspectives of usage are not yet completely determined.

There are some complex catalysts that lie outside of the above–mentioned classification. Among them a group of metal–containing catalysts should be noted, which allow obtaining of isobutylene polymers of the highest MM at temperatures of about 250–255 K [102–105]. These catalysts represent a combination of metal compounds, for example, $C_pTi(CH_3)_3$, $(C_p)_2Zr(CH_3)_2$, $Zr(CH_3)[N(Si(CH_3)_3)_2]_3$,

$(C_p)_2Hf(CH_3)_2$, $(C_p)_2AlCH_3$, where C_p is 1,2,3,4,5–pentamethylcyclopentadienyl, combined with tris(perfluorophenyl)boron. Formation of ionic intermediates with active metal cation and low–nucleophilic non–coordinating anion of $[B(C_6F_5)_3CH_3]^-$ is supposed. Water may activate individually chosen components forming polymers. In this case the mechanism of action of catalytic systems is supposed to be complex. These catalysts are active also in copolymerization of isobutylene with isoprene [105].

In concluding let us discuss one more type of cation–active catalyst — immobilized catalysts — which conceptually generalize all the above–mentioned systems. Interest in immobilized cationic catalysts is defined as in the case of metal–complex systems by the factors of technological effectiveness, purity of final products, combination of advantages of homogeneous and heterogeneous systems, etc. [106].

Immobilized forms of catalysts for cationic transformations of isobutylene are known for both individual and complex Lewis and Brönsted acids (Table 2.6).

Table 2.6.

Classification of immobilized cationic catalysts.

Type of acid	Type of carrier	References
Brönsted acids		
H_2O, HCl, HF, H_2SO_4, H_3PO_4, FSO_3H, heteropolyacids – $H_4SiW_{12}O_{40}$, $H_3PW_{12}O_{40}$, $H_3PMo_{12}O_{40}$, $H_4SiMo_{12}O_{40}$	Oxides, alumosilicates, zeolites, salts (sulfates, etc.), silicagels, organomodified silicagels, styrene copolymers with divinylbenzene, polyphenylene ketone, perfluorolefins and their copolymers, net copolymers of alcoxy– and alcoxy–organosilanes	54–57, 92, 107–125
Lewis acids		
R_nAlCl_{3-n}, $AlCl_3$, $AlBr_3$, BF_3, BCl_3, BBr_3, $ZnCl_2$, $CuCl_2$, $FeCl_3$, $TiCl_4$, TiF_4, $SnCl_4$, SbI_5, SbF_5, NbF_5	Oxides, mixed oxides, alumosilicates, salts, silicagels, metals, polystyrene, styrene copolymers with divinylbenzene and 4–vinylpyridine, acryl monomers copolymers, coal, graphite, fluorinated and modified graphite	93, 126–156
Complex acids		
HX (H_2O, HCl, HF) — MeX_n (R_nAlCl_{3-n}, $AlCl_3$, $FeCl_3$, BF_3)	Oxides, alumosilicates, polystyrene, styrene copolymers with divinylbenzene, sulfonated styrene copolymers with divinylbenzene	129, 157–172
Carbenium (carboxonium) ions from groups with labile halogen and MeX_n — R_nAlCl_{3-n}, $AlCl_3$, $TiCl_4$, VCl_4, $SnCl_4$, $AgClO_4$	Polyvinyl chloride, chloro(bromine)butyl rubbers, polychloroprene, modified carbon black	25, 173, 174

The conditional sequence of immobilized catalysts starts from the simplest H–acid — water. It is interesting to note in this connection that although the concept of applied catalysts was formed as a direction for metal–complex systems [106], applied catalysts had been known long before. Solid acids of mineral origin with surface Lewis and Brönsted centers conceptionally are the first example of immobilized systems, on which the dependence of acidic–catalytic properties on the carrier nature has been stated [40]. Although various methods of the oxygen component fixation is used at the formation of immobilized catalysts — starting from impregnation and intercalation to chemical covalent linking — the general character of the carrier influence on the catalyst behavior can be sought. Being the base or possessing places of different basicity in its composition, the support enters acid–base interactions with the catalyst, which depend on the chemical and physical nature of the components. Owing to the presence of a spectrum of acidic (basic) properties of the components, selection of acid–base interactions happens in conformity with the known concept of "hard" and "soft" acids and bases (HSAB) [107]. Finally, the support serves as a microligand, which increases sizes of ions and stabilizes them. Consequently, an increase of acidity and catalytic activity of systems takes place.

Since catalytic systems under consideration principally display the H–activity, some aspects of the proton catalysis will be remembered. Direct connection between the acidity function H_0 of applied acids and the amount of linked acids was set [175]. In its turn, the activity (lg$K$) of such catalysts is in direct dependence on H_0: $\lg K = (1 - \delta)H_0 + const$, where $\delta = 0$ for homogeneous reactions, and for catalytic reactions on solid acids the value $(1 - \delta)$ characterizes proton mobility on the surface. There are data that the increase of the number of protons in a solid acid affects its activity non–linearly which, in its turn, marks the heterogeneity of acidic centers [176]. It will be remembered that, as in solutions, a pure proton does not exist in solids even during a negligibly short time and easily enters electron covers of atoms and molecules [177]. Thus initiation of a catalytic system by linked protons reflects their transfer to a stronger acceptor (to a monomer in the case of polymerization). Obviously, a fine mechanism of such competitive interactions for possessing proton defines both the ability of the process initiation and the balance of activity and selectivity of the action of catalysts [107]. This is confirmed by the analysis with the help of quantum–chemical calculations [178]. Some specific features of applied Brönsted acids behavior are shown below, which differentiate them from both other immobilized systems and low–molecular analogs.

In relation to low acidity, not all molecules of H_2O linked to the solid support are active. For the latter, most probably, the molecules complexly linked to cations are responsible. The behavior of these molecules depends on the nature of cation and is characterized by high sorption heats (up to chemosorption) [179, 180]. In the limiting case hydroxonium ions and especially substituted hydroxonium ions correspond to a catalytically active state of H_2O [181].

In a similar manner, another traditionally used catalyst — sulfuric acid — displays catalytic properties as a complexly linked compound, for example, on metal sulfates [109, 110], as well as in the form of sulfogroups covalently attached to the matrix, i.e. polymeric sulfoacids [114–117]. In both cases, the greater is the amount of linked acid (SO_3H–groups) and the stronger is its connection with the matrix, the higher is the acidic–catalytic activity. General ideas on the character of action of such catalysts can be illustrated by the example of sulfonated styrene copolymers with divinylbenzene. As for any solid matrix, in this case a significant role is also played by polymer net permeability, defined by the linking degree, swelling degree, granule size, and other factors, too. The chemical aspect of the catalytic action of sulfocations is connected with the presence of a hydrogen bond network, cooperative effects and formation of associates — centers of increased local concentration of acidic groups [182, 183]. The presence of residual water provides necessary mobility of protons, dynamic character of the network and the ratio of action activity and selectivity observed in the experiment. Building–in a substrate into the network is favorable before its simple interaction with the surface [184–186]. Taking into account low polarity of olefins, isobutylene, for example, it may be suggested that its electrophilic transformations in the presence of sulfoacids proceed via intermediate formation of an alcohol with the further building–in into the matrix network. Possible structural elements of polymeric sulfoacids are shown below:

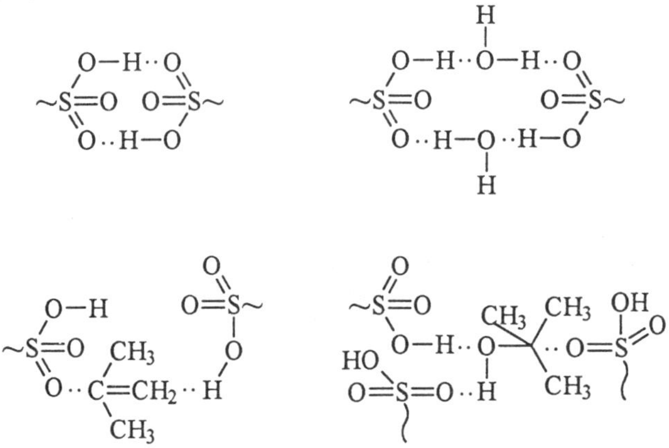

In the general case, polymerization of spatial monomers — styrene, isobutylvinyl ester, etc. — in the presence of polysulfoacids, obviously, may proceed through initiation in the volume with diffusion limits or from the surface. This is marked by low values of molecular masses of products or effectiveness of the initiation. Additional sulfonation, fluorination of the polymeric matrix, application of thermoresistant polymers (sulfopolyphenylene ketone) and inorganic polymers with grafted sulfobenzyl groups increase stability of catalysts, their acidity and activity [117–120, 187]. Among other things, fluorinated polystyrene divinylbenzene sulfoacids adhere properties of superacids. This allows performance of not only polymerization of simple olefins — propylene, butenes, isobutylene, etc. — but also oligomerization of olefins with sterically hindered double bonds, for example, diisobetylenes — $(CH_3)_3CCH_2{-}C(CH_3){=}CH_2$ and $(CH_3)_3C{-}CH{=}C(CH_3)_2$ [188].

As indicated earlier, the simplest Brönsted acid HCl, inactive in polymerization of olefin and vinyl monomers, retains its catalytic properties under immobilization, which leads to an increase in the proton concentration and the initiation rate owing to the coordination effect, and also to slowing down the reaction of Cl^- ions with the growing polymeric carbocation, i.e. deceleration of the chain termination.

Among other known Brönsted acids, immobilization of HF (doubtless facilities of the work owing to elimination of its high volatility and toxicity) and especially of heteropolyacids should be noted [108, 122–125]. Having a rigid framework of heteropolyanions, composed from metal oxide (Mo—O, W—O) octahedrons with included heteroatoms (P, Si, As, etc.) in their structure, they are hard proton acids and are active (selective) in many homogeneous and heterogeneous reactions, exceeding typical mineral acids (H_2SO_4, $HClO_4$) in these points and simultaneously favorably differing from the latter by high heat resistance and low corrosion activity. Taking into consideration the presence of spatial polyanions with low charge surface density, which in HSAB terms are soft bases, heteropolyacids can be considered as immobilized H–acids with a weak electrostatic bonding between proton and anion in the ionic pair and high acidity. This is confirmed by typical structures of proton centers in homogeneous, heterogeneous (water) and heterogeneous (water–free) variants of heteropolyacids, respectively [124]:

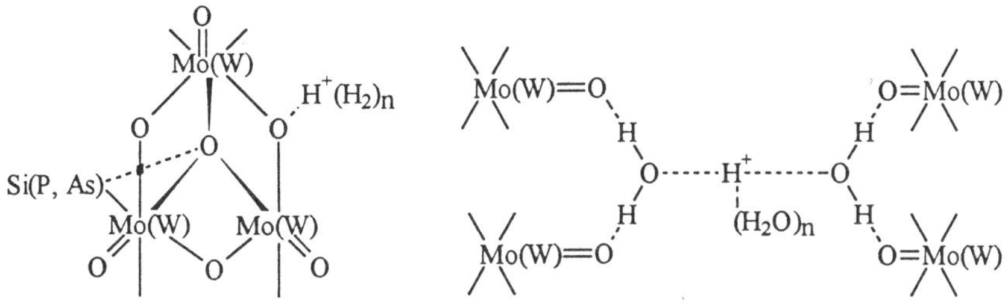

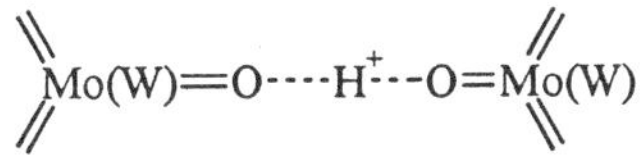

The activity of heteropolyacids in isobutylene oligomerization is defined by the total acidity, and the part of the most high–molecular products in acidic strength measured in the sequence $H_3PW_{12}O_{40}$ > $H_4SiW_{12}O_{40}$ > $H_3PMo_{12}O_{40}$ > $H_4SiMo_{12}O_{40}$ [125]. This sequence roughly coincides with some changes in softness parameters of anions of the mentioned acids. Immobilization of heteropolyacids on activated carbon, silicagel and other inert carriers increases selectivity of catalyzing reactions with sacrifice of the interaction of heteropolyanions with the carrier, a change of properties of the carrier itself (in the case of its functional character) being indicated also. Obviously, such an effect may be expected at complexing of heteropolyanions with appropriate ligands with formation of more complex structures containing 150 or more atoms.

The spectrum of acidic–catalytic properties of mobile Lewis acids is more diversified than that of Brönsted acids. Methods for their immobilization are more representative, too, from simple mixing of disperse acids and carriers (salts) or introduction of them into cavities (graphite) to complexing with organic polymers (styrene divinylbenzene copolymer), chemical interaction with inorganic supports or formation of Lewis acids directly at the sacrifice of a reaction on the carrier surface (fluorination of aluminum or aluminum oxide, aluminum oxide chlorination) [128–132, 136–138, 142]. In the latter case a surface phase of the Lewis acid with strong acidic centers ($H_0 < -14.5$) is formed. Consequently, the direction and selectivity of reactions are defined either by the nature of the Lewis acid and carrier or steric hindrances, or by all these factors. Of special note in this relation are compounds of Lewis acids' inclusion into graphite (layered graphite compounds with $MeHal_n$ or intercolation graphite compounds) representing ionized compounds of $C_m^+[MeHal_{n+1}^- \cdot xMeHal_n]$ gross formula (m and x are defined by the nature of the halide and formation conditions) with a positive charge on the graphite network [150]. A wide variety of different types of bonds in $MeHal_n$ compounds inclusion into graphite and features of their structures allows supposition of several probable ways (marked by arrows) of catalytic action of the compounds [150]:

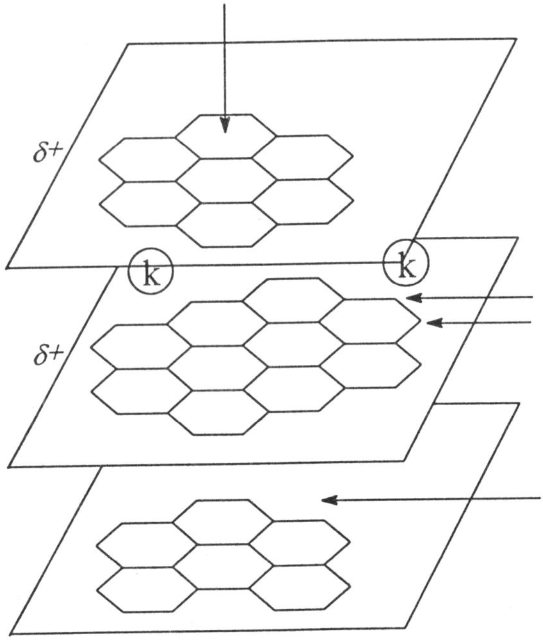

Reactions may proceed in surface layers of graphite, and the role of the introduced acid is in creation (regulation) of the positive charge on carbonic graphite networks. If molecules of a substrate are introduced into the free interlayer space of graphite, the reaction will proceed without direct contact of substrate molecules with molecules of introduced acids, and the reaction direction is generally defined by steric hindrances created by carbonic layers of graphite. If the process is localized in the interlayer space of graphite filled by acids, the process is affected by both the nature of acid–catalysts and mentioned steric hindrances. Introduced acids disposed on the margins of graphite crystals can also be catalytic centers. In this case, the role of steric hindrances created by the graphite network must be insufficient. The most distinctive method of catalytic action involves washing off substances, introduced into the solution, and proceeding of the reaction outside the graphite. To put it differently, layered graphite compounds are "internal" dispensers of a catalyst. From the standpoint of excitation of monomer polymerization reactions, moderate temperatures of the process (~20°C) are preferable, which increase the influence of both the nature of introduced acid and parameters of spatial graphite network. This is indicated by dependencies of the catalyst efficiency on nature of the Lewis acid and inactivity of individually chosen graphite (acid) [154, 155]. Usually, low rates of transformations define insufficient technological ability of catalysts by inclusion of compounds into the graphite, though they also possess obvious advantages: stability in air, resistance to hydrolysis, selectivity in some processes.

Certain advantages, such as long (up to several years) stability in air, multiple application without noticeable decrease of activity, low or decreased sensitivity to humidity are peculiar for Lewis acids, fixed on styrene polymers and copolymers, used as powders or spherical particles dispersion [139–145]. Such catalysts — "polymer–protected aluminum chlorides" or "polymer–protected adducts of aluminum chlorides" (for $AlCl_3$ complexes with ester or nitromethane) — are less active in the reactions, than aluminum chloride itself. However, these catalysts were found to be relatively insensitive to changes of the reaction conditions (ratio of components, nature of solvent, presence of admixtures, etc.). Selective formation of phenyl–*tert*–butyl ester in the reaction of isobutylene with phenol, catalyzed by $AlCl_3$ on styrene copolymer with divinylbenzene, proves in accordance with [189] a decreased acidity of the mentioned system compared with $AlCl_3$.

If application of these catalysts is necessary, it is worthwhile to implement more "rigid" fixation of Lewis acids at the sacrifice of introduction of monomers with electron–donor groups (4–vinylpyridine, akrylonitrile) into styrene polymers (copolymers) [142, 146, 147]. In this respect, gel–immobilized Lewis acids seem to be most promising. For example, BF_3 fixed on butylacrylate rubber (butylacrylate copolymer with acrylic acid, acrylonitrile and methylolacrylamide) is highly active in polymerization of N–vinylbutyl ester and N–vinylcarbazole under moderately negative temperatures [190, 191]. Alongside significant crystallinity, the polymers obtained are characterized by high polydispersion testifying to the significant role of diffusion processes in the synthesis owing to the effect of the gel matrix. Covalently fixed Lewis acids ($AlCl_3$, $AlBr_3$, SbF_5, AsF_5) on solid carriers are mostly efficient in catalysis of alkylation and isomerization reactions [129], but not in polymerization.

The nature of the support affects acidity and selectivity of the fixed Lewis acid action. For example, $(C_2H_5)_2AlCl$ applied to silica gel is a weak acid and serves as a catalyst of selective polymerization of isobutylene (from C_4 hydrocarbon fraction) [93]. At the same time, $C_2H_5AlCl_2(BF_3)$||sulfocationite is a typical superacid representing a low–selective catalytic system [127]. It should be noted that even the simplest way of mixing of traditional Lewis acids ($AlCl_3$, $AlBr_3$) in 1:1 molar ratio with metal salts (preferably sulfates of Ti, Fe, Ni, Cu, Mn, etc., or metal chlorides inactive in Friedel–Crafts reactions — $BiCl_3$, $CuCl_2$, $NiCl_2$) leads to formation of strongly acidic catalytic systems [130–132]. Synergic effects of mixtures in cationic reactions indicate formation of active complexes or monomeric forms of a Lewis acid.

The general tendency of increasing catalytic properties of cationic systems of polymerization in the presence of appropriate cocatalysts is also a property of applied Lewis acids. For example, this is displayed by significant increase of acidic properties (up to superacid level) and activity of compounds SbF_5, BF_3, etc., included into graphite, in the presence of proton–donor additives [154]. Real active centers of applied acids represent a dynamic combination of Lewis and Brönsted centers appearing during thermal treatment of catalytic systems in the presence of water, for example, for AlF_3||carrier (H_2O) system [128]:

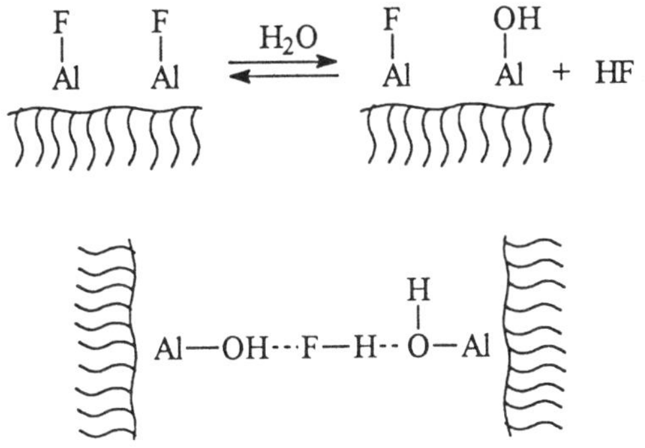

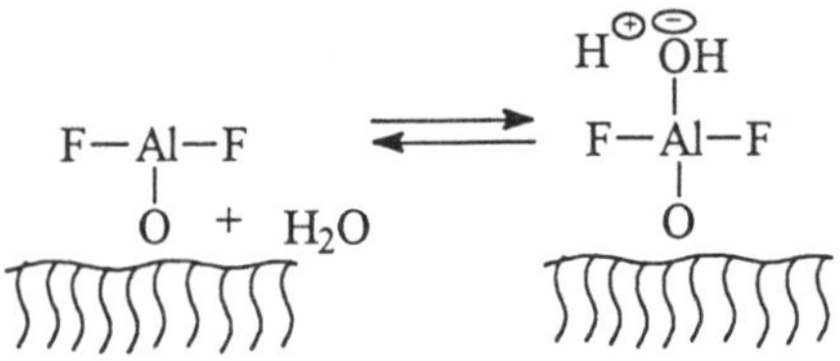

In the general case, in the absence of special conditions, catalysis with the help of applied individual Lewis acids is rather problematic, because the influence of co-catalytic admixtures can be hardly excluded and, consequently, formation of several complex systems can hardly be excluded either.

On the contrary, functioning of applied complex catalysts from Lewis and Brönsted acids may be agreed proved (Table 2.6).

First and foremost this relates to the cases when application of complex acids is not accompanied by side effects, for example, by formation of mobile acids, or when behavior of a carrier — polymeric ligand — can give additional information about its interaction with a complex acid. That is why, making no analysis of applied system on inorganic supports HF–BF_3||alumosilicate and FSO_3H–SbF_5||metal oxides, let us dwell on more informative immobilization of aluminum chloride complexes on styrene polymers and copolymers. For the $AlCl_3$–cross-linked polystyrene sulfoacid system after thermal treatment at 115°C, formation of the complex characterized by a ratio of about S:Al:Cl = 2:1:2 is indicated [162]. Superacidic properties (H_0 = –17 to 19), close to those for FSO_3H–SbF_5, are explained by an equilibrium existence of a number of structures, the latter two of which form an ionic pair:

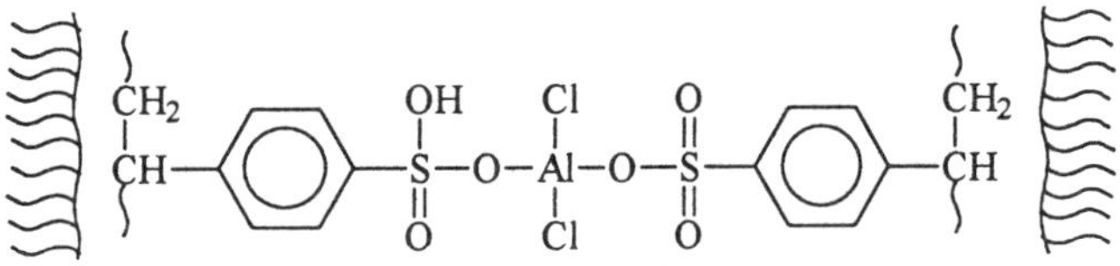

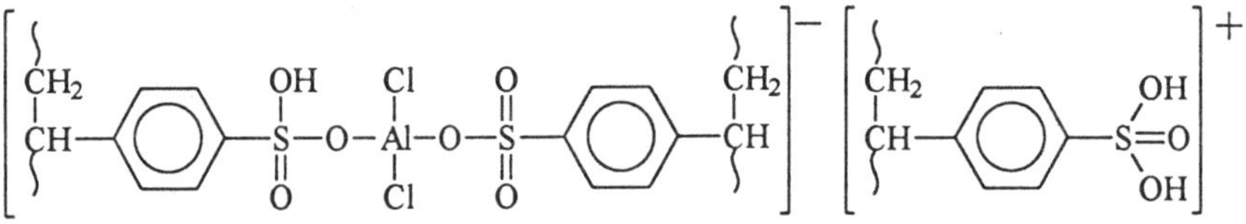

Extraction of HCl on the stage of complex formation and stabilization of aluminum in the 4–coordination state via linking SO_3H–groups, and the effect of cross–linked matrix are apparently responsible for the stability and technological ability (absence of the coordinating effect) of the catalytically active complex. In this respect it exceeds a low–molecular analog — $AlCl_3$ combined with H_2SO_4.

A stricter approach to immobilization of complex catalysts was undertaken in [164–172], where the principles of forming mobile complex acids were applied to creation of polymeric immobilized complexes. Taking into consideration that obtaining stable complexes of Lewis and Brönsted acids requires enough stability of a counterion and (or) accepting of a proton, two typical methods of immobilization have been suggested [166]:

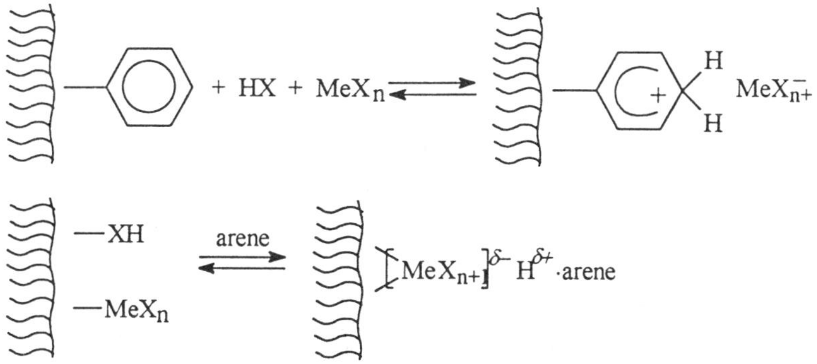

The first direction supposes reversible fixation of a complex acid by the polymer–ligand for the solution; the second, formation of a complex acid from Lewis and Brönsted groups chemically attached to the polymer.

Obtaining of polymeric complexes of the first type is based on the ability of arenonium ions in the composition of triple complexes (Gustavson's) to enter the exchange interaction with aromatic hydrocarbons of the solvation cover:

$$[ArH^{+}MeX^{-}_{n+1}]\cdot Ar^{1} \rightleftharpoons [Ar^{1}H^{+}MeX^{-}_{n+1}]\cdot Ar$$

With polymers used as Ar^{1}, containing aromatic fragments, analogous exchange interaction with formation of polymeric complexes at the sacrifice of the reversible transition of a mobile acid from aromatic hydrocarbons to the polymeric matrix takes place [165–167] (see Scheme 2.1 below).

During formation of a complex by adding mobile acid to polystyrene solution an increase of viscosity is observed and, at a definite critical ratio of components, there is structuring of the solution with complete loss of fluidity and formation of colored jelly. The presence of low–molecular aromatic compounds in the solvation cover provides easy transportation of protons in the system of aromatic nuclei and long lifetime (several weeks) of structured systems. The increase of polystyrene concentration makes structuring simpler, growth of values of "critical" molar ratios of the number of polymeric chains to Al being bound to strengthening of cooperative interactions of polymeric chains participating in formation of the solvation cover (Table 2.7) [166].

Temperature decrease to –78°C reduces the role of solvation interactions with aromatic fragments of the polymer and makes difficulties for the complex formation. On the contrary, application of polystyrene instead of polystyrene alkylated by isobutylene (into a benzene ring) involves strengthening of the cooperative interaction and increase of the "critical" mole of polymer/Al ratio up to 200.

Table 2.7.

Structuring of polystyrene (PS) solutions in the presence of Gustavson complexes $HCl\cdot 2AlCl_3\cdot 3C_6H_5CH_3$ (20% solution of the complex in CH_2Cl_2, 25°C).

PS concentration in CH_2Cl_2, % (mass)	Values of "critical" ratios of components, mol/mol		
	Al/PS	PS/Al chain	Aromatic fragment/Al
0.5	20,000	0.1	5
1.0	500	1.5	12
5.0	100	6.0	30
10.0	20	50.0	187

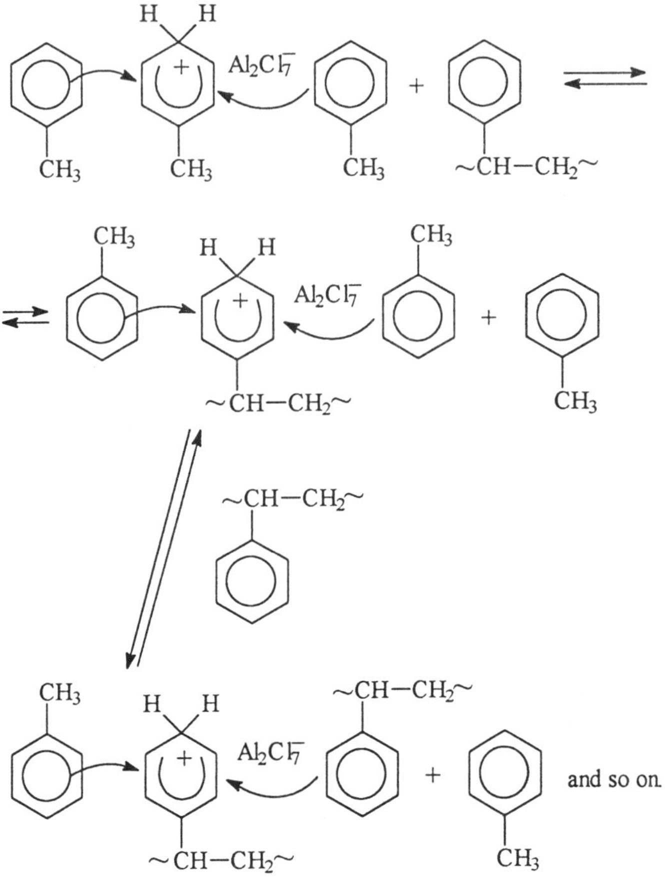

Scheme 2.1

When $HCl \cdot 2AlCl_3 \cdot 3C_6H_5CH_3$ complex contacts with styrene copolymer cross–liked with divinylbenzene, chemosorption of $AlCl_3$ and HCl by the polymeric matrix is observed with formation of an immobilized complex acid containing no toluene in its composition. The low number (n = 4) of the monomeric chain in the polymer in relation to Al points out limiting of cooperative interactions with sacrifice of limiting mobility of macrocarrier chains. The complex is stable and is characterized by constancy of Al and Cl content at long exposure in solvents (methylene chloride, toluene). Only a highly basic solvent, tetramethylbenzene (hemimellitol) causes gradual transition of linked $AlCl_3$ into the solution. Decomposition of the complex is also observed at heating above 50°C.

Contrary to the above–considered reversible immobilization of HCl–$AlCl_3$ by polymers and styrene copolymers (Scheme 2.1), irreversibly immobilized complexes represent Lewis and Brönsted acids, activated with the help of solvation by an aromatic solvent and preliminarily fixed on the polymeric matrix [164, 166]:

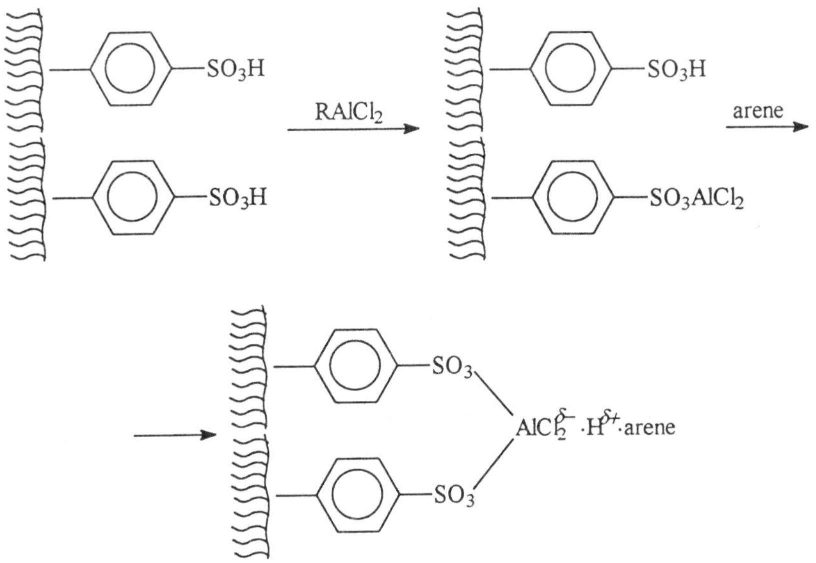

Scheme 2.2

Application of alkylaluminum chloride on the initial stage of pyrolysis provides simplicity and directivity of the reaction, the absence of HCl extraction (as in the case of $AlCl_3$ [162]) and the possibility of synthesizing products with the required ratio of SO_3H and SO_3AlCl_2 groups. Limiting the mobility of mentioned groups linked to the polymeric support creates higher heat resistance of catalysts (in the range of 70–100°C, in dependence on the groups ratio) compared with model low–molecular compounds — HSO_4AlCl_2 and $CH_3C_6H_4SO_3AlCl_2$ (50 to 60°C). The complex character of an immobilized acid is proved by the fact that according to Scheme 2.2, excitation of isobutylene polymerization, its copolymerization with vinylsilyl esters, and ionic hydrogenation of isobutylene oligomers require the presence of groups of two types in the catalyst — SO_3H^- and $SO_3AlCl_2^-$, the optimum ratio of which varies from 4 to 10 [169–172]. Individual sulfoacids in H^+– or Na^+–form or sulfopolyacids with sulfogroups completely substituted by SO_3AlCl_2 displayed no activity under conditions of the experiment. This conclusion correlates with the behavior of model compounds: activity of HSO_4AlCl_2 (contains groups of both types) and inertness of $ClSO_3AlCl_2$. Solvating function of aromatic hydrocarbons in immobilized catalyst follows from the twice higher value of exothermal heat effect (0.27 kJ/kg of carrier, 25°C) on the second stage of the above–considered scheme for the system of sulfocationite CU–2–8 with SO_3H^- and $SO_3AlCl_2^-$ groups-1,3,5-trimethylbenzene compared with the analogous system and cationite in the H–form [166].

It will be noted that electrophilic activity, too, is inherent to samples of immobilized catalysts only, prepared in the presence of aromatic compounds. It seems likely that, beside activation of linked complex acid, the presence of an aromatic solvent provides solvating (dissolving) ability in relation to a polymer forming on the catalyst surface, polyisobutylene in particular. In the case of highly polar monomers, for example, the above-mentioned trimethylvinyloxysilane, it appears deficient, and polymerization does not proceed [170, 171]. Despite significantly lower activity of isobutylene, an obvious decrease of chemisorption of polar polymer on the polar surface of immobilized catalyst and formation of a copolymer product happen during vinylsilyl ester reaction with isobutylene. As individually chosen components — sulfocationite and aluminum chloride — caused only separate polymerization of the mentioned ionomers, the immobilized complex system changes the ratio of their activities to the side of approaching.

Independent of the method of complex acid immobilization by ligand–carriers, the latter not only introduce a definite conditionality into the term of the complexes' stoichiometry, but also significantly change their properties as catalysts. The latter fact is an additional proof of the existence of immobilized catalysts. A decrease of activity in isobutylene polymerization typical for heterogeneous catalysts is observed as temperature decreases from 50 to –25°C; however, in this case, values of molecular masses of oligomeric products change insignificantly ($E_{\Delta MM} = 0$ compared with $E_{\Delta MM} = -25$ kJ/mol for soluble catalysts [192]) indicating the limiting role of reactions with the participation of medium components. It is also supported by low values of constants of immobilized catalyst poisoning by nuclei (esters, amines, sulfides) — 10 to 10^2 against 10^4 to 10^5 for soluble and model systems. Independently of the method of catalyst immobilization, their selectivity in polymeric division of isobutylene and α– and β–butylenes is low and does not practically depend on the reaction temperature, which is unusual for soluble systems. High acidic strength is also indicated by obtaining high–molecular products on the isobutylene basis at low temperatures (P ~ 1000 at –78°C).

Original "time lag" behavior is especially inherent in irreversibly immobilized acids. Complete absence of end aromatic groups in the polymer at isobutylene polymerization in toluene (xylene) testifies suppression of aromatic solvent alkylation inherent in soluble catalysts. In all probability, polymerization proceeds on the solid surface with the chain transmission to the catalyst and polymer chain propagation from the root similar to metal–complex Ziegler–Natta catalysts:

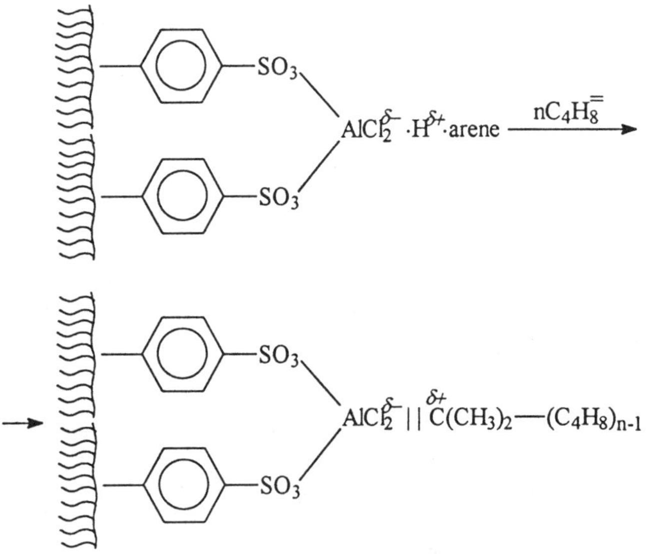

Scheme 2.3

Scheme 2.3 is also supported by independence of polyisobutylene molecular mass on the amount of catalyst used. The ratio of kinetic constants, more favorable than at polymerization in solution, gives a polymer with rather narrow MMD $\left(\overline{M}_w/\overline{M}_n\right)$ = 1.6–1.9). Some general signs of immobilized catalysts are inherent in polymeric sulfoaluminum chlorides, first of all, with respect to stability of isobutylene polymerization (the babbling regime) and formation of products containing no catalyst residues.

In conclusion, one more type of immobilized cationic catalysts should be mentioned, which can be assigned to the discussed variant and considered as generation of carbonium centers on a polymer (Table 2.6). They can be obtained by combining halogen–containing polymers with "soft" Lewis acids ($TiCl_4$, VCl_4, $(C_2H_5)_2AlCl$, etc.) [25, 173, 174]. "Elimination" of mobile halogen atoms from a polymer promotes formation of an ionic pair. Synthesis of grafted copolymers, for example, isobutylene on the polymer–carrier, proves existence of carbonium centers. Practical application of this immobilization method may result in stabilization of the initial polymer at the sacrifice of limiting or elimination of the negative effect of mobile halogen atoms promoting the polymer decomposition.

It should be pointed out that, contrary to the three above-discussed cases, immobilization of cationic catalysts, in which the carrier plays the role of ion stabilizer, for this type of system the carrier plays the role of a macrocation, and catalyst connection with the polymer–carrier is lost during excitation of polymerization.

This represents an overview of the current application of catalyst immobilization to cationic systems. A wide range of compounds, including those applied in industry and for other purposes (zeolites, silicagels, oxides, etc.), may be used as carriers of catalysts. The universal carriers are styrene polymers and copolymers, because, on the one hand, the physical parameters of the carrier are more easily regulated for them (permeability, mechanical strength, stability), and on the other hand, they are "saturated" by ligands which allow introduction of the whole spectrum of acidic agents. Immobilization of all types of acids was performed using polystyrene matrices, both individual Brönsted and Lewis acids and complex ones with variants, different from the chemical point of view. The mechanism of initiation of cationic processes by immobilized catalysts for the most part is reduced to redistribution of a proton in the acid–support–substrate system and finally, to accepting it by a substrate. That is why the problem of immobilized cationic catalysts is reduced to analyzing the problem of physicochemistry of a linked proton, one of the probable approaches to which is demonstrated in this chapter.

Allowing that the specificity of isobutylene polymerization processes are mostly low–temperature ones, there are limits of application of immobilized cationic catalysts as heterogeneous systems with respect to obtaining high-molecular products. However, for synthesis of low-molecular polymers of isobutylene they may be used as rather active and technologically applicable catalysts. Therewith advantages of solid acidic catalysts are usually inherent in them — easy separation of a product from catalyst, low ash content in products, multiple and long usage, decrease of device corrosion, lower sensitivity to admixtures, etc. This is demonstrated by the data of numerous native and foreign patents. Both general signs of applied systems (stable, stationary type of action) and some differences from the latter, which was discussed in the analysis of particular works, are inherent in the processes applying immobilized cationic catalysts.

Analysis of many examples of excitation of isobutylene cationic polymerization by combinations of Lewis and Brönsted acids of different strength testifies to the role of complex formation during acid–base interaction between system components. Typical features of isobutylene transformation, among them selective polymerization of one monomer mixed with the others, obtaining products with high molecular masses, etc., are directly or indirectly bound to them. Formation of Lewis and Brönsted acids complexes significantly increases abilities of new catalysts and their application to objects, which were not yet exciters of electrophilic processes, for example, isobutylene polymerization, polyisobutylene depolymerization. At the same time, application of the principles of acid–base interaction and complex formation in a wider sense, i.e. not only to the excitation stage, but also to the whole polymerizing system, allows better understanding of the mechanism of the polymer formation.

2.2. Polymerization mechanism.

In the general case, polymerization of isobutylene as well as of many other monomers, can be subdivided into four elementary stages: initiation (excitation of polymerization) — formation of ACs carrying positive charge; chain propagation — attaching of a monomer to AC; limiting of the chain propagation — material and (or) kinetic chain termination; and chain transfer. Each of the stages may include more than one of the elementary acts. High sensitivity of the process to the presence of alien admixtures, as well as extremely high rates of the polymerization, under real conditions which create instability of the reaction process (a consequence of the exothermic nature of the polymerization reaction), make it difficult to obtain reliable information about the mechanism of cationic polymerization.

At the present time, there is no general theory of electrophilic polymerization of cation–active monomers. The picture is not clear yet, even if the range of considered monomers is added to by olefins, and that of catalysts, by Lewis acids of Friedel–Crafts halides with proton-containing compounds. Nevertheless, the complexity of the problem does not exclude, but rather supports and stimulates investigations in the field of isobutylene polymerization mechanism. In recent years new fundamental results have been obtained that give the possibility to make a step forward in regulation of polymerization processes and molecular characteristics of polymer, and to create a more rigorous theory of polymerization as a super-fast process. Discussion of these results is foregone by the classic analysis of separate stages of polymerization.

2.2.1. Initiation

Initiation of isobutylene polymerization proceeds at the interaction of cationic polymerization exciters with the monomer. The positively charged particles so formed are able to initiate chain propagation.

All known cases of initiation can be reduced to one of several types differing by the number and type of components participating in the formation of initiating parts [193]: 1) catalyst (without a monomer) — initiation by proton acids in rather polar media; 2) catalyst–cocatalyst complex (without a monomer) — initiation by

Friedel–Crafts reagents in the presence of proton donors or alkyne halides; 3) catalyst–monomer complex realized only under the condition of the cocatalyst role played by the monomer in relation to MeX_n; 4) catalyst–cocatalyst–monomer complex, usually realized during formation of coordination complexes based on MeX_n in media with low dielectric permeability, in which ionization of the complex formed proceeds only under participation of the monomer.

Regardless of the mechanism by which the initiating particles are formed several types of primary active particles can be formed: 1) cations K^+, which react regardless of anion A^- (free cations) or in the composition of ionic pair $[K^{\delta+} A^{\delta-}]$; 2) Zwitterions $[K^{\delta+}—A^{\delta-}]$; 3) ion–radicals $K\bullet^+$; 4) coordination complexes possessing no clearly expressed ionic character. Clearly, all these primary initiating particles possess different activities. In one and the same system in relation to conditions of initiation, simultaneous formation of primary particles of various structure or transition of one into another during polymerization is possible.

Initiation by cations is typical for isobytulene polymerization under real conditions (catalysts BF_3, R_nAlX_{3-n} at $n \leq 2$). The process proceeds gradually by several stages. First, K^+ or Zwitterions $[K^{\delta+}—A^{\delta-}]$ are formed at the interaction of an initiator with a single or several components of the polymerization system. Subsequently this cation reacts with the monomer molecule M forming AC:

$$K^{\delta+} A^{\delta-} + M \rightarrow KM^{\delta+} A^{\delta-}.$$

Ionic pairs are always formed, if chemical catalysts are used in media of a moderate polarity. In cases when the first stage of initiation is significantly slower than the second one, interaction of the ionic pair $K^{\delta+}A^{\delta-}$ with the monomer molecule M is often correspond to the chain propagation stage.

In the general case, formation of primary cations at excitation of polymerization by metal halides can be represented as follows: 1) formation of the catalyst–cocatalyst $MeX_n{\cdot}RX$ or catalyst–cocatalyst–monomer $MeX_n{\cdot}RX{\cdot}M$ complex; 2) reaction between complex components with formation of an intermediate compound with clearly expressed ionic character, for example, $MeX_n{\cdot}RX \rightarrow R^+MeX^-_{n+1}$, which reacts with the monomer. The role of cocatalyst is reduced to MeX_n transformation into the active form of anionic fragment:

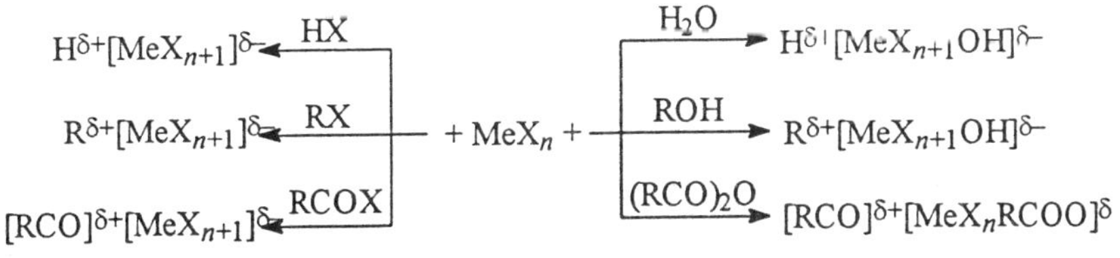

Not all stages of formation of ionic pairs are unequivocal [193]. Even identification of the simplest complexes $MeX_n{\cdot}RX$ or acids of $H^{\delta+}[MeX_n]^{\delta-}$ type usually gives no positive results. It is supposed that their formation requires traces of humidity. The reaction with RX often proceeds in systems in which RX is the medium. This case was named "cocatalysis by the solvent", although it might be expected that here ACs can also be formed with the participation of traces of humidity.

In media with low polarity, which is typical for many industrial synthetic processes, the above-mentioned reactions become much more complicated, because products of catalyst and cocatalyst interaction are incapable of ionization. Formation of primary cations is possible only in the presence of a monomer, however information about the reaction mechanism of such processes is evidently insufficient:

$$MeX_n \cdot RX \cdot M \rightarrow RM^{\delta+}[MeX_{n+1}]^{\delta-}.$$

Additional difficulties in understanding the mechanism of polymerization initiation by Lewis acids occur due to differences in stoichiometric composition of catalyst–cocatalyst complexes. For example, metal halides (BF_3, $AlCl_3$, R_nAlCl_{3-n}) with alkyl halides usually form complexes of 1:1 composition, but with H_2O and other oxygen–containing compounds formation of several types of complexes of different stoichiometric composition is possible, which differ significantly in the initiating activity; but 1:1–complexes are still the most active.

It should be noted that in systems of this type, the change of the initial catalyst–cocatalyst molar ratio may cause a major change in the role of cocatalysts, specifically of Lewis bases [193]. Outside some range of the value of catalyst–cocatalyst ratio (2 or 3 most often), an excess of the base is formed, which is usually the agent of the chain termination. Stoichiometric composition of initiating complexes may vary during the cationic polymerization.

On the final stage of initiation, at the primary K^+ cation interaction with the monomer M and AC formation, the structure KM^+ depends on the nature of the primary particle and the monomer, and its reactivity is determined by its stability.

Usually, attachment of the primary K^+ cation to the monomer happens at the position of the highest electron density in the molecule, which correlates with quantum–chemical calculations of relative electron density on carbon atoms of the monomer alkenic group.

It should be taken into consideration that acid–base interactions of various types and first of all formation of π-complexes or charge transfer complexes between the monomer and Lewis and Brönsted acids contribute to the initiation reaction

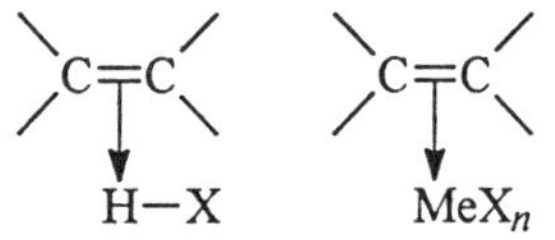

On account of the low polarity and dielectric permeability of solvents used in isobutylene polymerization, Lewis and Brönsted acids exist mostly in the form of covalent compounds or associates. Existence of free ions is of low probability because of the ease with which homo– and heteroconjugation processes may occur:

$$HA \leftrightarrow H^{\delta+} + A^{\delta-} \xleftrightarrow{2HA(2HA')} H_2^{\delta+}A + HA_2^{\delta-} \text{ or } H_2^{\delta+}A' + HA'A^{\delta-}$$

$$2MeX_n \leftrightarrow MeX_{n-1}^{\delta+}MeX_{n+1}^{\delta-}$$

$$MeX_n + MeX_{n+1} \leftrightarrow Me_2X_{2n+1}^{\delta-}$$

$$MeX_n + MeX'^{\delta-}_m \leftrightarrow Me_2X_nX'^{\delta-}_m$$

$$MeX_n + Me'X_m^{\delta-} \leftrightarrow MeMe'X_{n+m}^{\delta-}$$

Because of their equilibrium character, these reactions will affect each other in accordance with values of equilibrium constants. At a very high value of the complex formation constant k' of the monomer with an acid, polymerization may not proceed at all. For the isobutylene–AlB_3 complex constant k' = 620 cm^3/mol, and polymerization does proceed [194]. Linking of a Lewis acid into a complex anion decreases its concentration and decelerates the initiation. At rather high basicity the monomer is able to substitute the Lewis acid from the conjugated anion. Quite obviously, a simple change of the monomer concentration (initial or during the reaction) also affects the equilibrium. Such interactions indicate the importance of various factors in the polymerization initiation, in particular of the composition and degree of the catalytic system formation, solvent nature, temperature, etc.

In all cases, the acidic strength of the initiator H_0 apparently plays a significant but not the decisive role. Both the rate of isobutylene polymerization and the polymerization degree of the product increase with H_0 value of the initiator (H_2SO_4—H_2O); however, this rigorous dependence is broken at high H_0 [58]. This suggests that there are side reactions competing with the chain propagation. This is also the reason for the ineffectiveness of H–acids with significantly different acidity functions (H_0) in the synthesis of rather high–molecular polyisobutylene: HCl (–3), H_3PO_4 (–5), HF (–10.2), H_2SO_4 (–11), $ClSO_3H$ (–12.8), FSO_3H (–13.9), $HClO_4$ (–15) [13]. Even in the presence of a "superacid", only isobutylene oligomers are formed in low–nucleophilic media ($-H_0 > 15$) [34]. Consequently, the protonating ability stated for HX acid when using an appropriate acceptor (not a monomer), is not an adequate criterion for determining usefulness of the Brönsted acid as an initiator of the cationic polymerization.

Another favorable condition is formation of a stable weak–nucleophilic anion from the acid by interaction with the monomer. In the case of application of mineral acids, anions formed are related to the category of hard bases easily reacting with carbonium ions appearing in the form of heavy (hard) acidic centers.

Table 2.8.

Parameters of acidic strength for models of HCl complexes with R_nAlCl_{3-n}.

Complex	$\Delta E^{H^+}_{theor}$, kJ/mol	q_{H^+}	Δq_{H^+}	Activity/ Selectivity
HCl	–440	+0.09	—	low/high
$HCl\cdot Al(CH_3)_3$	+280	+0.18	0.09	—
$HCl\cdot Al(CH_3)_2Cl$	+280	+1.18	0.09	high/low
$HCl\cdot AlOHCl_2$	+340	+0.20	0.11	—
$HCl\cdot AlCH_3Cl_2$	+260	+0.18	0.09	high/low
$HCl\cdot AlCl_3$	+300	+0.19	0.10	high/low
$HCl\cdot AlC_2H_5Cl_2$	+270	+0.18	0.09	high/low

The most favorable situation for providing chain propagation is created with a counterion for complex compounds formed from Lewis and Brönsted acids. It is assumed that they are more stable and less nucleophilic than anions of mineral and organic salts. However, this conclusion was reached on the basis of indirect data, in particular, on the increase of acidic–catalytic properties of complex compounds compared with individual H–acids. For example, the complex acid $HClAlCl_3$

($HBrAlBr_3$) is many times stronger than H_2SO_4, and the complex H_2OAlR_3 is stronger than $AlCl_3$ [18].

Supporting information is also obtained from the calculation data, in particular for the case of Al complexes' compounds with HCl (Table 2.8) [94, 195, 196].

The value of the proton detachment energy $\Delta E^{H^+}_{\text{theor}}$ from complexes was calculated by the formula $\Delta E^{H^+}_{\text{theor}} = E_{0(R_nAlCl_{3-n}\cdot HCl)} - E^{-}_{0(R_nAlCl_{3-n}\cdot HCl)}$; activity and selectivity were estimated with respect to isobutylene polymerization from the C_4 fraction of hydrocarbons.

Note. The proton detachment is accompanied by a gain in energy, i.e. contrary to HCl, the process is energetically profitable. This is confirmed by growth of charge values of the hydrogen atom q_{H^+}. Although the calculation gives underestimated values of q_{H^+} due to the influence of the d-orbitals of the chlorine heteroatom, the change of the charge Δq_{H^+} gives a measure of the growth of the relative acidic strength. This is confirmed by the ratio between activity and selectivity of corresponding catalysts. Of interest is that, judging by values of $\Delta E^{H^+}_{\text{theor}}$ and Δq_{H^+}, differences in chemical structure of Lewis acids in the complex are not indicated and, consequently, appropriate anions of the complexes are approximately equivalent.

It is significant that the structure "polymeric carbonium ion-counterion" has no analogues in the chemistry of low-molecular compounds, excluding carbonium salts. But this analogy is also conditional, because only carbonium salts with a cationic component are stable, the latter being more stable than any component $C_n^{\delta+}$ of a propagating chain. That is why, in particular, $(C_6H_5)_3C^{\delta+}A^{\delta-}$ (A – BF_4, $AlCl_4$, etc.) cause no isobutylene polymerization [13]. Consequently, the absence of analogy in behavior of low-molecular and polymeric carbonium ions is connected with the specificity of complex counterion influence.

Growing carbonium ions represent original salts of hypothetic complex acids, not existing in the free state. That is why the condition of polymerization initiation provides a favorable way of counterion formation, which depends first of all on the catalyst and monomer nature. The general scheme of cationic polymerization initiation by complex acids, which takes into calculation the most probable acid–base reactions in the system at formation of a counterion, is quite plausible. The interconnection of the mentioned processes is reflected in the cooperative "concert" A_dE_3-mechanism [13], which successfully explains often observed high ($n > 1$) kinetic degrees by components of polymerizing systems:

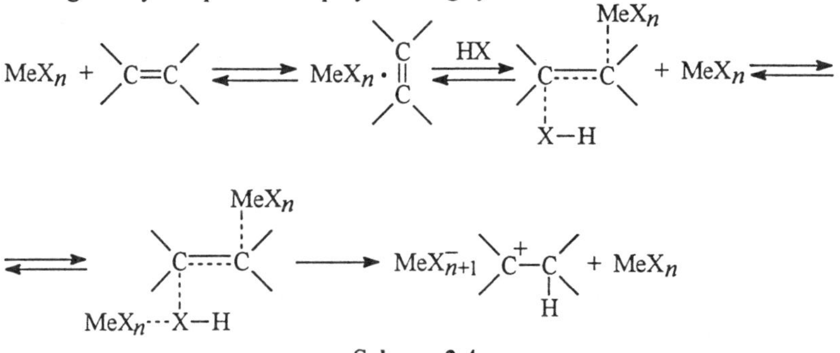

Scheme 2.4

The behavior of the HCl-$AlCl_3$ system during isobutylene polymerization is characteristic in dependence on the degree of component mixing [197]. HCl increases polymer yield if added to the monomer before catalyst is introduced. If HCl is added to the ready mixture of $AlCl_3$ with the monomer, the yield of polymer decreases. In accordance with Scheme 2.4, in the first case, HCl is the cocatalyst for free $AlCl_3$ and, in the second case, the agent binding $AlCl_3$ via conjugated base $AlCl_3 + AlCl_4^{\delta-} \rightarrow Al_2Cl_7^{\delta-}$.

In the case of H_2O as the cocatalyst, the formation degree of the system possesses a much higher value because of its high basic properties and ability of aqua-complexes to decay with generation of other active protogeneous particles — hydrogen halides, for example. If this possibility is excluded, as in the case of preparing $RAlCl_2 \cdot H_2O$ by a special method [71, 76], formation of a complex aqua-acid is the motive power in providing nucleophilic participation, necessary for formation and stabilization of an intermediate compound. That is why the second pathway among two of initiation by aqua-systems (MeX_n + H_2O):

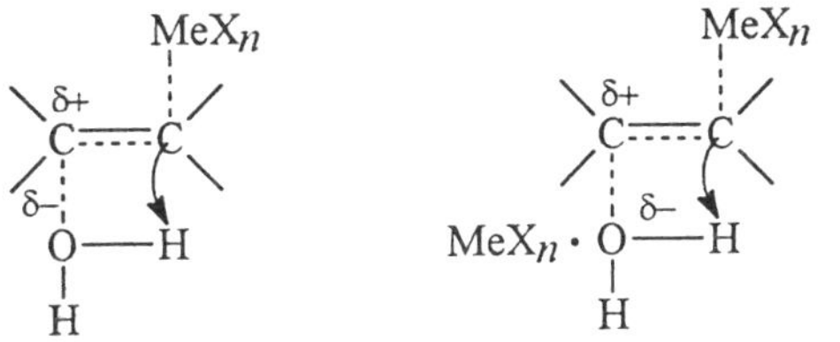

Scheme 2.5

is more probable because of increased (compared with H_2O) coordinating ability of the complex in relation to double bond of the monomer.

The presented scheme takes into account both the anion formation by the most favorable way and the expected order of the acid-base interaction reasoning from the HSAB concept. In keeping with this concept, the deterministic role in the system must be played by complexing of a hard acid (MeX_n) with the hardest base (H_2O).

The idea of the maximum activity of HX-MeX_n complexes, developed in some works [198–202], only in the presence of super-stoichiometric amounts of MeX_n, i.e. free Lewis acid, correlates with the above-mentioned Schemes 2.4 and 2.5. Obviously, the excess of MeX_n stabilizes the anion by reactions of homo- and heteroconjugation relating to the interaction of unitypical (hard) acids and bases. It should be noted that, however, from the point of view of reliability of such conclusions, not only the order of complex catalyst formation, but also fine details of interaction of components should be taken into account, the latter providing obtaining of complexes prearranged by the stoichiometry [71, 76].

The importance of formation of cooperative intermediate complexes and the connection of interaction processes of a complex aqua-acid with a monomer with formation of a counterion is confirmed by quantum-chemical calculations of C_4H_8–$R_nAlCl_{3-n} \cdot H_2O$ [203, 204], in which values of parameters characterizing the acidic strength of aqua-complexes of aluminum chlorides, the charge on H atom of the proton donor (q_{H^+}), energy of the proton detachment from the complex $\left(\Delta E_{theor}^{H^+}\right)$, and the acidity index ($\delta_p K_a$), were estimated. It is known that they correlate with the catalytic activity of complexes in electrophilic reactions [18, 205].

Preliminary calculations have shown (Table 2.9) that judging by values of charges q_{H^+} and values of $\delta_p K_\alpha$, the acidic strength of aqua-complexes is low, independent of the nature of the aluminum component. Negative values of the energy $\Delta E^{H^+}_{theor}$ indicates that the process of aqua-complexes' ionization may proceed only in the presence of accepting bases.

Table 2.9.

Calculated values of acidic strength parameters for aqua-complexes.

Aqua-complex	Charge on H atom, q_{H^+}, a.u.	Acidity indexes, $\delta_p K_\alpha$	Proton detachment energy, ΔE^{H^+}, kJ/mol
H_2O	+0.14	+15.7	–860
$H_2O{\cdot}Al_2Cl_2O$	+0.21	+4.4	–240
$H_2O{\cdot}Al_2Cl_3$	+0.21	+7.1	–390
$H_2O{\cdot}AlRCl_2$	+0.19	+7.7	–420
$H_2O{\cdot}AlR_2Cl$	+0.16	+14.4	–790
$H_2O{\cdot}AlR_3$	+0.16	+14.4	–790

Note: ΔE^{H^+} was calculated from the correlation equation $\delta_p K_\alpha = \beta{\cdot}\Delta E^{H^+}_{theor}$ [205], which was assumed true for both water and aqua-complexes; the coefficient β for H_2O equals –0.0182 mol/kJ ($\delta_p K_\alpha = 15.7$; $\Delta E^{H^+}_{theor} = -860$ kJ/mol).

At the same time, experimental studies of the behavior of R_nAlCl_{3-n} (Part 2.1) indicate that the process of AC formation requires calculation of all possible transformations of complexes under the effect of olefins [90, 91, 95]:

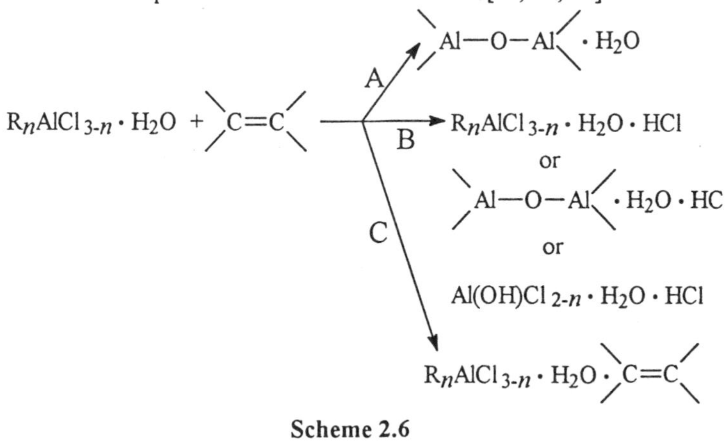

Scheme 2.6

Directions A and B in Scheme 2.6 reflect possible transformations of the aluminum–containing component of the system with further complexing with a proton donor, and the direction C shows formation of triple complexes with participation of the olefin.

Table 2.10

Parameters of alumoxane aqua-complex (CNDO/2).

Optimization parameters	E_0, kJ/mol	E_{CB}, kJ/mol	Dipole moment, D	$q_{Al_{(1)}}$	$q_{Al_{(2)}}$	$q_{O_{(1)}}$	$q_{O_{(2)}}$
Planar model							
0.21	-283580	-96550	2.48	-0.157	-0.157	+0.053	-0.135
0.24	-284010	-6970	3.51	-0.095	-0.095	+0.011	-0.159
0.27	-283940	-6920	3.88	-0.067	-0.067	+0.013	-0.205
0.20	-283500	-6470	2.47	-0.156	-0.156	+0.035	-0.120
0.22	-283590	-6550	2.50	-0.157	-0.157	+0.053	-0.135
0.23	-283527	-6500	2.48	-0.168	-0.168	+0.067	-0.140
Distorted model							
$\Theta_{(1)}* = 90°$	-284380	-7340	7.37	-0.194	-0.194	+0.001	-0.155
$\Theta_{(2)}* = 60°$	-284150	-7130	5.29	-0.132	-0.180	+0.014	-0.173
$\Theta_{(3)}* = 45°$	-284070	-7050	4.07	-0.105	-0.132	+0.015	-0.181
H_2O position							
"a"	-284380	-7340	7.37	-0.194	-0.149	+0.001	-0.132
"b"	-284410	-7390	7.00	-0.200	-0.160	+0.005	-0.167
"c"	-284400	-7370	6.52	-0.206	-0.140	+0.009	-0.193
Position at H_2O	-280910	-7130	5.55	-0.247	-0.197	+0.022	-0.326

As in some cases $R_nAlCl_{3-n}\cdot H_2O$ transforms into an alumoxane under the effect of electron donors [90, 95], an aqua–complex of alumoxane is chosen as one of the objects (direction A of the Scheme). Dialkylaluminum chloride and trialkylaluminum give unstable aqua-complexes even under negative centigrade temperatures but form appropriate alumoxanes or their aqua-complexes. Of importance is the fact that the activity of $R_3Al\cdot H_2O$ systems is usually prescribed to formation of aqua-complexes of alumoxane, which in accordance with [96] must possess increased Brönsted acidity. However, the role of alumoxanes in excitation of olefins' polymerization was stated only for the $C_2H_5AlCl_2\cdot H_2O$ system [9]. As hydrolysis of R_nAlCl_{3-n} in the presence of electron–donor compounds is usually accompanied by extraction of HCl — the acid stronger than water — cocatalytic action of HCl cannot be neglected, which may be displayed through formation of triple complexes (direction B of Scheme 2.6). In this case, different variants of AC structures possibly differ, for example, in their chemical structure.

There are experimental facts indicating that the electrophilic activity of $C_2H_5AlCl_2\cdot H_2O$ complex in arenes may be bound to formation of an appropriate alumoxane complex. It was stated by the IR–spectroscopy method that if heptane solutions of $C_2H_5AlCl_2$ are introduced into toluene, which is close in basicity to olefins, in particular to isobutylene, elimination of characteristic absorption bands of the initial complex (457, 500 and 605 cm^{-1}) and occurrence of alumoxane absorption bands (485, 575, 626, 1630 and 3200—3500 cm^{-1}) are observed. When oligomerization of a hardly polymerizing olefin — diisobutylene — is excited by $C_2H_5AlCl_2\cdot D_2O$, besides formation of the alumoxane complex, weak absorption bands at 2137 and 2183 cm^{-1} (ν_{C-D}) are observed for the products of the olefin transformation, which probably indicates protonation of the olefin by water from the complex with alumoxane:

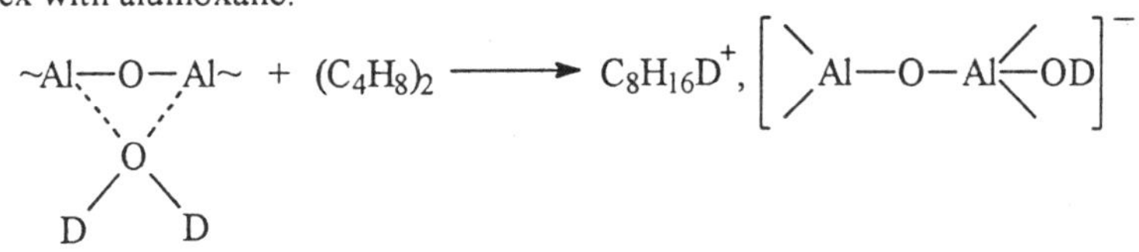

In this connection of interest are the data of quantum-chemical calculations of model structures (Figure 2.1) of alumoxane aqua-complexes.

It is typical that oxygen atoms $O_{(1)}$ and $O_{(2)}$ are nonequivalent. Taking account of the nonequivalence of these atoms is performed by two methods. In the first case the structure of the alumoxane aqua–complex was taken, in which the fragment

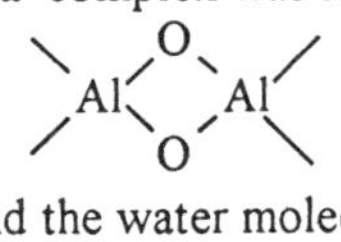

was considered equishouldered, and the water molecule was coordinated in relation to Al atoms. In response to partial optimization, a stable configuration of the complex was found (Tables 2.10 and 2.11), for example, the position of the water molecule (δ). The angle $\Theta = 90°$, $r_Y = 0.22$ nm, $r_Z = 0.24$ nm, in this case showing that the group $Al_{(1)}Cl_{(1)}Cl_{(2)}$ possesses sp^3-hybridization, and $Al_{(2)}Cl_{(3)}Cl_{(4)}$ possesses sp^2-hybridization (Figure 2.1, structure II).

The calculation has indicated that transition from the planar model of a symmetric complex (both $AlCl_2$ groups possess sp^3-hybridization) to a coplanar one is accompanied by an energy gain of 380 kJ/mol. Break of the equishouldered bridge is energetically favorable. Charges on hydrogen atoms of water are similar (+0.2) and low by the absolute value. Judging by the energy value $\Delta E^{H^+}_{theor}$ = –240 kJ/mol, detachment of a proton from the alumoxane aqua-complex is energetically unfavorable.

Table 2.11.

Charges on atoms in alumoxane aqua-complexes.

$q_{C_{(1)}}$	$q_{Cl_{(3)}}$	$q_{Cl_{(2)}}$	$q_{Cl_{(4)}}$	$q_{H_{(1)}}$	$q_{H_{(2)}}$
Planar model					
+0.062	–0.148	+0.162	–0.148	+0.283	+0.283
+0.039	–0.106	+0.039	–0.106	+0.236	+0.236
+0.029	–0.049	+0.029	–0.049	+0.191	+0.191
+0.071	–0.167	+0.071	–0.167	+0.293	+0.293
+0.068	–0.161	+0.068	–0.161	+0.289	+0.289
+0.062	–0.148	+0.062	–0.148	+0.283	+0.283
+0.063	–0.142	+0.063	–0.142	+0.284	+0.284
Distorted model					
+0.005	–0.081	+0.035	+0.106	+0.197	+0.215
+0.006	–0.021	+0.031	+0.052	+0.197	+0.215
+0.007	–0.016	+0.033	+0.030	+0.195	+0.213
+0.005	–0.081	+0.035	+0.106	+0.197	+0.215
+0.002	–0.023	+0.034	+0.107	+0.202	+0.201
+0.001	–0.002	+0.039	+0.111	+0.193	+0.189
-0.142	+0.026	-0.031	+0.070	+0.170	-

Note

$E^{H_2O}_{B3} = E_{0(Al_2Cl_4O \cdot H_2O)} - \left(E_{0(Al_2Cl_4O)} - E_{0(H_2O)}\right)$; values of $E^{H^+}_{B3}$ and $E^{H_2O}_{B3}$ for the alumoxane aqua-complex are 3240 and 220 kJ/mol, respectively; *Θ_1, Θ_2, Θ_3 are values of angles (see Figure 2.1).

In the second case, the position of the water was fixed, and the position of the oxygen atom $O_{(1)(r_1)}$ changed. For the found minimum energy, values of r_2 and α were varied, a more stable nonequishouldered structure III was found (E_{CB} = –7390 kJ/mol), break of the bridge of which is energetically unfavorable, and both possess sp^3-hybridization (see Figure 2.1).

Charges on hydrogen and oxygen atoms for the structure III were: $q_{H_{(1)}}$ = +0.21; $q_{H_{(2)}}$ = +0.21; $q_{O_{(2)}}$ = -0.21; $q_{O_{(1)}}$ = 0, respectively. In spite of the high stability of the structure III and nonequivalence of oxygen atoms, the acidity of the alumoxane aqua-complex, judging by charges on hydrogen atoms in the water molecule, hardly changed compared with the equishouldered model — the structure II (see Figure 2.1). Symmetric hypothetic structure III (a) was found energetically less favorable than the structure III, and was not considered further on. The structure IV (see Figure 2.1) is formed from the structure III as a result of proton detachment. The system charge becomes equal to –1 (at $H^+ \to \infty$, and multiplicity remains equal to 2S + 1 = 1). Optimization of the structure IV was performed in an analogous way to that of the structure III, i.e. first the position of $O_{(1)}$ atom was fixed, and then the position of OH-group was varied for the found minimum of the energy. The most stable structure - IV (sp^3-hybridization) — is characterized by parameters: r_3 = 0.20 nm, r_4 = 0.20 nm, r_5 = 0.19 nm, E_{CB} = –7280 kJ/mol. Comparison of obtained parameters correlating with the acidic strength of aqua-complexes: $E^{H^+}_{detach}$ = –420 kJ/mol, q_{H^+} = +0.19 for $RAlCl_2{\cdot}H_2O$ and $E^{H^+}_{detach}$ = -240 kJ/mol, q_{H^+} = +0.21 for $Cl_4Al_2O{\cdot}H_2O$, shows that although charges on hydrogen atoms in both complexes are similar, the proton detachment is simplified in the case of the alumoxane aqua-complex. However, even this, judging by the value pK_a = +4.4, must display low catalytic activity.

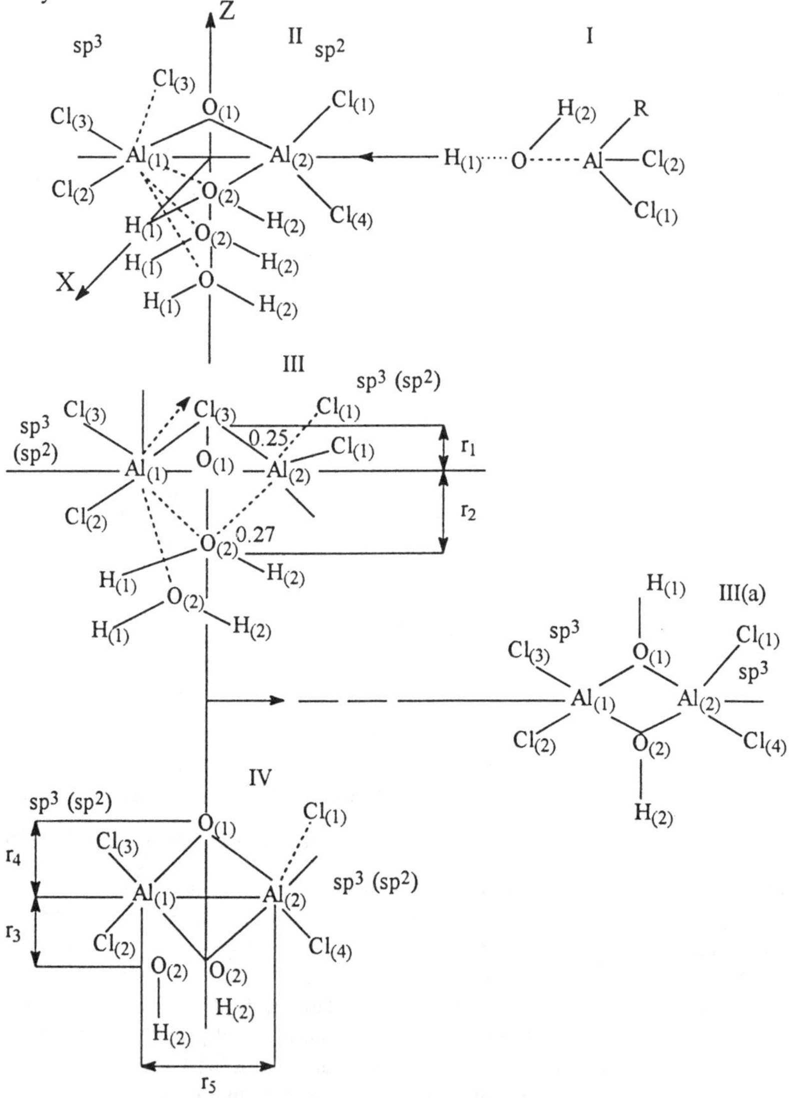

Figure 2.1. Model structures of alumoxane aqua-complexes (CNDO/2).

Another possible route of AC formation (direction B in Scheme 2.6) is connected to appearance of hydrogen chloride in the catalytic system and formation of complexes of the type $R_nAlCl_{3-n}\cdot H_2O\cdot HCl$. Values of $E_{\mathrm{detach}}^{\mathrm{H}^+}$ in the energy minimum for the main state of triple complexes $AlCl_3\cdot H_2O\cdot HCl$, $RAlCl_2\cdot H_2O\cdot HCl$, $Al_2Cl_4O\cdot H_2O\cdot HCl$ equal –85, –70, and –60 kJ/mol, respectively. Comparing these values with appropriate ones for double complexes (see Table 2.11) shows that the proton detachment in the case of formation of triple complexes is simplified by 3–5 times. However, negative values of $E_{\mathrm{theor}}^{\mathrm{H}^+}$ indicate that spontaneous ionization typical of $AlCl_3\cdot HCl$ complexes [94, 195], is not relevant in this case. Consequently, activity of the initial $R_nAlCl_{3-n}\cdot H_2O$ is not connected with an intermediate formation of hydrogen chloride. This important point is confirmed experimentally in the comparative study of selective isobutylene polymerization in the presence of catalysts $C_2H_5AlCl_2\cdot H_2O$ (Figure 2.2). Selectivity of the isobutylene polymerization from a mixture of C_4-olefins was determined in unit parts as the ratio of transformation degree of isobutylene to the total transformation of all olefins with an accuracy up to 15% (mass). In the case of adding HCl to $C_2H_5AlCl_2\cdot H_2O$, selectivity of the process decreases and approaches selectivity of the process initiated by $C_2H_5AlCl_2\cdot HCl$ complexes. This indicates that there is no connection of $C_2H_5AlCl_2\cdot H_2O$ activity in selective polymerization of olefins with the formation of triple $C_2H_5AlCl_2\cdot H_2O\cdot HCl$ complexes.

By this means, transformation of $AlCl_{3-n}\cdot H_2O$ into alumoxane and possible formation of triple complexes with participation of HCl do not respond probable AC, and suggested transformations of A and B directions in Scheme 2.6 are not significant from the point of view of initiation of the cationic polymerization of olefins.

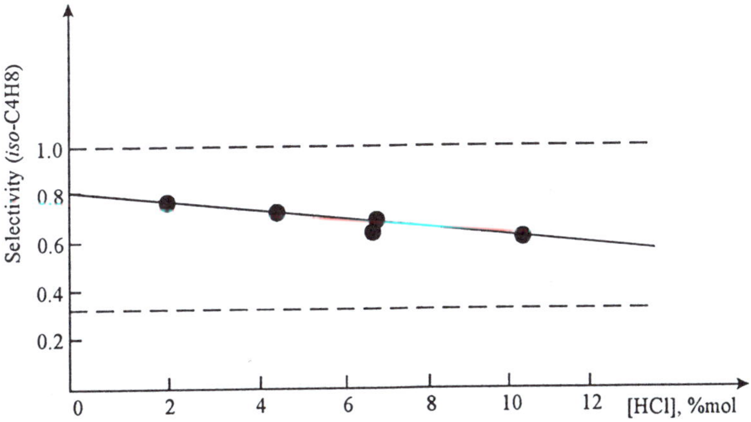

Figure 2.2. Dependency of isobutylene polymerization selectivity on adding of HCl to $C_2H_5AlCl_2\cdot H_2O$ complex ($3\cdot 10^{-6}$ m^3 – C_4 fractions of hydrocarbons, $[C_2H_5AlCl_2] = 3\cdot 10^{-3}$ mol/l, 243 K).

For the purpose of taking into account the olefin contribution to initiation of the process (direction B in Scheme 2.6), the dynamics of the complex formation in the $R_nAlCl_{3-n}\cdot H_2O$—olefin system was considered. For this purpose, calculation of the interaction of the $AlCl_3{\cdot}H_2O$ complex with isobutylene was conducted by the method suggested in [206]. Bond lengths and bond angle values in the initial model (Figure 2.3) were taken the same as in the work [203]. The water molecule in the complex $AlCl_3{\cdot}H_2O$ was rigidly fixed by sacrificing coordination Al—O and hydrogen O—$H_2{\cdots}Cl_{(1)}$ bonds. The position of the O—$H_{(1)}$ bond was determined by interaction with the acceptor, isobutylene, and was changed during the reaction.

The process of $AlCl_3{\cdot}H_2O$ interaction with isobutylene may be subdivided into three stages (coordination, π–bond scission in the olefin, carbocation formation), which differ by loosening the degree of the olefin double bond up to its rupture and formation of AC. Optimization of structures by the reaction coordinate $R_{OC_{(2)}}$ was performed by six parameters for the mentioned stages: length of the double bond $\left(R_{C_1C_2}\right)$, length of O—$H_{(1)}$ bond $\left(R_{OH_{(1)}}\right)$, the angle characterizing orientation of the $AlCl_3{\cdot}OH_{(2)}$ fragment in relation to isobutylene (α), the angle of the proton attack on olefin $\left(\varphi_{OH_{(1)}}\right)$, displacement of $AlCl_3{\cdot}H_2O$ molecule in relation to isobutylene along the $Z(R_Z)$, axis and the angle of the tetrahedron formed (γ) (Figure 2.3). Parameters and changes of the total energy (E_0) of $AlCl_3{\cdot}H_2O–C_4H_8$ system are shown in Table 2.12.

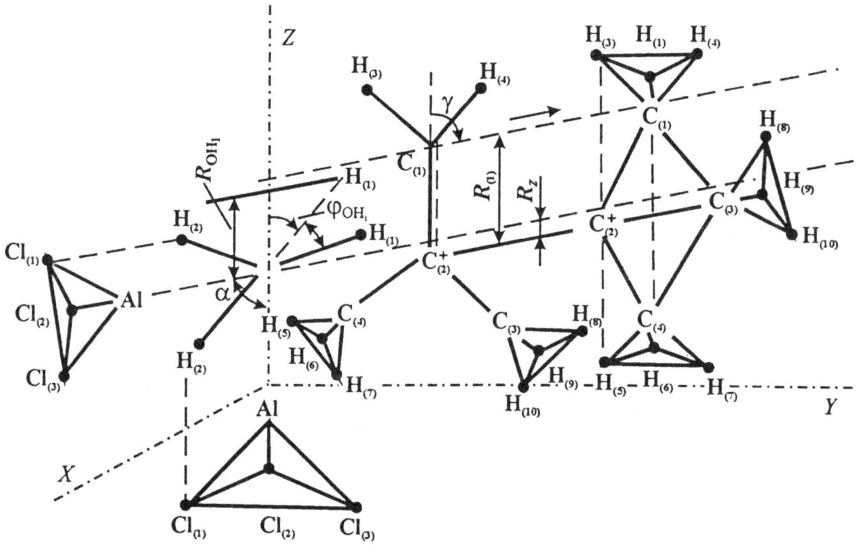

Figure 2.3. Interaction of $AlCl_3$ aquacomplex with isobutylene.

The true number of varied parameters of the system exceeds six. However, introduction of additional optimization parameters, for example, the length of the Al—O bond and others characterizing positions of atoms of the complex, having not participated directly in the interaction, appeared insufficient, when the reaction pathway was studied. The stage of the olefin coordination includes four steps. The initial value of $R_{OC_{(2)}}$ was taken equal to 0.300 nm, and the energy of the first step put at zero. Characteristic features of the coordination stage are rotation of the $AlCl_3{\cdot}OH_{(2)}$ fragment of the complex molecule by the angle α = 90° (it does not change in further stages) and orientation of O—$H_{(1)}$ bond to the side of isobutylene. On the second and the third stages, the value of E_0 decreases to –20 and –40 kJ/mol, and the length $R_{OH_{(1)}}$ grows up to 0.108 and 0.111 nm, respectively. On the final stage of the coordination, E_0 of the system increases, parameters $R_{C_1C_2}$, $R_{OH_{(1)}}$ and $\dot{\gamma}$ remaining unchanged, and the values of R_z and $\varphi_{OH_{(1)}}$ decrease. By this means the stage of isobutylene coordination with aluminum chloride aqua–complex is characterized by the minimum value of the energy (Figure 2.4).

Table 2.12.

Varied parameters and change of the total energy of the system in interaction of $AlCl_3$ complex with isobutylene.

Step number	Reaction coordinate $R_{OC_{(2)}}$, nm	Optimization parameters						E_0, kJ/mol
		$R_{C_1C_2}$, nm	$R_{OH_{(1)}}$, nm	α, degr.	$\varphi_{OH_{(1)}}$, degr.	R_z, nm	$\dot{\gamma}$, degr.	
1	0.300	0.134	0.105	0	70	0.4	0	0
2	0.280	0.134	0.108	90	70	0.4	0	–20
3	0.260	0.134	0.111	90	70	0.4	0	–40
4	0.240	0.134	0.111	90	60	0.2	0	–35
5	0.220	0.136	0.111	90	40	0.2	0	–5
6	0.180	0.136	0.111	90	0	0.2	0	0
7	0.170	0.136	0.111	90	0	0.2	0	+5
8	0.165	0.136	0.111	90	0	0.2	0	+5
9	0.163	0.136	0.111	90	0	0.2	0	+10
10	0.160	0.146	0.158	90	20	0.2	0	–420
11	0.155	0.148	—	90	—	0.2	7	–480
12	0.150	0.148	0.177	90	—	0.2	19	–500
13	0.142	0.148	0.212	90	—	0.1	—	–620
14	0.140	0.148	0.200	90	—	0.1	—	–610

The second stage of excitation of the electrophilic reaction begins with $C_{(1)}$=$C_{(2)}$ π–bond fission in the olefin (Table 2.12, the fifth step). E_0 reaches its maximum at the value of 10 kJ/mol. Consequently, the energetic barrier of the reaction is 50 kJ/mol. At $R_{OC_{(2)}}$ = 0.160 nm and the attack angle $\varphi_{OH_{(1)}}$ = 20°, the length of the π–bond increases from 0.136 to 0.146 nm, and the bond O—$H_{(1)}$ is practically broken (Table 2.12, the tenth step), parameters $R_{C_1C_2}$, $R_{OH_{(1)}}$ and $\dot{\gamma}$ becoming independent. The particle $H_{(1)}^{\delta+}$ attacks the most hydrated atom $C_{(1)}$ forming a $C_{(2)}H_{(1)}H_{(3)}H_{(4)}$ tetrahedron ($\dot{\gamma}$ = 19°). On the eleventh step, the π–bond $C_{(1)}$—$C_{(2)}$ is broken ($R_{C_1C_2}$ = 0.148 nm), and E_0 is reduced to –480 kJ/mol. The stage of carbocation formation completes the π–bond rupture and attachment of a $H_{(1)}^{\delta+}$ particle to C_1 atom. In this case, carbocation $C_{(2)}^{+}$ is formed, hybridized by the sp^2-

atom. Atoms $C_{(1)}$, $C_{(2)}$, $C_{(3)}$ and $C_{(4)}$ form a tetrahedral structure (steps 12—14) because of the interaction with negatively charged $[AlCl_3{\cdot}OH_{(2)}]^{\delta-}$ fragment of the complex. This process is accompanied by an additional gain in energy (E_0 reaches –620 kJ/mol). Consequently, formation of AC with total energy gain of 580 kJ/mol happens.

Changes of charges on atoms of the $AlCl_3{\cdot}H_2O$ complex by the reaction coordinate are displayed in Figure 2.4. As isobutylene approaches, the charge on $H_{(1)}$ atom in the active bond O—$H_{(1)}$ increases from +0.22 to +0.29, which by the correlation dependence $\delta pK_a - q_{H^+_{(1)}}$ [207] corresponds to an increase of δpK_a from 0 to –15. Detachment of the $H^{\delta+}_{(1)}$ particle (drop on the energetic curve, Figure 2.4) happens at a somewhat lower value of the acidic strength of the complex (q_{H^+} = +0.27, $\delta pK_a = -14$), but at the maximum charge on the $C_{(1)}$ atom of the monomer (–0.23). The positive charge on the α–carbon atom $C_{(2)}$ grows gradually in reaction coordinate from 0.06 to 0.2 at the sacrifice of the π–electron density shear of the double bond to the atom $C_{(1)}$. The oxygen atom of H_2O, although it acquires additional electronic density at the stage of the monomer coordination, is not significantly changed, which is probably connected with a partial charge transmission from oxygen to Al atom on the stage preliminary to the π-bond break. Approximate equality of the positive charge on the carbon atom $C_{(2)}$ and the negative charge on the oxygen atom indicates formation of polarized intermediate $[Cl_3AlOH]^{\delta-}$---$(CH_3)_3C^{\delta+}$.

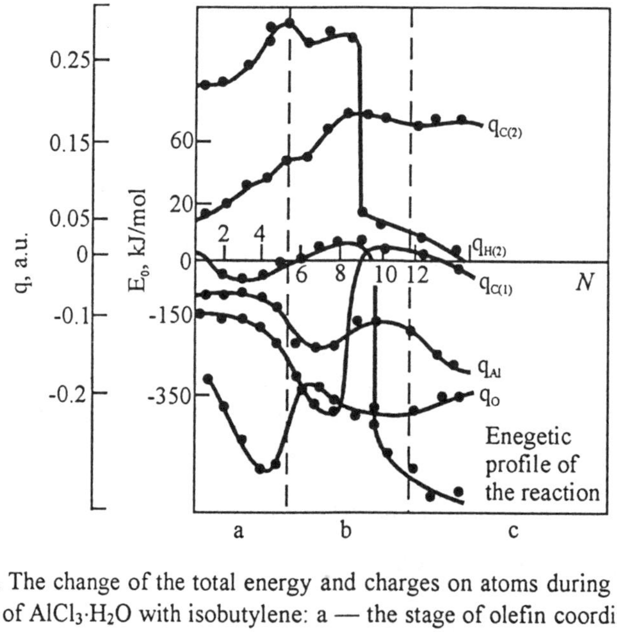

Figure 2.4. The change of the total energy and charges on atoms during the reaction of $AlCl_3{\cdot}H_2O$ with isobutylene: a — the stage of olefin coordination; b — the stage of π–bond fission; c — the stage of carbocation formation.

Displaying of hard Brönsted acid properties by $AlCl_3$ aqua-complexes and formation of AC are promoted by the monomer coordination with a catalyst, weakening (breaking) of O-$H_{(1)}$ and $C_{(1)}$=$C_{(2)}$ bonds proceeding simultaneously.

Preliminary complexing of water with $AlCl_3$ concludes not only in its partial activation and providing direction of the attack of the monomer O-$H_{(1)}$, but also in "rigid" linking of O-$H_{(2)}$ group to $AlCl_3$.

Consequently, acidic-catalytic properties of soluble R_nAlCl_{3-n} aqua-complexes, characterized by low acidic strength, are displayed at direct participation of a monomer (substrate) as a coordinating additive. In the coordination process there occurs the mutual orientation of the aqua-complex with the monomer with further simultaneous activation of O-$H_{(1)}$ bond of water and loosening of the olefin π-bond. Substantially in the process of cooperative interaction of a monomer with the aqua-complex, a significant increase of Brensted acidic strength of the complex occurs, which reaches the properties of a hard acid. However, the proton transfer is not the limiting stage and is conformed in time to counterion formation, which is promoted by preliminary coordination of water with the Lewis acid. It can be stated that formation of a counterion, in its turn, assists in the proton transfer, i.e. cooperative mechanism of excitation is also possible in the case of soluble cationic systems [18, 208–210].

The data obtained by calculation on regularities of initiation by aqua-complexes R_nAlCl_{3-n} were borne out also for the case of complexes of another soft acid — an alcohol [84, 211]. This concerns, first of all, the participation of a monomer in the stage of AC formation. A feature of systems with alcohols is their possibility of displaying proton-donor or carbocationic activity by appropriate complexes:

$$R_1OH + R_nAlCl_{3-n} \xrightarrow{M} H^{\delta+}R_nAlCl_{3-n}OR_1^{\delta-}$$
$$R_1OH + R_nAlCl_{3-n} \xrightarrow{M} R_1^{\delta+}R_nAlCl_{3-n}OH^{\delta-}.$$

As follows from the calculation of alcoholic complexes of Lewis acids, the fission of C—O and O—H bonds of alcohols by the processes described above is approximately equally probable (close degrees of corresponding bonds in the complexes). Growth of the Lewis acid strength and existence of substituents at the α-carbon atom of alcohol assists in carbocationic activity, for example, the $AlCl_3$ complex with *tert*-C_4H_9OH. Low values of counterion nucleophility and the possibility of linking hydroxygroups into a stable form are among a number of factors promoting this direction of the alcohol decay; for example, [18]

$$(CH_3)_3C-OH \xrightarrow{SbF_5 \cdot FSO_3H} (CH_3)_3C, SbF_5 \cdot FSO_3^- + H_3O^-.$$

In spite of the high energy of the OH bond compared with that of the C—O bond, at a decrease of the acidic strength of the complex forming acid and especially application of alcohols with non-branched radicals (CH_3OH, C_2H_5OH, etc.), the latter preferentially display proton-donor properties. Additional specific interactions between components of the complexes and other factors affect the direction of the alcohol decay. For example, in the model reaction with diisobutylene considered in part 2.1:

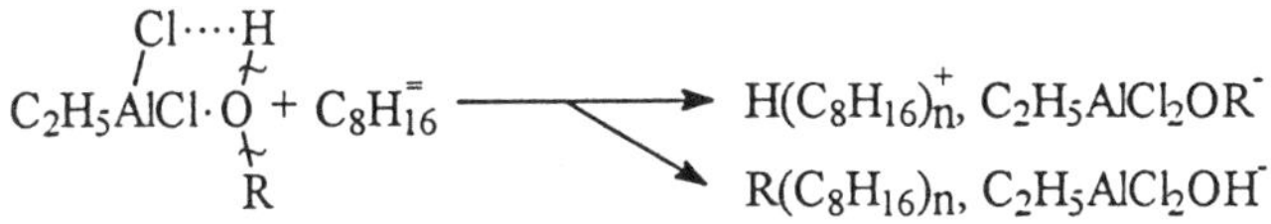

the latter of these reactions is assisted by formation of hydrogen bonds Cl·······H-O hindering the olefin assess to the proton.

In the case of olefins and sterically hindered phenols (combined with $SnCl_4$, $C_2H_5AlCl_2$, etc.) H-initiation of isobutylene polymerization is assisted by both the proton acidity, higher than in aliphatic alcohols, and p_π–p_π-conjugation of oxygen atoms with an aromatic nucleus hindering the fission of the C_{arom}—C bond [211]. Steric effects at the interaction of components should also be also taken into consideration.

For adamantols in the presence of $AlBr_3(SnCl_4)$, formation of adamantyl cations is bound to the stabilization effect at the sacrifice of secondary reactions [209].

The outlined features of the behavior of complexes with alcohols should be taken into account in their application as initiators of isobutylene polymerization.

A definite contribution into understanding of the acidic–catalytic mechanism of initiation is made by modeling of the proton transfer reaction. Such a possibility is given by the analysis of the behavior of proton attachment complexes, the formation of which is mostly considered as a model of the process initiation under the influence of, for example, the hypothetic acid $HAlCl_4$:

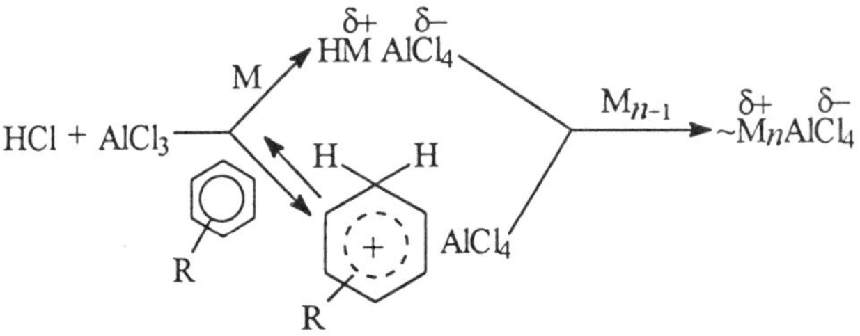

The general feature in behavior of ionic particles in the reaction is the dependency of activity on factors of external (solvation of ions, electrostatic effect of a counterion) and internal (the influence of electron–donor substituents) stabilization, and the difference in reversible and irreversible type of formation of arenonium and carbonium ions, respectively. Consequently, in the presence of a stronger than arene π–acceptor in the system, its protonation must happen. This is confirmed by application of various complexes of the proton attachment based on substituted arenes, including the composition of styrene polymers, for initiation of the isobutylene poymerization.

Whatever the nature of the complexes (of the Lewis acid and aromatic component type) or their composition (the ratio of aromatic fragments to metal halide is from 3 to 200), the H–initiation of isobutylene and other monomer polymerization proceeds [54, 166, 212]. As no products of the monomer attachment to the arenonium ion in the form of a σ–complex was found, the proton transfer is obviously performed via non-classic arenonium ion (π–complex) with regeneration of the initial aromatic structure:

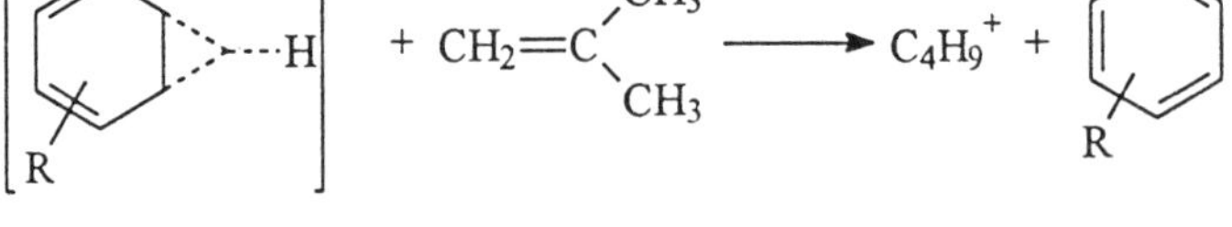

The motive force of the process is a higher π–basicity of olefin compared with arene, i.e. initiation of polymerization is of a competitive type for capturing a proton between isobutylene and bases participating in the complex.

In this manner, excitation of the isobutylene polymerization as a typical cation-active monomer by various complex catalysts represents a specific variety of acidic-catalytic reactions. Although the fine mechanism of proton transfer is not clear yet, the comprehensive analysis of dependent interactions in the monomer-catalyst system may assist in solving a complex problem of initiation of the cationic isobutylene polymerization.

2.2.2. Chain propagation.

Chain propagation is the main stage of macromolecular product formation:

$$\ldots - M_n^{\delta+}A^{\delta-} + M \rightarrow \ldots - M_{n+1}^{\delta+}A^{\delta-}.$$

The reaction proceeds at a relative rate noticeably higher than of other reactions of the growing center. It is suggested that end reactionary ionic particles exist in the cationic polymerization as free ions, ionic pairs, ionic triples [+ – +] and [– + –], quadrupoles [± ∓] and more complicated aggregates [193]. Reactions of these formations proceed in the liquid phase, in which most of them can coordinate with the solvent molecules. In some media, one and the same ionic particles exist in two or more different coordination states. Consequently, in the presence of the solvent S in the system, a series of different types of AC may exist in the equilibrium:

$$\ldots - MA \leftrightarrow \underset{\text{I}}{\ldots - M^{\delta+}A^{\delta-}} \leftrightarrow \underset{\text{II}}{\ldots - M^{\delta+}S_nA^{\delta-}} \leftrightarrow \underset{\text{III}}{\ldots - M^{\delta+} + A^{\delta-}},$$

where I, II and III are contact ionic pairs, ionic pairs separated by a solvent ($n \geq 1$), and free ions, respectively. However, there are no direct proofs of participation of particular structures I–III in the cationic polymerization. Various ionic forms of growing polymer are usually in dynamic equilibrium with each other. The equilibrium significantly depends on temperature and concentration of reagents. For the same reason, kinetics of the cationic polymerization can be extremely complicated and change suddenly with the external conditions [193].

Sometimes during polymeric chain propagation the stoichiometric structure of its cationic end changes, which also raises the number of different AC. Existence of many types of AC seriously complicates kinetic analysis of the polymerization process, especially in those cases when the equilibrium between different ions shifts during the polymerization process, the more so since no methods of reliable identification of various types of active particles exist yet. That is why measured elementary constants of the chain propagation are effective.

Kinetic behavior of cations is defined by their activity. The sequence of activity of carbocations coincides with that of appropriate free radicals — the consequence of the induction effect of substituents. The greater is the positive induction effect, the smaller becomes the positive charge on the valently unsaturated carbonic atom of the cation and the smaller is its reactivity.

The *tert*–butyl cation is the least reactive in the sequence of alkyl cations with respect to the electron density at the reactive center. Cations disposed in the sequence to the left of the *tert*–butyl one easily enter both reactions of the chain propagation and the side ones. Cations disposed to the right react too slowly with the monomer. In both cases no long non-branched chain molecules are formed. The optimal combination of stability and activity of the $(CH_3)_3C^+$ cation appears extremely favorable for the chain propagation reaction; that is why linear products of high molecular mass are easily formed during the isobutylene polymerization.

In the most of the systems, ionic pairs are active particles of the process. The specific feature of the chain propagation on such AC is dependency of the reaction rate of the chain propagation k_p on the process conditions, nature of the medium and catalyst. The reactivity of the active chain end in the ionic pair depends on the counterion nature and the distance between ionic pair components, which is determined by the medium polarity and its capability for specific solvation of ionic pair components, the nature of solvent or monomer, and temperature.

Quantitative data reflecting the counterion effect on the kinetics of the cationic polymerization are scarce [193]. Obtaining them is difficult, because the concentration of growing ionic pairs and the constant of their dissociation to free ions are usually unknown. The latter circumstance may cause sufficient errors in measured values of k_p due to high reactivity of free ions formed from ionic pairs.

Activation energy E_a of the chain propagation significantly depends on the degree of dissociation of growing pairs. The dissociation reaction is exothermic. Consequently, the concentration of free ions possessing high activity grows with temperature reduction. This is one of the main reasons for the existence of negative values of measured E_a.

Although the difference in activity of free cations of those in the structure of ionic pairs is obvious (for radiation polymerization of isobutylene $k_p = 1.5 \cdot 10^8$ $l \cdot mol^{-1}s^{-1}$, which is by many degrees higher than that for the substance initiation [213]), great differences in kinetics of such behavior of ionic agents in solutions may be hardly expected. Studying styrene polymerization, which is close in activity to isobutylene, under the effect of CF_3SO_3OH (303 K) by the method of stopped flow, the values of k_p were obtained only 6–24 times exceeding the values of $k_p^{\pm}$ [214], which confirms the data published before [215, 216]. This is significantly smaller compared to the common great difference between k_p^{+} and $k_p^{\pm}$. The reason should be sought in properties of carbonium ions. Forming as a result of the total process of acid–base interactions in the system, carbonium ions as rather reactive particles exist as a result of the effects of internal (the influence of substituents in the monomer) and external (effects of the medium, monomer, active additives) stabilization. The role of these two factors is obvious, because both high stability of carbocations (triphenylmethyl salts) and their low resistance (aliphatic cations in pair with strongly nucleophilic anions) are unfavorable from the point of view of initiation and propagation of polymeric chains.

The amphoteric type carbonium ion in the HSAB concept supposes ability to interaction by soft acid–soft base and hard acid–hard base types on the propagation stage. The limiting cases are the reactions of free cations in the gas phase, where solvation can be performed by a substrate only, and the chain propagation proceeds at the ether bond, for example, M—$OClO_3$. For the relatively stable carbonium ion from isobutylene, efficient chain propagation is provided by preference for the reaction with a soft base–monomer comparing with harder ones (counterion and others). Of importance is that competition conditions change as the polymerization processes on account of the monomer expenditure, change of the catalyst state and other processes. The reaction of carbocation which is unfavorable at the beginning, for example with a counterion or its fragment, becomes favorable at the end of the process. Evidently, this is the reason why deactivation of AC happens, the consequence of which is that in many cases isobutylene polymerization does not reach complete exhaustion of the monomer. This is the reason why the ionic pair state has not been established and raises ambiguities that imply a manifold role for the counterion during the polymeric chain propagation. Consequently, the terms "free carbonium ions" and, respectively, "free counterion", applied to the chain propagation at the substance initiation of the cationic polymerization, are rather provisional. The known low ability of volumetric counterions to solvate during cationic polymerization explains why solvent polarity has little effect on the chain propagation stage. In a similar manner, judging by the comparatively small increase of polyisobutylene molecular masses [217], complexing of a counterion with electron accepting compounds or introduction of salty additives with analogous (to the catalyst) anion insignificantly change the ionic pair behavior. Suitable information on the role of counterions on the propagation stage is obtained from quantum-chemical calculations of carbocation interaction with the monomer [218]. Taking into account the competitive character of the monomer and counterion reaction with AC, the transition state of the propagation stage can be represented by nucleophilic substitution reactions of the S_N2 type:

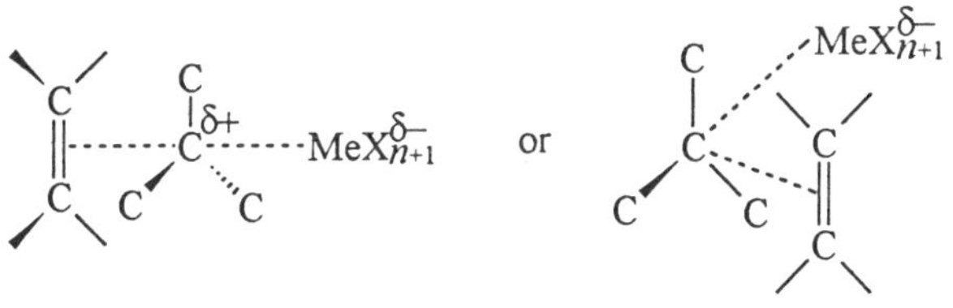

Scheme 2.7

With lack of carbocation screening by a counterion, the attack by olefin by Scheme 2.7A should be more favorable than by Scheme 2.7B. At the same time, in the transition state of the reaction S_N2, an important role is played by steric factors. The presence of volumetric substituents at the carbocationic center (the case of growing carbonium ions) makes difficulties for the monomer approach, and makes the direction A less favorable than B (Scheme 2.7).

To select the most probable reaction direction, models of olefin interactions with the ionic pair have been discussed. The initial state of the ionic pair $\ldots - C^{\delta+}MeX^{\delta-}_{n+1}$ corresponds to the most reactive structure, and is obtained at the stage of process initiation. The calculation was conducted in a similar manner to the one for model interactions of the $AlCl_3{\cdot}H_2O$ complex with isobutylene [203, 204].

Figure 2.5 displays the attack of the monomer on AC from the side opposite to the complex anion. Optimization of the monomer position (ethylene was chosen to simplify the case) in relation to the cation was performed by six parameters: length of the double bond $R_{C_1C_2}$, angles of the monomer rotation in planes zOy (θ_1) and zOx (φ_1), the distance between carbocation and counterion R_{OC_2}, rotation angles θ_2 and

φ_2 in planes *zOy* and *zOx* in pair for atoms H_2, H_3 and H_4, H_5, respectively. As the monomer approached the carbocation (Scheme 2.7A, decrease of $r_{C_1C_5}$), the system energy always increased as the chosen parameters were varied up to the distance $R_{C_1C_5}$, which equals the covalent bond length (0.148 nm). Changes of charges on atoms of the reactive center are shown in Table 2.13. The positive charge on the carbon atom C_1 (q_{C_1}) and the negative one on the oxygen atom (q_O) increase. Atoms of C in ethylene also change charges, growth of the negative charge on atom C_5 being more noticeable than of the positive one on C_6. In spite of these, the double bond C_5C_6 is not changed significantly ($P_{C_5C_6}$ = 1.8–1.9; $R_{C_5C_6}$ = 0.135 nm). When $R_{C_1C_5}$ is reduced from 0.30 to 0.148 nm, the distance R_{OC_1} remains practically constant. Absence of a minimum at E_0 change and loosening of the monomer π–bond verify the difficulty of the ethylene attack on the carbocation from the side opposite to the anion.

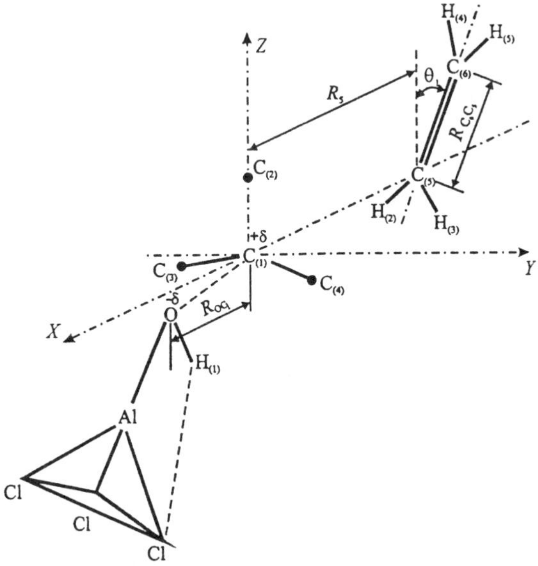

Figure 2.5. Olefin attack on the ionic pair by Scheme 2.7A.

The model illustrating the monomer approach to carbocation by Scheme 2.7B and formation of the end structure is shown in Figure 2.6.

At optimization of the ethylene position at the AC, first, at the monomer molecule rotation around O—C_1 bond (rotation radii $R_{C_1C_5}$ vary from 0.250 to 0.500 nm) the most stable configuration of the system was determined ($R_{C_1C_5}$ = 0.30 nm, Figure 2.4*a*), the value of E_0 of which equaled zero. Furthermore, angles of the ethylene rotation in *zOx* (θ_5, θ_6) and *zOy* (φ_5, φ_6) planes, length of the double bond $R_{C_5C_6}$, rotation angles of atoms H_2, H_3, H_4, H_5 in the *zOx* plane ($\varphi_{2\text{-}5}$), the angle of counterion rotation in the *zOx* plane (α), the length of forming the C_1C_5 bond ($R_{C_1C_5}$). The distance R_{OC_1} from carbocation to counterion was selected as the reaction coordinate.

Table 2.13.

The change of charges on atoms and total energy of the system at ethylene attack on a carbocation by Scheme 2.7A.

$R_{C_1C_5}$, nm	Charge, a.u.				E_0, kJ/mol
	q_{C_1}	q_O	q_{C_5}	q_{C_6}	
0.30	+0.18	-0.20	-0.04	-0.03	0
0.25	+0.18	-0.20	-0.06	-0.02	+10
0.20	+0.22	-0.21	-0.06	-0.01	+160
0.19	+0.22	-0.23	-0.14	+0.03	+220
0.185	+0.25	-0.24	-0.20	+0.06	+350
0.18	+0.26	-0.24	-0.20	+0.0	+620
0.16	+0.27	-0.25	-0.22	+0.10	+850
0.148	+0.28	-0.26	-0.24	+0.11	+960

It is useful to consider the change of parameters and the system energy during the reaction (Table 2.14), as in the case of initiation, by conditional stages: coordination of olefin, π–break and formation of the end structure.

For the coordination stage (R_{OC_1} grows from 0.142 nm to 0.160 nm) rotation of the counterion by the angle $\alpha = 100°$ (which does not change on further stages), the monomer outlet from the *zOy* plane ($\varphi_6 = 190°$) and decrease of the system energy to –175 kJ/mol are typical.

Table 2.14

Change of parameters and total energy of the system at the ethylene attack on carbocation by Scheme 2.7B.

R_{OC_1}	α, degr	$R_{C_1C_5}$ nm	θ_5, degr	φ_5, degr	$R_{C_5C_6}$ nm	θ_6, degr	φ_6, degr	$\varphi_{2,3}$, degr	$\varphi_{4,5}$, degr	E_0, kJ/mol
0.142	0	0.300	45	200	0.135	90	170	±75	±80	0
0.145	100	0.270	55	190	0.137	80	180	±75	±80	-125
0.150	100	0.270	55	190	0.137	80	180	±65	±80	-135
0.155	100	0.260	65	190	0.139	75	180	±65	±80	-145
0.160	100	0.250	65	190	0.141	75	180	±65	±75	-175
0.165	100	0.240	68	190	0.143	78	180	±65	±65	-150
0.170	100	0.230	68	190	0.145	80	180	±65	±65	-135
0.180	100	0.200	50	190	0.147	93	180	±65	±65	-210
0.260	100	0.170	17	190	0.148	107	180	±65	±65	-390
0.360	100	0.148	0	190	0.148	109	180	±60	±60	-720

On the second stage (R_{OC_1} = 0.160-0.180 nm) the system energy increases to –135 kJ/mol. Consequently, the energetic barrier is about 40 kJ/mol. It is remarkable that in spite of initiation (see Table 2.7), loosening of the monomer double bond proceeds gradually during all stages of the interaction. At the stage of the end structure formation, the double bond C_5C_6 scission ends, and the monomer is introduced at the O—C_1 bond (Figure 2.6b).

Changes of indexes of reactivity (charges on atoms and bond orders) are shown in Table 2.15.

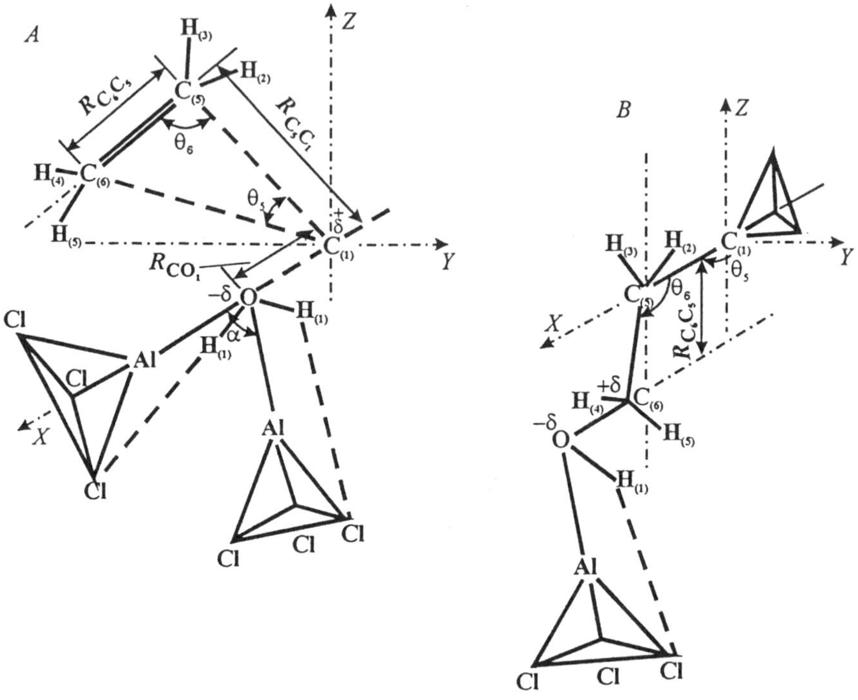

Figure 2.6. Olefin attack on ionic pair by Scheme 2.7, B.
a — the initial structure, b — the final structure after olefin introduction.

Analysis of reactivity indexes with regard to the reaction coordinate shows that the positive charge on carbon atom C_1 of the carbocation decreases, and the negative charge on atom C_5 of the monomer is increased. The deficit of electrons on the α-carbonic atom C_6 grows also gradually at the sacrifice of the shift of double bonds' π-electrons in the direction to the atom C_5. The total charge on the counterion $q_{(AlCl_3OH)^{\delta-}}$ is not significantly changed. The order of the double bond C_5—C_6 decreases from 2 to 1. The bond OC_1 is weakened simultaneously (up to complete scission), and bonds C_6—O and C_1—C_5 are formed. A small change of the positive charge on atom C_1 of the carbocation and simultaneous significant weakening of the double bond C_5C_6 indicate a significant role of orbital overlapping at this stage of the interaction. Obviously, introduction of the next monomeric chain is performed by newly formed ionic bond C_6—O, because its parameters (length of 0.142 nm, order of 0.89) are close to those of the initial bond C_1—O of the ionic pairs. The introduction condition should be the planar disposition of atoms C_6, C_5, H_4 and H_5.

Table 2.15

Reactivity indexes of the system at the ionic pair interaction with ethylene by Scheme 2.7B.

$p_{C_6C_5}$	Charge, a.u.				Bond order				
	q_{C_1}	q_{C_5}	q_{C_6}	$q_{(AlCl_3OH)^{\delta}}$	(q_O)	$p_{C_6C_5}$	p_{OC_1}	p_{C_6O}	$p_{C_1C_5}$
0.142	+0.09	–0.10	+0.04	–0.39	(–0.16)	1.97	0.83	0.03	0.02
0.145	+0.07	–0.26	+0.16	–0.36	(–0.12)	1.66	0.81	0.27	0.07
0.150	+0.06	–0.27	+0.17	–0.36	(–0.11)	1.62	0.79	0.31	0.08
0.155	+0.04	–0.30	+0.18	–0.33	(–0.09)	1.54	0.72	0.39	0.09
0.160	+0.03	–0.32	+0.20	–0.31	(–0.09)	1.30	0.65	0.61	0.15
0.165	–0.02	–0.28	+0.19	–0.35	(–0.11)	1.26	0.49	0.65	0.19
0.170	–0.07	–0.22	+0.19	–0.37	(–0.12)	1.18	0.31	0.76	0.27
0.180	–0.07	–0.16	+0.18	–0.38	(–0.14)	1.11	0.23	0.80	0.44
0.260	–0.08	–0.07	+0.17	–0.39	(–0.16)	1.07	0.17	0.87	0.69
0.360	–0.09	+0.02	+0.19	–0.39	(–0.16)	1.01	0.01	0.89	0.96

By this means, calculation indicates the important role of the counterion in the interaction of olefin with a carbocationic center by Scheme 2.7B (see Figure 2.6). The anionic fragment of the ionic pair directly participates in electron transitions bound to disclosure of the olefin double bond, although it does not generally change the result of the reaction. The behavior of charges on carbon atoms of olefin, carbocation and oxygen atom of the anion shows the coincident character of the interaction between carbocation and olefin. Weakening of the carbocation-counterion bond (dissociation of ionic pairs) is not accompanied by expected charge growth of charges on the ionic pair components and increase of the total energy of the system. Unfavorable separation of charges is compensated by the olefin coordination in relation to the active bond $\equiv C^{\delta+}$---$O^{\delta-}$. The process is accompanied by the counterion disposition in space, at which minimal obstacles for the olefin attack are made.

Of interest is that if the break and formation of the $C^{\delta+}$—anion bond proceed simultaneously, formation of the $C_1^{\delta+}$—monomer bond is a little delayed in relation to the aliquot bond break (see Table 2.15). Accompanying small charges on the C atom of the carbocation indicates a significant role of p-orbital overlapping at the coordination stage. Here the specificity of carbocation behavior as an amphoteric acidic agent is displayed: it participates in reactions with a hard base (a counteranion) and with a soft base (an olefin). Consequently, the reaction of carbocations with an olefin represents an unusual combination of electrostatic and orbital interactions controlled by the presence of a counterion. Impeding the approach to a carbocationic center at the coordination stage (growth of the system energy), the counterion provides simultaneously the directivity of the further attack displayed in the mutual orientation of system components. Their further interaction is accompanied by energy gain.

Dependency of the processes of olefin π–bond breaking and formation of a new ionic pair, parameters of which are close to the initial one, give grounds to suppose that attachment of the next olefin molecule to the newly formed AC proceeds by a similar scheme. As initiation with the help of soft acidic complex catalysts is, the chain propagation is characterized by coordinated interaction of the system components. The above–said is probably true also for the reaction of olefins with solvately separated ionic pairs of $[AlCl_3{\cdot}OH]^{\delta-}||C^{\delta+}(CH_3)$ type, i.e. when the medium effect is taken into account.

Calculation results coincide with the experimental data on the role of the counterion in the reaction of growing polymeric carbocation termination, which proceeds though a cyclic transition state in the carbonium ion–monomer–counterion system [219]. At the same time, the low probability of the reaction at the olefin attack on the carbocation from the side opposite to the counterion found experimentally contradicts the conclusions of the theoretical work [220], according to which this very reaction truly models the chain propagation. It appears to be that the main argument — no barrier character of the olefin coordination with the carbocationic center of the ionic pair — cannot be the reason to suppose that the whole reaction is preferable, because it must cause unfavorable separation of charges of the stage of the olefin π–bond fission.

Considering the experimental data on the role of ionic pairs at the stage of the chain propagation during polymerization, the unsuccessful attempt to perform non–terminating chain propagation under substance initiation should be noted [34]. In spite of using media with extremely low nucleophility (1,1,2-trichloro-1,2,2-trifluoro-ethane, etc.), favorable to stabilize carbonium ions, typical features of "living polymerization" were absent in the isobutylene polymerization under the effect of SbF_5: linear dependency of the polymer molecular mass on conversion and formation of a block-copolymer at styrene addition to preliminarily polymerized isobutylene. Obviously, besides its main function, the monomer acts as a nucleophilic agent able to detach a proton from AC and hence help in the chain transmission. This must be promoted by AC hardness growth due to Scheme 2.8:

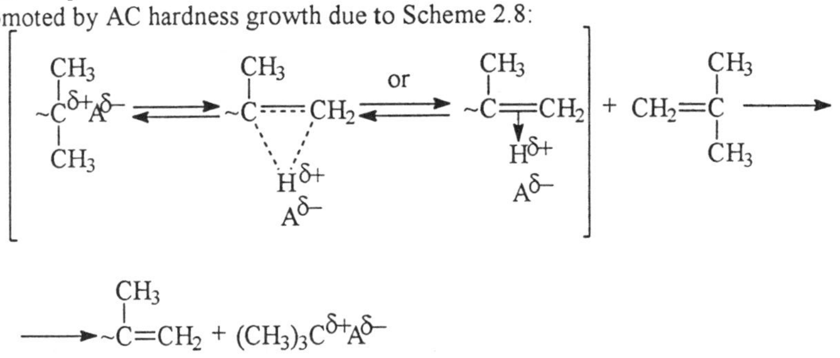

Scheme 2.8

In accordance with the HSAB concept, growth of the coordination number of AC at the sacrifice of complexing and increasing the charge transmission to acid enlarges the role of the softness factor. Judging by the behavior of model compounds [221], these processes in cationic polymerization can proceed with the participation of the monomer. Consequently, depletion of the monomer during polymerization or sufficient decrease of its initial concentration must cause a deviation of the role of soft acid (...-$C^{\delta+}$)-soft base (a monomer) interactions at the sacrifice of growth of the contribution of hard acid interactions (...-$C^{\delta+}$) with negatively charged particles (anion, its fragment, etc.). To put it differently, in "pure" systems, polymers with the highest molecular mass should be obtained at the initial stages of polymerization (equilibrium of the reactions by Scheme 2.8).

The experimental data which confirm this statement were obtained for the case of isobutylene and styrene polymerization on syncatalytic systems $R_3Al\text{-}Cl_2$

characterized by slow initiation and absence of the chain transmission to the monomer [222, 223]. Therefore, a condition of limiting the role of processes of polymeric chain termination as the reaction of hard acids and bases can be formulated — introduction of an agent into a system coordinating with AC by the principle of the monomer, but not participating in the irreversible interaction with it. In greatest degree, this requirement is fulfilled by polyunsaturated or conjugated aromatic systems of nonalkylating type, i.e. coordinating matrices able to limit secondary reactions of carbocations. As is obvious, a definite analogue with realization of the propagation stage by the principle of "living chains" in free-radical polymerization of vinyl monomers are considered [224].

In this regard, encouraging results were obtained. Of great interest is the effect of the molecular mass increase and narrowing of the polymer MMD in polymerization of isobutylene and other monomers in the presence of "soft" complex catalysts ($BCl_3{\cdot}H_2O$, $TiCl_4{\cdot}H_2O$, $SnCl_4{\cdot}H_2O$, etc.) with the introduction of small additives of sterically hindered amines (2,6-di-*tert*-butylpyridine, 4-methyl-2,6-di-*tert*-butyl-pyridine, etc.) [25]. Amines mostly suppress the reaction of the chain transmission to the monomer by linking a proton detached from the growing carbocation in the complex to substituted pyridine:

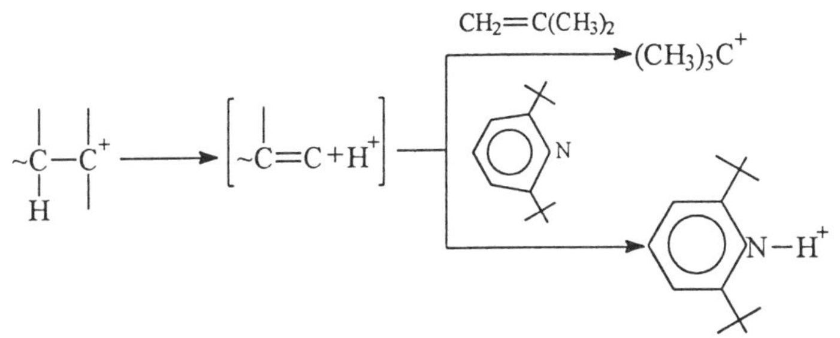

Reduction of the monomer conversion and formation of only exo-olefin [$\sim CH_2C(CH_3)=CH_2$] in isobutylene polymerization may be related to this reaction.

Amine additives play the role not only of proton traps, but also of coordinating agents in relation to growing carbocations,

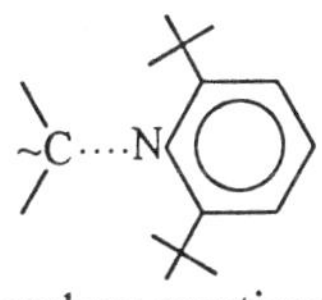

that limits their inclination to secondary reactions and assists in increase of MM and narrowing of MMD. This should be favored by low nucleophility at high basicity of sterically hindered pyridine compounds [225, 226]. Simultaneously, adding these compounds serves as unusual test agents. The absence of their influence on the stage of initiation and transmission (by the second order reaction) indicates a high probability of the concert mechanisms of the appropriate stages.

Coordination with a carbocationic center, which increases its soft character, must positively influence the selectivity of the system's action, which differs by increased sensitivity to small differences in the basicity of monomers. Specifically, an exceptionally high selectivity of isobutylene polymerization was observed in the presence of four–fold excess of α– and β–butylenes on adding to $C_2H_5AlCl_2$ catalyst of equimolar amounts of symmetric trimethylbenzene (253 K). At practically complete conversion of isobutylene the concentration of butenes remained constant. A similar picture — growth of the role of the competition factor in the reaction with unitypical bases, i.e. increase of sensitivity to the change of their nucleophilic strength — was observed at the stability increase (softness growth) of low–molecular carbonium ions [227].

Isobutylene polymerization in which products of very high molecular mass are obtained when some heterogeneous systems are used [228, 229], and radiation [219] and nuclear–chemical initiation [230, 231] are performed, shows the possibility of sufficient limiting chain termination reactions at the sacrifice of either steric factors or the absence of anions in the system. The latter case is especially interesting, because it demonstrates amazing stability of polyisobutylene carbocations under low temperatures (195 K) and the role of the monomer as a stabilizing factor. These outstanding examples only support the general picture of the propagation stage in the cationic polymerization representing a specific reaction of carbocations in the composition of ionic pairs with the monomer.

The role of medium in the cationic polymerization is generally reduced to two effects: stabilization of forming charged particles and change of AC reactivity. In the first case, energetic losses for heterolysis of chemical bonds during formation of initiating ions are compensated. In the second case, the change of AC reactivity in various media happens at the sacrifice of: a) influence of the medium polarity; b) its catalytic effect; c) specific solvation; d) formation of complexes with system components. Besides the chemical effect as a cocatalyst, the polarity of the medium is the dominating factor. Usually, the process rate and the molecular mass of the polymeric products formed grow during cationic polymerization with the medium polarity. This is indicated by growth of the initiation rate and slowing down of the chain termination.

It is hard to tell anything certain about the growth stage, because growth of the average distance between ionic pair components must cause an increase of k_p; however, the degree of the reactivity change of ionic pairs and its dependence on the distance between cation and anion is unknown. The solvating ability of a solvent also affects the mechanism and kinetics of the cationic polymerization. In particular, solvent molecules capable of complex formation with the molecules of polymerization initiators can change significantly, and in some cases completely suppress their activity. Paraffin hydrocarbons and benzene possess similar dielectric permeability; however, polymerization in benzene usually proceeds at higher rates than in saturated hydrocarbons due to better solvation of growing particles.

At the present time the influence of solvent on polymerization should be considered not only from the point of view of polarity, but also in the frames of the "coordination model" developed for the chemistry of ionic reactions in non–water solvents and characterizing the medium as donor (*DN*) and acceptor (*AN*) numbers [78, 232]. The optimum combination of donor (cation solvation) and acceptor (anion solvation) properties of the medium with calculation of monomer properties as a solvent will favor the separation and stabilization of charges. Hence, for example, nitromethane is a better solvent than methylene chloride ($DN_{CH_3NO_2} = 2.7$, $DN_{CH_2Cl_2} = 0$, $AN_{CH_3NO_2} = 20.4$, $E_{CH_3NO_2} = 35.8$, $E_{CH_2Cl_2} = 9.8$) at the sacrifice of specific coordination and unspecific solvation.

Viscosity is an important property of the medium. It can affect the fastest stages of polymerization (growth, scission). Though reliably measured kinetic constants of isobutylene polymerization lie below the diffusion border, accumulation of a gel–like product around a solid catalyst may represent a case of diffusion control of the reaction.

In most cases, cationic polymerization is non–isothermic and very sensitive to small amounts of admixtures, which complicate obtaining of reproductive results. At low concentrations of admixture the total rate of the process often increases, which is usually connected to its cocatalytic action. Growth of the admixture content in the reaction zone causes the opposite effect, which is exhibited by reactions limiting chain propagation.

2.2.3. Limiting of the chain propagation.

Limiting of chain propagation in the cationic polymerization proceeds at the expense of transmission and chain termination reactions. Transmission and chain termination can often be separated just conditionally, because regeneration of the catalyst takes place at the chain termination. This catalyst is able to cause further polymerization of the monomer.

In the general case, limiting of the chain propagation can be observed in growing AC interaction with a counterion, monomer, solvents, admixtures, including deactivators (stoppers) specially introduced at the end of the process, or a polymer.

Reactions with the counterion can proceed: a) with neutralization of charge on the chain end (mono– and bimolecular process)

$$\ldots\text{—}M^{\delta+}\,A^{\delta-} \rightarrow \ldots\text{—}MA$$
$$\ldots\text{—}M^{\delta+}\,BA^{\delta-} \rightarrow \ldots\text{—}MB + A$$
$$\ldots\text{—}M^{\delta+}\,A^{\delta-} + BX \rightarrow \ldots\text{—}MX + BA$$
$$\ldots\text{—}M^{\delta+}\,A^{\delta-} + X^{-} \rightarrow \ldots\text{—}MX + A^{-},$$

where B is the base, which is a part of the catalyst; or b) with the counterion protonation, for example,

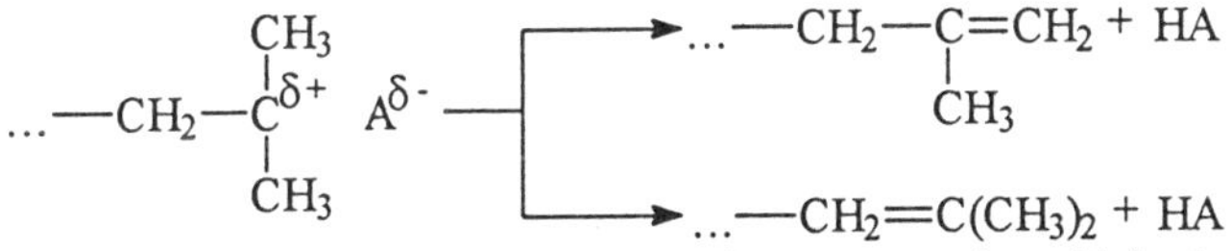

The reactions (a) preferably proceed in systems, in which the process is excited by HX proton acids, limiting of the chain propagation proceeding the more frequently, the stronger is newly formed chemical bond Me—X and the less stable is X^{-} counterion. Reactions with BX and X^{-} are also observed at application of aprotic acids

$$\ldots\text{—}M^{\delta+}\,[BCl_3OH]^{\delta-} \rightarrow \ldots\text{—}M\text{—}Cl + BCl_2OH.$$

This very type of scission was proved for the isobutylene polymerization with several metal halides and proton acids.

A very important role in cationic polymerization is played by chain transmission to the monomer:

$$\ldots\text{—}M^{\delta+}\,A^{\delta-} + M \rightarrow \ldots\text{—}M^{=} + HM^{\delta+}\,A^{\delta-}.$$

In most cases, during cationic polymerization of isobutylene, the chain transmission to the monomer determines the polymerization degree $\overline{P}$ of the polymeric products formed.

The feature of the chain transmission is the dependency of the rate on the process conditions, which is obviously bound to the role of counterion A^-. Specific rate of the chain transmission to the monomer changes with the catalyst acidity in the sequence $TiCl_4 > SnCl_4 > FeCl_3 > BF_3{\cdot}O(C_2H_5)_2$ and decreases with increase in the medium polarity [193]. The chain transmission to solvents, arenes for example, is also typical and proceeds via the stage of protonation or alkylation of an aromatic nucleus with further proton detachment from the product of the primary interaction:

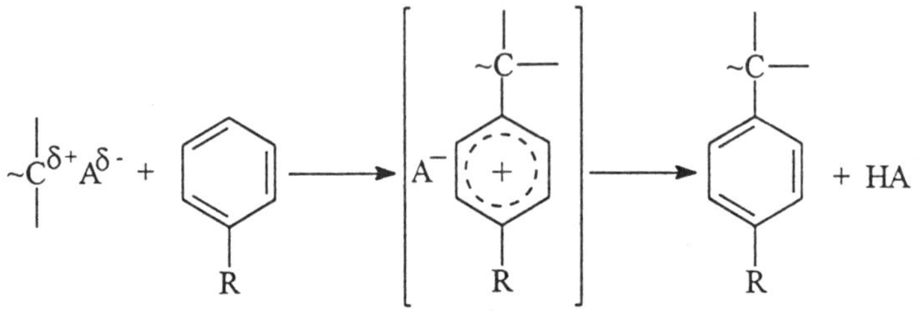

Chain transmission proceeding in the medium of alkyl halides is also possible:

$$\ldots\text{—}M^{\delta+} + RX \rightarrow \ldots\text{—}MX + R^{\delta+}.$$

In the presence of active admixtures (water, alcohol and others) a great part of the chains are degraded, for example,

$$\ldots\text{—}\overset{|}{\underset{|}{C}}{}^{+} \xrightarrow{2H_2O} \ldots\text{—}\overset{|}{\underset{|}{C}}OH + H_3O^+$$

The cases of chain transmission under AC interaction with a monomer, solvent or polymer are rather widespread involving detachment of hydrogenated (H^-) or methide ($^-CH_3$) ions. In these cases branched polymeric products, typical of cationic polymerization are formed.

The specific contribution of different types of chain propagation limiting depends on the nature of the catalytic system. Consideration of this stage, taking into account the deterministic role of acid–base interactions, will allow us to point out some general tendencies. The most typical reactions of polymeric chain limiting with the application of complex catalysts (excluding transmission to a monomer) are:

a) for combinations $TiCl_4{\cdot}H_2O$, $TiCl_4{\cdot}CCl_3COOH$, $SnCl_4{\cdot}C_6H_5OH$ [233 — 236]

$$HM_m^{\delta+}MeX_{n+1}^{\delta-} \rightarrow HM_mX + MeX_n$$

b) for the combination $BCl_3{\cdot}H_2O$ [237], and

c) for the combination $BF_3{\cdot}H_2O$ [62]

$$HM_m^{\delta+}MeX_nX'^{\delta-} \rightarrow HM_mX + MeX_{n-1}X'$$

$$HM_m^{\delta+}MeX_n^{\delta-} \rightarrow HM_m^{=} + H'MeX_n,$$

which reflect the dependence of the process direction on strength of the donor–acceptor interaction of initial Lewis and Brönsted acids. At its high value ($BF_3 \cdot H_2O$), the scission with regeneration of a complex catalyst or transmission to a monomer while keeping the counterion $BF_3OH^{\delta-}$ (reaction c) is preferable. When stability of complex catalysts (Ti and Sn compounds) is reduced, initiation and termination proceed with the participation of HX activated by the Lewis acid (reaction a), i.e. HX serves as a peculiar "telogen". As detachment of ligands from MeX_nX' counterion (reaction b) depends on the energy of their linking to metal and stability of the Lewis acid formed, the case of the chain termination at elimination of ligand X, initially participated in the composition MeX_m, is probable. This type of termination is especially typical for initiation of isobutylene polymerization by HX(RX)—R_nAlCl_{3-n} and proceeds though alkylation of polymeric carbocation and formation of higher chlorinated aluminum compounds [60]. Such reactions allow introduction of various functional groups into polyisobutylene: aromatic, vinyl, cyclopentadiene, hydrocyanic, etc. As these complex catalysts respond to the systems with controlled type of initiation, they can be used for introduction of assigned "head" and "tail" groups into the polymer, i.e. for the synthesis of α,ω–bifunctional polyisobutylenes or obtaining of block– and grafted copolymers (see Chapter 3).

Modeling of limiting polymeric chain termination reactions (a) and (b) in studying kinetics of triphenylmethyl cation death as a result of halide ion detachment from the complex counterion $MeX_{n+1}^{\delta-}$ (Me — Al, B, Sn, Sb, etc.; X — F, Cl, Br) [238] points to the linear connection of observed rate constant to donor numbers of ligand X^- (the average donor number in the case of anions with different halogen ligands), which is well interpreted from the HSAB statements.

From the point of view of the nature of end groups in the polymer, considered effects become of special interest in reactions of termination on bases assuming different directions of the process, alcohols, for example [211]. Comparing experimental and calculated (quantum–chemical analysis of $\equiv C^+$ or MeX_n complexes with an alcohol by the PPDP/2 method under Santri–Pople–Siegal parametrization) data indicate that the fragmentation direction of complex–linked ROH depends on the nature of R and the acid strength. Growth of acidity of a complexing agent and branching degree of the alcohol radical promotes favorable fragmentation at the C—O bond (introduction of hydroxy groups into polymer). The role of special interactions is important for $MeX_n \cdot ROH$ complexes. For example, formation of hydrogen bonds H····Cl between ligands in addition to the donor–acceptor interaction of $RAlCl_2$ with R'OH assists in displaying carbocationic activity of alcohol and leads to the absence of expected hydroxy end groups in isobutylene polymers [36, 88]. For $SnCl_4 \cdot C_6H_5OH$ [236] the break via detachment of phenoxy groups is indicated first by the effect of $(p\text{-}p)_\pi$-conjugation of oxygen atom with an aromatic nucleus, which hampers the breaking of the C(phenyl)–oxygen bond.

Generally, limiting of the chain propagation in the ionic pair

$$\ldots—\overset{|}{\underset{|}{C}}{}^{\delta+}MeX_{n+1}^{\delta-}$$

should be considered as its stabilization by the mechanism of hard (electrostatic) interactions resulting in weakening of the coordination effect of the monomer (in media with low dielectric permeability).

Another feature of chain limiting reactions in cationic polymerization is the formation of intermediate ionic structures of reduced acidic activity. One example may be the allyl scission in the isobutylene polymerization [69]:

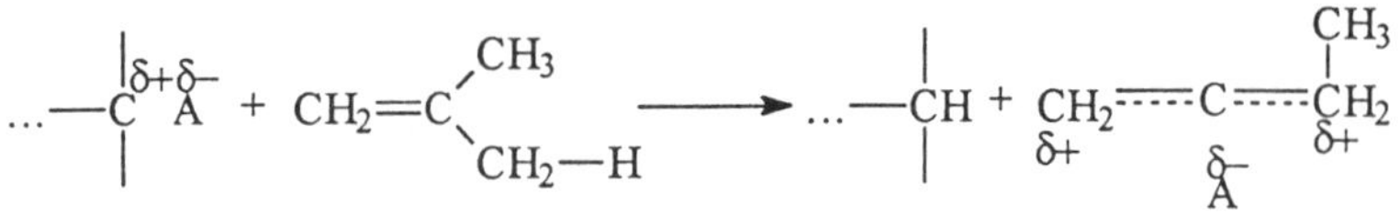

Growth of the softness factor (stability growth) of cations suppresses their ability to react with counterion and inherent monomer, but does not exclude interaction with harder basic monomers, such as styrene and cyclopentadiene. Occurrence of similar cationic structures is typical of isobutylene polymerization in the presence of admixtures of unsaturated compounds (α–olefins or diene monomers) and is responsible for the process inhibition. Arenes also react by the same scheme. In more detail, these interactions were considered in the analysis of conjugated electrophilic reactions of polyisobutylene carbocations and kinetic aspects of isobutylene polymerization. Here it should be mentioned also that traditional admixtures (water, etc.), first, may cause catalyst poisoning with reduction of the polymeric product yield, and after that they block centers of macromolecules growth limiting their length [239]. Therein lies the display of favorable hard interactions of admixtures with an initiator compared with growing carbocations. On the contrary, due to weakness of the effect of external stabilization by free cations in radiation polymerization, their death on admixtures is highly probable. Though the fine mechanism of chain termination on admixtures has not yet been studied, formation of structures of the type

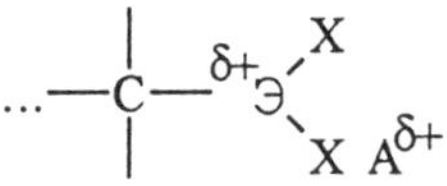

(where Э = O, S; X = H, R)

stable under polymerization conditions, should be expected. Obviously, under favorable conditions, they may be macromolecular initiators for polymerization of higher basic monomers, heterocyclic ones, for example. In the case of arenes, ease of the proton elimination from arenonium ions allows performance of repeated initiation of the isobutylene polymerization at the introduction of fresh monomer portions with formation of polyisobutylene aromatic compounds [54, 212, 240].

Reaction of low-temperature alkylation of aromatic hydrocarbons by propagating polymeric chains appears worthy of special consideration, because it represents a possibility to analyze regularities of the chain termination (transmission) reaction, which rarely appear in the cationic polymerization of isobutylene, and gives us methods by which we may regulate the fundamental reactions. If the catalytic nature of the reaction is neglected, the process formally represents isobutylene telometization, during which the arene plays the role of telogen:

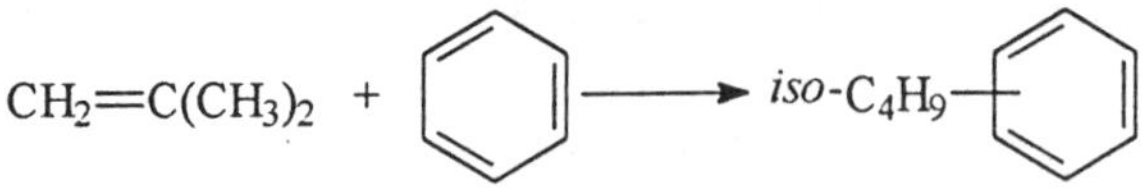

In accordance with the catalyst nature and ratio of components, temperature telomer homologues can be obtained up to the highest alkyl aromatic compounds, formation of 1:1–adduct in the reaction of direct arene alkylation by isobutylene being the limiting case of inhibited olefin polymerization, for example, in the isobutylene–$C_6H_5CH_3$, $FeCl_3$ system (273 K). In a similar manner, diisobutylene and toluene (with

1:3 molar ratio) in the Gustavson complex composition (CH_2Cl_2, 243 K) give the telomer $C_8H_{17}C_6H_4CH_3$ with a yield over 70% [212].

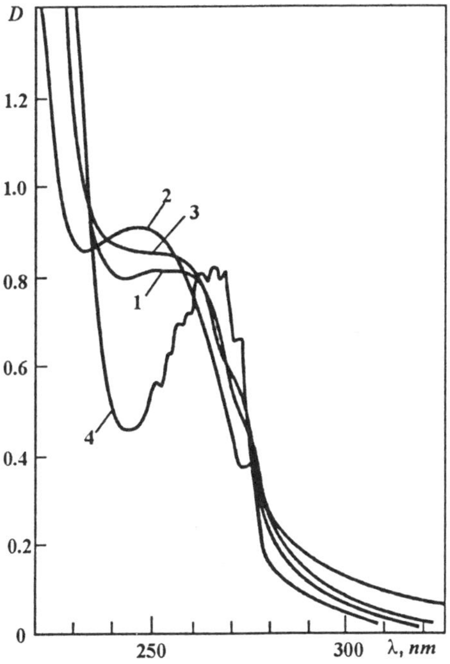

Figure 2.7. Electron spectra of polyisobutylene solutions in hexane at l = 1 cm and various polymer concentrations: 1, 4 — 7.8; 2 — 2.6; 3 — 3.51 kg/m^3; samples were produced in the presence of catalysts: 1, 4 — $3CH_3C_6H_5$; 2 — $4(CH_3)_3C_6H_3{\cdot}HCl{\cdot}Al_2Cl_6$; 3 — $3(CH_3)_3C_6{\cdot}HCl{\cdot}AlCl_3$; 1, 3 — in CH_2Cl_2 solutions; 4 — in toluene solution.

Selectivity of the alkylation reaction in isobutylene polymerization in the presence of arenes regarding the growing chain transmission to the monomer, which determines the ratio of polyisobutylene with unsaturated and aromatic fragments, significantly depends on the catalyst nature (Table 2.16) [241]. Moving from HAOC to complex catalysts, besides isobutylene conversion increasing the content of aromatic fragments as polyisobutylene grows, its molecular mass and non–saturation degree reduce on account of the conjugated alkylation of toluene by the growing carbonium ion. The feature of isobutylene polymerization in toluene medium in the presence of a heterogeneous catalyst based on sulfocationite modified by $C_2H_5AlCl_2$ is a selective formation of unsaturated polymer, which is probably connected to steric obstacles of the alkylation reaction. The maximum content of aromatic fragments in polyisobutylene is reached, if Gustavson complexes based on mesitylene and toluene are used as catalysts. The existence of aromatic fragments in thoroughly reprecipitated samples of the polymer is indicated by the data of UV-spectroscopy [absorption in the range of 240–280 nm] (Figure 2.7).

Isobutylene conversion decreases with the relative increase in the amount of toluene in the reaction mixture at the sacrifice of the low inhibiting effect of methylbenzene (Figure 2.8, curve 5). Dependence of the monomer conversion on the methylbenzene/isobutylene ratio during polymerization in the presence of $C_2H_5AlCl_2$ is more complicated (curves 1–4). It is well known that methylbenzenes can play the role of ligands regarding $C_2H_5AlCl_2$ [59, 242]. Simultaneous display of the properties of weak inhibitors and activators of $C_2H_5AlCl_2$ (at the sacrifice of the complex formation) by methylbenzenes is evidently responsible for occurrence of maximums on the corresponding curves. Shear of the maximums to the left with growth of methylbenzenes' basicity is bound to the increase of their ability to form complexes with the catalyst.

Table 2.16

Isobutylene polymerization in the toluene medium (isobutylene — $4{\cdot}10^2$ mol/m^3, catalyst — 15 mol/m^3, 195 K, 600 s).

Catalyst	Isobutylene conversion, %	Relative content of end groups in polymer, %		$\overline{M}_n \cdot 10^{-4}$
		Unsaturated	Aromatic	
$C_2H_5AlCl_2$ — KU-2-8*	50	100	0	5.0
$C_2H_5AlCl_2$	23	50	50	3.5
$C_2H_5AlCl_2{\cdot}HCl$	40	40	60	5.9
$C_2H_5AlCl_2{\cdot}H_2O$	800	40	60	5.1
$2AlCl_3{\cdot}HCl{\cdot}3C_6H_3(CH_3)_3$	80	25	75	1.0
$2AlCl_3{\cdot}HCl{\cdot}3C_6H_5CH_3$	70	20	80	0.9

* Heterogeneous catalyst based on non–aqueous sulfocationite KU-2-8[4] activated by $C_2H_5AlCl_2$ (SO_3H/Al 6:1 mol/mol).

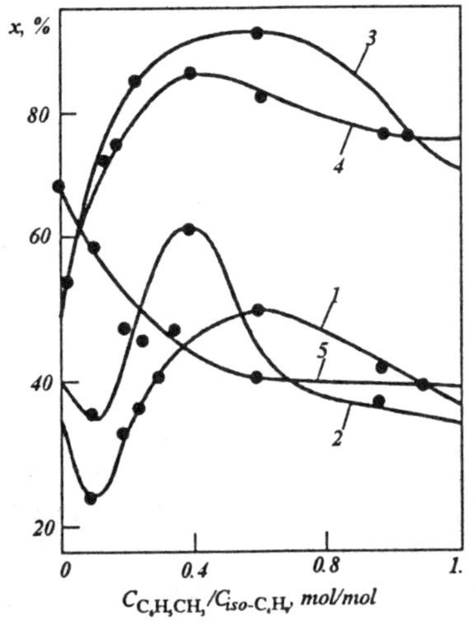

Figure 2.8. Dependence of isobutylene conversion on the methylbenzene/isobutylene ratio at polymerization in the presence of toluene (1, 5), μ–xylene (2), hemimellitol (3), mesitylene (4); solvent — CH_2Cl_2, catalyst — $C_2H_5AlCl_2$ (1–4) and $3C_6H_5CH_3{\cdot}HCl{\cdot}Al_2Cl_6$ (5); 195 K, 10 min.

[4] Russian trademark.

Contrary to HAOC, Gustavson complexes are chemically inert to arenes, which allows rather strict estimation of the contribution of the conjugated alkylation reaction as a function of temperature [240] (Table 2.17). Activity of the Gustavson complex in isobutylene polymerization grows with temperature decrease; in this case each molecule of the polymer produced in a non-aromatic solvent (tests 1–3) contains one or more C=C-bond.

Table 2.17

Influence of temperature and aromatic solvent on the isobutylene polymerization in the presence of Gustavson complexes [solvent 40cm^3, $C_6H_5CH_3{\cdot}HCl{\cdot}Al_2Cl_6$ — $8.5{\cdot}10^{-2}$ mol/l, the monomer transmission rate is 2.5 h^{-1}, bubbling regime (tests 1, 2, 5–12), monomer concentration — 1.3 mol/l (tests 3, 4, 13)].

No.	Tempera-ture, K*	Solvent	k**, min^{-1}	Polyisobutylene yield regarding light alkylate, %	Molecular mass	$n_{C=C}$***	C/H, mol/mol
1	316	Hexane	—	100	410	—	—
2	320	Methylene chloride	—	100	690	1.04	0.50
3	272		—	100	1,800	1.27	0.50
4	213		—	100	125,000	—	—
5	338	Toluene	0.044	6	370	0.35	—
6	328		0.032	5	370	0.32	—
7	318		0.028	11	380	0.26	0.53
8	308		—	15	390	0.27	—
9	290		0.018	19	350	0.13	0.54
10	278		0.020	42	370	0.10	0.53
11	268		0.020	57	380	0.08	0.51
12	243		0.014	90	750	0.09	0.51
13	212		—	100	9,000	0	—

* Kinetic measurements were performed under isothermal conditions.

** k —rate constants of the toluene consumption calculated by the first order reaction equation; relative error of k determination is 10%.

*** The number of double bonds in polyisobutylene molecule was determined iodometrically.

When toluene is used as a solvent, the molar ratio C/H in polyisobutylene exceeds 0.5, and the concentration of C=C-bonds, molecular mass and the yield of polymers are reduced (tests 5–12) because of formation of alkyl-aromatic compounds. Isobutylene polymerization in toluene above 273 K degenerates into arene alkylation by the monomer with formation of a light alkylate with the boiling point below 523 K/98 kPa, which consists of 31.6% of μ- and 43.4% of *n-tert*-butyltoluenes (molecular mass of 148), and of 25% of alkyltoluenes with molecular masses of 204, 260 and 306, respectively (composition for the test 5).

In the range of negative temperatures the yield of polyisobutylene grows regarding the light alkylate (tests 9–12). Noticeable toluene consumption under low-temperature polymerization conditions in the absence of *tert*-butyltoluene and light alkylate, and the formation of oligomeric products with low content of C=C-bonds indicate toluene introduction into polyisobutylene macromolecules as end groups. Existence of aromatic fragments in polymeric products is confirmed by the data from

PMR spectra (signals with chemical shifts at 6.90 and 2.15 m.d.), IR-spectra (bands at 1900, 820, 725 cm^{-1}) and UV-spectra (bands at 257, 265, 272 nm; ε_{272} = 420 l/mol·cm), which characterize aromatic nuclei, *meta*- and *para*-substituted, regarding the methyl group.

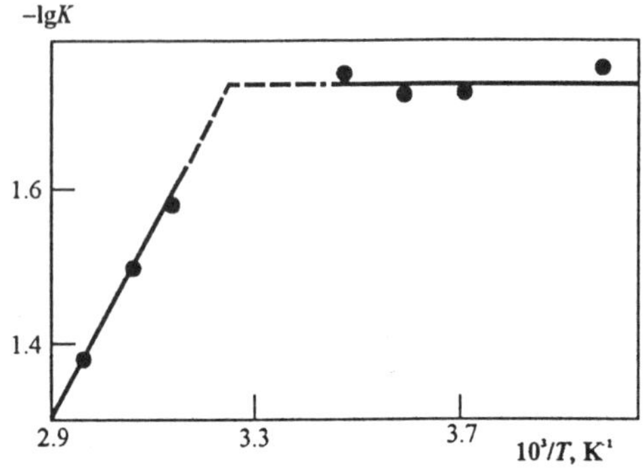

Figure 2.9. Temperature dependence of rate constants of toluene expenditure (experimental conditions are shown in Table 2.17).

Figure 2.9 shows the temperature dependence of the change of toluene consumption rate constants, displayed in Table 2.17. The left branch of the curve (temperature range above 298 K) is characteristic for the typical arene alkylation by a monomer (E_a = 26 kJ/mol). The break at the transition to low-temperature range (E_a value is close to zero) points out the predominant role of the isobutylene polymerization. Dissociation of ionic pairs at the temperature reduction increases activity of carbonium ions in relation to both the chain propagation reaction and its limiting, if toluene is applied. Hence the value $E_{\Delta M}$ = –25.1 kJ/mol (activation energy of the isobutylene polymerization degree change) in the presence of toluene is close to $E_{\Delta M}$ of the process proceeding in the absence of arene, i.e. in the range of negative temperatures isobutylene polymerization with the chain limiting by toluene or a peculiar alkylation reaction by Friedel-Crafts takes place.

In this manner, proceeding of reactions of conjugated alkylation of arenes during low-temperature isobutylene polymerization supposes their participation in parts of the material chain transmission (reactions A and C, Scheme 2.9):

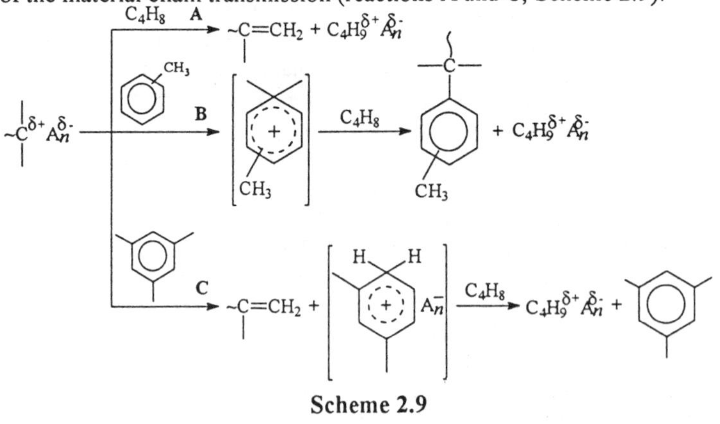

Scheme 2.9

In line with the change of character of the chain propagation limiting reaction at adding methylbenzenes (Scheme 2.9), MMD curves of polymer samples are widened at the sacrifice of obtaining low-molecular fractions (Figure 2.10). In this case, the $\overline{M}_w/\overline{M}_n$ ratio rises from 2 (for polyisobutylene produced in the absence of methylbenzene) to 3.

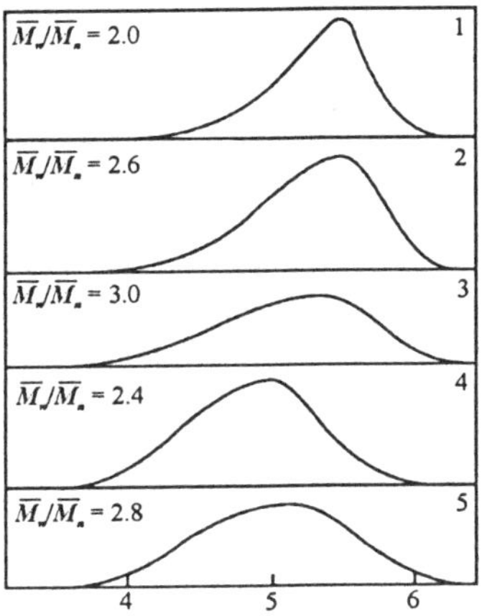

Figure 2.10. GPC[5] of polyisobutylene samples; isobutylene polymerization (catalyst $C_2H_5AlCl_2$, CH_2Cl_2, 195 K) in the presence of toluene (2), *m*-xylene (5), and in the absence of methylbenzene (1); CL-1302 device, columns with styrogel by Waters Co., the pore size from $3.5\cdot10^{-8}$ to $7\cdot10^{-5}$ m, eluant – chloroform, 298 K.

Conduction of the process at 263-273 K in toluene causes bimodal MMD of isobutylene at the sacrifice of the presence of a light alkylate – a mixture of products of arene mono- and polyalkylation by isobutylene and its oligomers. Relative constants of the polymeric chain transmission to methylbenzenes, calculated by the Mayo-Lewis equation, are shown in Table 2.18 (the data on styrene polymerization are presented for comparison). It is typical that k_t/k_p values grow with π–basicity of methylbenzenes (for styrene). Existence of such correlations complies with the electrophilic nature of the conjugated alkylation reaction and is implied by differences in activity of growing carbocations of polyisobutylene and polystyrene. Analogous to the reaction of electrophilic substitution in the aromatic sequence it may be suggested that the intermediate state with the maximum energy in the reaction of conjugated akylation of methylbenzenes by an electrophilic agent (growing carbocation) is similar structure to the π–complex [18, p. 460].

Higher electrophilic activity of the growing polyisobutylene carbocation compared with the polystyrene one in reactions with appropriate monomers (values of k_s differ by 2-3 decimal degrees [62]) can also explain differences in values of k_p/k_s for isobutylene [241] and styrene [243] (see Table 2.18). Suggesting that k_p values for the given monomer do not depend on the methylbenzene nature, ratios of constants k_t/k_M, which characterize the competitive ability of studied bases in the reaction with polymeric carbocation, were calculated on the basis of relative constants of the chain

[5] GPC — gel-penetrating chromatography.

transmission to methylbenzene k_t/k_p and to monomer k_M/k_p. Contrary to the propagating polystyrene displaying selectivity in the reaction of conjugated alkylation of methylbenzenes (high values of k_t/k_M), a more active carbocation of polyisobutylene is less selective (k_t/k_M values are close to 1). That is why there is no possibility for complete elimination of the growing chain transmission to isobutylene by the reaction A. Beside aromatic fragments, polyisobutylene synthesized in the presence of methylbenzenes contains double bonds. In this case, content of double bonds in the samples obtained under different conversions of isobutylene (from 25 to 70%) is practically constant during the reaction and is 0.2 mole of C=C per 1 mole of polyisobutylene. This supports the competitive character of reactions of the propagating chain limiting by isobutylene and methylbenzenes (reactions A and B, C in Scheme 2.9, respectively).

The specific feature of the reaction B is the intermediate formation of quite stable σ–complexes. An indirect proof of their existence is the stability of the color of the reaction solution, when the isobutylene polymerization ends. As at 195–240 K σ–complexes are not inclined to intermolecular transformations, the possibility of proton attachment by arene with regeneration of the initial Gustavson complex is excluded. Decay of the σ–complex with chain transmission to isobutylene slows down with decrease of the monomer concentration and temperature. σ–Complexes, formed at 243 K by the reaction B, participate in reinitiation of polymerization of new portions of isobutylene (Figure 2.11). A similar but less marked effect is observed at 195 K during adding of isobutylene (polyisobutylene yield of 24%) and styrene (polystyrene yield of 11%) to the polymerizate. Constancy of polyisobutylene molecular mass at the repeated introduction of isobutylene and the absence of a block–copolymer at styrene addition (chain transmission to aromatic nuclei of polystyrene is absent under the experimental conditions) outline the specificity of polymerization in the presence of arene: "living" chains terminated by arene fragments are able to initiate transformation of new portions of the monomer, e.g. they are macroinitiators.

A specific feature of isobutylene polymerization in the presence of mesitylene is the absence of conjugated alkylation reaction as a consequence of steric shielding of an aromatic nucleus by symmetrically disposed methyl groups. Formation of polyisobutylene, which contains on average a single C=C-bond per macromolecule, points out the proton transmission to mesitylene with formation of a σ–complex and further reinitiation of the polymerization (reaction C, Scheme 2.9). Though unsaturated macromolecules of a polymer are formed through reactions A and B, they differ by the type of MMD curves (see Figure 2.10, curves 1 and 5), in particular, participation of mesitylene in the polymeric chain transmission to isobutylene causes formation of low–molecular fractions with molecular mass of about 10^4, not found in the process conducted in a non–aromatic solvent.

Application of substituted phenols in low–temperature polymerization of isobutylene confirms the possibility of the alkylation reaction proceeding conjugated with the chain propagation, accompanied by the chain transmission and phenol bonding to the polymeric chain as the end group [236, 244, 245]. As in the case with methylbenzenes, reduction of the molecular mass, polymer yield and growth of the amount of linked phenol with growth of its initial concentration in the system are observed. Decrease of molecular mass values of polymeric products is less evident for phenols with increased substitution degree. Concurrent formation of polyisobutylene with end phenolic and unsaturated groups indicates competition between alkylation

Table 2.18

Influence of methylbenzene nature on the reaction of polymer chain propagation limiting in isobutylene polymerization (195 K, $C_2H_5AlCl_2$, CH_2Cl_2) and styrene (273 K, $SnCl_4$, CCl_4-$C_6H_5NO_2$).

No.	Methylbenzene	Relative basicity of methylbenzene [157]		Isobutylene			Styrene	
		π(HCl)	σ(HF×BF₃)	$(k_p/k_s)\times10^3$	k_p/k_M	M_w/M_n	$(k_p/k_s)\times10^3$	k_p/k_M
1	—	—	—	(0.30)	1	2.0	(0.2)	1
2	Toluene	0.92	0.01	0.15	0.5	2.6	8.8	44
3	*o*-Xylene	1.13	2	—	—	—	830	415
4	*m*-Xylene	1.26	20	0.25	0.8	3.0	134.0	670
5	Hemimellitol	1.46	40	0.70	2.3	2.4	—	—
6	Mesithylene	1.59	2800	2.00	6.7	2.8	7.0	37
7	Methoxybenzene*	—	—	5.30	17.7	—	1620	8100

* Catalyst $TiCl_4 \cdot H_2O$, solvent CH_2Cl_2, 183-264 K [1, p. 171].

conjugated with propagation and other chain termination (transmission) reactions. Their ratio and, consequently, the amount of included phenol depend on the catalyst nature, arene, solvent and conditions of the reaction.

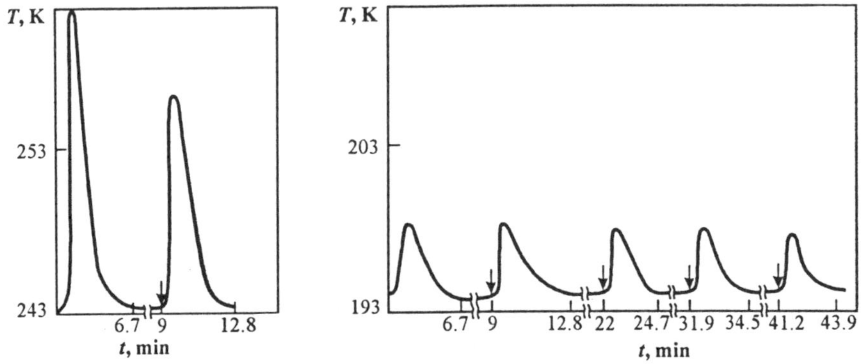

Figure 2.11. The change of polymerization mixture temperature with time in the reactor during isobutylene polymerization:
a — 243 K; b — 195 K; arrows mark moments of repeated introduction of isobutylene.

In accordance with the electrophilic nature of alkylation of phenols and the orienting effect of *ortho-para*-substituents in the aromatic nucleus (electron density increase in positions 2, 4 and 6 of phenol), favorable substitution of H atom on polyisobutylene in the position 4 for 2,6–disubstituted phenols, as well as for *ortho*–alkylation in the case of 4–alkylsubstituted phenols are observed. Steric factorsare important, for example, 2,4,6-tri-*tert*-butylphenol does not essentially affect isobutylene polymerization.

Reduction of polyisobutylene molecular mass and the amount of linked phenol with polymerization temperature growth reflect an unfavorable effect of the ions⇔ionic pairs equilibrium shifted to the right in both the propagation stage and the stage of phenol alkylation — chain limiting — conjugated to the former one. For fractionated samples, the concentration of the polyisobutylene fraction with end phenolic groups weakly depends on the molecular mass of the polymer [245], which also correlates with co–existence of the propagation reaction and chain termination as a result of conjugated alkylation.

The sequence of experimental facts [246], in particular, domination of conjugated alkylation with temperature decrease, growth of $\overline{M}_n$ values for polyisobutylene fractions containing phenolic groups (compared with the samples that do not contain them), different temperature effect on the molecular mass of polymeric products formed with end phenolic fragments and without them, point out that competition of chain propagation and termination reactions, including the ones with participation of phenols, can proceed differently depending on the AC type ($SnCl_4$, CH_2Cl_2, heptane, 203–263 K). The final result represents superposition of the processes proceeding on all ACs.

The possibility of polymeric chain termination with the help of OH–groups and their stronger inhibiting effect, comparing with methylbenzenes, must be emphasized. This applies restrictions to the amount of phenols used (about 0.1–1%

mol). Besides the transmission, the chain termination observed for anisole is explained by formation of oxonic ions incapable for growth due to their high stability [1, p. 171]:

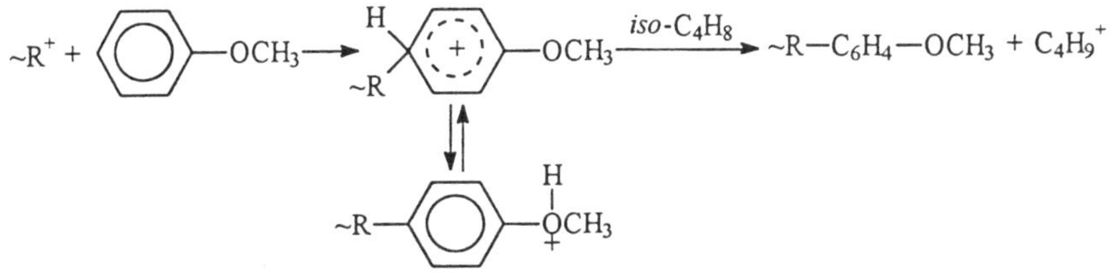

Thus, all the details of the mechanism of the cationic polymerization of isobutylene have not been cleared up yet. From the point of view of the chemistry, the cationic polymerization of isobutylene is a relatively simple process. In particular, the possibility of competitive side reactions occurring, such as hydride-ion transmission and isomerization, is reduced to the minimum of possible variants due to the absence of H atoms at tertiary C atoms in the monomer and polymer. Tertiary carbonium ions are formed as processes of chain propagation. These ions have higher stability than the secondary ones. That is why polyisobutylene is formed as linear macromolecules with "head to tail" chain dislocation. Macromolecules with a favorable content of end groups (one per macromolecule) of two types are formed in the presence of Brönsted acids or complex catalysts on the basis of BF_3, $AlCl_3$, as well as HAOC [68, p. 110] to give continuous endo- and exo-forms of olefins ...—CH_2—$C(CH_3)=CH_2$ (I) and ...—$CH=C(CH_3)_2$ (II). The factors determining the ratio of exo– and endo–forms in the chain termination are not clear yet. It is suggested [68, p. 111] that the chain termination in a particular system may be controlled by steric factors, because the transition state at formation of an endo–olefin is somehow compressed due to the interaction of valently non–linked hydrogen atoms in groups

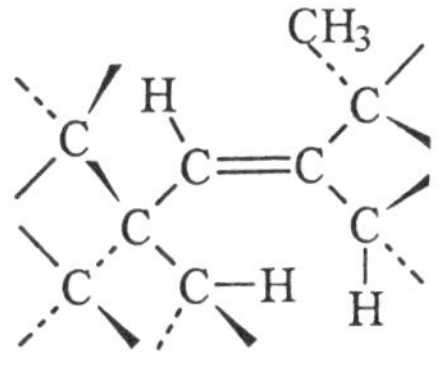

Two atoms of H (at the bottom) strive to approach each other at the moment of the double bond formation. However, the state corresponding to the exo-form is significantly less deformed than the exo-form that compensates the difference in heats of formation of the endo- and exo-olefins. If the internal potential cannot be eliminated by rotation of interacting groups, an exo-olefin is obtained as the main product.

Based on the existing experimental data, the process of isobutylene polymerization in the presence of metal (Lewis acids) halides combined with a Brönsted acid generally includes the following elementary stages:

formation of a catalytic complex

$MeX_n + H_2O \rightarrow [MeX_n{\cdot}H_2O] \rightarrow H^{\delta+}\, MeX_nOH^{\delta-}$

initiation

$$H^{\delta+}\,MeX_nOH^{\delta-} + CH_2 = C(CH_3)_2 \xrightarrow{k_i} (CH_3)_3C^{\delta+}\,MeX_nOH^{\delta-}$$

chain propagation

$$(CH_3)_3C^{\delta+}\,MeX_nOH^{\delta-} + CH_2 = C(CH_3)_2 \xrightarrow{k_p}$$
$$\rightarrow (CH_3)_3C - CH_2 - C^{\delta+}(CH_3)_2MeX_nOH^{\delta-}$$

chain transmission to a monomer

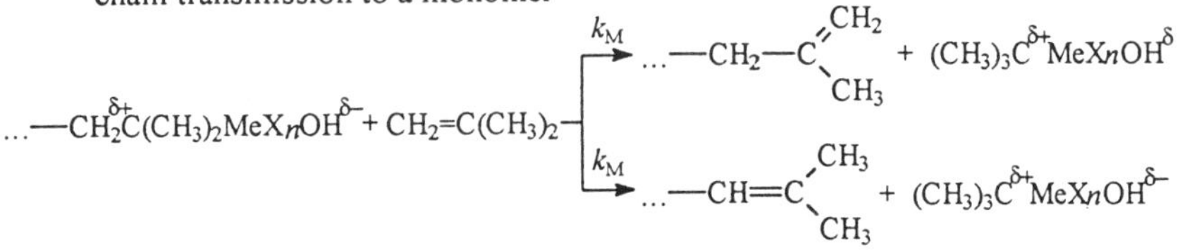

proton transmission to a counterion (probable reaction)

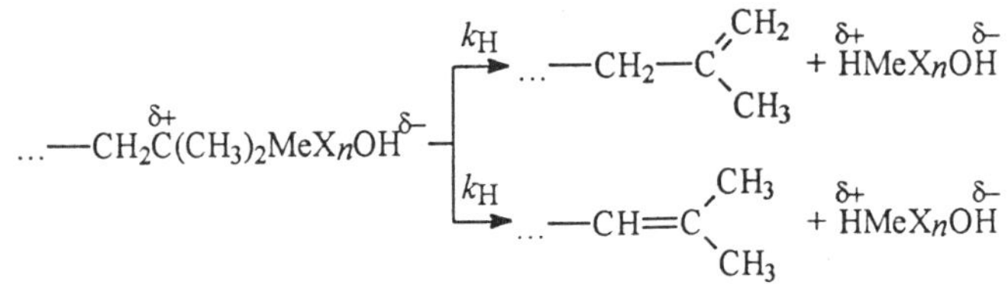

chain termination with a counterion fragment capture

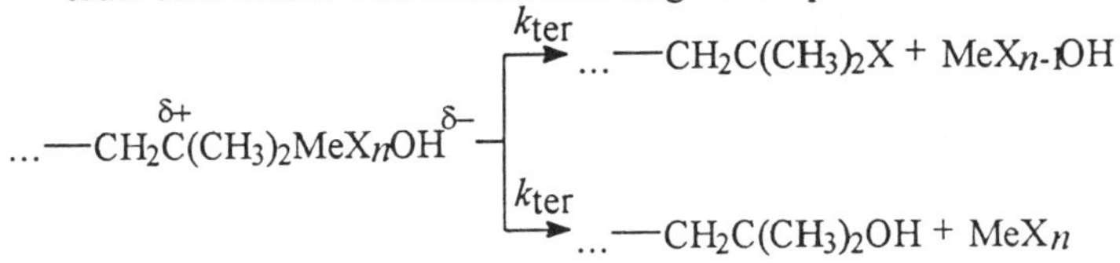

The chain termination on admixtures is also possible.

Thus, already when we consider the general picture of isobutylene polymer formation by cationic polymerization, difficulties in understanding the fine mechanism of separate elementary stages are obvious. First of all, it relates to the stages of initiation and propagation of polymeric chains in low-polar hydrocarbonic media, which reflect the high specificity of reactions of formation and growth of polymeric carbocations absent in the chemistry of their low-molecular analogues. In spite of low values of dielectric permeability (ε = 2–3), the weak effects of the internal stabilization of growing carbonium ions with isobutylene structure indicate a significant role for the solvating (electrostatic) function of the monomer due to the absence of strong electron-donor substituents. In the universally adopted sense, poor solvation of carbonium ions due to the large sizes of associates does not exclude this, but suggests efficient external stabilization with the help of the monomer, bound to its nucleophilic function. Of importance is to note the interconnection of electrostatic and catalytic binding (or hardness and softness factors) in the reaction of carbonium ions with olefin. Stabilization of carbocations by a monomer, determined by the orbital coordination, is connected with the reverse regulating action of the ion on the monomer molecule in its close vicinity, which orients them in accordance with the electrostatic factor. The totality of these factors explains the fast growth of cations in non-polar media and the observed kinetic features of the polymerization reaction.

2.3. Problems of cationic polymerization of isobutylene.

2.3.1. Molecular engineering.

The material on catalysts and the elementary stages of the isobutylene polymerization, discussed in Parts 2.1 and 2.2, demonstrate one of the approaches, the most general and acceptable, to the analysis of the process regularities. The concept of the deterministic role of acid–base interactions in the cationic polymerization is based on the fact that the process considered represents a variety of a wide class of cationic reactions in non-aqueous media with all main features inherent in them. In the frame of this concept and as an addition to it, other features of the cationic polymerization of isobutylene should also be considered, which distinguish it from reactions of low-molecular compounds and other reactions of polymer formation. In the generalized formulation, achievements in regulation of the cationic polymerization of isobutylene and constructing polymeric molecules have been called macromolecular (or molecular) engineering [25, 247]. Extension of this multivalued term first occurred with consideration of radical and anionic polymerization and, during 1975–1980, in cationic polymerization also. Macromolecular engineering means the regulated construction of head and end groups, repeated chains, MM and MMD microstructure, nature of branchings, frequency of the network, block-, graphite- and star-like structures. Most of these features are also applicable to polyisobutylene. Elements of the macromolecular engineering are controlled elementary processes (initiation, scission, transmission) and the "quasi-living" mechanism of chain propagation. As the problem is considered in detail in reviews [25, 247], the present discussion can be made in brief. Realization of elements of the macromolecular engineering is connected to two initial moments: directed selection of complex catalytic systems determining the character of initiation, transmission and chain termination, and similarity of properties of the initial monomer to those of formed polymeric compounds of the olefin class:

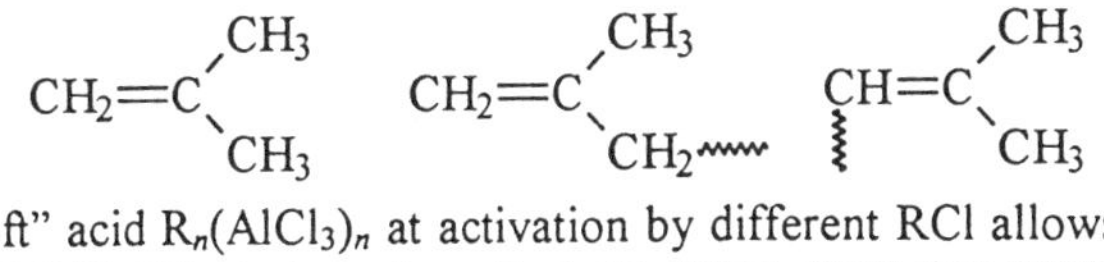

The basic "soft" acid $R_n(AlCl_3)_n$ at activation by different RCl allows introduction of silyl ($H_3SiCH_2C_6H_4CH_2$–), bromine–alkyl ($Br(CH_3)_2CCH_2CH_2C(CH_3)_2$–), unsaturated ($CH_3$–CH=CH–$CH_2$–) and other groups used for further reactions of the synthesis of grafted polymers.

Inifers — bifunctional initiators and agents of chain transmission — perform functions of controlled initiation and controlled chain termination with introduction of desired end groups.

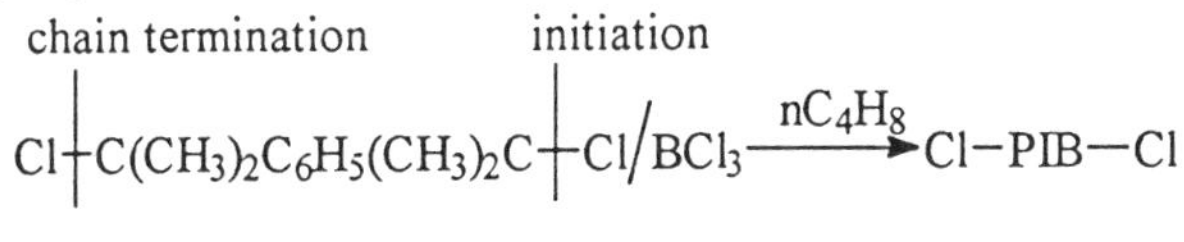

From the point of view of polyisobutylene chain engineering, the "ballast" reaction is the one of the chain transmission to the monomer. In accordance with HSAB principles, complexing of AC with proton traps of the type

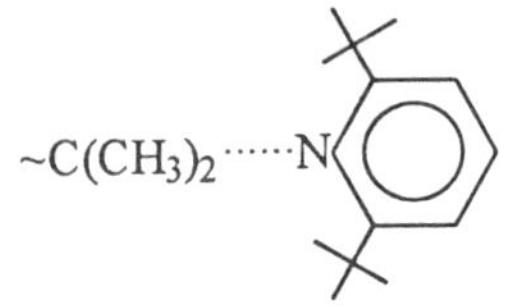

and other coordinating matrices suppresses direct reaction of the chain transmission to the monomer and assists in MM growth and narrowing of the polymer MMD. If the motive force of the chain transmission reaction by proton elimination remains, then in accordance with the increase of AC rigidity in the elementary process, proton accepting by a trap is preferable. Reduction of the monomer conversion coordinates with this scheme. Absence of the chain transmission to the monomer allows an increase of further polymer "grafting" up to ~100%.

In such a manner, controlled reactions of initiation, termination and transmission of the chain create premises for "quasi-living" polymerization. Contrary to classic "living" polymerization, in which all ACs are kinetically stable and can participate simultaneously in the chain propagation (chain termination and transmission are absent), in cationic "quasi-living" polymerization of isobutylene, termination and transmission of the chain may be present, but they are reversible, and the monomer is expended in acts of chain initiation and propagation only. Propagating macromolecules are not always active, but only under necessary conditions, in particular, reversibility of kinetic chain termination, loss of a proton and bimolecular transmission to a monomer. These conditions are realized, if the condition $[M] < [P^{=}]$ is postulated, i.e. at a monomer concentration below that of unsaturated potentially reactive polymeric chains and shift of the reaction equilibrium to the left:

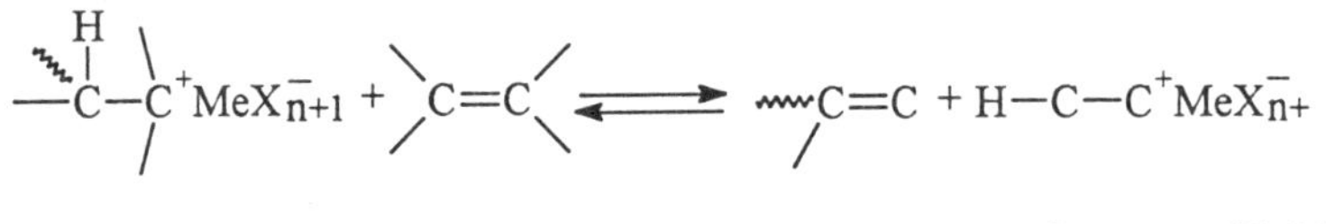

Experimentally, these conditions are fulfilled for the sequence of systems ($H_2O/TiCl_4$, $C_6H_5C(CH_3)_2Cl/TiCl_4$, etc.) [248] under slow addition of isobutylene at a constant rate providing the condition $[P^{=}] > [M]$ [25, 247]. For these systems signs of "quasi-living" polymerization are evident — constancy of AC concentration, growth of MM with adding of a monomer, $1 < M_w/M_n < 2$ in the presence of chain termination and transmission reactions. Another proof is the formation of block-copolymers (the list of realized structures is shown in the Section on copolymerization of isobutylene: see Chapter 4).

On account of reversibility of chain transmission and termination reactions in polymerization of isobutylene, the works [249, 252, 253] should be mentioned, conclusions from which appear unexpected and controversial. It is maintained that the polymerization–depolymerization equilibrium occurs at a certain stage of the polymerization process under the effect of complex catalysts and negative centigrade temperature. Experimentally, this is displayed in the appearance of maximums of the conversion (MM)–time curves. This fact is amazing, but the suggested explanation about the connection of this effect with a local temperature increase seems to be likely and once more reminds us about the difficulties of behavior and, consequently, investigation of rather fast high-energy polymerization processes by traditional methods (see Parts below for details, especially Chapter 7).

However, the details of low-temperature (20–60°C) polyisobutylene degradation under the influence of the same catalysts in solution, discussed in later works [249–253], suggests that the true reason for the above-mentioned events may be much more complex. Degradation is accompanied not only by decrease of MM, but also by formation of branched structures, e.g. by deep rebuilding of polyisobutylene macromolecules. Reasons that suggest a free-radical degradation mechanism do not correlate with the low E_a (9 kcal/mol). On the other hand, the more probable cationic mechanism (according to the data presented) is not confirmed.

2.3.2. Features of fast polymerization processes.

Chemistry possesses other examples of extremely fast processes proceeding in the liquid phase, when the rate of chemical reactions are comparable with or higher than those of reagent mixing, in particular chlorination, sulfurization of olefins, alkylation of alkanes by alkenes, chlorination of aromatic hydrocarbons, many processes of ionic and free-radical polymerization. Among rather fast reactions of polymerization, most attention for theoretical consideration of macrokinetic regularities of the process, generally, due to the total theoretical clearness and simultaneous great practical importance, is given to the cationic (electrophilic) liquid-phase isobutylene polymerization, which can be considered as the classic model of a fast chemical reaction.

In industry oligo- and polyisobutylenes with molecular mass (MM) of 112-50,000 and higher are obtained by cationic polymerization of isobutylene in the presence of $AlCl_3$ in the medium of hydrocarbons (butanes, etc.) or chlorinated hydrocarbons (chlorides of ethyl, methyl, etc.) in the range of 173-353 K in mixing reactors of 1.5-3.0 m^3 of a complex construction [107, 254].

Rate constants of chain initiation and propagation (k_i, k_p) are not lower than 10^5-10^6 l/(mol·s) [255-259] which, under real initial concentrations of the catalyst and monomer, determines the characteristic time of the chemical reaction of 10^{-1}-10^{-3} s. This means that isobutylene polymerization proceeds generally at a distance shorter than 1-10 cm from the point of catalyst introduction, and the process is limited by mixing of the catalyst, monomer and the reactionary mass. Heat accumulation in the reacting system under high turbulence of the reaction mixture motion causes osculation change of temperature in the reaction zone with time [255]. That is why thermostatic control of isobutylene polymerization processes, not only in their industrial production, but also in the laboratory, displays considerable difficulties. Because of non-isothermity, changes of MM and MMD of polyisobutylene are observed during the process proceeding along the reaction zone [107, 255–257, 259–262].

Of importance is that a macroscopic consequence of fast local operation of the polymerization is dependence of the average MM on concentrations of the catalyst and the monomer, as well as significant broadening of the MMD with increase of the initial monomer concentration. At the same time, the MM of polyisobutylene, according to the kinetic scheme, is completely determined by the chain transmission to monomer and predicts independence of MM and MMD on concentrations of the catalyst and the monomer in the isothermal regime. Moreover, usually, solutions of the catalyst and the monomer are introduced into the reactor irregularly and on account of high rates of polymerization it is difficult to achieve good mixing with the reactionary mass. This feature causes additional instability of the reactor operation, broadening of the MMD and decrease of the final product MM [252–257].

2.3.3. Problems of kinetics of fast polymerization processes.

The process of isobutylene polymerization in the presence of halides of metals (Lewis acids) combined with Brönsted acids proceeds (using the example of a catalyst based on MeX_n and H_2O) by the scheme given in refs. [107, 263, 264] and includes the following stages (Scheme 2.10):

$$MeX_n + H_2O \rightarrow MeX_n{\cdot}H_2O \rightarrow H^{\delta+},MeX_nOH^{\delta-};$$

I. initiation (k_i):

$$H^{\delta+},MeX_nOH^{\delta-} + CH_2{=}C(CH_3)_2 \rightarrow (CH_3)_3C^{\delta+},MeX_nOH^{\delta-};$$

II. chain propagation (k_p):

$$(CH_3)_3C^{\delta+},MeX_nOH^{\delta-} + CH_2{=}C(CH_3)_2 \rightarrow (CH_3)_3C\text{-}CH_2\text{-}(CH_3)_2C^{\delta+},MeX_nOH^{\delta-};$$

III. chain transmission (mostly to the monomer, k_M):

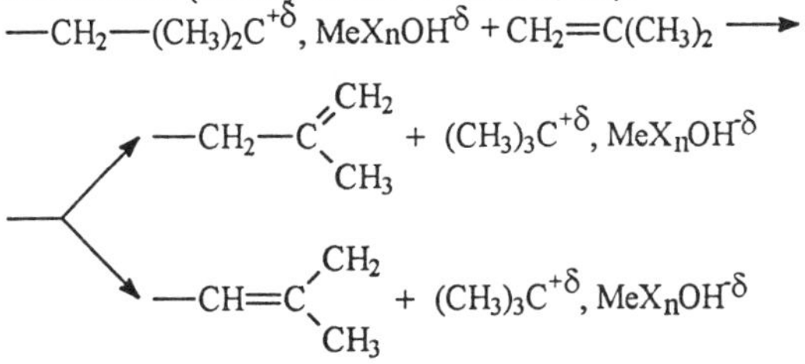

IV. reaction of the proton transmission to counterion (probable reaction), which can be neglected in many cases:

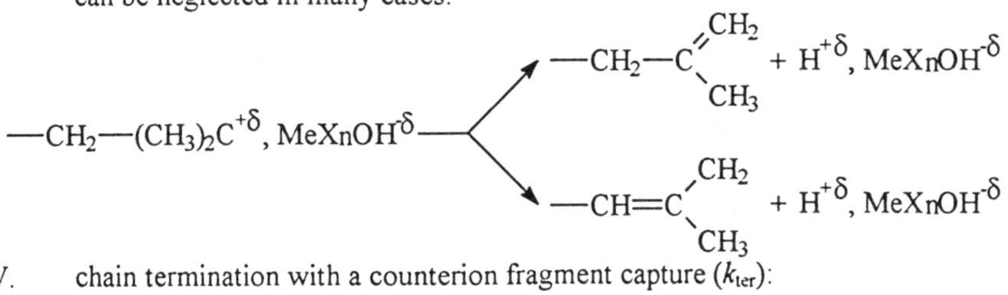

V. chain termination with a counterion fragment capture (k_{ter}):

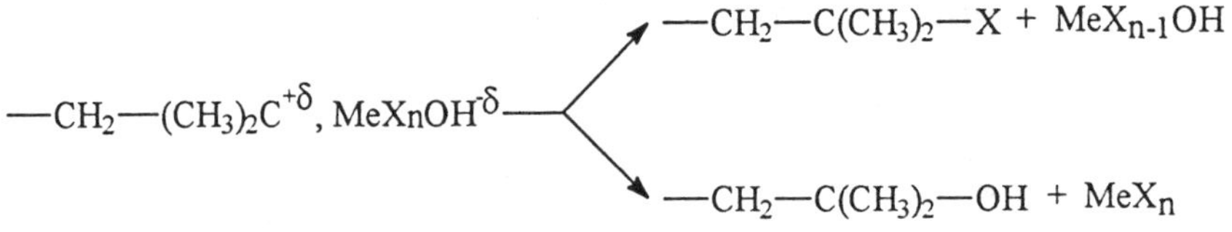

in this case regeneration of the catalyst owing to fast reinitiation transforms into the chain transmission to the counterion.

In such a way, kinetic Scheme 2.11 should be used for modeling of the chemical process of liquid-phase electrophilic polymerization of isobutylene:

VI.	Catalyst + M → A_1*	k_i
VII.	A_1* + M → A_2*	k_p
	....	
VIII.	A_{n-1} + M → A_n*	k_p
IX.	A_n* + M → A_1* + P	k_M
X.	A_n* → P	k_{ter}

Scheme 2.11

In accordance with the experiment, the rate of initiation is rather high ($k_i \geq k_p$); that is why under kinetic consideration concentration of active centers, $\Sigma A_n{*}$ can be identified with the catalyst concentration.

In accordance with Scheme 2.11 the average polymerization degree $\overline{P}_n$ depends on the monomer concentration:

$$\frac{1}{\overline{P}_n} = \frac{k_{ter} + k_M[M]}{k_p[M]}. \quad (1)$$

However, as the chain transmission to the monomer (III) is the main reaction determining MM and MMD in the isobutylene polymerization in a wide range of temperatures, $\overline{P}_n$ and MMD of the polymer formed in each rather small element of the reaction volume is defined only by temperature and does not depend on concentrations of the catalyst and the monomer:

$$\rho_n(j) = \frac{1}{\overline{P}_n}\exp\left(-\frac{j}{\overline{P}_n}\right), \quad (2)$$

where $\overline{P}_n = \dfrac{k_p}{k_M} = f(T) \neq \varphi([A^*],[M])$. (3)

A specific feature of isobutylene is its high reactivity in relation to cationic agents and, as a consequence, a rather high (over 10^5 l/mol·s) rate constant of the chain propagation [258, 259, 262]. This creates the extremely high rate of polymerization, accompanied by extraction of a significant heat amount, which is usually very difficult to remove from the reaction zone. This is the reason why rather accurate measurement of the cationic polymerization rate has hardly ever been made by anybody on the basis of the study of isobutylene polymerization in the kinetic range due to difficulties in setting correct quantitative experiments (first of all, on account of non-isothermity of the process) and the absence of strict experimental data on the concentration of active centers. That is why one can say with a fair degree of assurance that the data on kinetics of the isobutylene polymerization existing in the literature, obtained under application of classic methods of estimation of rate constants of the reactions, in the best case, are extremely approximate, even in cases when, performing experimental studies, the investigators were dealing with processes proceeding at a relatively moderate rate and forming polymeric products of low molecular mass. This very high order of numerical values of the chain propagation rate in electrophilic isobutylene polymerization (k_p) the main reason why the cationic isobutylene polymerization is regarded as a fast chemical process [107, 258, 259, 263, 265, 266].

It can be ascertained that very high rates of polymerization combined with exothermity of the process (q_{II} = 54 kJ/mol) create the situation in which even extremely slow injection of an initiator and fast mixing are not enough for heat disposal, extracted during the reaction. In the general case, isobutylene polymerization starts already before initiating particles manage to diffuse deep enough. Even with the help of fast recording (~3000 frame/s) it was found to be impossible to determine the time period between application of the $AlCl_3$ drop to the isobutylene surface (at 195 K) and appearance of the polymer. Hence it follows that in these and even many other fast ionic and non-ionic systems, uniformity of distribution of reagents and temperature in the reaction volume does not occur. This means that in practice, processes of cationic isobutylene polymerization and other similar rather fast chemical reactions are barely regulated. This circumstance requires research and development of new approaches to kinetic study of fast polymerization processes (and other fast reactions also), as well as regulation methods for these processes, necessarily using equations of chemical kinetics, heat transfer, diffusion and convection.

2.4. Kinetic features of isobutylene polymerization.

A specific feature of isobutylene is its high reactivity in relation to cationic agents and, as a consequence, very high rates of the process, accompanied by high heat release. Until now, it has not been possible to make fairly certain and accurate measurement of the rate of isobutylene polymerization due to difficulties in setting up correct quantitative experiments (influence of admixtures, non-isothermal type of the process) and the absence of experimental data on the nature and concentration of AC. In essence, the term "kinetics of isobutylene polymerization" does not exist for the following reasons: multiplicity of AC, absence of a stationary state, ambiguity of chain transmission reactions and order of reactions with respect to components (the first, second or higher), effects of solvent and counterion, not taken into account in the kinetic equations, and others. That is why data on the kinetics of isobutylene polymerization are of a peculiar type and are rather approximate even in study of processes proceeding at a moderate rate and with formation of products of low molecular masses. From relatively broad, but not always uniformly interpreted material, the following experimental data on kinetics of the isobutylene polymerization can be presented.

The equation for the degree of polymerization $\overline{P}$ taking into account characteristic reactions 2.I – 2.V of the cationic polymerization of isobutylene is the following [62, p. 110]:

$$\frac{1}{\overline{P}_n} = \frac{k_{\text{ter}} + k_n}{k_\text{p}[\text{M}]} + c_\text{M}.$$

The difference in activation energies of the propagation and chain propagation limiting reactions can be estimated by the temperature function of $\overline{P}$. It should also be noted that the inverse proportion of polyisobutylene molecular mass with respect to temperature is maintained (Figure 2.12), although slopes of linear dependencies of $\lg \overline{M}_v$ on $1/T$ may be different: for example, for series of straight lines conforming with the classic Friedel-Crafts metal halides ($AlCl_3$, $RAlCl_2$, BF_3) $E_{\overline{M}_v} = 27.6 \pm 4.2$ kJ/mol, and for systems R_nAlCl_{3-n}-*tert*-C_4H_9Cl (n = 2–3) $E_{\overline{M}_v} = 7.1 \pm 2.1$ kJ/mol and $C_2H_5AlCl_2 \cdot H_2O$ $E_{\overline{M}_v} = 7.5 \pm 1.7$ kJ/mol. Difference in values of preexponential multiplicands of the groups of initiating systems at one and the same $E_{\overline{M}_v}$ determines (in terms of the experimental scatter) differences in absolute values of molecular masses. In the range of 253–195 K maximum molecular masses are obtained in the presence of HAOC-based systems. Below these temperatures, polymers with high $\overline{M}_v$ are produced on traditional Friedel-Crafts catalysts ($AlCl_3$, BF_3), which is stipulated by a change of the relative role of some of the reactions that limit chain propagation. At the same time, closeness of the straight line slopes for one group of catalysts indicates the resemblance of termination mechanisms. In particular, as chain transmission processes via monomer with a counterion are typical for initiation by classic Friedel-Crafts halides, the dominant role for HAOC is played by the chain limiting in interaction with a fragment of the counterion. The observed reduction of $E_{\overline{M}_v}$ values under a change of polymeric chain termination mechanism is confirmed by calculations [267]. Closeness of the slope of lines and the difference in absolute values of parts of the ordinate axis intercepted by them means that energetic terms are approximately equal (the counterion influences E_M and E_p equally), and steric factors are different (different values of A_M/A_p) [68, p. 168]. It is supposed that differences in slopes of Arrhenius lines are brought about by the difference in "ionic degree" of growing ionic pairs. Initiators providing large slope of lines (large $E_{\overline{M}_v}$) form counterions possessing low nucleophility, which determines the probability of the isobutylene polymerization involving relatively free ions. Initiators displaying smaller slopes of the lines form counterions with a little higher nucleophility. As a consequence, the chain propagation can proceed from the proximity of ionic pairs.

According to Scheme 2.10 of the isobutylene polymerization, the difference in activation energies of chain propagation and propagation limiting equals

$$\Delta E = E_p - (E_M + E_d).$$

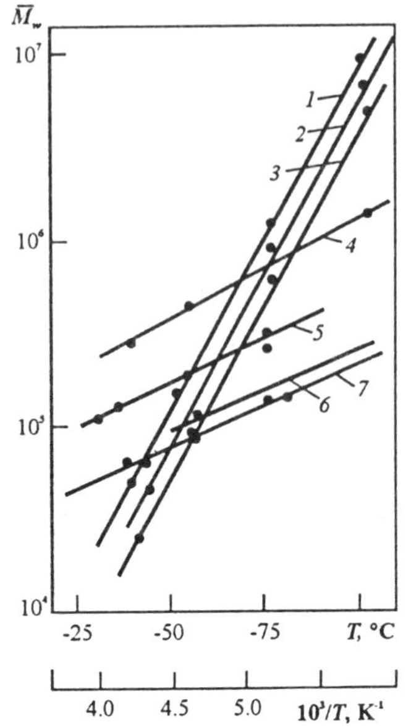

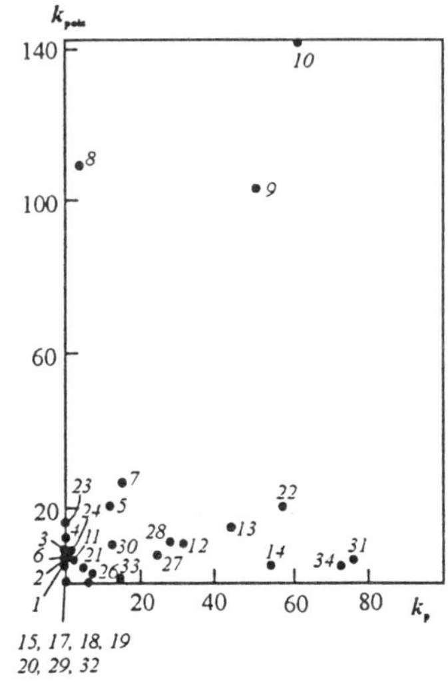

Figure 2.12. Dependence of the medium–viscous molecular mass of polyisobutylene, produced in the presence of catalysts, on the polymerization temperature:
1 — $C_2H_5AlCl_2$; 2 — BF_3; 3 — $AlCl_3$; 4 — $(C_2H_5)_2AlCl$-*tert*–C_4H_9Cl; 5 — $(CH_3)_3Al$-*tert*–C_4H_9Cl; 6 — $(C_2H_5)_3Al$-*tert*–C_4H_9Cl; 7 — $C_2H_5AlCl_2{\cdot}H_2O$.

Figure 2.13. Constants of chain poisoning (k_{pois}) and transmission (k_p) during isobutylene polymerization.

The relative role of chain termination and transmission reactions changes not only with chemical composition or nature of catalytic systems, but also with temperature range of polymerization in one and the same system [62, p. 114], as well as with the catalyst concentration in the reaction mixture. This is reflected by ΔE values which change during isobutylene polymerization (and for a number of other monomers as well) in a rather broad range, including also negative values. In particular, ΔE = –0.84 kJ/mol responds to the low–temperature branch of the Arrhenius curve (from 85 to 175 K), which characterizes isobutylene polymerization under the effect of $AlCl_3$ in CH_3Cl. In this case, the degree of polymerization $\overline{P}$ under these conditions does not depend on the monomer concentration. With temperature increase, $\overline{P}$ becomes more sensitive to the monomer concentration, and ΔE = –15.1 kJ/mol [268]. Besides the chain transmission to the monomer, this is also indicated by the existence of other reactions limiting chain propagation. *n*–Alkenes do not usually affect the molecular mass, but being poisons reduce the yield of polyisobutylene. Alkyl halides reduce the molecular mass causing no influence on the polymer yield, which is typical for agent–transmitters of the material chain. Many compounds display both factors in greater or lesser degree, for example, β–alkenes. Figure 2.13 generalizes the experimental data on kinetic parameters of the reactions of chain poisoning and transmission during isobutylene polymerization [68, p. 146]. Pure poisons (propylene, butene-1, pentene-1, etc.) and pure chain transmitters form the horizontal and vertical axes, respectively. *tert*-Butyl chloride and *tert*-butyl bromide are seen to be rather effective chain transmitters.

The poisoning effect of *n*-alkenes is explained by hydrogen transmission from an α–olefin to growing polyisobutylene cation and increase of stability of allyl cations

which is reflected by numerical values of poisoning constants (the catalytic system of $AlCl_3$ with CH_3Cl, solvent — *n*-pentane, 195 K): propylene 4.9, butene-1 6.2, pentene-1 9.7, hexene-1 11.8, octene-2 20.8.

Propylene is the weakest poison, because it forms the less stable non–substituted allyl cation. For compounds containing two secondary allyl hydrogen atoms, formation of allyl carbonium ion is simpler. Of distinct probability is that the deterministic role is played by basicity of secondary allyl hydrogen atoms, bound to the increase of electron–donor ability of CH_3^-, $C_2H_5^-$, $C_3H_7^-$ groups in the sequence of *n*–alkenes. Octane–2 is one of the strongest poisons, which is supposed [68, p. 147] to be connected with formation of a rather stable di-substituted secondary allyl ion

$$CH_3{-}\overset{\delta+}{C}H{\cdots}CH{\cdots}\overset{\delta+}{C}H{-}C_4H_9$$

and proceeding of the reaction

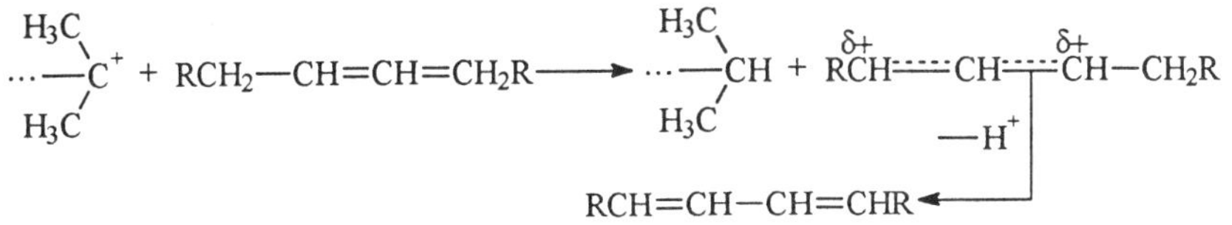

i.e. after hydride–ion transition to the growing cation, the forming allyl carbonium ion is stabilized by deprotonation, forming a conjugated diene; attacking a new monomer molecule, the proton propagates the kinetic chain. β–Alkenes influence both the yield and the molecular mass of polymeric products, and for this reason are poisons and transmitters of the chain simultaneously. This effect is generally indicated by the energetic efficiency of formation of stable di–substituted conjugated dienes [68, p. 148]. Branched olefins possessing no allyl H-atoms (for example, 3,3-dimethylbutene-1) or possessing a single allyl H-atom, but disposed unfavorably for attack by the growing carbocation (for example, 3-methylbutene-1), do not influence isobutylene polymerization.

To regulate the polymerization rate and/or molecular mass it is suitable to use diisobutylene [269]. 2,4,4-Trimethylpentene-1 is a strong poison and transmitter, and 2,4,4-trimethylpentene-2 is a strong poison, but weak transmitter:

	Poisoning constant	Transmission constant
2,4,4-Trimethylpentene-1	66.7	700
2,4,4-Trimethylpentene-2	66.7	34.6

Conjugated dienes significantly decrease the molecular mass of polyisobutylene. For example, pentadiene-1,3 (piperylene) is a strong poison (poisoning constant is 170) and chain transmitter (chain transmission constant is 327), which is explained by allyl chain termination: allyl cations formed at diene inclusion into the chain and hydride–ion transition are rather stable:

$$\cdots-CH_2-C(CH_3)_2-CH_2-\overset{\delta+}{CH}\!\!\cdots\!\!CH\!\!\cdots\!\!\overset{\delta+}{CH}-CH_3$$

$$\overset{\delta+}{CH_2}\!\!\cdots\!\!CH\!\!\cdots\!\!CH\!\!\cdots\!\!CH\!\!\cdots\!\!\overset{\delta+}{CH_2}$$

Allyl halides are not poisons, but represent strong transmitters of the chain. This is indicated by the fact that at hydride-ion loss, forming allyl halides possess similar or even higher stability compared to the allyl cation formed from olefin [270]. HCl is a promoter and chain transmitter in the $AlCl_3$–methyl chloride system [269].

In the most cases, arenes affect neither the yield, nor the molecular mass of the polymer. Such behavior contradicts the high π–basicity of arenes and ability to enter the alkylation reaction with carbocations [240]. It is more correct to consider them as weak inhibitors, in which their ability to enhance chain transmission is expressed more strongly than their ability to terminate. In its constants of poisoning and chain transmission (2 and 11, respectively) toluene, for example, is similar to isobutyl bromide. Discrepancy in data is connected with possible influence of arenes as bases on the catalytic system, displayed already at low concentrations. Only in the case of application of complex systems of $HCl\cdot AlCl_3$·arene type is a monotonous reduction of the monomer conversion with arene concentration growth observed. Relative constants of the chain transmission to arene (toluene) are close to that to isobutylene. For individual $MeCl_n$ (in the work [68, p. 146] $AlCl_3$ was used) the dependence of isobutylene conversion on the amount of introduced additive is complex in the range of low concentrations of additives, regarding to formation of a complex between them and $MeCl_n$. Consequently, the total effect of arenes on the isobutylene polymerization depends on both additive concentration and the nature of the catalyst.

Attempts were made to estimate rate constants of the chain propagation k_p in the isobutylene polymerization. For example, the probable rate constant of its polymerization on free ions (in the presence of solid ZnO) was calculated in [271]: $k_p = 1.5\cdot10^8$ l/(mol·s) (273 K). A value close to this k_p was calculated in [219] for the block polymerization of isobutylene under the influence of ^{60}Co γ–radiation, in the temperature range of 300–135 K the temperature dependence of the polymerization rate being significantly low at values of $k_p = 3\cdot10^2$ s^{-1} ($AlCl_3$–CH_3Cl in CH_3Cl, 195 K). Clearly that existing kinetic constants and k_p, first of all, are approximate, although their significantly high degree is established. At least, values of absolute rate constants of the isobutylene reaction with relatively stable cations $(C_6H_5)_2\overset{+}{C}H$ and $C_6H_5\overset{+}{C}H_2$, which can be considered as the lower limit of k_p, are approximately 10^7 l/(mol·s) (impulse radiolysis, 297 K) [272]. Values of k_p can be seen to lie below the diffusion threshold: for most monomers with viscosity of 0.5 kPa·s at 300 K, k_D is about 10^{10} l/(mol·s). Under conditions of a viscous product formation, especially during low-temperature isobutylene polymerization, k_D values decrease, for example, for the isobutane–polyisobutylene system $k_D \sim 10^6$ l/(mol·s) [273, p. 61], and kinetic constants k_p may appear higher than k_D. Extremely high polymerization rates combined with rather high exothermicity (71.9 kJ/mol) create the situation, in which even very slow addition of an initiator and fast mixing are not enough to eliminate the heat released during the reaction proceeding in the liquid phase, especially in microspheres adjacent to growing centers. This very conclusion was made from critical consideration of the experimental data from [68, p. 144]. Application of extremely diluted solutions of the monomer (about 0.01–0.02 mol/l), which are not of interest for the real isobutylene polymerization, is the only thing that can provide a regime close to the isothermal one. In the general case, when an initiator drop enters the liquid monomer, the reaction starts even before initiating particles manage to diffuse deep enough. The reaction proceeds so fast that isobutylene can be polymerized long before two drops – the monomer and the initiator are simply mixed up. Even with the help of fast recording (3000 frame/s) the time period between impact of $AlCl_3$ solution drop to the isobutylene surface and polymer occurrence could not be determined [68, p. 145]. Hence it follows that the stationary principle cannot be fulfilled in these and many other cationic systems during the isobutylene polymerization. This means that in practice molecular masses of the polymer appear much lower, MMD is wider, and viscosity of oligomeric products increases.

Features of isobutylene polymerization kinetics stipulated by instability of the process were studied in works [274, 275] using a flow-through device (Figure 2.14), which allows regulation of linear rates of reagent introduction and control of the temperature change in assigned points of the reaction volume (12% isobutylene solution and $C_2H_5AlCl_2$ solution with the concentration of $1.4 \cdot 10^{-2}$ mol/l in heptane was introduced simultaneously with similar constant rates of $1 \cdot 10^{-2}$ m/s and higher).

The temperature change in different points of the reaction zone (Figure 2.15) indicates that the highest rate lies at the point of the catalyst injection (point B, Figure 2.14). The lowest rate of the reaction is observed near reactor walls (point C). It is seen from the data of Table 2.19 that the reaction mainly (by some 50–70%) proceeds in the reagent injection zone during 1–2 s (point B).

The temperature of the reaction mass appears variable and dependent on the reagents' concentration, isobutylene in particular (Table 2.19, experiments 9–12). Reduction of the molecular mass of the polymer and decrease of isobutylene conversion with the increase of its concentration in the initial mixture proceed simultaneously.

The experimental observations are generated by increase of the heat amount releasing during polymerization, natural deviation of the heat exchange and increase of the role of the material chain termination by transmission to isobutylene. When the reaction zone is removed from the point of catalyst injection (conversion increases approximately up to 100%), the rate of polymerization changes at the sacrifice of the appearance another, different from that in the point B, temperature gradient and monomer concentrations, which indicate the dependence of the molecular mass (Table 2.19) and MMD (Figure 2.16) on the reaction length. Differential MMD curves are of various appearance also at the change of isobutylene concentration. MMD functions in $\lg \rho_n(j)$–j coordinates [276] for both cases are depicted in Figure 2.15. From the Figure is seen MMD broadening at the sacrifice of occurrence of significant amounts of a low-molecular fraction in the product both with remote from the catalyst injection place and increase of the isobutylene concentration in the reaction mixture. Consequently, appearance of the temperature gradient and fields of various rates in the polymerization zone determine increase of non-uniformity of polymeric product with respect to the molecular mass. To put it differently, in the topochemical aspect, isobutylene polymerization relates to significantly fast processes

Table 2.19

Influence of the reaction zone length and monomer concentration on the molecular mass of the product and monomer conversion.

No.	Probe picking point	Isobutylene concentration, % (mass)	Rate of reagents' introduction, cm/s	ΔT*	l, m	Conversion, %	Molecular mass	
							MCHE method (measurement of condensation heat effects)	DBA method (double bond analyzer)
1	A	12	1.0	35	0.02	56.2	960	950
2	B	12	1.0	36	0.10	72.0	940	935
3	C	12	1.0	37	0.30	86.3	700	690
4	D	12	1.0	40	0.48	98.4	510	500
5	A	12	1.05	32	0.02	48.9	1250	1210
6	B	12	1.5	35	0.10	70.4	1000	980
7	C	12	1.5	37	0.30	82.6	840	840
8	D	12	1.5	40	0.48	98.6	560	550
9	C	5	1.0	24	0.30	81.0	1500	1480
10	C	10	1.0	30	0.30	71.2	1210	1150
11	C	15	1.0	37	0.30	70.3	950	930
12	C	20	1.0	40	0.30	69.2	820	820

* Initial temperature is 273 K.

and represents a "flare", analogous to combustion, with various temperature zones, concentrations of reagents, process rates, etc. The above-mentioned facts should be taken into account in the kinetic analysis of such processes.

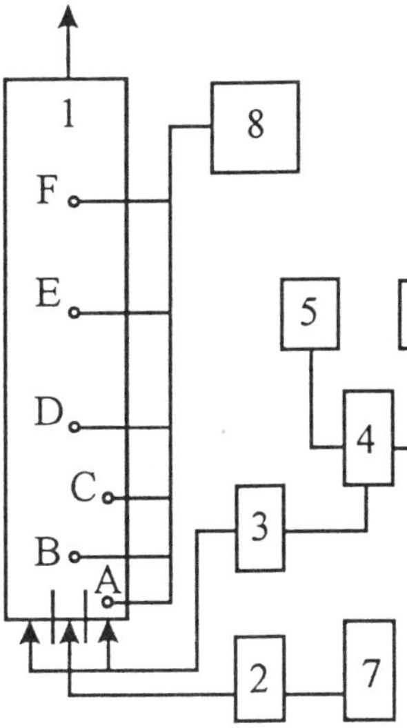

Figure 2.14. The scheme of a flow-through device for isobutylene polymerization: 1 – reactor; 2, 3 – pumps; 4 – volume for monomer and solvent mixing; 5, 6, 7 – volumes for monomer, solvent and catalyst; 8 – recorder EPR-09DMS[1]; A–E – points of the probe picking.

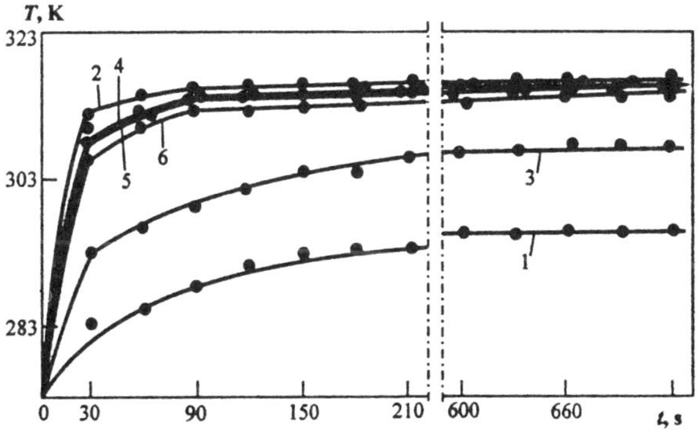

Figure 2.15. Dependence of the temperature change at various points of the reaction zone on time; curves 1–6 – correspond to probe selection points A–E in Figure 2.14.

[1] Russian trademark

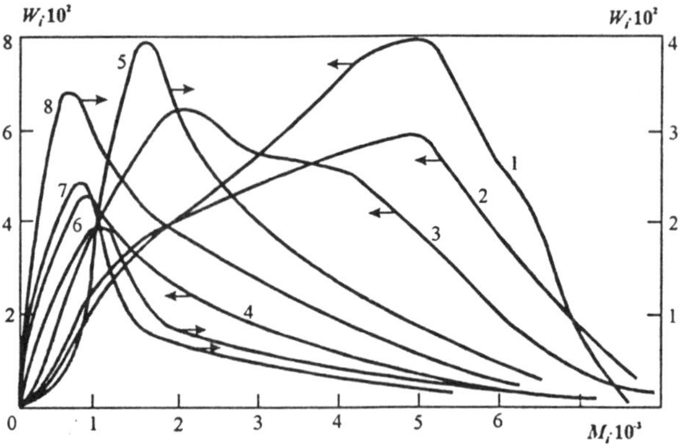

Figure 2.16. Dependence of polyisobutylene MMD on the reaction zone length (1–4 correspond to experiments 5–8 in Table 2.14) and the monomer concentration (5–8 correspond to experiments 9–12 in Table 2.14).

References to Chapter 2

1. Cationic Polymerization, Ed. P. Plesha, 1966, 584 p., Moscow, Mir. (Rus)
2. Hunter, W.H. and Yone, R.V., J. Amer. Chem. Soc., 1933, v. 55(3), pp. 1248–1252.
3. Cesca, C., J. Polym. Sci., Polymer Symp., 1976, No. 56, pp. 159–172.
4. Sigwalt, P., Makromol. Chem., 1974, Bd. 175(4), S. 1017–1038.
5. Olah, G.A., Makromol. Chem., 1974, Bd. 175(4), S. 1039–1064.
6. Korshak, V.V. and Lebedev, N.N., Doklady Akademii Nauk SSSR, 1947, v. 57(3), pp. 263–266. (Rus)
7. Marek, M. and Chmelir, M., J. Polymer Sci., 1968, Pt. C, No. 23, pp. 223–229.
8. Chmelir, M. And Marek, M., Coll. Czech. Chem. Comm., 1967, v. 32(8), pp. 3047–3049.
9. Lopour, P. and Marek, M., Makromol. Chem., 1970, Bd. 134(1), S. 23–31.
10. Kennedy, J.P., J. Macromol. Sci., 1972, Pt. A, v. 6(2), pp. 329–346.
11. Reactivity, Reaction Mechanisms and Structure in Chemistry of Polymers, Ed. A. Jenkins and A. Ledwith, (Transl. from English), 1977, 645 p., Moscow, Mir. (Rus)
12. Kennedy, J.P. and Mareshal, E., J. Polymer Sci., 1981, Pt. D, v. 16(16), pp. 123–198.
13. Candini, A. and Cheradame, H., Adv. Polymer Sci., 1980, v. 34/35, pp. 1–289.
14. Dallinga, G., Caaf, J., and Mackor, E.L., Rec. Trav. Chim., 1970, v. 80(10), pp. 1068–1078.
15. Gillespie, R.J., Peel, T.E., and Robinson, E.A., J. Amer. Chem. Soc., 1971, v. 93(20), pp. 5083–5087.
16. Palm, V.V., Doklady Akademii Nauk SSSR, 1956, v. 108(2), pp. 270–273.
17. Gillepsie, R.J., Acc. Chem. Res., 1968, v. 1(2), pp. 202–209.
18. Olah, G.A., Friedel-Crafts Chemistry, 1973, 581 p., N.V., etc.
19. Olah, G.A., White, A.M., and O'Brien, D.H., Chem. Rev., 1970, v. 70(5), pp. 561–591.
20. Brouwer, D.M. and Van Doorm, J.A., Rec. Trav. Chim., 1972, v. 91(7), pp. 895–905.
21. Faust R. and Kennedy, J.P., J. Polymer Sci., 1987, v. A25(7), pp. 1847–1869.
22. Zsuga, M. and Kelen, T., Polym. Bull., 1986, v. 16(4), pp. 285–291.
23. Nesmelov, A.I., Orlinkov, A.V., Murachev, V.B., Akhrem, N.S., Byirikhin, V.S., and Volpin, M.E., Isv. AN SSSR, Ser. Khim., 1988, No. 10, pp. 2232–2238. (Rus)
24. Byirikhin, V.S., Nesmelov, A.I., Murachev, V.B., Ezhova, E.A., and Chulochnikova, T.V., Highmolec. Comp., 1990, v. A33(7), pp. 1421–1427. (Rus)
25. Kennedy, J.P. and Mareshal, E., Carbocationic Polymerization, 1982, 510 p., N.Y.
26. Biswas, M. and Kennedy, J.P., Makromol. Chem., Makromol Symp., 1986, v. 3, pp. 113–127.
27. Toman, L., Spewacek, J., and Pokorny, S., J. Polym. Sci. A., 1989, v. 27(3), pp. 785–794.
28. Sangalov, Yu.A., Nelkenbaum, Ya.Ya., Romanko, T.V., and Elichev, A.A., Acta Polymerica, 1990, Bd. 41(3), pp. 173–177.
29. Sangalov, Yu.A., Nelkenbaum, Ya.Ya., and Ponedelkina, I.Yu., Kurzreferate 7 Internat. Makrosymp. Fortschritte aufdem Gebiet der ionischen Polymerisation, 1990, Post. 4, Berlin.

30. Erusalimskii, B.L., Ionic Polymerization of Polar Monomers, 1970, 288 p., Leningrad, Nauka. (Rus)
31. Korotkov, A.A. and Podolskii, A.F., Catalytic Polymerization of Vinyl Monomers, 1973, 284 p., Leningrad, Nauka. (Rus)
32. Jensen, W.B., Chem. Rev., 1978, v. 78(1), pp. 1–22.
33. Pearson, R.G., J. Amer. Chem. Soc., 1963, v. 85(22), pp. 3533–3539.
34. Kennedy, J.P., Melby, E., and Johnston, J., J. Macromol. Sci., 1974, Pt. A, v. 8(3), pp. 463–468.
35. Yatsimirskii, K.B., Zh. Anal. Khim., 1951, v. 6(4), pp. 211–217. (Rus)
36. Sangalov, Yu.A., Kirillov, A.P., and Minsker, K.S., In: Physico–chemical Grounds of Synthesis and Processing of Polymers, 1978, Iss. 3, pp. 3–14, Gorky. (Rus)
37. Topchiev, A.V., Zavgorodnii, S.B., and Paushkin, Ya.M., Boron Fluoride and Its Compounds as Catalysts in Organic Chemistry, 1956, 356 p., Moscow, Izd. AN SSSR. (Rus)
38. Ioffe, I.I. and Roginskii, S.Z., Zh. Fiz. Khim., 1957, v. 31(3), pp. 612–625. (Rus)
39. Zhorov, Yu.M., Panchenkov, G.N., and Volokhova, G.S., Olefin Isomerization, 1977, 204 p., Moscow, Khimia. (Rus)
40. Tanabe, K., Solid Acids and Bases (Transl. from English), 1973, 183 p., Moscow, Mir. (Rus)
41. Antipina, T.V., Chukin, G.D., and Kirina, O.F., Zh. Fiz. Khim., 1972, v. 46(11), pp. 2921–2922. (Rus)
42. Paushkin, Ya.M., Catalytic Polymerization of Olefins to Motor Oil, 1955, 184 p., Moscow, Izd. AN SSSR. (Rus)
43. Shephard, F.E., Roone, J.J., and Kemball, C., J. Catal., 1962, v. 1(4), pp. 379–388.
44. Holm, V.C.F., Baliev, G.C., and Clark, A., J. Phys. Chem., 1959, v. 63(1), pp. 129–133.
45. Hattori, H. And Shiba, T., J. Catal., 1968, v. 12(2), pp. 111–120.
46. Richardson, J.T., J. Catal., 1967, v. 9(2), pp. 182–194.
47. Eberly, P.E., J. Phys. Chem., 1968, v. 72(3), pp. 1042–1047.
48. Lapidus, A.L., In: Application of Zeolites in Catalysis, 1976, Iss. 1, pp. 53–60, Novosibirsk, Nauka. (Rus)
49. Eidus, Ya.T., Astoichiometric Components of Catalytic Reactions, 1975, 140p. Moscow, Nauka. (Rus)
50. Ben–Dor, L. and Margalith, R., Inorg. Chim. Acta, 1967, v. 1(1), pp. 149–153.
51. Bekker, G., Introduction into Electronic Theory of Organic Reactions (Transl. from German), 1977, 658 p., Moscow, Mir. (Rus)
52. Karatun, A.A., Sukhov, F.F., and Slovokhotova, N.A., Doklady AN SSSR, 1977, v. 234(3), pp. 632–635. (Rus)
53. Higashimura, T. and Kishiro, O., J. Polymer Sci., Polym. Chem. Ed., 1974, v. 12(5), pp. 967–984.
54. Razuvaev, G.A., Sangalov, Yu.A., Yasman, Yu.B., and Minsker, K.S., Doklady AN SSSR, 1978, v. 239(2), pp. 338–341. (Rus)
55. Sangalov, Yu.A., Yasman, Yu.B., Khudaiberdina, Z.I., and Ilkaeva, E.B., Highmolec. Comp., 1978, Ser. A, v. 20(1), pp. 185–190. (Rus)
56. Sangalov, Yu.A., Yasman, Yu.B., Odinokov, V.N., Ilkaeva, E.M., and Minsker, K.S., Neftekhimia, 1977, v. 17(6), pp. 865–872. (Rus)
57. Yasman, Yu.B., Sangalov, Yu.A., and Minsker, K.S., Zh. Obshch. Khim., 1976, v. 46(8), pp. 1665–1667. (Rus)

58. Wichterle, O., Kolinski, M., and Marek, M., Chem. Listy, 1958, v. 52(6), pp. 1049–1057.
59. Volkov, L.A., Chernova, I.D., Radushnova, I.L., Robyisheva, V.T., and Lyishova, V.I., Highmolec. Comp., 1976, Ser. A, v. 18(7), pp. 1486–1493. (Rus)
60. Kennedy, J.P. and Rengachary, S., In: Advances in Polymer Science, 1974, v. 14, pp. 1–14, Berlin, etc.
61. Kennedy, J.P. and Sivaram, S., J. Org. Chem., 1973, v. 38(12), pp. 2262–2264.
62. Erusalimskii, B.L. and Liubetskii, S.G., Ionic Polymerization Processes, 1974, 256 p., Leningrad, Khimia. (Rus)
63. Kennedy, J.P., Huang, S.Y., and Feinberg, S.C., J. Polymer Sci., Poly. Chem. Ed., 1977, v. 15(12), pp. 2801–2819.
64. Bogomolova, T.B. and Gantmakher, A.R., Doklady AN SSSR, 1974, v. 217(2), pp. 369–372; 1976, v. 230(1), pp. 117–120. (Rus)
65. Bogomolova, T.B. and Gantmakher, A.R., Highmolec. Comp., 1978, Ser. A, v. 20(6), pp. 1315–1324. (Rus)
66. Kucera, M., J. Macromol. Sci., 1973, Pt. A, v. 7(8), pp. 1611–1627.
67. Study and Development of Monomer and Synthetic Rubber Production Technology, Collection, 1979, Iss. 3, 216 p., Moscow, CNIITENeftekhim. (Rus)
68. Kennedy, J.P., Cationic Polymerization of Olefins. Critical review (Transl. from English), 1978, 430 p., Moscow, Mir. (Rus)
69. Petrova, V.D., Shcherbakova, N.V., and Minsker, K.S., Izv. AN SSSR, Ser. Khim., 1972, No. 7, pp. 1597–1600. (Rus)
70. Petrova, V.D., Rzhevskaya, N.N., Shcherbakova, N.V., Sangalov, Yu.A., and Minsker, K.S., Doklady AN SSSR, 1977, v.233(4), pp. 602–605. (Rus)
71. Petrova, V.D., Rzhevskaya, N.N., Shcherbakova, N.V., Sangalov, Yu.A., and Minsker, K.S., Izv. AN SSSR, Ser. Khim., 1978, No. 6, pp. 1373–1379. (Rus)
72. Petrova, V.D., Rzhevskaya, N.N., Shcherbakova, N.V., Sangalov, Yu.A., and Minsker, K.S., Zh. Obshch. Khim., 1978, v. 48(8), pp. 1854–1859. (Rus)
73. Tinyakova, E.I., Zhuravleva, T.G., Kuren'gina, T.N., Kirikova, N.S., and Dolgoplosk, B.A., Doklady AN SSSR, 1962, v. 144(3), pp. 592–594. (Rus)
74. Byirikhin, V.S., Ezhova, E.A., Yakovleva, I.M., Konovalenko, N.A., and Pravednikov, A.N., Doklady AN SSSR, 1977, v. 235(5), pp. 1079–1081. (Rus)
75. Araki, T.J., Polym. Sci., Polym. Chem. Ed., 1973, v. 11(4), pp. 699–712.
76. Razuvaev, G.A., Sangalov, Yu.A., Minsker, K.S., and Nelkenbaum, Yu.Ya., Izv. AN SSSR, Ser. Khim., 1975, No. 11, pp. 2547–2554. (Rus)
77. Sangalov, Yu.A., Nelkenbaum, Yu. Ya., Yasman, Yu.B., and Minsker, K.S., In: Thesis of All–Union Conf. on the Mechanism of Catalytic Reactions, 1978, pp. 158–166, Moscow, Nauka. (Rus)
78. Gutmann, V., Chemistry of Coordination Compounds in Non–water Solutions (Transl. from English), 1971, 218 p., Moscow, Mir. (Rus)
79. Medvedev, S.S., Uspekhi Khimii, 1968, v. 37(11), pp. 1923–1945. (Rus)
80. Volkov, L.A. and Medvedev, S.S., Doklady AN SSSR, 1969, v. 186(2), pp. 362–365. (Rus)
81. Sangalov, Yu.A., Petrova, V.D., and Minsker, K.S., In: III International Symp. On Ionic Polymerization, 1977, p. 21, Varna, Bulgaria. (Rus)
82. Ermakova, I.I., Kropacheva, E.N., Kol'tsov, A.I., and Dolgoplosk, B.A., Highmolec. Comp., 1969, Ser. A, v. 11(7), pp. 1639–1644. (Rus)
83. Ponomarev, O.A., Minsker, K.S., Sangalov, Yu.A., and Pshenichnikov, V.M., Highmolec. Comp., 1975, Ser. A, v. 17(2), pp. 309–317. (Rus)

84. Sangalov, Yu.A., Ponomarev, O.A., Romanko, T.V., and Minsker, K.S., Teor. and Eksper. Khim., 1977, v. 13(5), pp. 623–629. (Rus)
85. Sangalov, Yu.A., Babkin, V.A., Yasman, Yu.B., and Nelkenbaum, Yu.Ya., In: Thesis of III All–Union Conf. on Metal–organic Chemistry, Pt. II, 1985, p.277, Ufa. (Rus)
86. Betel, D. and Gold, V., Carbonium Ions (Transl. from English), 1970, 416 p. Moscow, Mir. (Rus)
87. Terlouw, J.K., Weiske, T., Schwartz, H., and Holmes, J.L., Org. Mass Spectrom., 1986, v. 21(10), pp. 665–671.
88. Sanganlov, Yu.A., Ponomarev, O.A., Nelkenbaum, Yu.Ya., Romanko, V.G., and Minsker, K.S., Highmolec. Comp., 1978, Ser. A, v. 20(6), pp. 1331–1338. (Rus)
89. Tsude, O., Tashiro, M., and Torii, A., Koge Kogaku Dzassi., 1967, v. 70(12), pp. 2287–2291; C.A., 1968, v. 68, 104300 t.
90. Sangalov, Yu.A., Nelkenbaum, Yu.Ya., Ponimarev, O.A., and Minsker, K.S., Highmolec. Comp., 1979, Ser. A, v. 21(10), pp. 2267–2274. (Rus)
91. Minsker, K.S., Sangalov, Yu.A., Kirillov, A.P., and Yudaeva, A.I., Chemistry and Technology of Fuels and Oils, 1977, No. 1, pp. 27–31. (Rus)
92. Pape, W., Hartung, H., Heublein, G., Staderman, D., Sangalov, Yu.A., Yasman, Yu.B., and Nelkenbaum, Yu.Ya., Acta Polymerica, 1984, Bd. 35(11), S. 677–680.
93. Sangalov, Yu.A., Yasman, Yu.B., Nelkenbaum, Yu.Ya., Minsker, K.S., Hartung, H., Heublein, G., and Staderman, D., Acta Polymerica, 1985, Bd. 36(11), pp. 618–621.
94. Sangalov, Yu.A., Yasman, Yu.B., Nelkenbaum, Yu.Ya., and Babkin, W.A., Acta Polymerica, 1987, Bd. 38(2), pp. 134–138.
95. Minsker, K.S., Sangalov, Yu.A., Nelkenbaum, Yu.Ya., Petrova, V.D., Rubina, S.A., and Kaluzhskii, A.A., Heftekhimia, 1977, v. 17(5), pp. 696–704. (Rus)
96. Nelkenbaum, Yu.Ya., Rubina, S.A., Sangalov, Yu.A., and Minsker, K.S., In: Synthesis and Transformations of Alkyl–aromatic Compounds, 1979, pp. 72–79, Irkutsk, IGU. (Rus)
97. Minsker, K.S., Sangalov, Yu.A., Churkin, Yu.V., and Nelkenbaum, Yu.Ya., Neftepererabotka i Neftekhimia, 1978, No. 2, pp. 21–22. (Rus)
98. Modern Problems in Chemistry of Carbonium Ions, Ed. V.A. Koptiug, 1975, 411 p. Novosibirsk, Nauka. (Rus)
99. Magnotta, V.L. and Gates, B.C., J. Catal., 1977, v. 46(3), pp. 266–274.
100. Yoke, K.M., J. Polym. Sci., Polym. Lett. Ed., 1975, v. 13(11), pp. 651–655.
101. Kunitake, T. and Takarabe, K., Polymer J., 1978, v. 10(1), pp. 105–110.
102. Gilles, D.J., Tudoret, M.J., and Baird, M.C., J. Amer. Chem. Soc., 1993, v. 115, p. 2543.
103. Shaffer, T.D. and Ashbaugh, J.R., J. Polym. Sci., Pt. A, Polym. Chem., 1977, v. 35, pp. 329–344.
104. Bochmann, M. and Dawson, D.M., Angew, Chem. Int. Ed. Engl., 1996, v. 35, pp. 3226–3228.
105. Carr, A.G., Dawson, D.M., and Bochmann, M., Macromol. Rapid Commun., 1998, v. 19, pp. 205–207.
106. Pomogailo, A.D., Polymeric Immobilized Metal–complex Catalysts, 1988, 303 p., Moscow, Nauka. (Rus)
107. Misker, K.S. and Sangalov, Yu.A., Isobutylene and Its Polymers, 1986, 224 p., Moscow, Khimia. (Rus)
108. Chosh, A. and Kydd, R.A., Catal. Rev. – Sci. Eng., 1985, v. 27(4), pp. 539–589.

109. Okamura, S., Higashimura, T., and Watanabe, T., Makromolek. Chem., 1961, v. 50, pp. 137–148.
110. Lai, J. and McGrath, J.E., J. Polym. Sci., 1964, v. A2, pp. 3369–3381.
111. Pape, W., Acta Polymerica, 1984, Bd. 25(11), S. 677–680.
112. Pat. 3340319 (USA); RZhKhim., 1969, 3S 226P. (Rus)
113. Pat. 1603104 (France); RZhKhim., 1973, 9S 319P. (Rus)
114. Ionites in Chemical Technology, Ed. B.P. Nikolskii and P.G. Romankov, 1982, 416 p., Leningrad, Khimia. (Rus)
115. Hodge, P. And Sherrington, D., Reactions of Polymeric Supports in Organic Synthesis, 1983, 604 p., Moscow, Mir. (Rus)
116. Moore, J.A. and Carraher, C.E., Polym. Sci. and Technol., 1983, v. 21, pp. 1–12.
117. Chaudhuri, B., Chem. Ind. News, 1987, v. 32(8), pp. 601–609.
118. Saus, A. and Schmidl, E., J. Catal., 1985, v. 94(1), pp. 187–194.
119. Yaretska, T., Zeleznaya, L., Meshcheriakov, S., and Yaskula, K., Thes. XIV Mendeleev's Congress, 1989, v. 2, p. 568, Moscow, Mauka. (Rus)
120. Sondheimer, S.J. and Bunce, J.N., J. Macromol. Sci., 1986, v. 26(3), pp. 353–413.
121. Suzuki, S., Tohmori, K., and Ono, Y., J. Molec. Catal., 1987, v. 43, pp. 41–50.
122. Chernyishkova, F.A., Heftekhimia, 1991, v. 31(5), pp. 579–591. (Rus)
123. Bada, T., Ono, Y., Ishimoto, T., Moritaka, S., and Tanooka, S., Bull. Chem. Soc. Jap., 1985, v.58(7), pp. 2155–2156.
124. Kozhevnikov, I.V., Uspekhi Khimii, 1987, v. 56(9), pp. 1417–1443. (Rus)
125. Cai Tain–xi, J. Catal. (China), 1985, v. 6(4), pp. 370–375.
126. Pat. 2442246 (France); RZhKhim., 1981, 16S 334P. (Rus)
127. Stocker, M. and Nilsen, B.P., Acta, Chem., 1984, Bd. 38(1), pp. 55–61.
128. Hugnes, T.R., White, H.M., and White R.J., J. Catal., 1969, v. 13(1), pp. 58–64.
129. Widdecke, H., Brit. Polym. J., 1984, v. 16, pp. 188–192.
130. One Y., Tanabe, T., and Kitajama, N., Chem. Lett. (Japan), 1978, No. 6, pp. 625–628.
131. Onon, Y., Sakuma, S., and Tanabe, T., Chem. Lett. (Japan), 1978, No. 10, pp. 1061–1064.
132. Ono, Y., Tanabe, T., and Kitajama, N., J. Catal., 1979, v. 56(1), pp. 47–51.
133. Kozorezov, Yu.I. and Erofeeva, I.P., Zh. Fiz. Khim., 1975, v. 49, pp. 666–668. (Rus)
134. Pat. 2402478 (France); RZhKhim., 1981, 7N 6P. (Rus)
135. Fadel, A., Yefsah, R., and Salaun, J., React. Polym., 1987, v. 6(2–3), pp. 93–97.
136. Ayame, A., Ohta, K., Izumizawa, T., Sawade, G., Sato, H., and Kakizaki, H., Catalyst (Japan), 1988, v. 30(2), pp. 72–75.
137. Ayame, A. and Izumizawa, T., J. Chem. Soc. Commun., 1989, No. 10, pp. 645–646.
138. Atamanenko, O.P., Nelkenbaum, E.M., Yasman, Yu.B., Sangalov, Yu.A., and Wasserberg, V.E., Neftekhimia, 1987, v. 27(1), pp. 87–92. (Rus)
139. Klein, J. and Widdecke, H., Chem. Jag. Techn., 1982, v. 54(6), pp. 595–596.
140. Ran, R., Rei, W., Jia, X., Shen, J., and Giang, S., Chem. J. Chin.Univ., 1986, v. 7, pp. 645–650.
141. Ran, R., Rei, W., Jia, X., Shen, J., and Giang, S., Polym. Commun. (China), 1986, No. 5, pp. 379–385.
142. Ran, R. and Mao, G., J. Catal., 1989, v. 10(1), pp. 94–97.
143. Ran, R., Pei, W., Jia, X., and Wu, X., Org. Chem. (China), 1987, No. 4, pp. 268–278.

144.Ran, R. and Giang, S., Polym. Commun. (China), 1985, No. 5, pp. 376–379.
145.Sivaram, S. and Naik, K.M., J. Appl. Polym. Sci., 1978, v. 22(11), pp. 3293–3300.
146.Ran, R., Huang, J., and Shen, J., J. Appl. Chem. (China), 1987, v. 4(5), pp. 49–54.
147.Sket, B. and Zupan, M., J. Macromol. Sci., 1983, v. A19(5), pp. 643–652.
148.Neckers, D., Kooistra, D.A., and Green, G.W., J. Amer. Chem. Soc., 1972, v. 94(26), pp. 9284–9285.
149.Blossey, E.C., Turner, L.M., and Neckers, D.C., J. Org. Chem., 1975, v. 40(7), pp. 959–960.
150.Volpin, M.E. and Novikov, Yu.N., Catalysis Mechanism. Part 2. Methods of Studying Catalytic Reactions, Ed. G.A. Boreskov, 1984, p. 100, Novosibirsk, Nauka. (Rus)
151.Laali, K., Muller, and M., Sommer, J., J. Chem. Soc. Chem. Commun., 1980, No. 22, pp. 1088–1089.
152.Latanette, J.M. and Lafontaine, J., J. Chem. Soc. Chem. Commun., 1973, No. 21, pp. 815–817.
153.Ola, G.A., Kaspi, J., and Bukala, J., J. Org. Chem., 1977, v. 42, pp. 4187–4191.
154.Kakulia, Ts.V., Khanashvili, L.M., Tsomaya, N.I., Novikov, Yu.N., and Volpin, M.E., AN GSSR News, 1988, v. 130(3), pp. 557–560. (Rus)
155.Rachkov, I., Gitsov, I., and Panayotov, I., Polym. Bull., 1983, v. 10(11–12), pp. 487–490.
156.Ebert, J., J. Mol. Catal., 1982, v. 15(3), pp. 275–296.
157.Kolesnikov, I.M. and Mirgaleev, I.G., Zh. Prikl. Khim., 1981, v. 54(7), pp. 1751–1753. (Rus)
158.Hattori, H., Takahashi, O., Takagi, M., and Tanabe, K., J. Catal., 1981, v. 68(1), pp. 132–143.
159.Pat. 3862258 (USA); Chem. Abstr., 1975, v. 82, No. 19.124696.
160.Whang, K.J., Lee, K.I., and Lee, Y.K., Bull. Chem. Soc. Jap., 1984, v. 57(8), pp. 2341–2342.
161.Dooley, K.M. and Gates, B.C., J. Polym. Sci., 1984, v. 22(11), pp. 2859–2870.
162.Magnotta, V.L. and Gates, B.C., J. Polym. Sci., Poly. Chem. Ed., 1977, v. 15(6), pp. 1341–1347.
163.A.C. 896880 (USSR); B.I., 1982, No. 26. (Rus)
164.Sangalov, Yu.A., Gladkikh, I.F., and Minsker, K.S., Highmolec. Comp., 1982, v. B24(5), pp. 356–358. (Rus)
165.Sangalov, Yu.A., Yasman, Yu.B., and Minsker, K.S., Highmolec. Comp., 1983, v. B25(5), pp. 317–320. (Rus)
166.Sangalov, Yu.A., Yasman, Yu.B., Gladkikh, I.F., and Minsker, K.S., Doklady AN SSSR, 1982, v. 265(3), pp. 671–674. (Rus)
167.Sangalov, Yu.A., Yasman, Yu.B., and Nelkenbaum, Yu.Ya., Thes. V International Symp. On Coonection Between Homogeneous and Heterogeneous Ctalysis, 1986, p. 217, Novosibirsk. (Rus)
168.Sangalov, Yu.A., Yasman, Yu.B., Nelkenbaum, Yu.Ya., Murinov, Yu.I., and Prokofiev, K.V., Thes. I All–Union Conf. on Problems of Catalyst Deactivation, 1985, p. 88, Ufa. (Rus)
169.Sangalov, Yu.A., Yasman, Yu.B., Gladkikh, I.F., and Minsker, K.S., Doklady AN SSSR, 1981, 261(2), pp. 428–432. (Rus)
170.Sangalov, Yu.A., Gladkikh, I.F., Komarov, N.V., and Lisovin, E.G., Izv. Vyssh. Uch. Zav., 1984, v. 27(6), pp. 707–710, Khimia i Khim. Tekhnol. (Rus)

171.Sangalov, Yu.A. and Nelkenbaum, Yu.Ya., Collection of Synthesis and Processing of Polymers, 1984, Iss. 7, pp. 23–26, Gorky. (Rus)
172.Sangalov, Yu.A., Gladkikh, I.F., Komarov, N.V., and Lisovin, E.G., Int. Polym. Sci. and Technol., 1984, No. 10, pp. 41–43.
173.Pomogailo, A.D., Tosrosyan, A.A., Raspopov, L.N., Diachkovskii, F.S., and Enikolopov, N.S., Doklady AN SSSR, 1980, v. 252(5), pp. 1148–1151. (Rus)
174.Heunlein, G., Hartung, H., Helbig, M., and Staderman, D., J. Macromol. Sci. Chem., 1982, v. A17(5), pp. 821–846.
175.Kobayashi, D., Morioka, J., Sandzaki, X., and Masui, M., Shokubai Catalyst, 1980, v. 22(4), pp. 235–238.
176.O'Connor, C.T., Kojima, M., and Schumann, W.K., Appl. Catal., 1985, v. 16(2), pp. 197–207.
177.Ernsberger, F.M., J. Non–Cryst. Solids, 1980, Pt. 2, v. 39, pp. 557–561.
178.Sangalov, Yu.A., Yasman, Yu.B., Nelkenbaum, Yu.Ya., and Babkin, V.A., Izv. AN SSSR, 1986, No. 12, pp. 2661–2666. (Rus)
179.Vucelic, D. and Vucelic, V., Thes. IV International Symp. On Heterogeneous Catalysis, 1979, Pt. 2, pp. 443–448, Varna. (Rus)
180.Ben–Dor, L. and Margalitt, R., Inorg. Chem. Acta, 1967, v. 1(1), pp. 149–153.
181.Grabowski, W., Misono, M., and Yoneda, Y., J. Catal., 1980, v. 61, pp. 103–108.
182.Thornton, R. and Gates, B.C., J. Catal., 1982, v. 34(2), pp. 275–287.
183.Widdecke, H., Klin, J., and Haupt, U., Macromol. Chem. Macromol. Symp., v. 4, pp. 145–155.
184.Gates, B.C. and Johanson, L.N., J. Catal., 1969, v. 14(1), pp. 69–76.
185.Dooley, K.M., Williams, J.A., Gates, B.C., and Albright, R.L., J. Catal., 1982, v. 74(2), pp. 361–372.
186.Diemer, R.B., Dooley, K.M., Gates, B.C., and Albright, R.L., J. Catal., 1982, v. 74(2), pp. 373–381.
187.Prokop, Z. and Setinek, K., Collect. Czech, Chem. Commun., 1982, v. 47(6), pp. 1613–1620.
188.Pat. 57–140729 (Japan); RZhKhim, 7N 3P. (Rus)
189.Widdecke, H. and Klein, J., Chem. Ing. Tech., 1981, v. 53(4), pp. 954–957.
190.Potapov, G.P., Politova, N.K., Malakhova, I.A., and Borodkina, M.A., Highmolec. Comp., 1987, v. A29(8), pp. 1689–1693. (Rus)
191.Potapov, G.P., React. Kinet. Catal. Lett., 1983, v. 23(3–4), pp. 307–312.
192.Yasman, Yu.B., Khudaiberdina, Z.M., Sangalov, Yu.A., and Minsker, K.S., Highmolec. Comp., 1979, v. B21(8), pp. 567–571. (Rus)
193.Encyclopedia of Polymers, Sov. Encyclopedia, 1972, pp. 974–991, Moscow, Sov. Encyclopedia. (Rus)
194.Chmelir, M., J. Polymer Sci., Polymer Symp., 1976, No. 56, pp. 311–321.
195.Sangalov, Yu.A., Yasman, Yu.B., Nelkenbaum, Yu.Ya., and Babkin, V.A., Izv. AN SSSR, Ser. Khim., 1986, No. 12, pp. 2661–2666. (Rus)
196.Sangalov, Yu.A., Babkin, V.A., and Nilkenbaum, Yu.Ya., Quantum–Chemical Calculation of Acidic Catalysts. Acidity and Catalytic Properties of Metal Compound Complexes with Proton–Donors, Report preprint, 1989, 46 p., BNC UrO AN SSSR, Ufa. (Rus)
197.Kennedy, J.P. and Squires, R.G., J. Macromolec. Chem., 1967, Pt. A, v.1(6).
198.Byirikhin, V.S., Nesmelov, A.I., Murachev, V.B., Belousov, S.I., Ezhova, E.A., and Pravednikov, A.N., Doklady AN SSSR, 1984, v. 274(3), pp. 626–629. (Rus)
199.Byirikhin, V.S., Nesmelov, A.I., and Ezhova, E.A., Highmolec. Comp., 1988, v. A30(10), pp. 2046–2051. (Rus).

200.Nesmelov, A.I., Orlinkov, A.V., Akhrem, I.S., Murachev, V.B., Byirikhin, V.S., and Volpin, M.E., Izv. AN SSSR, Ser. Khim., 1990, No. 7, pp. 1496–1499. (Rus).
201.Murachev, V.B., Nesmelov, A.I., Byirikhin, V.S., Ezhova, E.A., Orlinkov, A.V., Akhrem, I.S., and Volpin, M.E., Izv. AN SSSR, Ser. Khim., 1996, No. 5, pp. 1175–1179. (Rus)
202.Nesmelov, A.I., Murachev, V.B., Ezhova, E.A., Tregubenkov, S.I., Byirikhin, V.S., Orlinkov, A.V., and Akhrem, I.S., Izv. AN SSSR, Ser. Khim., 1996, No. 5, pp. 1180–1183. (Rus)
203.Minsker, K.S., Sangalov, Yu.A., Ponomarev, O.A., Babkin, V.A., and Lomakin, G.S., Teor. Eksper. Khim., 1980, v. 16(2), pp. 181–188. (Rus)
204.Sangalov, Yu.A., Minsker, K.S., Babkin, V.A., and Ponomarev, O.A., Teor. Eksper. Khim., 1982, v. 18(5), pp. 535–542. (Rus)
205.Schastnev, P.V. and Bogdanchikova, G.A., Teor. Eksper. Khim., 1975, v. 11(5), pp. 667–671. (Rus)
206.Novaro, O., J. Chem. Phys., 1978, v. 68(5), pp. 2337–2351.
207.Grabowski, W., Misono, M., and Yoneda, Y., J. Catal., 1980, v. 61(1), pp. 103–108.
208.Kazanskii, V.B., Kinetica i Katalis, 1977, v. 18(5), pp. 1179–1191. (Rus)
209.Shchokova, E.A., Knopova, S.I., and Kazanskii, B.A., Doklady AN SSSR, 1973, v. 212(2), pp. 386–388. (Rus)
210.Nesmelov, A.N., Murachev, V.B., Ezhova, E.A., and Byirikhin, V.S., Highmolec. Comp., 1988, Ser. A, v. 30(9), pp. 1957–1961. (Rus)
211.Sangalov, Yu.A., Babkin, V.A., Nelkenbaum, Yu.Ya., and Minsker, K.S., Teor. Eksper. Khim., 1983, v. 19(5), pp. 615–620. (Rus)
212.Sangalov, Yu.A., Yasman, Yu.B., Valeev, F.A., and Minsker, K.S., Highmolec. Comp., 1978, Ser. A, v. 20(6), pp. 1339–1344. (Rus)
213.Pikaev, F.K., Modern Radiation Chemistry. The Solid and Polymers. Applied Aspects, 1987, 446 p., Moscow, Nauka. (Rus)
214.Kinitake, T. and Takarabe, K., Macromolecules, 1979, v. 12(6), pp. 1067–1071.
215.Lorimer, J.P. and Pepper, D.C., Proc. Rev. Soc., 1976, Ser. A, v. 351(1667), pp. 551–568.
216.Cotrel, R., Macromolecules, 1976, v. 9(5), pp. 931–936.
217.Heublein, G., Zum Ablauf ionischer polymerizations–reaktionen, 1975, 316 S., Berlin, Akademie–Verlag.
218.Minsker, K.S., Sangalov, Yu.A., and Babkin, V.A., Teor. Eksper. Khim., 1984, v. 20(6), pp. 669–675. (Rus)
219.Kennedy, J.P., Shinkawa, A., and Williams, F., J. Polymer Sci., 1971, Pt. A–1,v. 9(6), pp. 1551–1561.
220.Hallpap, P. and Heublein, G., Z. Chem., 1976, Bd. 16(1), S. 21–22.
221.Hoiblan, G., In: IV International Macrosymp. Advances in Ionic Polymerization, 1979, p. 4, Ufa. (Rus)
222.Di Maina, M., Makromol. Chem., 1977, Bd. 178(8), S. 2223–2234.
223.Kennedy, J.P. and Sivaram, S., J. Macromol. Sci., 1973, Pt. A, v. 7(4), pp. 969–989.
224.Encyclopedia of Polymers, 1977, v. 3, pp. 260–271, Moscow, Sov. Encyclopedia. (Rus)
225.Bernasconi, C.F. and Carre, D.J., J. Amer. Chem. Soc., 1979, v. 101, pp. 2707–2709.

226.Jasinski, J.M. and Brauman, J.L., J. Amer. Chem. Soc., 1980, v. 102, pp. 2906–2913.
227.Swain, C.G., Scott, C.B., and Lohmann, K.H., J. Amer. Chem. Soc., 1953, v. 75(1), pp. 136–140.
228.Wichterle, O., Marek, K., and Trekoval, J., J. Polymer Sci., 1961, v. 53(158), pp. 281–287.
229.Imanishi, Y., Yamamoto, R., Higashimura, T., Kennedy, J.P., and Okamura, S., J. Macromol. Sci., 1967, Pt. A, v.1(5), pp. 877–890.
230.Akulov, G., Geller, N., Kropatchev, V., Nefedov, V., Sonotova, E., Skorokhodov, S., Stepanov, V., and Toropova, M., Makromol. Chem., 1978, Bd. 179(11), S. 2775–2777.
231.Akulov, G.P., Highmolec. Comp., 1979, Ser. B, v.21(4), p. 243. (Rus)
232.Day, L. and Selbin, V., Theoretic Inorganic Chemistry, 1976, 568 p., Moscow, Khimia. (Rus)
233.Biddulph, R.H., Plesch, P.H., and Rutherford, P.P., In: Intern. Symp. On Macromol., 1959, Paper III, A10, Weisbaden.
234.Flett, M.St.C. and Plesch, P.H., J. Chem. Soc., 1952, pp. 3355–3358.
235.Bauer, R.F., Laflair, R.T., and Russel, K.E., Canad. J. Chem., 1970, v. 48(8), pp. 1251–1262.
236.Bauer, R.F. and Russel, K.E., J. Polym. Sci., 1971, Pt. A–1, v. 9(6), pp. 1451–1458.
237.Kennedy, J.P., Huang, S.Y., and Feinberg, S.C., J. Polym. Sci., Polym. Chem. Ed., 1977, v. 15(12), pp. 2869–2892.
238.Heublein, G., Spange, S., and Hallpap, P., Makromol. Chem., 1979, Bd. 180(8), S. 1935–1949.
239.Sangalov, Yu.A., Nelkenbaum, Yu.Ya., Ishmuratova, N.M., and Minsker, K.S., Highmolec. Comp., 1979, Ser. B, v. 21(11), pp. 875–878. (Rus)
240.Yasman, Yu.B., Khudaiberdina, Z.I., Sangalov, Yu.A., and Minsker, K.S., Highmolec. Comp., 1979, Ser. B, v. 21(8), pp. 567–571. (Rus)
241.Sangalov, Yu.A., Yasman, Yu.B., Neleknbaum, E.M., Gataullin, R.F., and Kozlov, V.G., Highmolec. Comp., 1983, Ser. A, v. 25(2), pp. 227–232. (Rus)
242.Sangalov, Yu.A., Yasman, Yu.B., Nelkenbaum, E.M., Bedretdinova, V.Ya., and Minsker, K.S., Highmolec. Comp., 1980, Ser. A, v. 22(7), pp. 1588–1591. (Rus)
243.Overberger, C.G. and Newton, M.G., J. Amer. Chem. Soc., 1960, v. 82(14), pp. 3622–3626.
244.Russel, K.E. and Vail, L.G., J. Polymer Sci., Polymer Symp., 1976, No. 56, pp. 183–189.
245.Russel, K.E., Vail L.G., and Woolston, M.E., Europ. Polymer J., 1979, v. 15(11), pp. 969–974.
246.Rooney, J.M., J. Appl. Polymer Sci., 1980, v. 25(7), pp. 1365–1372.
247.Kennedy, J.P., J. Macromolec. Sci., Chemistry, 1982, v. A18(1), pp. 1–152.
248.Kennedy, J.P., J. Macromolec. Sci., Chemistry, 1982–83, v. A18(9), pp. 1185–1417.
249.Khalafov, F.R., Novrusova, F.A., Ismailov, E.G., and Krentsel, B.A., Acta Polymerica, 1990, Bd. 41(4), pp. 226–229.
250.Khalafov, F.R., Novrusova, F.A., Ismailov, E.G., and Krentsel, B.A., Highmolec. Comp., 1990, v. 32(3), pp. 165–167. (Rus)
251.Khalafov, F.R., Novrusova, F.A., Faradzhev, G.M., Abramov, F.R., and Krentsel, B.A., Highmolec. Comp., 1991, v. A33(12), pp. 2668–2673. (Rus)

252.Khalafov, F.R., Novrusova, F.A., Krentsel, B.A., Jafarov, O.I, Mekhtiev, S.A., and Kuliev, A.M., Highmolec. Comp., 1990, v. 32(3), pp. 194–196. (Rus)
253.Khalafov, F.R., Novrusova, F.A., Krentsel, B.A., Agaeva, E.A., Jafarov, O.I., Mekhtiev, S.A., and Kuliev, A.M., Acta Polymerica, 1992, Bd. 43(5), pp. 102–106.
254.Prokofiev, K.V., Verbitskii, B.G., Rogov, S.A., and Kirichenko, P.N., Low–molecular Polybutenes. Thematic Review, 1982, p. 51, Moscow, CNIITENeftekhim. (Rus)
255.Minsker, K.S., Berlin, Al.Al., Svinukhov, A.G., Prochukhan, Yu.A., and Enikolopian, N.S., Doklady AN SSSR, 1986, No. 5, pp. 1171–1173. (Rus)
256.Berlin, Al.Al., Minsker, K.S., Prochukhan, Yu.A., Karpasas, M.M., and Enikolopian, N.S., Highmolec. Comp., 1986, v. B28(6), pp. 461–465. (Rus)
257.Berlin, Al.Al., Minsker, K.S., Prochukhan, Yu.A., Karpasas, M.M., and Enikolopian, N.S., Internat. Polym. Sci. Technol., 1986, v. 13(11), pp. 95–97.
258.Prochukhan, Yu.A., Berlin Al.Al., Minsker, K.S., and Enikolopian, N.S., Doklady AN SSSR, 1986, v. 287(3), pp. 682–685. (Rus)
259.Minsker, K.S., Berlin, Al.Al., Prochukhan, Yu.A., and Enikolopian, N.S., Internat. Polym. Sci. Technol., 1986, v. 13(12), pp. 90–93. (Rus)
260.Berlin, Al.Al., Minsker, K.S., Sangalov, Yu.A., Svinukhov, A.G., Kirillov, A.P., and Enikolopian, N.S., Highmolec. Comp., 1980, v. A22(3), pp. 566–574. (Rus)
261.Berlin, Al.Al., Minsker, K.S., Prochukhan, Yu.A., Kaikolopianrpasas, M.M., and Enikolopian, N.S., Doklady AN SSSR, 1986, v. 287(1), pp. 145–148. (Rus)
262.Minsker, K.S., Berlin, Al.Al., Prochukhan, Yu.A., and Enikolopian, N.S., Highmolec. Comp., 1986, v. B28(6), pp. 466–469. (Rus)
263.Minsker, K.S. and Berlin, Al.Al., Fast Polymerization Processes, 1996, 146 p., Gordon & Breach Publ. Inc.
264.Minsker, K.S., Babkin, V.A., and Zaikov, G.E., Polym. Plast. Technol. Eng., 1990, v. 29(1–2), pp. 43–72.
265.Berlin, Al.Al., Minsker, K.S., Prochukhan, Yu.A., and Enikolopian, N.S., Highmolec. Comp., 1989, v. A31(9), pp. 1779–1798. (Rus)
266.Berlin, Al.Al., Minsker, K.S., Pochukhan, Yu.A., and Enikolopian, N.S., Polym. Plast. Technol. Eng., 1991, v. 30(2–3), pp. 253–297.
267.Kennedy, J.P., J. Polymer Sci., Polymer. Symp., 1996, No. 56, pp. 1–11.
268.Kennedy, J.P. and Thomas, R.M., Adv. Chem. Ser., 1962, v. 34(1), pp. 111–119.
269.Kennedy, J.P. and Squires, R.G., J. Macromol. Chem., 1967, Pt. A, v. 1(6), pp. 805–845.
270.Kennedy, J.P., Bank, S., and Squires, R.G., J. Macromol. Sci., 1967, Pt. A, v. 1(6), pp. 977–993.
271.Taylor, R.B. and Williams, F., J. Amer. Chem. Soc., 1969, v. 91(14), pp. 3728–3732.
272.Dorfman, L.M. and de Palma, V.M., In: Proc. 30–th Intern. Meet., Thiais, 1977; Amsterdam c.a., 1978, pp. 215–226, Discuss. 227.
273.Reitlinger, S.A., Permeability of Polymeric Materials, 1974, 269 p., Moscow, Khimia. (Rus)
274.Berlin, Al.Al., Minsker, K.S., Sangalov, Yu.A., Novikov, D.D., and Pozniak, T.I., Hoghmolec. Comp., 1979, Ser. B, v. 21(6), pp. 468–471. (Rus)
275.Pozniak, T.I., Lisitsin, D.M., Novikov, D.D., Berlin, Al.Al., Diachkovskaya, F.S., Prochukhan, Yu.A., Sangalov, Yu.A., and Minsker, K.S., Highmolec. Comp., 1980, Ser. A, v. 22(6), pp. 1424–1427. (Rus)

276.Berlin, Al.Al. and Volfson, S.A., Kinetic Method in Polymer Synthesis, 1973, 341 p., Moscow, Khimia. (Rus)

Chapter 3. Macrokinetics of isobutylene polymerization

3.1. Modeling of isobutylene polymerization as a fast chemical reaction.

Experimental data of fast cationic polymerization of isobutylene have provided evidence for calculation and mathematical modeling of the process [1]. The reaction zone of the equipment was chosen from the laboratory model by the scale transfer method, e.g. the same ratios of geometrical sizes of the device, rates of reagents' feeding were taken, as well as the principle of catalyst feeding [2]. High rates of flow in the reaction zone (1–10 m/s) provided turbulent mixing of the catalyst solution ($c_0 = 10^{-4}$–10^{-1} mol/l) and mixture of the monomer (M_0 = 0.01–1 mol/l), polymer and solvent. The criterion Re, calculated for the present linear flow rate, its density (0.5–1 g/cm^3), dynamic viscosity coefficient [(5–10)$\cdot 10^{-3}$ g/(cm·s)] and the pipe diameter (10 cm), equaled 10^4. That is why the coefficient of turbulent diffusion, equal to the coefficient of temperature conductivity, can be used as coefficients of mass and heat transmission: $D_t = D_T = \lambda \cdot c/\rho$ (where λ, c, and ρ are average heat conductivity, heat capacity, and density of the reaction medium, respectively).

In keeping with the existing experimental data, the following kinetic model was chosen:

$$\text{Catalyst} + \text{M} \xrightarrow{k_i} \text{A}^* \qquad \text{A}^* + \text{M} \xrightarrow{k_p} \text{A}_p^*$$

$$\text{A}_p^* \xrightarrow{k_{ter}} \text{P}_p \text{A}_p^* + \text{M} \xrightarrow{k_M} \text{A}^* + \text{P}_p,$$

where M is the monomer; A* and A_p^* are active monomeric and polymeric centers, respectively; P_p is the terminated chain; k_i, k_p, k_{ter}, and k_M are rate constants of initiation, propagation, termination and chain transmission to the monomer.

It is suggested that the initiation rate is rather high ($k_i >> k_p$), although initial equations and their solutions will not change, if relatively slow initiation and stability of AC are expected. The main reaction determining the value of the molecular mass and MMD in a wide range of temperatures, and especially at high temperatures, is the chain transmission to the monomer. That is why the polymerization degree $\bar{\nu}_p$ and MMD in each specific element of the reaction volume $\rho_p(j)$ are determined by temperature only: $\bar{\nu}_p = k_p / k_M$ and $\rho_p(j) = 1/\nu_p e^{-j/\nu_p}$. Changes in the molecular mass and MMD of the product reflect temperature field of the reaction zone.

Based on the above-considered and in accordance with experimental data [1–8], the kinetic equations describing changes in concentrations of the monomer, and active centers, as well as variations in temperature during fast isobutylene polymerization in a flow which has axial symmetry, are the following:

$$\frac{\partial[\text{M}](x,r)}{\partial \tau} = D_t \nabla[\text{M}](x,r) - k_p^0 \exp\left(-\frac{E_p}{RT}\right) \cdot [\text{M}](x,r)[\text{A}^*](x,r) - V\text{grad}[\text{M}](x,r)$$

$$\frac{\partial[\mathrm{A}^*](x,r)}{\partial\tau}=D_{\mathrm{t}}\nabla[\mathrm{A}^*](x,r)-k_r^0\exp\left(-\frac{E_r}{RT}\right)\cdot[A^*](x,r)-V\mathrm{grad}[\mathrm{A}^*](x,r) \tag{3.1}$$

$$\frac{\partial[T](x,r)}{\partial\tau}=\chi_{\mathrm{t}}\nabla T(x,r)+\frac{q_{11}}{\rho\cdot c}k_p^0\exp\left(-\frac{E_{\mathrm{p}}}{RT}\right)\cdot[\mathrm{M}](x,r)[\mathrm{A}^*](x,r)-V\mathrm{grad}T(x,r)$$

under limiting conditions:

$$T(0,r)=T_0,$$

$$\mathrm{M}(0,r)=\begin{cases}\mathrm{M}_0,\ \text{if } R>r>r_0, r_0<r<R\\ 0,\ \text{if } r<r_0\end{cases}$$

$$\mathrm{A}^*(0,r)=\begin{cases}\mathrm{M}_0,\ \text{if } R>r>r_0, r_0<r<R\\ \mathrm{A}_0,\ \text{if } r<r_0\end{cases}$$

$$\partial\frac{\partial\mathrm{M}(x,r)}{\partial r}=\frac{\partial\mathrm{A}^*(x,R)}{\partial r}=\frac{\partial\mathrm{A}^*(x,0)}{\partial r}=\frac{\partial\mathrm{M}(x,0)}{\partial r}=\frac{\partial T(x,0)}{\partial r}=0,$$

$$\frac{\partial T(x,R)}{\partial r}=\alpha\{T(x,R)-T_1,$$

where x and r are length and radial coordinates of the reaction zone; T is temperature; T_0 and T_1 are temperatures of introduced reagents and external thermostatic control; α is the coefficient of heat transmission though the reactor wall; q is the heat effect of polymerization in the liquid phase, equal to 72 kJ/mol; $k_{\mathrm{p}}=k_{\mathrm{p}}^0\exp(-E_{\mathrm{p}}/RT)$, $k_{\mathrm{ter}}=k_{\mathrm{ter}}^0\exp(-E_{\mathrm{ter}}/RT)$, $k_{\mathrm{M}}=k_{\mathrm{M}}^0\exp(-E_{\mathrm{M}}/RT)$ are rate constants of chain propagation, termination and transmission to the monomer, respectively; k_{p}^0, k_{ter}^0, k_{M}^0 and E_{p}, E_{ter}, E_{M} are pre-exponents and activation energies of these reactions.

Taking into account the stationary character of the process, i.e.

$$\frac{\partial\mathrm{M}(x,r,t)}{\partial t}=\frac{\partial\mathrm{A}^*(x,r,t)}{\partial t}=\frac{\partial T(x,r,t)}{\partial t}=0,$$

and neglecting the diffusion transfer along the x axis compared with the linear one at the rate V, solution of quasi-linear parabolic equations is simplified by neglecting the summands

$$D_{\mathrm{t}}\cdot\frac{\partial^2\mathrm{M}}{\partial x^2},\ D_{\mathrm{t}}\cdot\frac{\partial^2\mathrm{A}^*}{\partial x^2},\ \lambda_T\cdot\frac{\partial^2 T}{\partial x^2}.$$

Consequently, calculation of the mathematical model is reduced to solution of the limit situation for the system of Equations (3.1). The numerical solution was performed with the help of a somewhat equivocal differential scheme in the stationary approximation. It was suggested that D_{t} and X_{t} are constant in the reactor space, and the plane front of the flow is realized according to the ideal displacement (V = Const). Fluctuations of concentrations and temperatures were also neglected. It should be noted that usage of the assumption about D_{t} being constant along the flow length is generally wrong. However, as calculations by the more accurate Navier–Stokes model have shown, the results obtained for the diffusion model quite accurately qualitatively describe all the main macrokinetic regularities of fast polymerization processes. Approximation of differential equations by difference ones is of the order $0(\tau + h)^2$, where τ and h are increments in x and r, respectively. Exclusion conditions are set accurately, and the limit ones are approximated with the accuracy of $0(h)$. The mass distribution function $\rho_w(j)$ by polymerization degrees at the exit from the reactor of l length and moments ($I_0 - I_3$) were calculated by the formulae:

$$\rho_w(j) = \int_0^l \int_0^R \frac{j}{\bar{v}_n^2}\left(-\frac{j}{\bar{v}_n}\right) k_p \cdot A^* \cdot M \cdot \exp\left(-\frac{E_p}{RT}\right) \cdot 2\pi r dx dr, \tag{3.2}$$

where $\bar{v}_n = \dfrac{k_p}{k_M} = \dfrac{k_p^0}{k_M^0} \exp\left[-\dfrac{E_p - E_M}{RT}\right]$;

$$\begin{aligned} I_0 &= \int_0^\infty \frac{\rho_w(j)}{j} dj = \int_0^R \int_0^l k_i^0 \cdot A^* \cdot M \cdot \exp\left(-\frac{E_M}{RT}\right) \cdot 2\pi r dx dr = \\ &= \int_0^R \int_0^l \frac{1}{\bar{v}_n} k_p^0 \cdot A^* \cdot M \cdot \exp\left(-\frac{E_p}{RT}\right) \cdot 2\pi r dx dr, \end{aligned} \tag{3.3}$$

$$I_1 = \int_0^\infty \rho_w(j) dj = \int_0^R \int_0^l k_p^0 \cdot A^* \cdot M \cdot \exp\left(-\frac{E_p}{RT}\right) \cdot 2\pi r dx dr, \tag{3.4}$$

$$\begin{aligned} I_2 &= \int_0^\infty j \rho_w(j) dj = \int_0^R \int_0^l 2\bar{v}_n k_p^0 \cdot A^* \cdot M \cdot \exp\left(-\frac{E_p}{RT}\right) \cdot 2\pi r dx dr = \\ &= \int_0^R \int_0^l \frac{\left(k_p^0\right)^2}{k_M^0} \cdot A^* \cdot M \cdot \exp\left(-\frac{2E_p - E_M}{RT}\right) \cdot 4\pi r dx dr, \end{aligned} \tag{3.5}$$

$$\begin{aligned} I_2 &= \int_0^\infty j^2 \rho_w(j) dj = \int_0^R \int_0^l 6\bar{v}_n^2 k_p^0 \cdot A^* \cdot M \cdot \exp\left(-\frac{E_p}{RT}\right) \cdot 2\pi r dx dr = \\ &= \int_0^R \int_0^l 6 \frac{\left(k_p^0\right)^3}{\left(k_M^0\right)^2} \cdot A^* \cdot M \cdot \exp\left(-\frac{3E_p - E_M}{RT}\right) \cdot 2\pi r dx dr, \end{aligned} \tag{3.6}$$

where $\bar{v}_n = \dfrac{I_1}{I_0}$; $\bar{v}_w = \dfrac{I_2}{I_1}$; $\bar{v}_z = \dfrac{I_3}{I_2}$ are average numerical, average mass and z-average degrees of the product polymerization.

Integrals $\int_0^R\int_0^l f(x,r)dxdr$ with respect to (x, r) are calculated by the existing network.

Numerical values of E_p and E_0 for fast processes of isobutylene cationic polymerization are low [3, 4, 9]. That is why calculations were made, which neglect temperature dependencies of k_p, k_0 and D_t. The system of equations (3.1), if the limit diffusion is neglected, is reduced to dimensionless coordinates by substitutions:

$$\rho = \sqrt{\frac{k_p M_0}{D}} \cdot r, \quad x = \frac{k_p M_0}{V}, \quad \theta = \frac{\rho \cdot c}{qM} \cdot T,$$

$$\bar{\alpha} = \alpha \sqrt{\frac{D}{k_p M_0}}, \quad \frac{M}{M_0} = m, \quad \frac{A^*}{A} = a, \quad \frac{A_0}{M_0} = c_0, \quad \frac{k_{ter}}{k_p M_0} = \chi.$$

Then we obtain that

$$\frac{\partial^2 m}{\partial \rho^2} + \frac{1}{\rho} \cdot \frac{\partial m}{\partial \rho} - \frac{\partial m}{\partial x} - mac_0 = 0, \tag{3.7}$$

$$\frac{\partial^2 a}{\partial \rho^2} + \frac{1}{\rho} \cdot \frac{\partial a}{\partial \rho} - \chi a = 0, \tag{3.8}$$

$$\frac{\partial^2 \theta}{\partial \rho^2} + \frac{1}{\rho} \cdot \frac{\partial \theta}{\partial \rho} - \frac{\partial \theta}{\partial x} - mac_0 = 0 \tag{3.9}$$

with appropriate border conditions. In this case, growth of the rate V means just a change of the scale along the axis x; a change of coefficient D is a change of the scale by r and x and corresponding change of the coefficient α. Changes of k_p and M_0 are also equivalent to some distortion of scales by both axes. Moreover, it is the ratio of the kinetic constants which is important rather than their absolute values.

Transformations performed allow a generalization of calculation results for a wider range of parameter values.

Figure 3.1 shows temperature fields[1], monomer and catalyst concentrations, and Figure 3.2 — fields of polymerization rates for various values of D, k_p and α. As in [2], we notice that the process generally proceeds at the catalyst feeding the reactor at its mixing with the monomer solution. Temperature and rate in the reaction zone are variables depending on the reagent concentrations, and coefficients D and α (coefficient of heat transmission through walls). Although the maximum of the polymerization rate is observed near the catalyst feeding zone, the reaction continues also along the x axis, which leads to a change of yield and properties of polymer as the maximum moves from the zone. Formation of a polymer in different points of the reaction volume (under different temperatures, respectively) leads to widening of the

[1] The field term implies changing of a parameter in conditional units by length and width of the reaction zone.

polymer MMD compared with the most probable MMD $\left[\rho_n(j) = 1/\overline{P}_n \cdot \exp\left(-j/\overline{P}_n\right)\right]$, typical for isothermal conditions. As mentioned above, as the average molecular mass and MMD of the polymeric product is determined by the chain transmission to the monomer, the main factor affecting these characteristics is the difference of activation energies of the chain transmission and propagation reactions ($E_M - E_p$). Figure 3.3 shows dependencies of $\overline{v}_w/\overline{v}_n$ and $\overline{v}_z/\overline{v}_w$ on $E_M - E_p$. Figure 3.4 displays the view of the MMD function and its anamorphosis in semi-logarithmic coordinates $\lg\rho_n(j)$ on j (curves 5 and 6). As seen from the Figure, a noticeable deviation of MMD from the most probable value (straight line in Figure 3.5) demands a difference in values of E_M and E_p at least by several units, that does indeed occur.

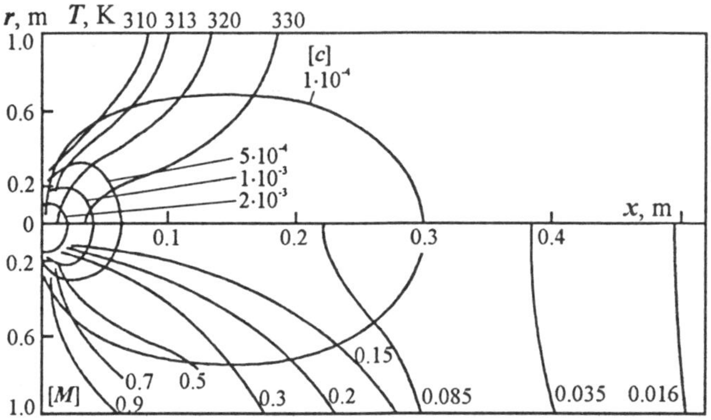

Figure 3.1. Fields of temperatures (T), concentrations of monomer (M) and catalyst (c): $k_p = 10^5$ l/(mol·s), $k_{ter} = 1$ s^{-1}, $M_0 = 1$ mol/l, $c_0 = 10^{-2}$ mol/l, $D = 1$ m^2/s, $\alpha = 0$; numbers on curves indicate values of corresponding parameters.

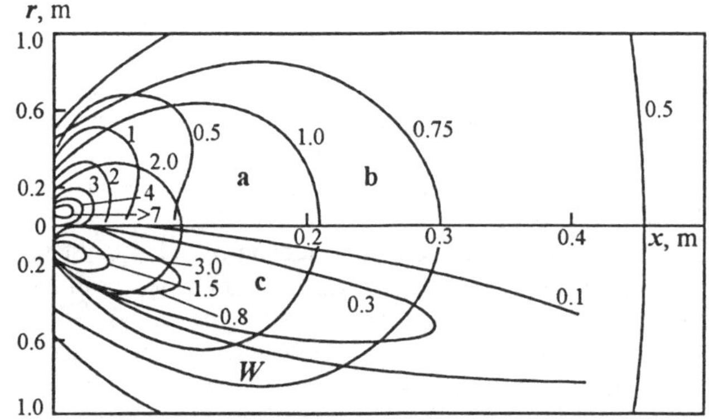

Figure 3.2. Fields of isobutylene polymerization rates for various values of indexes: a – $k_p = 10^5$ l/(mol·s), $(E_M - E_p)R = 5000$ K, $D = 1$ m^2/s, $\alpha = 0$; b – $k_p = 10^4$ l/(mol·s), $(E_M - E_p)R = 5000$ K, $D = 1$ m^2/s, $\alpha = 0$; c – $k_p = 10^4$ l/(mol·s), $(E_M - E_p)R = 2000$ K, $D = 0.1$ m^2/s, $\alpha = 10$, $k_{ter} = 1$ mol/l, $[M]_0 = 1$ mol/l, $c_0 = 10^{-2}$ mol/l. Numbers at curves mark values of rate parameters in conditional units.

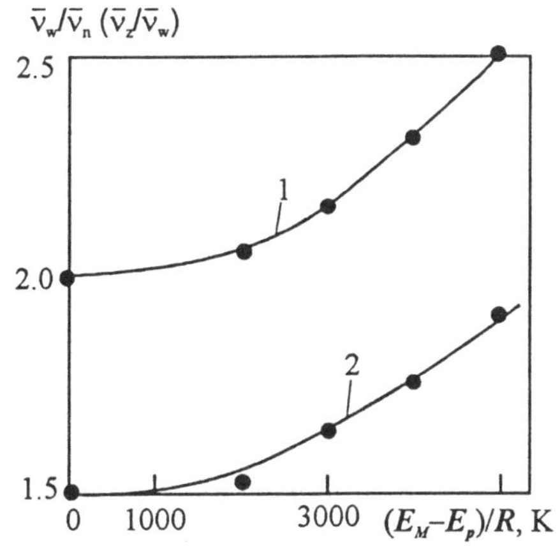

Figure 3.3. Dependence of $\bar{\nu}_w/\bar{\nu}_n$ (1) and $\bar{\nu}_z/\bar{\nu}_w$ (2) on $(E_M - E_p)R$ for $k_p = 10^5$ l/(mol·s), $k_{ter} = 1$ s^{-1}, $[M]_0 = 1$ mol/l, $c_0 = 10^{-2}$ mol/l, $D = 1$ m^2/s, $\alpha = 0$, l = 1 m.

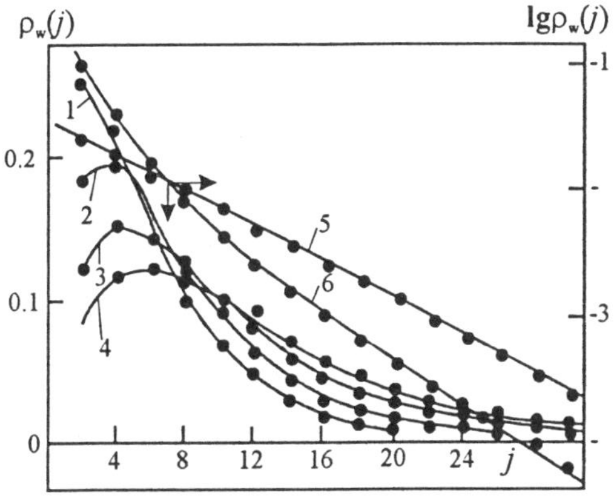

Figure 3.4. Functions of polyisobutylene MMD for various $(E_M - E_p)R$ (1 – 5000, 2 – 4000, 3 – 3000, 4 – 2000) and their semi-logarithmic anamorphoses (5 – 2000, 6 – 5000); calculated parameters are the same as in Figure 2.18.

Growth of the coefficient α under constancy of other parameters of the processes causes some narrowing of MMD, which corresponds to smoothing of the temperature field. However, a noticeable effect of the heat transmission may appear only in the case of small size of the reaction volume and growth of coefficients D and λ, respectively. Temperature in the reaction zone and its gradient are determined by the process rate and quantity of isolated heat and, consequently, by concentrations of monomer and catalyst. Influence of the monomer concentration on functions and parameters of MMD in $\lg\rho_n(j)$—j coordinates is displayed in Figures 3.5 and 3.6. Hence it follows that MMD is significantly widened with the monomer concentration, generally, at the sacrifice of occurrence of large amount of a low-molecular fraction. Analogous regularities are observed when the catalyst content is varied (Table 3.1).

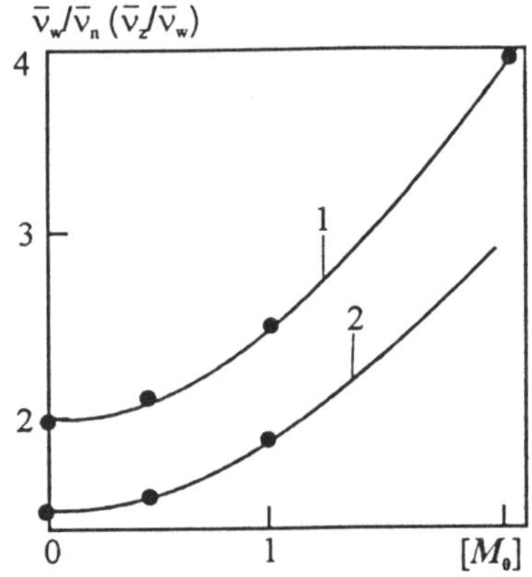

Figure 3.5. Dependence of $\bar{v}_w/\bar{v}_n$ (1) and $\bar{v}_z/\bar{v}_w$ (2) on the monomer concentration: for $k_p = 10^5$ l/(mol·s), $k_{ter} = 1$ s^{-1}, $c_0 = 10^{-2}$ mol/l, $D = 1$ m^2/s, $\alpha = 0$, $(E_M - E_p)R = 5000$ K, $l = 1$ m.

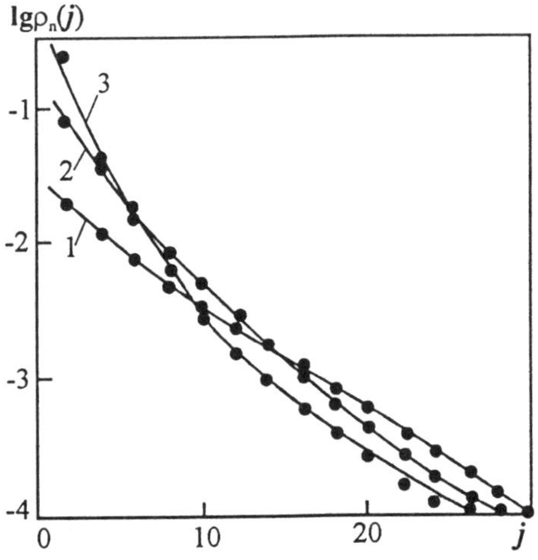

Figure 3.6. Semi–logarithmic anamorphoses of polyisobutylene MMD functions for various monomer concentrations: 1 – 0.5; 2 – 1; 3 – 2 mol/l.

As was found, besides expected growth of the polyisobutylene yield, growth of the catalyst concentration leads to increase of temperature and temperature gradient in the reaction zone and, consequently, to decrease of the mean molecular mass and widening of MMD (the ratio $\bar{v}_w/\bar{v}_n$ increases). The catalyst concentration only below 10^{-4} mol/l provides production of the polymer with MMD close to the calculated one (see Table 3.1). Influence of the process rate on the temperature gradient and, consequently, on the MMD width is observed from the data in Table 3.2.

Table 3.1

Influence of the catalyst concentration on the monomer conversion and polyisobutylene MMD parameters*.

No.	Catalyst concentration, mol/l	Conversion, %	$\bar{v}_n$	$\bar{v}_w$	$\bar{v}_w/\bar{v}_n$	$\bar{v}_z/\bar{v}_w$
1	$1\cdot10^{-1}$	99.7	3.6	8.9	2.46	1.94
2	$6\cdot10^{-2}$	98	3.8	9.3	2.46	1.93
3	$3\cdot10^{-2}$	89.5	4.2	10.2	2.41	1.87
4	$1\cdot10^{-2}$	60	5.6	13.3	2.24	1.70
5	$1\cdot10^{-2}$	99.9	4.1	10.3	2.50	1.90
6	$3\cdot10^{-3}$	89.2	5.1	12.4	2.42	1.80
7	$1\cdot10^{-3}$	53.6	8.0	17.3	2.17	1.62
8	$1\cdot10^{-4}$	8.5	13.8	27.6	2.00	1.50

* $k_p = 10^5$ l/(mol·s), $k_d = 10$ s^{-1} ($k_{ter} = 1$ s^{-1} for tests 5 – 8), $[M]_0 = 1$ mol/l, $c_0 = 10^{-2}$ mol/l, $(E_M - E_p)R = 5000$ K, $D = 1$ m^2/s, $\alpha = 0$.

Table 3.2

Influence of k_p and k_{ter} on the monomer conversion and MMD parameters.

k_p, l/(mol·s)	k_{ter}, s^{-1}	$\bar{v}_z/\bar{v}_n$	$\bar{v}_z/\bar{v}_w$	Conversion, %
10^5	1	2.50	1.90	99
10^5	10	2.24	1.70	60
10^6	1	2.46	1.94	100
10^6	10	2.26	1.94	99.7
10^4	1	2.17	1.62	53.6

Table 3.3

Influence of the reaction zone length l on the monomer conversion and MMD parameters.

Calculation parameters: $c_0 = 10^{-3}$ mol/l, $[M]_0 = 2$ mol/l, $(E_M - E_p)R = 5000$ K, $D = 1$ m^2/s, $\alpha = 0$.

l, m	Conversion, %	$\bar{v}_n$	$\bar{v}_w$	$\bar{v}_z$	$\bar{v}_w/\bar{v}_n$	$\bar{v}_z/\bar{v}_w$
0.1	10	9.3	19.8	31.3	2.11	1.58
0.3	27	7.6	16.2	26.2	2.14	1.62
0.7	45.7	5.1	12.4	22.5	2.44	1.81
1.0	53.6	4.2	11.2	21.5	2.66	1.92
3.0	68.6	3	9.4	20.2	3.13	2.17

Growth of the k_p/k_{ter} ratio induces simultaneous increase of $\bar{v}_w/\bar{v}_n$ and $\bar{v}_z/\bar{v}_w$ values, as well as of the monomer conversion. In this case, the ratio of constants, but not of their absolute values is most important, which is confirmed by closeness of values of MMD parameters and polymeric product yield for k_p/k_{ter} equal to 10^5:1 and 10^6:10, and 10^5:10 and 10^4:1, respectively.

Changes of the monomer conversion and molecular characteristics of polymer by the reaction zone length are shown in Table 3.3.

Hence MMD parameters change in the whole zone of the reaction volume as the conversion increases, being mostly noticeable at the distance of 1 m from the catalyst injection point. Change of the character of mass differential MMD curves [$\rho_w(j)$] with respect to the reaction zone length (Figure 3.7) indicates a gradual appearance of the low-molecular fraction at the sacrifice of temperature growth.

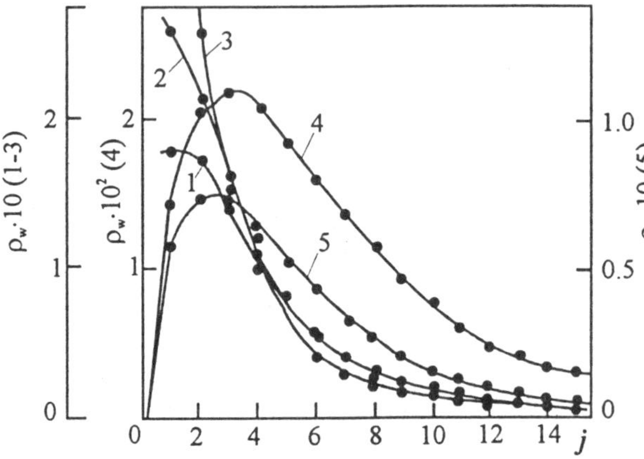

Figure 3.7. Functions of polyisobutylene MMD for various lengths of the reaction zone, l (m): 1 – 0.7; 2 – 1; 3 – 2; 4 – 0.1; 5 – 0.3.

The analysis of fast polymerization reactions performed have shown that effects marked in the mathematical modeling are identical to these experimentally observed using the example of isobutylene polymerization [2]. Comparison of calculated and experimental data marks occurrence of gradients of concentration and temperature, i.e. fast reactions with local introduction of a catalyst proceed in separate zones as flares with different temperatures and kinetic parameters. An important consequence of non-isothermity of the process is in increase of the product polydipersion by mean molecular masses, i.e. in deviation of its properties. Existence of a flare in fast polymerization processes, of isobutylene, in particular, determined specific method and practical techniques for their performance. For example, external thermostatic control is not effective and limits application of dilatometry and many other experimental methods of process kinetics investigation. The only parameters that can provide isothermal character of the process, and production of a polymeric product with MMD close to the calculated one, are low catalyst concentrations (below 10^{-4} mol/l) under the condition of effective reaction mass mixing.

It follows from the above-mentioned features that production of a product with MMD close to the calculated one in a single device requires limiting of the reaction zone by an area of comparatively low monomer conversions of 50–70%. This, in its turn, applies restrictions on size of the zone itself, i.e. the device size, which must be selected on grounds of kinetic parameters of the reaction. Analysis of operation of industrial polymerization reactors has shown that artificial limiting of the reaction zone causes a specific influence on the character of the reaction process of isobutylene polymerization. The polymer yield is maximum (100% approximately) at small radii of the reaction zone, and fronts of spreading of temperature and concentrations of all reagents is planar; this determines the regime of a quasi-ideal superceding.

Formation of a "flare" is displayed only under high values of the reaction zone radius. There is a zone outside the flare, in which polymerization does not practically proceed, which leads to a monomer breakthrough and, consequently, to a reduction of the polymer yield.

The intermediate regime is characterized by heterogeneous concentration fields and the absence of a precise border between the reaction area and zones of its absence.

One of the most significant results is the effect of the geometry of the reaction volume on molecular characteristics of polymeric products formed. In this case, heat elimination becomes ineffective, and coolers are able just to cool down reacting substances approaching the point of catalyst and monomer feeding, which is not an effective or sufficient factor.

3.2. Macroscopic features of isobutylene polymerization. Novel fundamental regularities.

3.2.1. On existence of several macroscopic regimes.

Analysis of experimental results and theoretical calculations of fast polymerization processes in turbulent flows (the diffusion model) using the example of isobutylene polymerization brought out a significant influence of geometric parameters of the reaction zone — radius R and length L — on kinetic parameters of the processes, as well as on the depth of monomer transformations. In this case, apparently, one of the most significant results is influence of the reaction volume geometry on molecular–mass indexes of resulting polymeric products (Figures 3.8, 3.9, and Table 3.4). In topochemical aspect, three macrokinetic types of process are clearly distinguished: A, B and C (Figure 3.10).

At small values of radius R (type A) mixing of reagents is quite effective, active centers A* are comparatively uniformly distributed in the reaction zone (Figure 3.10b). Consequently, the reaction temperature is uniformly distributed through the reaction zone radius R (Figure 3.10a). Surfaces of equal concentrations of the monomer, active centers and temperatures represent planes perpendicular to the reactor axis. This determines high (up to 100%) conversion of the monomer and the regime of quasi-ideal displacement in highly turbulent flows.

Another limiting case (local flare regime) is realized at relatively large values of R (type C). Active centers A* die not managing to diffuse to peripheral areas of the reaction volume which, consequently, plays the role of breakthrough zones of unreacted monomer (Figure 3.10f). As a consequence, specific fields of complex configuration of concentrations of the monomer, active centers and temperatures appear (Figure 3.10e, f). The flare radial size is determined by the ratio of two competing processes: mixing of flows (in the present case, at the sacrifice of the turbulent diffusion) and death of active centers. The reaction does not reach the reactor walls, and the monomer conversion at the sacrifice of its breakthrough between the flare border and the reactor wall is significantly decreased (see Table 3.4).

Table 3.4

Dependence of the monomer transformation depth β, polymerization degree P_n and polydispersion index P_w/P_n on the reaction zone radius R ($k_p = 10^5$ l/(mol·s), $k_{ter} = 20$ s^{-1}, $[M]_0 = 1$ mol/l, $[A^*]_0 = 0.0045$ mol/l, $D_t = 0.025$ m^2/s) [9].

Reaction type	R, m	Conversion, β mass%	P_n**	P_w/P_n**
A	0.01	100	13	2.0
	0.03	100	13	2.0
	0.05	100(97.7)*	12 (30)	2.1 (3.1)
	0.08	99.3	10	2.1
B	0.10	90.0 (90.0)	8 (21)	2.2 (3.7)
	0.25	65	6	2.4
C	0.50	32 (29.7)	6 (17)	2.4 (4.0)

Notes. * Values shown in brackets are experimental results obtained during isobutylene polymerization in the presence of $AlCl_3$ in ethyl chloride at 243 K.

** Differences between calculated and experimental data of P_n and P_w/P_n are defined by selected value of k_p/k_M and can be easily neglected.

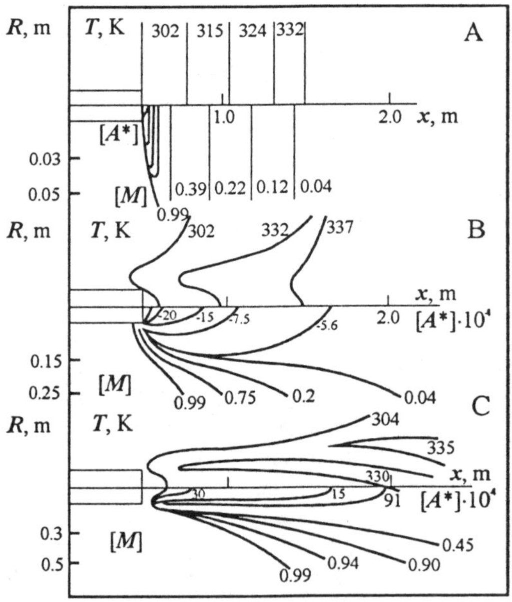

Figure 3.8. Dependence of temperature fields (T, K), concentrations of the monomer [M] and active centers [A*] on the reaction zone radius R ($[M]_0 = 1$ mol/l, $[AlCl_3]_0 = 0.0045$ mol/l, $D_t = 0.025$ m^2/s, $R/r = 5$, $k_p = 10^5$ l/(mol·s), $k_{ter} = 20$ s^{-1}, $T_0 = 300$ K).

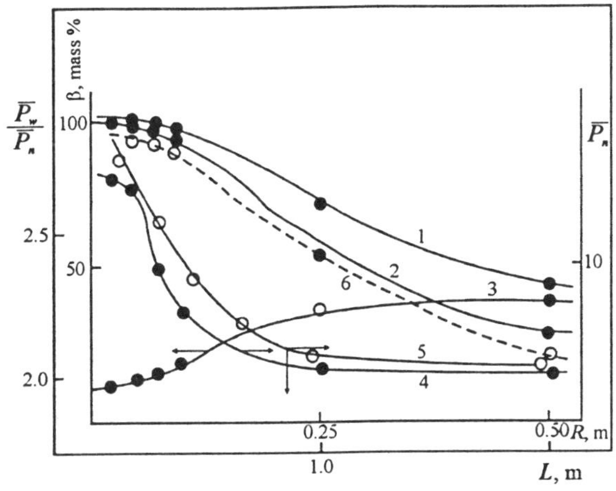

Figure 3.9. Dependence of the monomer transformation depth β (1, 2, 6), polydispersion index P_w/P_n (3) and polymerization degree P_n (4, 5) on radius R and length L of the reaction zone (conditions of Figure 3.8) for k_{ter} (s^{-1}) equaling: 1 – 15; (2 – 5) – 20; 6 – experimental curve for $[M]_0$ = 1.9 mol/l, $[AlCl_3]_0$ = 0.0037 mol/l (in C_2H_5Cl and T_0 = 248 K), $k_p = 10^5$ l/(mol·s), V = 5 m/s, R/r = 2.5.

Type B (the third macroscopic type) determines an intermediate regime of the process, defined by formation of a flare with no monomer breakthrough zones. The regime B (as well as the regime C), characterized by existence of high temperature gradients, as well as by high concentrations of active centers and the monomer along both x axis and R (Figure 3.10c, d), significantly affect the uniformity of the forming polymeric product (MMD is widened). In this case, the monomer conversion is somewhat increased compared with the limit case, when the macroscopic regime C is realized (see Table 3.4).

Occurrence or absence of the temperature gradient in fast polymerization processes, its change at transition from one macroscopic regime to another (of A, B, C types) noticeably influence the molecular-mass characteristics of forming polymer (see Table 3.4). The reason is that temperature in the reaction zone at low values of R (under macroscopic regime of the type A) is distributed rather uniformly, whereas occurrence of the temperature gradient as a flare by reaction volume coordinates (macroscopic regimes B and C) at radii values above some critical value R_{cr} (R_{cr} is the value of R stipulating transition from the A type regime to the B one) leads to MMD widening at the sacrifice of accumulation of a part of low-molecular fraction. It should be taken into account simultaneously that MMD of the polymeric product is widened at points along the x axis remote from the catalyst injection point, which is the consequence of temperature growth and formation of macromolecules under different temperature conditions along the x axis. Calculations adequately reflect the tendency of the effect of the reactor geometrical size on molecular-mass characteristics of forming polymeric products, when liquid-phase fast polymerization processes are conducted, and correlate with the experimental ones (see Table 3.4) [9].

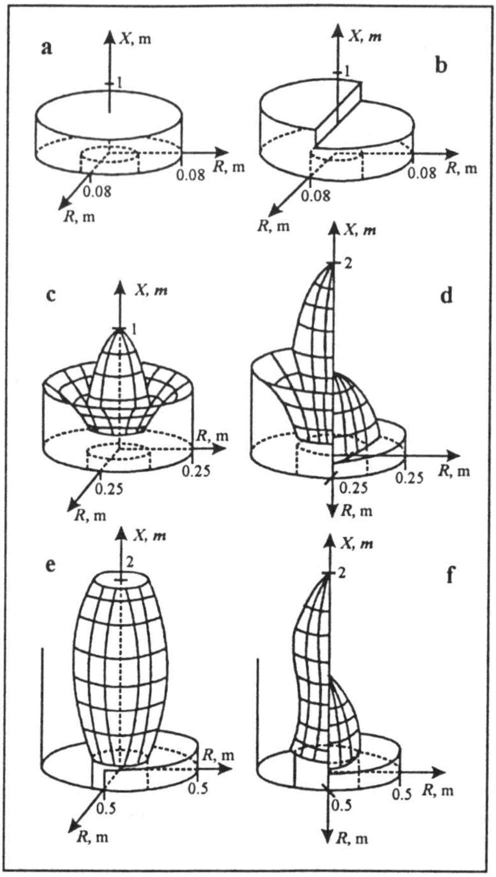

Figure 3.10. Fields of temperature (a, c, e), concentrations of active centers [A*] and the monomer [M] (b, d, f) being formed during fast polymerization in a flow for R = 0.08 m (a, b); 0.25 m (c, d); 0.5 m (e, f); the maximum monomer conversion β, mass%, is 99.3 (a, b), 65.0 (c, d), and 32.0 (e, f) at ΔT_0 = 28 (a, b), 22 (c, d), and 9 (e, f) ($[M]_0$ = 1 mol/l, $[AlCl_3]_0$ = 0.0045 mol/l, $D_t = 0.025$ m^2/s, $R/r = 2.5$, $k_p = 10^5$ l/(mol·s), $k_{ter} = 20$ s^{-1}, $T_0 = 300$ K).
Macroscopic regimes correspond to types: A — a, b; B — c, d; C — e, f.

Hence influence of the reaction volume geometry on molecular-mass characteristics of forming polymer and possibility of transition from one macroscopic regime to another (regimes A, B, C) at the sacrifice of the radius R of change of the reaction zone are clearly displayed in the example of fast liquid-phase electrophilic polymerization of isobutylene. MMD width (P_w/P_n) grows with the radius R. The mean MM of the polymeric product decreases simultaneously. Consequently, a new phenomenon is observed in fast liquid-phase polymerization processes — the influence of geometrical parameters of the reaction zone on the yield of final products and their molecular-mass characteristics [8–12]. Contrary to classic chemical processes, growth of the reaction zone volume and injection of greater amounts of ingredients at the same linear rate will usually cause a significant reduction of the monomer conversion and hence in productivity of reactors. In this case, three different macroscopic regimes can be observed:

1. At small radius — plane front of concentrations and temperatures — the regime of quasi-ideal displacement. The inclusion of lengthwise shear differentiates this regime from the ideal displacement.
2. At large radii — the reaction proceeds in the central zone of the reactor, as active centers do not reach the external wall. The monomer breakthrough happens on the periphery — the flare regime.
3. The intermediate regime with significant gradients of concentrations and temperature along the reactor radius.

Obviously, the highest yield of the product and uniformity by MMD can be obtained in the first regime.

3.2.2. On connection between geometrical size of the reaction zone and kinetic and hydrodynamic parameters.

Transition from the quasi-ideal displacement regime (type A) to the flare one (types B and C) is accompanied by monomer conversion reduction and deterioration of molecular–mass characteristics of the polymer (reduction of P_n and widening of MMD). The critical radius R_{cr} determining transition from the flare regime of reactor operation (types B and C) to a macroscopic regime of displacement in turbulent vortexes (type A), depends on the ratio of the monomer diffusion (D_t) and death of active centers A* (k_{ter}), i.e. is determined by the ratio of characteristic times of mixing τ_{mix} and chemical reaction τ_{ch}. From the dimensionality ratio R_{cr} is presented as follows [10]:

$$R_{cr} = m \cdot \sqrt{\frac{D_t}{k_{ter}}}, \tag{3.10}$$

where m is a numerical coefficient.

In the general case, conditions of reaching the quasi-ideal displacement regime in turbulent flows are determined by the ratio:

$$\tau_{mix} = \frac{R^2}{D_t} < \tau_{ch} = \frac{1}{k_{ter}}.$$

Calculations have displayed that if the reaction zone radius is assumed as R_{cr}, at which the polymer yield reduces by 10% compared with the yield at $R \to 0$, then $m = 2 \pm 0.2$ (Figure 3.11) [10 – 12].

Under real conditions of fast polymerization reaction processes of any monomer, correlation (3.10) also allows solution of the reverse problem. Induced change of the reaction zone (reactor) radius and measuring of dependence of the monomer conversion on the parameter R give the possibility of calculating the ratio D_t/k_{ter}. Of importance is that a general dependence between the relative yield of polymeric products and the ratio $R\sqrt{k_{ter}/D_t}$ exists in a wide range of k_{ter}, D_t and R values (Figure 3.12).

Notice that, at the change of radius R (at constant linear rate V of the flow) three different regimes of liquid flow can be separated: laminar (small R), transitional and turbulent ones. Each regime is characterized by individual values of mean characteristic times of mixing ($\tau_{mix} \approx R^2/D_t$) (Figure 3.13).

Restrictions with regard to the minimum and maximum diameter of a pipe reactor for selected rates of the flow are seen to exist.

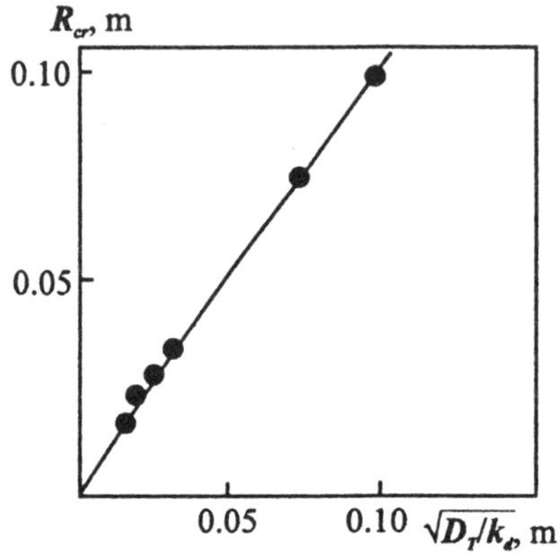

Figure 3.11. Dependence of R_{cr} on $\sqrt{D_t/k_{ter}}$ ($[M]_0 = 1$ mol/l; $[A^*]_0 = 0.0045$ l/mol; $V = 5$ m/s; $k_p = 10^5$ l/(mol·s); $T_0 = 300$ K).

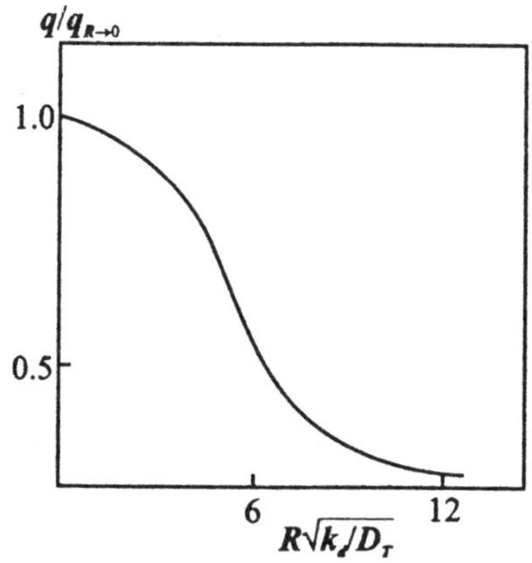

Figure 3.12. Dependence of relative monomer conversion in fast polymerization on $R\sqrt{k_{ter}/D_t}$ for various combinations of R, k_{ter} and D_t.

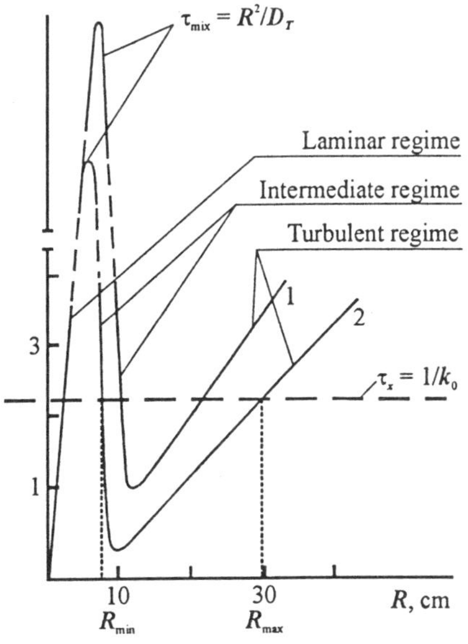

Figure 3.13. Dependence of the effective time of mixing ($\tau_{mix} = R^2/D_t$) on the reactor radius R for the flow rate of 2.5 (1) and 5 m/s (2).

In the laminar regime, the diffusion coefficient is very small ($D = 10^{-9}$ m^2/s). As a consequence, mixing times are large and grow with R changing as R^2. In the transitional regime, efficiency of mixing increases abruptly and τ_{mix} decreases. In the turbulent regime, D_t values are rather high ($\approx 10^{-3}$ m^2/s) and grow approximately by the law $D_t \sim R$.

Comparison of τ_{mix} with the typical time of chemical reaction τ_{ch} shows that conditions of the quasi-ideal displacement regime ($\tau_{mix} < \tau_{ch}$) are fulfilled in the ranges $0 < R < R_1$ and $R_2 < R < R_3$. The range $0 < R < R_1$ corresponds to capillary reactors ($R_1 \approx 10^{-5}$ m, $D = 10^{-9}$ m^2/s), which are of no practical interest. Just rather narrow (depending on τ_{ch} and V) range of the reaction zone radii remains, which provides turbulent flow ($R_2 < R < R_3$), with rather high productivity (Figure 3.13). For example, at $V = 1$ m/s and $R_2 = 3$ cm, productivity of the turbulent reactor is ≈ 20 m^3/h. When the radius changes to $R_3 = 15$ cm, productivity of the pipe turbulent reactor increases to 500 m^3/h. Growth of the flow rate causes widening of the possible sphere of reactor exploitation (in the quasi-ideal displacement regime in turbulent flows R_2 decreases, and R_3 grows). However, in this case, hydrodynamic resistance of the reactor can grow (in proportion to V). Consequently, in some cases, pressure in the system should be increased. However, this may cause undesirable effects (growth of boiling temperature of the reaction mass, requirements of the process equipment become more demanding, loads of pumps increase, etc.).

An important aspect of the problem connected with carrying out fast polymerization processes is determination of the reaction front spreading length L along the x axis, defined not only by transversal mixing of reagents in the reaction cross-section, but also by flowing, as well as lengthwise mixing, which, in its turn, is defined by the flow rate V, the rate of direct chemical reaction. In this case, washing of the reagents' concentration is determined by the turbulent diffusion D_t. As a consequence, there is an interconnection between efficient exposure time of the reaction mass in the reaction zone (L/V), necessary for reaching high product yield, and effective time of the chemical reaction (τ_{ch}), which depending on the process conditions is defined by lower of the values $1/k_{ter}$ or $1/k_p[A^*]_0$ [10–12].

Calculations have shown that reagents manage to move by $S = v\tau_{ch}$ along the flow under developed turbulent flow during a characteristic reaction time, and their concentrations distribute by an average distance equal to

$$S_D = \sqrt{D_t \tau_{ch}} \sim \sqrt{v}. \tag{3.11}$$

Consequently, the reactor working regime at high rates will approach the ideal displacement, i.e. the role of lengthwise diffusion will be decreased. In this case, the reactor length L can be calculated by comparing exposure time in the reactor $\tau = L/v$ with characteristic reaction time τ_{ch}.

At low catalyst concentrations $\tau_{ch} = 1/k_0$. At high catalyst concentrations ($C_0 >> k_0/k_p$) $\tau_{ch} = 1/k_pC_0$. To reach the maximum yield the condition $\tau \geq \tau_{ch}$ must be fulfilled, wherefrom $L \geq v\tau_{ch}$.

Based on this analysis of the model and experiment in the pipe reactor, determining polymer yield in dependence on the radius, reactor length and catalyst concentration, effective rate constants of the reaction k_0, k_p, as well as D_t for particular catalytic systems were successfully determined.

With flow rate decrease in the frames of the turbulent regime, "washing" of temperature in the reaction zone owing to lengthwise turbulent diffusion leads to a constant temperature in the whole reaction volume (quasi-isothermal regime). It is the principal difference between this regime and the ideal displacement one.

Polymer yield β for the reactor with ideal displacement is determined in the analytic form by the correlation:

$$\beta = 1 - \exp\left\{-\frac{k_p[A^*]_0}{k_{ter}}\left[1 - \exp\left(-k_{ter}\cdot\frac{L}{V}\right)\right]\right\}. \tag{3.12}$$

If the parameter $(L/V)_{eff}$, the effective time of exposure determining the reaction zone length, at which the polymer yield reaches 90% of the maximum possible one at $L \to \infty$, is introduced, then:

$$\left(\frac{L}{V}\right)_{eff} = -\frac{1}{k_{ter}}\ln\left\{1 + \frac{k_d}{k_p[A^*]_0}\ln\left[0.1 + 0.9\exp\left(-\frac{k_p[A^*]_0}{k_{ter}}\right)\right]\right\}. \tag{3.13}$$

Dependence of $(L/V)_{eff}$ on $1/k_p[A^*]_0$ is depicted in Figure 3.14. Typically, each temperature corresponds to a particular effective time of raw material exposure in the reactor $(L/V)_{eff}$. For example, when great amounts of catalyst are used (243 K), it is obtained that $(L/V)_{eff} \approx 0.005 \pm 0.0005$ s. In the case when there is lack of catalyst in the system, and the reaction is not finished, the value of L/V may change by a degree or more (in particular, at 226 K $(L/V)_{eff} = 0.02 \pm 0.005$ s).

Two limiting cases are of importance. At high catalyst concentrations, when $k_{ter}/k_p[A^*]_0 << 1$ the polymer yield approaches the quantitative one (100 mass%). In this case, $(L/V)_{eff} \approx 2.3/k_p[A^*]_0$. At low catalyst concentrations $k_{ter}/k_p[A^*]_0 \geq 1$and $(L/V)_{eff} \approx 2.3/k_{ter}$.

Calculation of fast polymerization processes with consideration for the lengthwise diffusion, but under the condition $R < \sqrt{D_t/k_{ter}}$, has indicated that curves of $(L/V)_{eff}$ dependence on $1/k_p[A^*]_0$ are close to those calculated by Equation (3.13) (Figure 3.14).

It is well known that determination of elementary constants of fast polymerization processes is extremely difficult. The data shown give a pathway for experimental determination of elementary constants in processes of similar type. Studying change of the polymer yield with the reaction zone length and/or the flow rate at different catalyst concentrations under the condition $R < R_{cr}$, rate constants k_p and k_{ter} could be estimated, if two ranges of ratios $k_{ter}/k_p[A^*]_0 < 1$ and $k_{ter}/k_p[A^*]_0 > 1$ are used, or even one of them, when both ranges of ratios could not be used [10–12].

The method suggested for calculation of kinetic constants allowed experimental estimation of k_p and k_{ter} for cationic liquid-phase polymerization of isobutylene in operating industrial reactors, which at 243 K ($[AlCl_3]_0$ = 0.0037 mol/l, $[M]_0$ = mol/l) were found: $k_p = 1.0 \cdot 10^6$ l/(mol·s), which generally corresponds to literary data [1 — 3], and $k_{ter} = 17.5 \pm 5$ s^{-1} (was estimated for the first time).

One more feature of fast polymerization processes can be indicated. Polymer MM decreases as concentrations [M] and [Cat] grow, obviously, because of heat release growth with $[M]_0$ and $[Cat]_0$ and, as a consequence, noticeable increase of temperature in the reaction zone. Additional widening of MMD of the polymeric product and MM reduction are always observed under realization of macroscopic regimes of B and C types due to formation of temperature fields (a "flare").

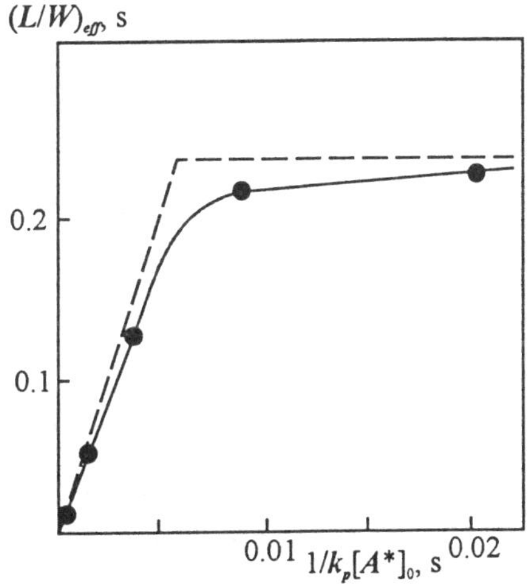

Figure 3.14. Dependence of $(L/V)_{eff}$ on $1/k_p[A^*]_0$ for the reactor of ideal (continuous curve) and quasi-ideal (points on the curve) displacement (V = 2.5 m/s; L = 0.08 m; $k_p = 10^5$ l/(mol·s); $k_{ter} = 10$ s^{-1} ; T_0 = 300 K).

Hence kinetic parameters of fast polymerization processes (k_p and k_{ter}) and linear rate of the reagents' flow determine geometrical sizes (R, L) and configuration of the reaction zone. In this case, new possibilities and methods for process control appear, which permit regulation of the monomer conversion degree and molecular–mass characteristics of resulting polymeric products, in particular, at the sacrifice of induced change (limiting) of geometrical parameters of the reaction zone (R and L).

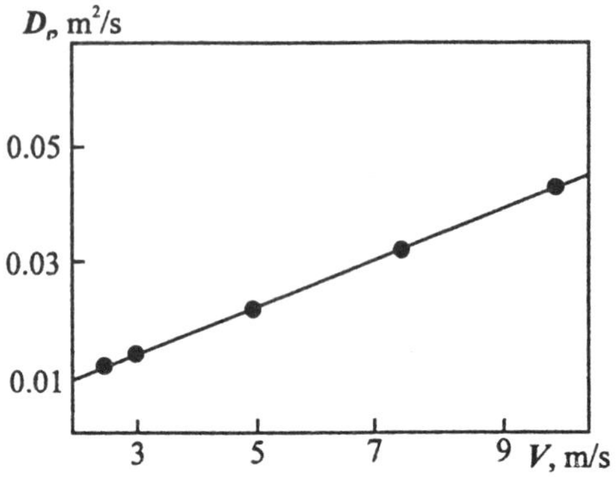

Figure 3.15. Dependence of the turbulent diffusion coefficient D_t on the linear rate of the monomer and catalyst flow motion V.

3.2.3. Influence of linear flow motion rate on molecular-mass characteristics of resulting polymer.

The turbulent diffusion coefficient D_t can be varied in rather wide ranges at the sacrifice of preliminary turbulization, changing the way of mixing, direction and rate of reagents' flows motion, etc. [13–17]. Figure 3.15 shows dependence of the turbulent diffusion coefficient D_t on the flow motion rate at accompanying injection of reagents, determined in "q–ε" frames of the Navier–Stokes model with regard to real values of the solution viscosity, its temperature, heat release rate, etc. Growth of the turbulent diffusion coefficient is practically linear depending on the motion rate of reaction flows (monomer, catalyst). Substitution of numerical values of D_t obtained from calculations by the Navier-Stockes equation (Figure 3.15) into the system of equations (3.1), which describes change of concentrations of M and active centers A*, and temperature in the reaction zone for the pipe turbulent reactor with accompanying injection of reagents as well, the influence of D_t and V on main parameters of fast polymerization processes can be easily estimated [13–17].

Conversion of the monomer during polymerization proceeding in reactors of the radius $R > R_{cr}$ does not usually reach 100%. Growth of the turbulent diffusion coefficient several-fold, including with the motion rate of flows, causes a significant increase of the monomer transformation depth, despite shortening of time of raw material exposure in the reaction zone in this case (the reaction zone length L is constant) (Figure 3.16, curve 2).

If the reaction area is limited to a single exposure time for all values of the flow rate, for example, $\tau = 0.2$ s (Figure 3.16, curve 1), then the monomer conversion in fast polymerization processes can grow by more than three times with the flow rate increase from 2.5 to 10 m/s, all other factors being the same. Of great importance is that molecular-mass characteristics of the formed product change simultaneously with growth of the monomer transformation depth (with growth of V and, consequently, D_t). Increase of the flow motion rate leads to a growth of mean MM (P_n), MMD of the product narrowing simultaneously (Figure 3.17). Figure 3.18 depicts temperature changes by R for various linear rates of the reagent flow motion (V) (turbulent diffusion coefficient D_t). Despite growth of the total polymer yield, smoothing of temperature maximums in the reaction volume appears with V and consequent D_t growth. Washing of temperature leads to increase of the average MM and narrowing of MMD of resulting polymeric product.

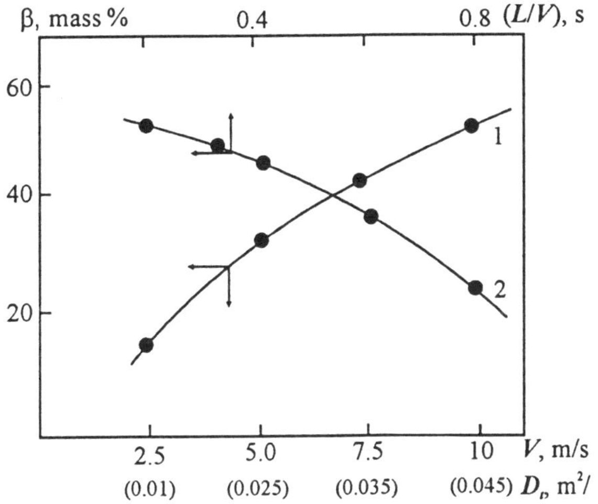

Figure 3.16. Dependence of the polymer yield on the linear rate of reaction mixture flow motion (1) at τ_{ch} = 0.2 s and on characteristic time of the chemical reaction τ_{ch} (2) at L = 2 m ($[M]_0$ = 2 mol/l, $[AlCl_3]_0$ = 0.0045 mol/l, R = 0.25 m, T_0 = 300 K).

Of special importance is that with no dependence on the turbulent diffusion coefficient D_t in the range of radii over R_{cr} the external heat elimination does not practically affect MM and MMD of resulting polymeric products. This is bound to insufficiency of reagent diffusion for reaching the thermostatically controlled wall. Active centers decay as a result of not reaching it, i.e. the process proceeds in a local zone, the size of which is determined by a ratio of two competing processes: diffusion of active centers and their death (macroscopic type C, Figure 3.10). Growth of the mean molecular mass and narrowing of MMD of the resulting polymer is explained not by improving the heat elimination, but by improving distribution of reacting components in the volume. This leads to a more uniform distribution of the temperature front, i.e. no clearly displayed zones of high temperatures exist in this case.

Sensitivity of isobutylene polymerization to temperature (dependence of the mean MM) is determined by difference in activation energies of the chain transmission and propagation (E_t – E_p). Then the condition of existence of quasi-isothermal regime is defined by the correlation

$$\frac{k_0 D_t R T^2 c}{v^2 Q \Delta M (E_t - E_p)} >> 1, \tag{3.14}$$

where R is the gas constant; T is the mean temperature in the reaction zone; ΔM is the part of reacted monomer.

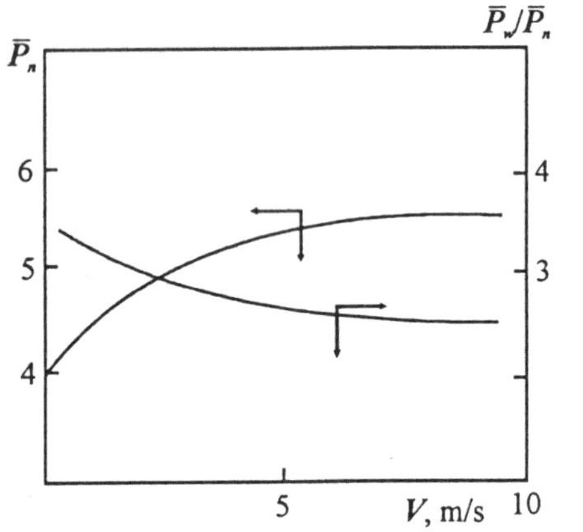

Figure 3.17. Dependence of P_n (1) and P_w/P_n (2) on the linear rate of reaction mixture flow motion at the monomer transformation depth $\beta = 30$ mass% ($[M]_0 = 2.5$ mol/l, $[AlCl_3]_0 = 0.01$ mol/l, $T_0 = 200$ K).

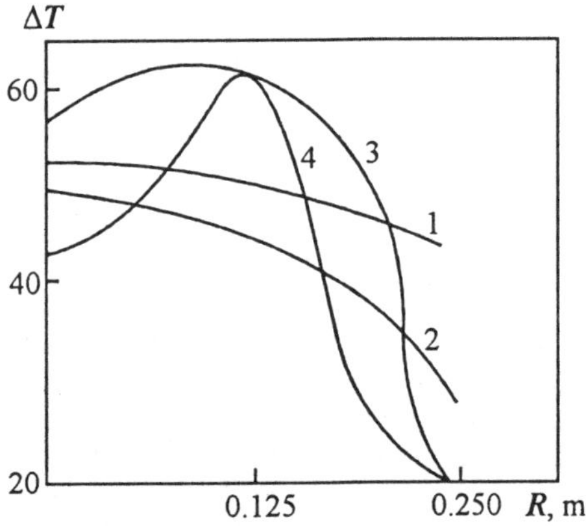

Figure 3.18. Radial profiles of temperatures for D_t, m^2/s: 1, 2 – 0.0045; 3, 4 – 0.010 ($V = 2.5$ m/s) at various times of monomer and catalyst contacting: 1, 3 – 0.2 s ($\beta = 90$ mass%); 2, 4 – 0.1 s ($\beta = 60$ mass%).

Under this regime a polymer with the narrowest MMD must be produced. In the present case, it is the most probable distribution, confirmed experimentally, $\bar{P}_w/\bar{P}_n \cong 2$.

On the other hand, the condition of low sensitivity of MM and MMD to temperature in the reaction zone, more precisely to temperature gradient (which determines the quasi-isothermal regime of the process), is the correlation [5, 6]:

$$\Delta T \leq \frac{RT^2}{E_M - E_p}, \tag{3.15}$$

where T is the average temperature; ΔT is the difference of temperatures in the reaction zone. The temperature gradient ΔT in the reaction zone is defined by equality of heat release rates at polymerization (W_n) and its distribution at the sacrifice of turbulent transfer (W_t):

$$W_t = \frac{\lambda_t \cdot \Delta T}{\delta} = \frac{c_p \rho D_t k_{ter} \Delta T}{V} = W_n = \rho V q_{11} \Delta M, \tag{3.16}$$

where $\delta = V/k_{ter}$ is the reaction zone length at $k_p[A^*]_0 \leq k_{ter}$.

The generalized criterion determining formation of a quasi-isothermal regime of the process proceeding in displacement reactors in turbulent flows is obtained by combination of equations (17) and (18) [3, 4]:

$$\frac{k_{ter} \lambda_t R T^2}{\rho V^2 q_{11} (E_{ter} - E_p)} \geq 1. \tag{3.17}$$

Hence observance of conditions of the quasi-isothermal regime formation in displacement reactors in turbulent flows requires fulfillment of two criteria presented by correlations (3.10) and (3.17).

Typical size of washing (averaging) of the reaction temperature is proportional to $\sqrt{\lambda_t} \propto \sqrt{D_t} \propto \sqrt{V}$, as in the turbulent regime $D_t \sim \lambda_t \sim V$, and temperature shear along the reaction axis and, consequently, growth of the reaction zone length is proportional to V. That is why growth of the flow rate breaks the condition (3.17), and at $V \to \infty$ the regime becomes close to the ideal displacement (in the classic variant) with the characteristic rather wide MMD of the product.

Abrupt decrease of the flow rate V causes a transition from the turbulent regime to a laminar one, the criterion (3.17) being broken due to decrease of λ_t, which also means deviation from the quasi-isothermal regime. To put it differently, a limited range of flow rates exists for real performance of the quasi-isothermal regime in pipe reactors.

Hence the yield of resulting polymeric product can be increased simultaneously with MM and improvement of its quality (narrowing of MMD) by changing the flow rate, i.e. reaction rate of polymerization and molecular characteristics of resulting polymers can be effectively controlled. This is connected with the change of the temperature field profile in the reaction zone by the reactor coordinates. Despite shortening of the contact time of the reacting mixture with thermal controlling reactor wall ($\tau_{mix} = L/V$), growth of V will lead to a noticeable improving of the efficiency of external heat elimination, and, consequently, to increase of the molecular mass and narrowing of MMD as well as quantitative yield of the polymeric product, which is unexpected from the common kinetic grounds.

3.2.4. Influence of the method of catalyst transmission on molecular-mass characteristics of resulting polymer.

A strict hydrodynamic model was developed on the basis of solution of the Navier-Stockes equation for turbulent flow together with chemical reaction equations (q,ε-model) [13, 14]. By this model the real profile of rates was calculated, allowances are made for concentration and temperature fluctuations, and the turbulent diffusion coefficient at each point of the reactor was numerically calculated. Naturally, the volume of calculations is significantly greater in this case.

The q,ε-model allowed calculation and comparison with experiment of absolute values of the turbulent diffusion mean coefficient and determination of effects of the reactor geometry and the catalyst injection method on the reactor efficiency. It turned out that, for example, a reactor with constant diameter is not the optimal one, and its efficiency and the product quality can be raised by a change in the geometry. Analogously, changing the method of reagent feeding into the reaction zone significantly influences the character of a fast polymerization reaction (isobutylene polymerization, in particular) [13–17].

In the frames of the united mathematical model the influence of the feeding method of the reaction zone by monomer (M) and catalyst (A*) during isobutylene polymerization in flow was studied [15, 17]. Two methods of reagent feeding were chosen: a) model I — with central single-point catalyst injection; b) model II — with catalyst injection at the external part of the pipe device (Figure 3.19). The system of kinetic equations (3.1) with border conditions was used in the calculation (at central injection of the catalyst into the monomer solution flow with axial symmetry).

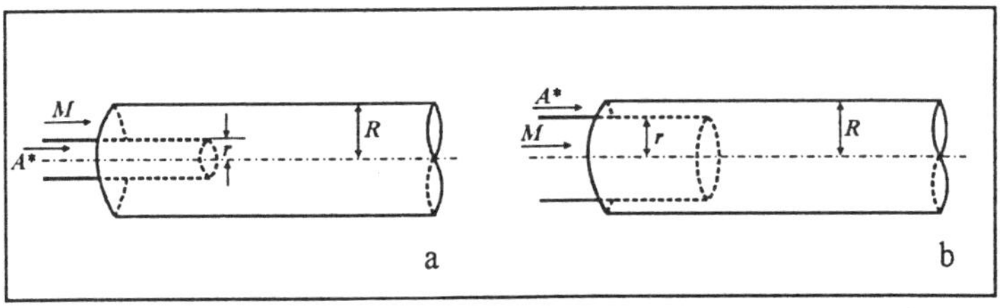

Figure 3.19. Models I (a) and II (b) of the catalyst feeding method of the pipe turbulent reactor.

Numerical solution of the system of equations (3.1) with assumed border conditions was performed with the help of implicit difference scheme according to the sweep method; MM and MMD were calculated according to [1].

At small radii of the reaction zone (Figure 3.20) in the absence of external heat elimination the method of reagent feeding causes no effect on P_n and P_w/P_n of resulting polymeric product, whereas MM and MMD of products differ in respect to the reagent feeding method with growth of efficiency of the external heat elimination. For example, if P_n grows at realization of the model I, and in this case P_w/P_n is gradually decreased practically approaching the exponential one at $\alpha = 50$, then P_w/P_n in the model II tends to two already at $\alpha = 20$. P_n of the model II is characterized by higher value than that of the model I (under similar conditions).

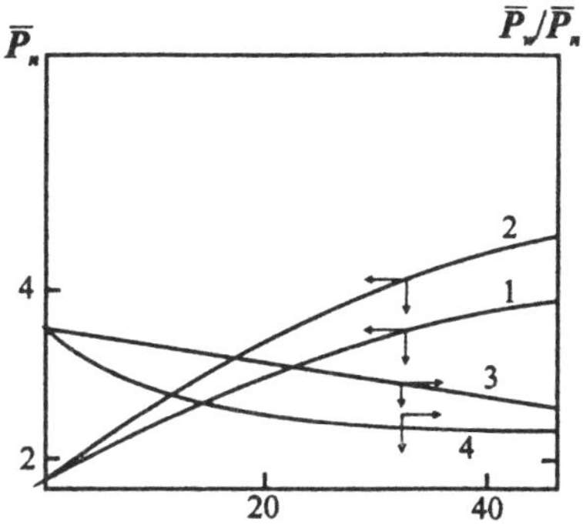

Figure 3.20. Dependence of P_n (1, 2) and polydispersion index P_w/P_n (3, 4) on the heat elimination efficiency (α) at accompanying (1, 3) and radial (2, 4) methods of the catalyst feeding ($[M]_0 = 2$ mol/l; $[AlCl_3]_0 = 0.0045$ mol/l; $R = 0.66$ m; $T_0 = 300$ K).

Model I

$T(-d, r) = T_0$

$$M(-d,r) = \begin{cases} [M]_0 & \text{at } R > r > r_0 \\ 0 & \text{at } r_0 > r > 0 \end{cases}$$

$$A^*(-d,r) = \begin{cases} [A^*]_0 & \text{at } R > r > r_0 \\ 0 & \text{at } r_0 > r > 0 \end{cases}$$

$$k_{ter}(x,r) = \begin{cases} k_{ter} & \text{at } x \geq 0 \\ 0 & \text{at } x < 0 \end{cases}$$

Model II

$T(-d, r) = T_0$

$$M(-d,r) = \begin{cases} 0 & \text{at } R > r > r_0 \\ [M]_0 & \text{at } r_0 > r > 0 \end{cases}$$

$$A^*(-d,r) = \begin{cases} 0 & \text{at } R > r > r_0 \\ [A^*]_0 & \text{at } r_0 > r > 0 \end{cases}$$

$$k_{ter}(x,r) = \begin{cases} k_{ter} & \text{at } x \geq 0 \\ 0 & \text{at } x < 0 \end{cases}$$

$$\frac{\partial[M](x,R)}{\partial\tau} = \frac{\partial[A^*](x,R)}{\partial\tau} = \frac{\partial[A^*](x,0)}{\partial\tau} = \frac{\partial[M](x,0)}{\partial\tau} = \frac{\partial T(x,0)}{\partial\tau} = 0,$$

$$\frac{\partial T(x,R)}{\partial\tau} = \alpha \cdot \{T(x,R) - T_1\}$$

Here x and r are coordinates of length and radius of the reaction zone.

The effect of change of polymer molecular-mass characteristics becomes more noticeable as the radius of the reaction zone increases (Figure 3.21), for the model I: P_n therewith changes only slightly with growth of α, while P_w/P_n changes significantly and tends to the exponent with simultaneous growth of P_n.

The effect found is connected to the character of temperature fields in the reaction volume. If at small radii, i.e. when a plane front of the reaction is formed, the thermostatic control effect is bound to the reaction temperature reduction in the initial stage of the process, and it changes similarly with reaction zone length in both models (Figure 3.22); the temperature field in the whole reactor changes at large radii (Figure 3.23). For example, realization of the model II indicates formation of the maximal temperature zone near the thermostatically controlled wall, which greatly simplifies heat elimination and smoothes off the temperature field. This leads to narrowing of the MMD of resulting polymeric product.

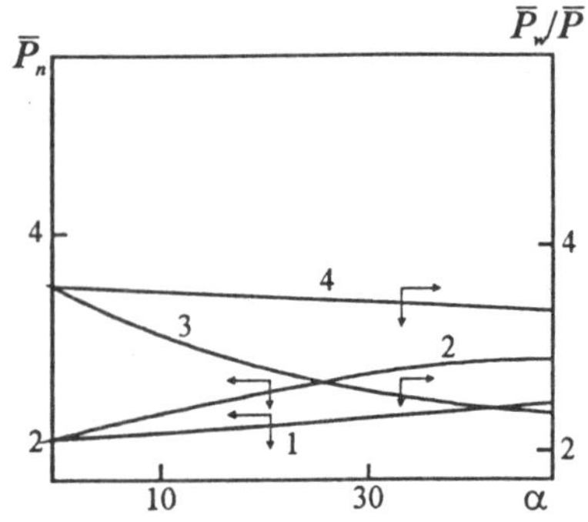

Figure 3.21. Dependence of P_n (1, 2) and P_w/P_n (3, 4) on the heat elimination efficiency α for models I (1, 4) and II (2, 3) at R = 0.264 m ($[M]_0$ = 2 mol/l; $[AlCl_3]_0$ = 0.0045 mol/l; T_0 = 300 K).

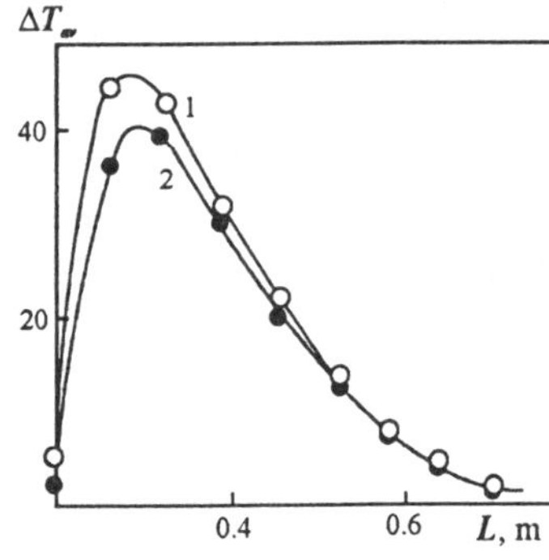

Figure 3.22. Changing of the mean cross-section temperature along the reactor length for models I (1) and II (2) at R = 0.066 m ($[M]_0$ = 2 mol/l; $[AlCl_3]_0$ = 0.0045 mol/l; T_0 = 300 K).

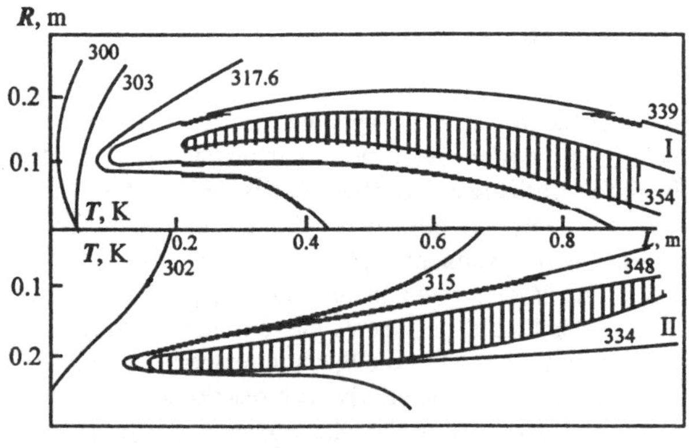

Figure 3.23. Fields of temperature distribution during isobutylene polymerization with external heat elimination for models I and II (R = 0.264 m). The field of maximal temperatures is shaded. Numbers near curves indicate T (K).

3.3. Heat regime of fast isobutylene polymerization.

To reach specific production aims it is necessary to be able to direct control of the product MMD to obtain narrow or wide distributions. In the synthesis of polyisobutylenes, control of this parameter it is easier to achieve by changing polymerization temperature. This requires efficient heat elimination from the reaction zone.

In high-power industrial reactors, this task is achieved by introduction of external and internal (inside) surfaces of heat exchange, through which refrigerants are circulated. The larger the reactor volume and the higher the exothermal effect of the reaction, the more difficult is the solution of this task by this method.

Feeding of the reactor by preliminarily cooled reagents allows the process transition into autothermal regime, when the reaction heat is completely or partially spent for heating the reaction mixture up to the reaction temperature. This principle formed the basis of ethylene polymerization under high pressure in an autoclave with mixer employed by the ICI Company. One of the variants of polyisobutylene production in the solution of methylene chloride was also introduced, when deep cooling (to –100°C) of the raw material feeding the reactor was possible due to existence of liquid ethylene at the plant.

Another method for heat elimination utilizes the principle of solvent boiling at a given temperature by pressure selection (or rarifying).

As heat capacity of organic liquids is low, the cooling regime of the raw material flow usually provides no required heat elimination, and it is combined with other methods.

In its turn, evaporation heat is rather high; moreover, boiling of the solvent in the volume intensifies the mixing process, stabilizes the temperature field, and excludes local heating. Such techniques are applied to technological processes.

3.3.1. Heat regime of polymerization in the absence of heat elimination.

The most important factor defining the efficiency of fast polymerization processes is temperature field in the reaction zone, which depends on the polymerization reactor geometry, concentrations of monomer and catalyst, rates of motion and turbulence of flows, method of component mixing, etc. [3, 5, 6, 9, 16, 18–21].

Intensive lengthwise and transversal mixing during conduction of fast polymerization processes, when $R < R_{cr}$ [9, 21], averages temperature in the reaction zone in such a way that MMD and mean MM become close to those typical for isothermal conditions at temperatures corresponding to adiabatic heat-up of the medium. However, polymerization abilities under adiabatic conditions are usually limited by significant decrease of MM and initiation of secondary processes at high temperatures, for example, degradation and cross-linking of macromolecules.

The main limit for temperature in the cationic polymerization (of isobutylene, for example), under which the process can be conducted, is the chain transmission to the monomer leading to formation of oligomers with the polymerization degree of several units already at temperatures above 300 K. That is why it is necessary to understand possibilities of isobutylene polymerization conduction under adiabatic conditions and to determine limiting factors of possible heating of the initial reaction mixture, as well as to eliminate the heat expense at the sacrifice of external thermostatic control or internal heat elimination.

The possibility of cooling the initial mixture in the case of the process realization, during which the boiling point of the reaction mass is not reached, is limited by the refrigerant boiling point, liquid ethylene, for example (b.p. = 183 K). Before the solvent boiling begins, temperature of the mixture will rise in proportion to the polymer yield in accordance with the correlation [20]:

$$\Delta M = \frac{c_p}{q_{ll}}\left(T - T_0\right) \approx 0.00175\left(T - T_0\right). \qquad (3.18)$$

Here c_p is the mean heat capacity of the reaction mixture (1.68 J/(g·grad)). If the range of temperature change of 90° is applied (liquid ethylene is the cooling agent), about 15 mass % of dissolved isobutylene can be polymerized at the sacrifice of the reaction mixture heating. If possibilities of the monomer solution cooling are lower, for example, the cooling agent is liquid ammonia (working temperature is 243 K), less than 10 mass % of the polymer can be produced during the time of heating the reaction mixture. Figure 3.24 displays dependence of the amount of resulting polymer on the temperature difference at the reactor entry and the reaction mixture exit from the reactor, which allows determination of either the mean reaction temperature or amount of resulting polymer, if one of parameters is known.

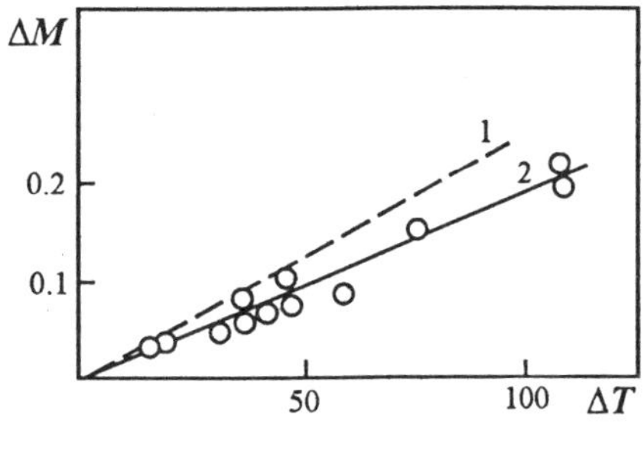

Figure 3.24. Theoretical (1) and experimental (2) dependencies of change of the reaction flow temperature at the reactor entry and exit by the reaction mixture on the amount of resulting polymer ($[M]_0$ = 2.15 mol/l; $[AlCl_3]_0$ = 0.0045 mol/l, T_0 = 213 K).

3.3.2. Heat regime of polymerization with reaction mass boiling.

One of the methods of elimination of chemical reaction heat is keeping of the assigned temperature in the reaction volume at the sacrifice of boiling of a solvent or (and) monomer. Boiling takes place in some end rather narrow range of temperatures, the b.p. depending, in a complex way, on the amount of polymerized monomer, MM of resulting polymer, etc. To accommodate strictly this dependence it is necessary to know the change of the boiling temperature and evaporation heat with the composition of the reaction mixture.

In the first approximation it may be considered that boiling proceeds at a constant temperature, but the heat of solvent evaporation for some processes is quite enough to end polymerization (initial concentration of the monomer in a chlorinated solvent is $[M]_0$ = 1.5-3.0 mol/l). Complete analysis of the heat balance was discussed in [20]. The reaction temperature increases until it reaches the b.p., and then the process proceeds at practically constant temperature until boiling ends after complete boiling out of the solvent and/or some part of the monomer. When liquid is completely boiled out, the temperature in the system begins to grow fast as a consequence of polymerization of liquid (if the monomer in the system is not boiling) or gaseous monomer [20, 21].

Simple calculation shows that if the isobutylene–isobutane fraction of hydrocarbons is used (50:50 mass%), only 70 mass% of isobutylene can be polymerized calculated for the initial polymer (b.p. = 261.3 K for isobutane; b.p. = 266 K for isobutylene; boiling heat of isobutane χ_s = 87.5 cal/g and of isobutylene χ_m = 95 cal/g) at the sacrifice of the isobutane boiling.

The amount of polymer obtained at the sacrifice of heating of the initial mixture of the solvent and monomer from the initial temperature to its b.p. can also be estimated [20]. At the moment, when the solvent and the non-polymerized monomer are converted into the gas phase, the heat balance equation becomes the following:

$$q_{11}\cdot\Delta M^* = c_p\cdot(T_{boil} - T_0) + \chi_s\cdot S_0 + \chi_m\cdot(M_0 - \Delta M). \qquad (3.19)$$

Therefrom

$$\Delta M^* = \frac{c_p\cdot(T_{boil} - T_0) + \chi_s\cdot S_0 + \chi_m\cdot M_0}{\chi_m + q_{11}} = \frac{c_p\cdot(T_{boil} - T_0) + \chi_s + (\chi_m - \chi_s)\cdot M_0}{\chi_m + q_{11}} =$$
$$= 0.37 + 0.023 M_0 \qquad (3.20)$$

Here S_0 = (1 – M_0) and M_0 are initial mass parts of the solvent and the monomer in the mixture, respectively.

Figure 3.25 displays dependencies of ΔM^* on M_0 calculated by Equation (3.20) for different cases and the real system. If

$$M_0 \le \frac{c_p\cdot(T_{boil} - T_0) + \chi_s}{q_{11} + \chi_s},$$

the monomer is polymerized completely at the sacrifice of the initial mixture heating and solvent boiling ($\Delta M^* = M_0$). If

$$M_0 > \frac{c_p\cdot(T_{boil} - T_0) + \chi_s}{q_{11} + \chi_s},$$

a part of the monomer will evaporate and pass into the gas phase, the greater this part is, the lower is χ_m. If heats of monomer and solvent are equal, an equal amount of the polymer can be obtained both at the expense of the monomer and solvent evaporation (Figure 3.25, curve 2).

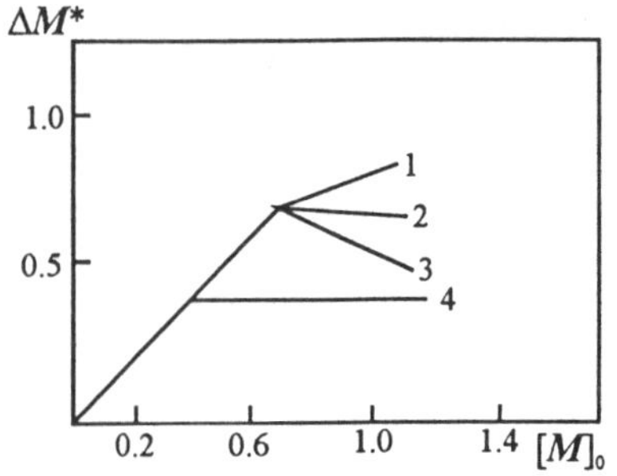

Figure 3.25. Dependence of the polymer yield ΔM* on the mass part of the monomer $(M)_0$ in the reaction mixture: 1 - $\chi_m > \chi_s$; 2 - $\chi_m = \chi_s$; 3 - $\chi_m < \chi_s$; 4 – the experimental curve for isobutane–isobutylene mixture (1:1); q_{ll} = 230 cal/g; c_p = 0.4 cal/g; χ_m = 95 cal/grad; χ_m = 87.5 cal/grad; T_0 = 186 K; T_{boil} = 266 K).

When boiling proceeds at constant temperature, the quantitative criterion of approximate constancy of temperature in the analysis of MM and MMD of the polymer is the correlation [20]:

$$T_{boil}^{max} - T_{boil}^{min} << \frac{RT_{boil}^{2}}{E_M - E_p}. \tag{3.21}$$

Notice that at the analysis of other characteristics of the resulting polymer, the quantitative criteria will be different.

If it is assumed that temperature in the reaction zone changes linearly with the amount of the resulting polymer (ΔM), two cases can be distinguished (Figure 3.26): when the solvent boils, and when both the solvent and the monomer boil.

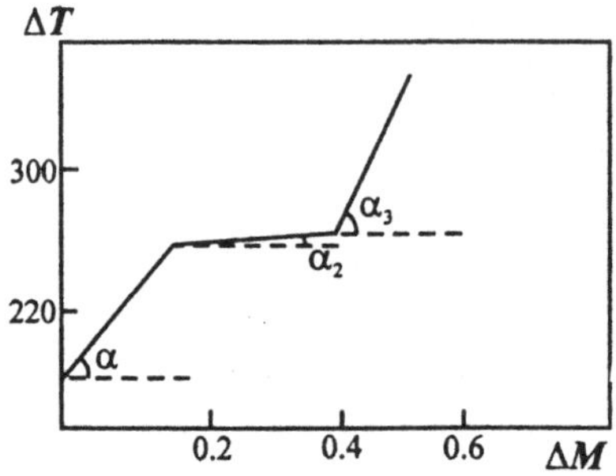

Figure 3.26. Typical curve of temperature change on the polymer yield in the reactor without external heat elimination with the reaction mass boiling.

a) If only the solvent boils:

$$\Delta T = \alpha_{boil}\Delta M \text{ at } T_{boil}^{min} < T < T_{boil}^{max}, \tag{3.22}$$

where $\alpha_{boil} = \alpha^{s}_{boil} = \alpha_2 = \left(T^{max}_{boil} - T^{min}_{boil}\right)\cdot q_{11}/\chi_s S_0$;

b) If both solvent and monomer boil:

$$\alpha_{boil} = \frac{T^{max}_{boil} - T^{min}_{boil}}{\Delta M^* - M} = \frac{\left(T^{max}_{boil} - T^{min}_{boil}\right)\cdot q_{11}\cdot\left(q_{11} - \chi_M\right)}{q_{11}\cdot\left(\chi_s S_0 + \chi_M M_0\right) - \chi_M \cdot c_p \cdot \left(T^{min}_{boil} - T_0\right)}. \tag{3.23}$$

In this case, if the assumed model is used, the validity criterion in the analysis of MM and MMD of the resulting polymer in the first approximation is the correlation (3.15), ΔT being the maximum deviation of the real curve $T = f(\Delta M)$ from a straight line.

If the monomer gas remained in the system, heating of the whole system will proceed after boiling ends, the mean heat capacity of the gas-liquid system being somewhat changed and polymerization heat rising significantly $q_{gl} = q_{ll} + \chi_m$. Consequently, the slope of the curve ΔT on ΔM in this range will equal: $\alpha_3 \cong q_{ll}/c_p$ (at $T > T^{max}_{boil}$, if the monomer remained liquid) or $\alpha_3 \cong (q_{ll} + \chi_m)/c_p$ (at $T > T^{max}_{boil}$, if the monomer passed into the gas phase). Here, c_p is the mean heat capacity of the liquid mixture. The schematic curve of ΔT versus ΔM in the whole possible range of temperature change is presented in Figure 3.26. Graphical dependence allows determination of distribution of mean temperatures by reactor zones with the multi-stage feeding of catalyst, if the amount of the resulting polymer in each zone is known, which is unequivocally connected with the amount of fed catalyst. In a particular case, when rate constants of propagation and termination of active polymerization centers do not depend on temperature, and the kinetic scheme of polymerization correlates with fast initiation and the first order of the chain propagation reaction rate by monomer and active center termination by their concentration, the expression for the polymer yield is the following:

$$\sum_{i=1}^{k} \Delta M_i = M_0\left[1 - \exp\left(-\frac{k_p}{k_{ter}}\cdot\sum_{i=1}^{k} A^*_i\right)\right]. \tag{3.26}$$

In the general case for the same scheme, but under temperature dependence on the ratio k_p/k_{ter}, it becomes:

$$\sum_{i=1}^{k} \Delta M_i = M_0\left[1 - \exp\left(-\sum_{i=1}^{k}\frac{k_p}{k_{ter}}\cdot A^*_i\right)\right]. \tag{3.27}$$

Hence, in the general case, the internal heat released as reaction mixture components get used up is a rather effective way of achieving thermostatic control of chemical reactions. Limiting of temperature of the reaction mixture during fast polymerization in the liquid phase if only a part of the monomer or solvent boils differently influences the reaction proceeding with dependence on the reaction zone radius R. In the area of small radii ($R < R_{cr}$), when a plane front of the reaction is formed and temperature is quite uniformly distributed in the reaction zone along the reaction zone radius R, the process rate and molecular–mass characteristics of the resulting polymer become independent of the initial temperature of raw materials in a definite temperature range (Figure 3.27) [21].

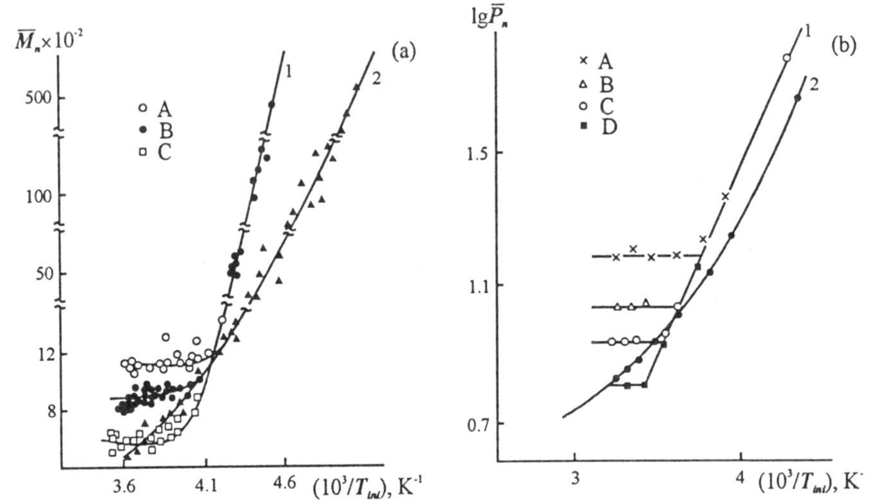

Figure 3.27. Dependencies of $\overline{M}_n$ (a) and $\lg \overline{P}_n$ of polyisobutylene (b) on raw material temperature at the reactor entry: 1 — the plane front regime $\left(R << \sqrt{D_t/k_0}\right)$; 2 — the flare regime $\left(R >> \sqrt{D_t/k_0}\right.$; a — experimental data, T_{boil} = 297 (A), 305 (B) and 313 K (C); b — calculation results, T_{boil} = 300 (A), 305 (B), 310 (C) and 320 K (D).

Polymerization of the monomer proceeds under conditions close to isothermal ones, the polydispersion index of the polymer approaching the most probable value, but depends on the difference between the initial temperature of raw materials (T_0) and the boiling point of the reaction mass (see Table 3.5), and the polymerization degree logarithm linearly depending on the reverse boiling point of the reaction mass (Figure 3.28). The data are shown for small (curves 1, 3) and large (curves 2, 4) diameter of the pipe reactor. Points of inflection correlate with the beginning of solvent boiling (isobutane). It is impossible to separate vapors of boiling solvent from the liquid reaction volume in the pipe reactor, as it is made in volumetric mixing reactors, where the reverse cooler is often used to return solvent into the reactor.

Foam is formed in the pipe reactor in the turbulent regime under boiling conditions. However, in this case it promotes agitation of the flow increasing the effective value of D_t and linear rate of the flow.

If the reaction zone radius R rises above R_{cr}, when the monomer breakthrough is observed in boundary areas of the reaction zone, the typical break in the curve of lgP_n dependence on $1/T_0$ is smoothed (see Figure 3.27a, curve 2), as zones with a broad selection of temperatures along reaction coordinates do exist. As a consequence, macromolecules of various sizes are formed, and the zone of reagent boiling is limited by the "flare" epicenter. As the "flare" zone includes less than 30% of the reaction volume, the inefficiency of internal heat elimination is obvious at rather great volumes of the reaction zone by virtue of local boiling of the reagents.

Calculation and modeling of fast polymerization with internal heat elimination at the sacrifice of reagents' boiling correlate with the experiment (Figure 3.28).

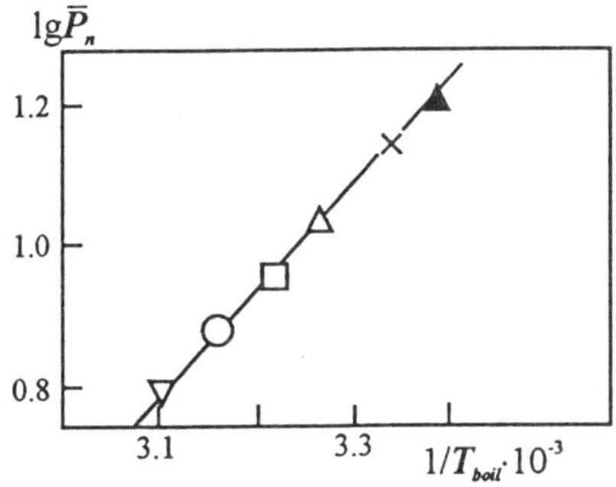

Figure 3.28. Dependence of $\ln P_n$ on $1/T_{boil}$ ($[M]_0 = 0.5$ mol/l; $[AlCl_3]_0 = 0.0045$ mol/l; $R = 0.08$ m; $\alpha = 0$).

Table 3.5

Dependence of the polydispersion index (P_w/P_n) of polymer and the boiling point of reagents ($\Delta T = T_{boil} - T_0$) on the difference of initial temperature of raw materials ($[M]_0 = 0.5$ mol/l; $[A^*]_0 = 4.5\cdot10^{-3}$ mol/l; $\alpha = 0$; $R = 0.08$ m) [21].

T_0, K	T_{boil}, K	ΔT	P_w/P_n
270	310	40	3.10
280	310	30	2.90
290	310	20	2.40
300	310	10	1.95
305	310	5	1.75
290	300	10	1.95
280	300	20	2.40
280	290	10	1.95

Hence formation of volumetric temperature gradient, i.e. polymerization conduction in the flare regime, complicates the process control. The thermostatic control effect on the reaction mass at the sacrifice of local boiling of reagents is obviously not enough in this case, which leads to widening of MMD of the resulting product.

3.3.3. Efficiency of external heat elimination.

Analysis of temperature change in the reaction flow during fast polymerization indicates that growth of the reaction zone radius over R_{cr} in the absence of external heat elimination leads to a significant change of temperature regime of the reaction. For example, with small radii of the pipe, the turbulent reactor difference in temperatures of central (T_0) and peripheral (T_R) flows of the reaction mixture is small, less than $\Delta T' = 2\text{-}5°$, just at the initial stage of polymerization (Figure 3.29).

When R grows (up to $R = 0.5$ m), $\Delta T'$ reaches the maximum value (50–60°), despite low (below 30 mass%) degrees of the monomer transformation. In this case, the temperature profile of the monomer polymerization is distorted and characterized by the area of maximal temperatures in the flow center in the absence of peripheral changes of temperature (Figure 3.29b). The latter means that the reaction does not practically approach the wall, i.e. formation of a volumetric temperature gradient takes place (along the radius R and length L of the reaction zone). Clearly, this, first of all, influences quality of the polymeric product, with widening of the MMD, in particular (Table 3.6). Growth of $\Delta T'$ in the reaction volume, as well as total growth of the flow temperature bound to heat release rising with the monomer content in the system ($\Delta T''$), also lead to widening of the polymer MMD at the sacrifice of accumulation of the low-molecular fraction.

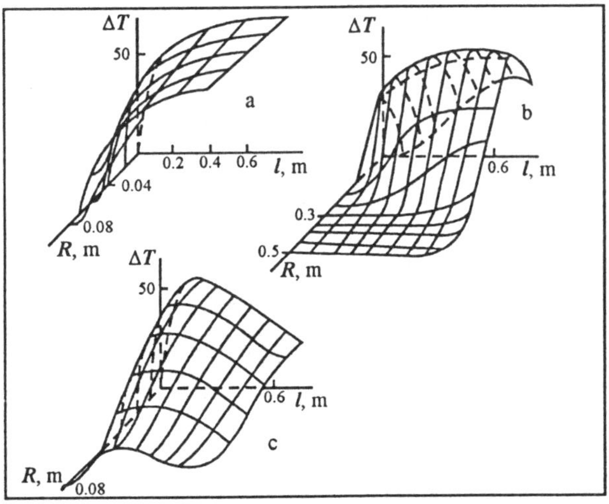

Figure 3.29. Change of the reaction flow temperature during fast polymerization in the absence of heat elimination (a, b) (at $R = 0.08$ m (a) and $R = 0.5$ m (b)) and in the presence of external heat elimination ($\alpha = 50$) at $R = 0.08$ m ($[M]_0 = 2.5$ mol/l; $[AlCl_3]_0 = 0.0045$ mol/l; $k_p = 10^5$ l/(mol·s); $k_{ter} = 20$ s^{-1}; $T_0 = 300K$).

During fast polymerization processes polymer is also formed, although with lower intensity, along the reaction zone length. This fact is remarkable, because, in spite of very high rates of isobutylene polymerization, in particular ($k_p \geq 10^5\text{–}10^6$ l/(mol·s)), the reaction also proceeds rather far from the place of injection of accompanying monomer and catalyst flows. Different temperature conditions along the radius and length of the reaction zone cause widening of the product MMD and formation of a noticeable amount of the low-molecular fraction in the polymeric product. Temperature reduction in the reaction zone at the sacrifice of heat elimination through the wall under medium transformation degrees (of about 50 mass%) decreases the amount of the low-molecular fraction contained in the final polymeric product (Figure 3.30).

Table 3.6

Influence of the reaction zone radius on $\Delta T'$, $\Delta T''$ and MMD of the product ($k_p = 10^5$ l/(mol·s); $k_{ter} = 20\ s^{-1}$; $\alpha = 0$) [22].

No. of experiment	R, m	$\Delta T'$ ($\Delta T''$)	P_w/P_n
1	0.08	0 (10)	2.0
2	0.10	7.0 (9.7)	2.5
3	0.125	9.0 (9.5)	2.9
4	0.25	11.0 (9.8)	3.4
5	0.5	9.3 (15.0)	4.2
6	0.08	1.0 (20.0)	2.2
7	0.5	19.0 (34.0)	5.9
8	0.08	1.5 (55.0)	2.5

Note. For No. 1-5 $[M]_0 = 0.5$ mol/l; $[A^*]_0 = 0.0045$ mol/l; $T_0 = 300$ K; No. 6-7 $[M]_0 = 1.0$ mol/l; No. 8 $[M]_0 = 2.5$ mol/l.

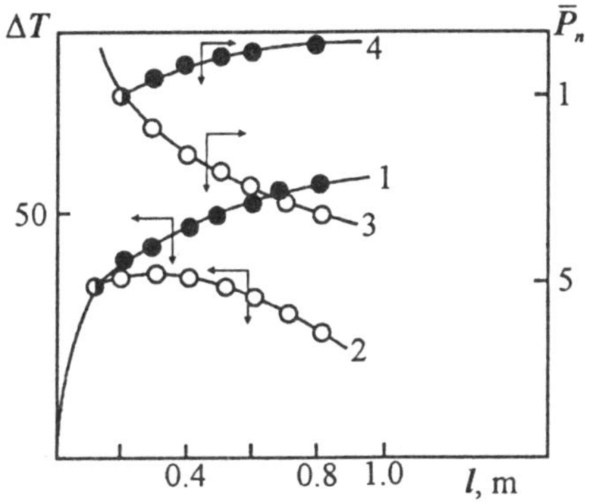

Figure 3.30. Change of reaction flow temperature and average polymerization degree of resulting polymer along the reaction zone length: 1, 3 — in the absence of heat elimination ($\alpha = 0$); 2, 4 — at external heat elimination ($\alpha = 20$) (under conditions from Figure 3.29).

Change of the average polymerization degree P_n along the reactor length under external heat elimination is characterized by a minimum on the curve, which corresponds to the maximum of the heat release during polymerization. If P_n decreases in the absence of the heat elimination, indicated by a noticeable increment of low-molecular fraction in the polymeric product, the process is stabilized already at $\alpha = 10$ and growth of the low-molecular fraction does practically stop.

The effect of external thermostatic control is smoothed with the growth of the reaction zone volume, and at $R = 0.5$ m the heat elimination becomes insufficient for the process character even at α growth up to 100. This is bound to the existence of the monomer breakthrough in the boundary area that plays the role of heat insulator.

3.4. Optimization of molecular–mass characteristics of polymer in fast isobutylene polymerization. The zone model.

Increase of MM and narrowing of MMD of the resulting polymer with conduction of fast polymerization processes in the liquid phase under adiabatic heating is a significantly complex problem. Of importance in this connection is the problem of obtaining polymeric products with maximal values of P_w, P_n and P_z and possibly narrowest MMD.

One possible method of controlling MM and MMD of the polymer is to change the method of catalyst feeding to the reaction zone, in particular, effecting multistage catalyst feeding to PTR. The zone model used to analyze such a reactor represents consecutive connection of several independent reaction zones, where the reaction mixture enters after the process end in the previous zone and to where a new portion of catalyst and/or monomer is delivered.

Realization of the zone model in the real PTR requires fulfillment of several conditions: 1) reaction zones must not intersect each other; that is why the distance between neighboring places of catalyst injection must be $l_i > V/k_{ter}$; 2) the reactor must fulfill the condition of formation of homogeneous temperature and concentration fields along the radius at the turbulent reacting flow in the reaction zone, i.e. the correlation $R \le R_{cr} = \sqrt{D_t/k_{ter}}$; 3) intensity of the heat release in each zone in accordance with the polymerization rate and amount of the monomer reacting in the reaction zone is low, and heat exchange coefficients are such that the requirement of the temperature constancy in each reaction zone must be fulfilled. To put it differently, a polymer must appear in the reaction zone with exponential MMD and MM correlating to temperature defined by the heat balance.

Notice that separate catalyst injection itself leads to a growth of the mean degree of the polymer molecular mass, all other factors being the same; growth of the number of catalyst injection zones along the pipe reactor raises possibilities of polymer MMD regulation with the monomer transformation degree.

3.4.1. Estimation of molecular-mass polyisobutylene characteristics in the absence of heat elimination at the multistage catalyst delivery into the reacting turbulent flow.

Calculations have shown that such conditions are fulfilled if [19]:

$$\Delta T_i \cong \frac{\rho \delta V q_{11} \Delta M_i}{\lambda_t} << \Delta T_2 = \frac{RT_i^2}{E_M - E_p}, \qquad (3.26)$$

where ΔT_i is the characteristic temperature drop in the i-th reaction zone; $\rho V q_{11} \Delta M_i$ is the rate of heat release in the i-th zone; ΔM_i is the amount of reacted monomer in the i-th zone; $\lambda_t \Delta T_i/\delta$ is the rate of heat "smearing" along the reactor; δ is the characteristic size of the reaction zone; ΔT_2 is the characteristic range of temperatures for MM change; T_i is the mean temperature in the i-th zone.

If heat released during polymerization is spent for heating the reaction mass, and the heat transmission through the external wall is absent, an unequivocal connection between the mean temperature in the i-th zone T_i, the initial temperature T_0 and amount of the monomer reacted in all previous zones, including the i-th one, exists.

Expressions for mean polymerization degrees appear as:

$$\left(P_n^i\right)^{-1} = \sum_{k=1}^{i} \Delta M_k \cdot \left(P_n^k\right)^{-1} \Big/ \sum_{k=1}^{i} \Delta M_k , \qquad (3.27)$$

$$P_w^i = \sum_{k=1}^{i} \Delta M_k \cdot P_w^i \Big/ \sum_{k=1}^{i} \Delta M_k , \qquad (3.28)$$

$$P_z^i \cdot P_w^i = \sum_{k=1}^{i} \Delta M_k \cdot P_z^i \cdot P_w^k \Big/ \sum_{k=1}^{i} \Delta M_k . \qquad (3.29)$$

Taking into consideration that reaction in each zone proceeds practically at constant temperature, we obtain:

$$\rho_w^k(j) = \frac{j}{\left(P_n^k\right)^2} \cdot \exp\left(-\frac{j}{P_n^k}\right) \text{ and}$$

$$\frac{1}{3} P_z^k = \frac{1}{2} P_w^k = P_n^k = \frac{k_p^0}{k_M^0} \cdot \exp\left(-\frac{E_p - E_M}{RT_k}\right). \qquad (3.30)$$

Hence there is a locked system of equations (3.27–3.30), which allow calculation of MMD and mean MM for arbitrary distribution of polymer yields in each zone (ΔM_i). To determine the connection between the polymer yield in a particular zone and the amount of fed catalyst, using the scheme of isobutylene cationic polymerization [4], we obtain:

$$\sum_{k=1}^{i} \Delta M_k = M_0 \cdot \left\{1 - \exp\left(-\frac{k_p}{k_{ter}} \cdot \sum_{k=1}^{i} [A^*]_k\right)\right\}. \qquad (3.31)$$

3.4.2. Two-zone model of pipe turbulent reactor.

Two-stage catalyst feeding to the reactor (the two-zone PTR model) transforms Equation (3.28) as follows:

$$P_w = 2\frac{\Delta M_1}{\Delta M}\cdot\frac{k_P^0}{k_M^0}\cdot\exp\left(-\frac{E_P - E_M}{RT_1}\right) + 2\frac{\Delta M_2}{\Delta M}\cdot\frac{k_P^0}{k_M^0}\cdot\exp\left(-\frac{E_P - E_M}{RT_2}\right). \quad (3.32)$$

Henceforward ΔM is the total amount of the polymer yield in the reactor; $\varepsilon = \frac{E_M - E_P}{R}$, $P_w^0 = \frac{k_P^0}{k_M^0}\cdot\left(-\frac{E_P - E_M}{RT_0}\right)$ is the polymerization degree of the polymer formed under isothermal conditions at temperature T_0 of fed monomer solution.

After transformation of Equations (3.27)–(3.29) and (3.32) we obtain:

$$\frac{P_w}{P_w^0} = \frac{x_1}{x}\cdot\exp(-x_1) + \frac{x_2}{x}\cdot\exp(-x), \quad (3.33)$$

$$\frac{P_n}{P_n^0} = \frac{x_1}{x}\cdot\exp(-x_1) + \frac{x_2}{x}\cdot\exp(-x), \quad (3.34)$$

$$\frac{P_z}{P_z^0}\cdot\frac{P_w}{P_w^0} = \frac{x_1}{x}\cdot\exp(-2x_1) + \frac{x_2}{x}\cdot\exp(-2x), \quad (3.35)$$

Here $x = \frac{\varepsilon\alpha\Delta M}{T_0}$, $x = x_1 + x_2$ [23].

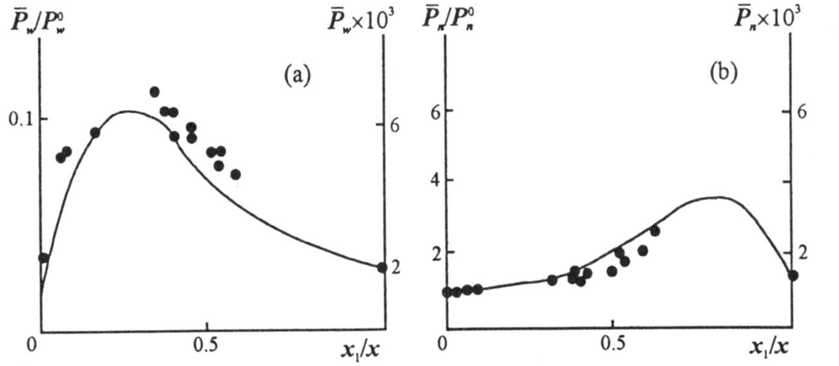

Figure 3.31. Dependence of the weighted mean (a) and average (b) MM of polyisobutylene on the part of resulting polymer in the first zone (x_1/x) in the two-zone reactor; continuous curves depict calculation results, points mark experimental data.

The parameter x at the end of the reaction (at the reactor exit) always significantly exceeds unity ($x >> 1$). Figure 3.31 displays the example of dependence of the mean polymerization degree of the products on the part of polymer resulting in the first zone under the two-zone regime of catalyst feeding. Consecutive catalyst feeding of each zone allows reaching of the maximum value of one or another mean MM. Reduction of the catalyst part in the first reactor causes decreasing of temperature in the reaction zone which leads to growth of the mean MM of resulting polymer in this zone. However, consequently, the amount of the resulting polymer in the first zone decreases. This leads to reduction of contribution of the polymer, formed in this zone, into the mean MM of the product. Competition of these factors leads to the extreme dependence of the mean MM of polymer on the part of catalyst, introduced into the first zone.

The two-zone method of catalyst feeding allows varying of two parameters: its overall amount (or gross-yield of the polymer ΔM) and catalyst distribution over two zones. Reduction of the total catalyst amount and, consequently, $\Delta M(x)$ always leads to increase of the mean MM and narrowing of MMD, the polymer yield and reactor productivity decreasing.

Growth of the number of catalyst feeding zones along the reactor length causes an increase of possibilities to regulate isobutylene polymer mean molecular masses, MMD growing with the monomer transformation degree.

3.4.3. Three-stage catalyst feeding.

Compared with the two-zone model, the three-zone model of the reactor (the three-stage catalyst feeding) gives great possibilities to regulate, control and optimize MMD, because one more independent controlled parameter occurs – the amount of catalyst feeding the second zone. The three-zone reactor gives the possibility of obtaining higher values of the mean MM. In this case, expressions for P_w and P_n will be represented as follows [23]:

$$\frac{P_w}{P_w^0} = \frac{x_1}{x} \cdot \exp(-x_1) + \frac{x_2}{x} \cdot \exp(-x_1 - x_2) + \frac{x - x_1 - x_2}{x} \cdot \exp(-x), \qquad (3.36)$$

$$\frac{P_n}{P_n^0} = \frac{x_1}{x} \cdot \exp(x_1) + \frac{x_2}{x} \cdot \exp(x_1 + x_2) + \frac{x + x_1 + x_2}{x} \cdot \exp(x), \qquad (3.37)$$

and $\left(\frac{P_w}{P_w^0}\right)_{max} \cong \frac{e^{1/e}}{e \cdot x} \cong \frac{1.5}{e \cdot x}$, which is approximately 1.5 times higher than the possible maximum value of P_w at the two-stage feeding of the catalyst.

Analogously, differentiating the expression (3.37) by x_1 and x_2 and putting partial derivatives $\frac{\partial P_n}{\partial x_1}$ and $\frac{\partial P_n}{\partial x_2}$ equal to zero, at $x > 1$ we obtain [23]:

$$\left(\frac{P_n}{P_n^0}\right)_{max} \cong \exp(-x) \cdot \frac{x}{1 + \ln(1 + \ln x)}. \qquad (3.38)$$

This value is also greater than that for the two-stage reactor. It is reached at

$$x_1 \cong x - \ln(x - \ln x) - \ln(1 + \ln x),\ x_2 \cong \ln(x - \ln x),\ x_3 \cong \ln(1 + \ln x). \qquad (3.39)$$

The three-stage catalyst feeding also gives other possibilities for optimizing MMD. For example, it is possible to obtain close to maximal values for two mean MM simultaneously by optimizing separately P_w by x_1 and P_n by x_2 at $x_1 \cong 1$, $x_2 \cong x - 1 - \ln x$. Substituting these values into equations (3.36) and (3.37), we obtain [19]:

$$(P_w)_{\max} \cong \frac{P_w^0}{e \cdot x}; \; (P_n)_{\max} \cong P_n^0 \cdot \exp(-x) \cdot \frac{x}{1+\ln x}. \tag{3.40}$$

In the general case, dependencies of P_w and P_n on two variables (x_1/x and x_2/x) represent surfaces in three-dimension space. Figure 3.32 shows an example of one of them. The absolute maximum of this surface for P_w is reached at $x_1 \cong 1 - 1/e$ and $x_2 = 1$. Analogous constructions can be made for P_n, P_z, P_w/P_n, P_z/P_w.

Surfaces of this type allow selection of a program of catalyst feeding and close to optimal values of catalyst concentrations, which should be fed to each of the zones to obtain a polymer with assigned (as much as possible) molecular-mass characteristics. Thus, a more or less narrow MMD is obtained near P_n maximums, or in the case when nearly all catalyst feeds the first zone.

3.4.4. The four-stage catalyst feeding.

Four-point catalyst feeding (four reaction zones) allows much higher increase of mean MM values or simultaneous optimization of three MMD parameters at constant x [19]. As an example the result of determination of x_1, x_2 and x_3 can be indicated for simultaneous obtaining of maximal P_w, P_n and P_z values. Values of optimal parameters x_1, x_2 and x_3 are really coincident with appropriate values for the two-zone reactor. The values $(P_n)_{\max}$, $(P_w)_{\max}$ and $(P_z)_{\max}$ also coincide. The only difference is that the two-zone reactor makes it possible to conduct only one of the variants and to reach the maximal value of either P_w, P_n or P_z, and in the four-zone reactor the same mean MM values can be reached, if necessary. Calculation formulae of optimal MM and MMD parameters and, respectively, results of numerical calculations are shown in Tables 3.7 and 3.8.

3.4.5. Multistage catalyst feeding (the reactor with infinite number of catalyst feeding zones without heat elimination).

The limiting case of catalyst feeding of the reaction zone is realized when there are many reactors and each of them is fed by an extremely small amount of the catalyst (the case of continuous catalyst feeding into rather long reaction zone with no intersection of the reaction zones). In kinetic respects, this model is identical to the ideal displacement reactor or to a periodical reactor with arbitrary catalyst feeding of the reaction zone, and expressions for the mean MM then become [19, 23]:

Table 3.7

Calculation formulae for optimal MMD parameters for different variants of catalyst feeding in the pipe adiabatic reactor.

Number of zones	$M_{\max}$	$\dfrac{\overline{P_w}}{P_w^0}$	$\dfrac{\overline{P_n}}{P_n^0}$	$\dfrac{\overline{P_z}}{P_z^0}$	$\dfrac{\overline{P_w}}{\overline{P_n}}$
1	—	e^{-x}	e^{-x}	e^{-x}	2
2	$\overline{P_n}$	$\sim xe^{-x}\left(1-\dfrac{2\ln x}{x}\right)$	$\dfrac{xe^{-x}}{\ln x+1}\left(1+\dfrac{2}{x}\right)$	$\sim xe^{-x}\left(1-\dfrac{\ln x}{x}\right)$	$\sim 2(\ln x+1)$
	$\overline{P_w}$	$\dfrac{1}{ex}\left(1+xe^{-(x+1)}\right)$	$\dfrac{xe^{-x}}{\ln x+1}\left(1+\dfrac{2}{x}\right)$	$\dfrac{1}{e}\left(1-xe^{-(x-1)}\right)$	$\sim 2\dfrac{e^{x-1}}{x}$
	$\overline{P_z}$	$\sim\left(e^{-x/2}/\sqrt{x}\right)\cdot\left(1-\dfrac{\sqrt{x}}{2}e^{-x/2}\right)$	$\sim e^{-x}\left(1-\dfrac{e^{-x/2}}{\sqrt{2}}\right)$	$1-2\sqrt{x}e^{-x/2}$	$\sim 2\dfrac{e^{x/2}}{\sqrt{x}}$
3	$\overline{P_w},\overline{P_n}$	$\sim\dfrac{1}{ex}$	$e^{-x}\dfrac{x}{\ln x+1}$	$\sim\dfrac{1}{e}$	$\sim 2\dfrac{e^{x-1}}{x^2}\cdot(\ln x+1)$
	$\overline{P_w}$	$\sim\dfrac{1}{ex}e^{1/e}$	$\sim e^{-x}$	$\sim\dfrac{1}{e}\left(1+\dfrac{1}{2e^2}\right)$	$\sim 2\dfrac{e^{x-1}}{x}e^{1/e}$
	$\overline{P_n}$	$-xe^{-x}(1+\ln x)\cdot\left(1-\dfrac{2\ln x}{x}\right)$	$\sim\dfrac{xe^{-x}}{1+\ln(1+\ln x)}$	$-xe^{-x}\left(1-\dfrac{\ln x}{x}\right)\cdot(1+\ln x)$	$\sim 2(1+\ln x)\cdot\left[1+\ln(1+x)\left(1-\dfrac{2\ln x}{x}\right)\right]$
4	$\overline{P_r},\overline{P_n},\overline{P_w}$	$\sim\dfrac{1}{ex}$	$\dfrac{xe^{-x}}{1+\ln x}$	~ 1	$\sim\dfrac{2e^{x-1}(1+\ln x)}{x^2}$
∞	—	$\dfrac{1}{x}\left(1-e^{x}\right)$	$\dfrac{x\left(1+e^{-x}\right)}{e^{x}}$	$\dfrac{1}{2}\left(1+e^{x}\right)$	$\dfrac{2e^{x}}{x^2}\left(1-2e^{-x}\right)$

Number of zones	$\frac{\overline{P_z}}{\overline{P_w}}$	x_1	x_2	x_3	x_4
1	3/2	x	—	—	—
2	$\frac{3}{2}\left(1+\frac{\ln x}{x}\right)$	$x-\ln(1+x-\ln x)$	$\ln(1+x-\ln x)$	—	—
	$-\frac{3}{2}x$	$\sim\left(1+e^{-(x+1)}\right)$	$\sim x - 1$	—	—
	$\sim\frac{3}{2}\sqrt{x}e^{x/2}$	$\sqrt{x}e^{x/2}\left(1-\frac{\sqrt{x}}{2}e^{-x/2}\right)$	$x-\sqrt{x}e^{x/2}$	—	—
3	$\sim\frac{3}{2}x$	~ 1	$x-\ln(1+x-\ln x)-1$	$\ln(1+x-\ln x)$	—
	$\sim\frac{3}{2}\frac{x}{e^{1/e}}\approx x$	$1-\frac{1}{e}$	1	$x-2+\frac{1}{e}$	—
	$\sim\frac{3}{2}\left(1+\frac{\ln x}{x}\right)$	$x-\ln(x-\ln x)-$ $-\ln(1+\ln x)$	$\ln(x-\ln x)$	$\ln(1+\ln x)$	—
4	$\sim\frac{3}{2ex}$	$\sim\sqrt{x}e^{-x/2}$	~ 1	$x-1-\ln(1+x-\ln x)$	$\ln(1+x-\ln x)$
∞	$\sim\frac{3}{4x}\left(1+2e^{-x}\right)$	$\ll\sqrt{x}e^{-x/2}$	—	$\ll 1$	$\ll \ln x$

Table 3.8

Numerical values of optimal MMD parameters for different variants of catalyst feeding in pipe adiabatic reactor ($x = 10$).

Number of zones	M_{max}	$\frac{\overline{P_n}}{P_n^0}\cdot 10$	$\frac{\overline{P_w}}{P_w^0}$	$\frac{\overline{P_z}}{P_z^0}$	$\frac{\overline{\overline{P_w}}}{\overline{P_n}}$	$\frac{\overline{\overline{P_z}}}{\overline{P_w}}$	x_1	x_2	x_3	x_4
1	—	0.5	$0.5\cdot 10^{-4}$	$0.5\cdot 10^{-4}$	2	3/2	10	—	—	—
2	$\overline{P_n}$	1.5	$4\cdot 10^{-4}$	$5\cdot 10^{-4}$	4.5	1.8	7.7	~ 2.3	—	—
	$\overline{P_w}$	0.55	0.035	0.35	$1.4\cdot 10^{-3}$	15	1	9	—	—
	$\overline{P_z}$	0.5	$0.22\cdot 10^{-2}$	1	90	$0.7\cdot 10^{-3}$		20	—	—
3	$\overline{P_w}, \overline{P_n}$	1.5	0.04	0.35	$0.5\cdot 10^{-3}$	13	1	6.7	2.3	—
	$\overline{P_w}$	~ 0.5	~ 0.042	~ 0.35	$2.3\cdot 10^{-3}$	10	0.65	1	—	—
	$\overline{P_n}$	~ 2.5	~ $1.3\cdot 10^{-4}$	$16.5\cdot 10^{-4}$	~ 10	1.8	~ 7	~ 2	~ 1	—
4	$\overline{P_r}, \overline{P_n}, \overline{P_w}$	~ 1.5	~ 0.04	1	$0.5\cdot 10^{-3}$	~ 40	$2.2\cdot 10^{-2}$	1	~ 6.7	2.3
∞	—	5.0	0.1	0.5	$0.4\cdot 10^{-3}$	7.5	10^{-3}	10^{-3}	10^{-3}	10^{-3}

Table 3.9

Calculation formulae for optimal MMD parameters for different variants of catalyst feeding in the pipe adiabatic reactor with boiling of the reaction mass.

$$\left(x = \frac{\alpha\varepsilon\Delta M}{T_0^2}; \frac{\alpha\Delta M}{T_0} << 1; x_k = \frac{\varepsilon(T_{boil} - T_0)}{T_0^2}; e^{x_k} = A >> 1; x_k \approx x \right).$$

Number of zones	M_{max}	$\frac{\overline{P_n}}{P_n^0}$	$\frac{\overline{P_w}}{P_w^0}$	$\frac{\overline{P_z}}{P_z^0}$	$\frac{\overline{P_w}}{\overline{P_n}}$
1	—	$1/A$	$1/A$	$1/A$	2
2	$\overline{P_n}$	$\frac{1}{A}\cdot\frac{x}{x-x_k+\ln x_k}\cdot\left(1-\frac{x}{x-x_k+\ln x_k}\right)$	$\frac{1}{A}\cdot\frac{x_k^2}{x}\cdot\left(1-\frac{2\ln x}{x_k}+\frac{x-x_k}{x_k^2}\right)$	$\frac{x_k}{A}\cdot\left(1-\frac{3\ln x}{x_k}-\frac{x-x_k}{x_k^3}\right)$	$2\frac{x_k^2(x-x_k+\ln x_k)}{x^2}$
	$\overline{P_w}$	$\sim\frac{1}{A}\cdot\left(1+\frac{1}{x}\right)$	$\frac{1}{ex}\cdot\left(1+e^{(-x_k-1)}\right)$	$\frac{1}{e}\cdot\left(1-xe^{-(x_k-1)}\right)$	$2\frac{A}{ex}$
	$\overline{P_z}$	$\sim\frac{1}{A}\cdot\left(1-\frac{e^{-x_k/2}}{\sqrt{x}}\right)$	$\frac{1}{\sqrt{xA}}\cdot\left(1-\frac{e^{-x_k/2}}{\sqrt{x}}\right)$	$1-2\sqrt{x}e^{-x_k/2}$	$2\sqrt{\frac{A}{x}}$
3	$\overline{P_w}, \overline{P_n}$	$\frac{1}{A}\cdot\frac{x}{x-x_k+\ln x_k}$	$\sim\frac{1}{ex}$	$\sim\frac{1}{e}$	$2\frac{A(x-x_k+\ln x_k)}{ex^2}$
4	$\overline{P_r}, \overline{P_n}, \overline{P_w}$	$\sim\frac{1}{A}\cdot\frac{x}{x-x_k+\ln x_k}$	$\sim\frac{1}{ex}$	~ 1	$2\frac{A(x-x_k+\ln x_k)}{ex^2}$
∞	—	$\frac{xe^{-x_k}}{x-x_k+1}\cdot\left(1+\frac{e^{-x_k}}{x-x_k+1}\right)$	$\frac{1}{x}\cdot\left[1-(x-x_k+1)e^{-x_k}\right.$	$\frac{1}{2}\left[1+(x-x_k+1)e^{-x_k}\right.$	$2\frac{x-x_k+1}{x^2}\cdot e^{-x_k}$

Number of zones	$\overline{P_z} / \overline{P_w}$	x_1	x_2	x_3	x_4
1	3/2	x	—	—	—
2	$\frac{3}{2} \cdot \frac{x}{x_k}$	$x_k - \ln(1 + x_k - \ln x_k)$	$x - x_k + \ln(1 + x_k - \ln x_k)$	—	—
	$\frac{3}{2x}$	$1 + e^{-(x_k - 1)}$	$\sim x - 1$	—	—
	$\frac{3}{2}\sqrt{Ax}$	$\sqrt{x}e^{-x_k/2} \cdot \cdot \left(1 - \frac{\sqrt{x}}{2} e^{-x_k/2}\right)$	$\sim x - \sqrt{x}e^{-x_k/2}$	—	—
3	$\frac{3x}{2}$	~ 1	$x - 1 - \ln(1 + x_k - \ln x_k)$	$x - x_k + \ln(x_k + 1 - \ln x_k)$	—
4	$\frac{3ex}{2}$	$\sim \sqrt{x}e^{-x_k/2}$	~ 1	$x - 1 - \ln(1 + x_k - \ln x_k)$	$x - x_k + \ln(x_k + 1 - \ln x_k)$
∞	$\frac{3x}{4}$	$x_i << \sqrt{x}e^{-x_k/2}$	—	$x_i << 1$	—

Table 3.10

Numerical values of optimal MMD parameters for different variants of catalyst feeding in pipe adiabatic reactor with boiling of the reaction mass.

Number of zones	M_{max}	$\frac{\overline{P_n}}{P_n^0}\cdot 10$	$\frac{\overline{P_w}}{P_w^0}$	$\frac{\overline{P_z}}{P_z^0}$	$\frac{\overline{P_w}}{\overline{P_n}}$	$\frac{\overline{P_z}}{\overline{P_w}}$	x_1	x_2	x_3	x_4
1	—	$0.7\cdot 10^{-2}$	$0.7\cdot 10^{-2}$	$0.7\cdot 10^{-2}$	2	3/2	10	—	—	—
2	$\overline{P_n}$	$1.4\cdot 10^{-2}$	$1.75\cdot 10^{-2}$	$3.5\cdot 10^{-2}$	2.5	3	3.2	6.8	—	—
	$\overline{P_w}$	$0.7\cdot 10^{-2}$	0.035	0.35	10	15	1	9	—	—
	$\overline{P_z}$	$0.7\cdot 10^{-2}$	$2.4\cdot 10^{-2}$	0.5	7	30	0.25	9.75	—	—
3	$\overline{P_w}, \overline{P_n}$	$1.4\cdot 10^{-2}$	0.035	0.35	5	15	1	2.4	6.6	—
4	$\overline{P_r}, \overline{P_n}, \overline{P_w}$	$1.4\cdot 10^{-2}$	0.035	1	5	40	0.25	1	2.15	6.6
∞	—	$1.2\cdot 10^{-2}$	0.1	0.5	~17	7.5	10^{-1}	—	—	—

$$\overline{P_w} = P_w^0 \cdot \frac{1-e^x}{x} \cong \frac{P_w^0}{x}; \ \overline{P_n} = P_n^0 \cdot \frac{x}{e^x - 1} \cong P_n^0 \cdot x \cdot e^{-x};$$

$$\overline{P_z} = \frac{1}{2} \cdot P_z^0 \cdot \frac{1-e^{-2x}}{1-e^{-x}} \cong \frac{1}{2} \cdot P_z^0 \cdot \left(1+e^{-x}\right). \tag{3.41}$$

This variant of the process conduction allows us to achieve maximal values of all distribution indexes.

The results of performed calculations are combined in Table 3.7, and Table 3.8 displays numerical values of appropriate values at $x = 10$.

It is important to know what amount of catalyst must feed each zone to make the deduced equation true. Most significant in this case are the initial zones, where the most high-molecular polymeric product is formed.

3.5. Molecular–mass characteristics of resulting polymer under the condition of internal heat elimination (at the sacrifice of the reaction mass boiling).

Boiling of a solvent and/or monomer permits a significant smoothing of the regime of fast polymerization conduction, obtaining of higher mean MM values and narrower MMD of resulting polymeric products [19, 21]. If the boiling point is reached in one of the zones, further splitting of zones is senseless. In this case, single catalyst feeding can cause the narrowest (exponential) MMD and the lowest mean value of MM of the resulting polymer [19]:

$$\frac{1}{3} \cdot \overline{P_z} = \frac{1}{2} \cdot \overline{P_w} = \overline{P_n} = \frac{k_P^0}{k_M^0} \cdot \exp\left(-\frac{\varepsilon}{T_{boil}}\right) = P_n^0 \cdot \exp(-x_{boil}), \tag{3.42}$$

where $x_{boil} = \frac{\varepsilon \cdot (T_{boil} - T_0)}{T_0^2}$.

3.5.1. The two-stage catalyst feeding under internal heat elimination.

The two-stage catalyst feeding allows an increase of the mean MM:

$$(P_w)_{max} \cong \frac{P_w^0}{e \cdot x} \cdot [1 + x\exp(1 - x_k)], \tag{3.43}$$

$$(P_n)_{max} \cong P_n^0 \cdot \exp(-x_k) \cdot \frac{x}{x - x_k + \ln x_k} \cdot \left(1 - \frac{x}{x - x_k + \ln x_k}\right), \tag{3.44}$$

$$(P_z)_{\max} \cong P_z^0 \cdot \left[1 - 2\sqrt{x} \cdot \exp\left(-\frac{x_k}{2}\right)\right]. \tag{3.45}$$

Maximal values of P_w and P_z are close to those obtained during conduction of polymerization without boiling, and P_n is much higher (e^{-x_k} instead of e^{-x}). The two-stage catalyst feeding of the reaction zone with boiling of solvent and/or monomer gives less gain in P_n compared with the one-stage model, as in the reactor without boiling. The greatest effect on narrowing of MMD of the resulting polymeric product is caused by boiling (Tables 3.9 and 3.10). Consequently, a much narrower MMD can be obtained if a polymer with the maximal P_w is not the objective. Thus, the width of the MMD for a polymer with the maximal P_n will equal [19]:

$$\frac{P_w}{P_n} \cong 2\frac{x_k^2 \cdot (x - x_k + \ln x_k)}{x^2} \cdot \left(1 - \frac{2\ln x_k}{x_k} + \frac{x - x_k}{x_k^2} + \frac{1}{x - x_k + \ln x_k}\right), \tag{3.46}$$

when $(x - x_k) << x_k^2 << \exp(x_k)$, $(x - x_k + \ln x_k)$,

$$\frac{P_z}{P_w} \cong \frac{3}{2} \cdot \frac{x}{x_k} \cdot \left(1 - \frac{\ln x_k}{x_k} - \frac{x - x_k}{x_k^2}\right), \tag{3.47}$$

which may already be quite close to the width of the exponential MMD (Table 3.9).

3.5.2. Three-stage catalyst feeding under internal heat elimination.

To optimize MMD somewhat additional freedom in action is presented by the three-zone reactor with boiling in the third zone. In this case, mean MM values are functions of two independent parameters (x_1/x and x_2/x):

$$\frac{P_w}{P_w^0} = \frac{x_1}{x} \cdot \exp(-x_1) + \frac{x_2}{x} \cdot \exp(-x_1 - x_2) + \frac{x - x_1 - x_2}{x} \cdot \exp(-x_k), \tag{3.48}$$

$$\frac{P_n}{P_n^0} = \frac{x_1}{x} \cdot \exp(x_1) + \frac{x_2}{x} \cdot \exp(x_1 + x_2) + \frac{x - x_1 - x_2}{x} \cdot \exp(x_k) \tag{3.49}$$

Appropriate surfaces in a graphical display of results are analogous to those depicted in Figure 3.32, with the only difference that absolute values of P_n are significantly higher in the reactor with boiling, and the distribution (MMD) is narrower. In the present case, maximums of P_z and P_w are obtained practically at the same comparatively low values of x_1 or at $(x_1 + x_2)$, the same as in the regime without internal heat elimination (boiling). Maximum values of the average MM of the polymer (P_n) are reached at $x_1 + x_2 \cong x_k - \ln x_k$ [19]. In this working regime, for example, maximum values of P_z and P_w can be reached simultaneously at

$x_1 = \sqrt{x} \cdot \exp\left(-\frac{x_k}{2}\right)$ and $x_2 \cong 1$, or P_w and P_n at $x_1 \cong 1$ and $x_2 = x_k - \ln x_k$, Tables 3.9 and 3.10.

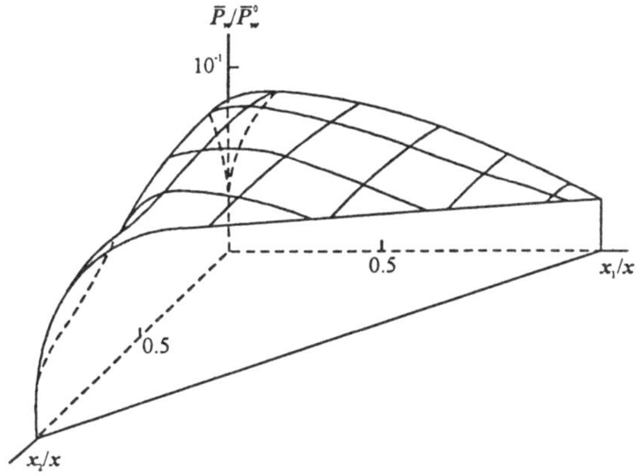

Figure 3.32. The surface of weight–average polymerization degree $\overline{P_w}$ on the resulting polymer distribution by zones.

3.5.3. Four-stage catalyst feeding under internal heat elimination.

In the four-zone reactor, as well as in the regime without boiling, there are possibilities either to increase much more the maximal values of one of mean MM or to reach simultaneously high values of P_z, P_w and P_n. For $x = 10$ and $x = 5$ numerical examples are presented in Table 3.10.

3.5.4. Multistage catalyst feeding under internal heat elimination.

In conclusion, let us discuss the example of continuous catalyst feeding by small portions up to T_{boil}, after which the reaction zone is fed by the remaining catalyst. Then expressions for the mean MM are as follows [19]:

$$P_w = \frac{x - x_k}{x} \cdot P_w^1 + \frac{x_k}{x} \cdot P_w^2, \tag{3.50}$$

$$\frac{1}{P_n} = \frac{x - x_k}{x} \cdot \frac{1}{P_n^1} + \frac{x_k}{x} \cdot \frac{1}{P_n^2}, \tag{3.51}$$

$$P_z \cdot P_w = \frac{x - x_k}{x} \cdot P_z^1 \cdot P_w^1 + \frac{x_k}{x} \cdot P_z^2 \cdot P_w^2, \tag{3.52}$$

where P_n^1, P_w^1, and P_z^1 are average polymerization degrees obtained in all zones up to boiling, and P_n^2, P_w^2, and P_z^2 are those in the boiling zone.

Expressions for P_n^1, P_w^1, and P_z^1 can be found while solving the task without boiling based on Equation (3.41), and for the second zone we obtain:

$$\frac{1}{3}P_z^2 = \frac{1}{2}P_w^2 = P_n^2 = P_n^0 \cdot \exp(-x_k). \tag{3.53}$$

A polymer with rather wide MMD and absolutely maximal values of all distribution indexes is formed in the multi-zone catalyst feeding method. The minimal values of all distribution indexes of the mean MM and their ratios are possessed by the resulting polymer using the single catalyst injection. Hence both boundary cases limit possible variations of MMD function indexes from above and from below [19].

To realize fast polymerization with indexes close to the multi-zone process, as well as for the regime without boiling, it is necessary to observe definite restrictions on volume of catalyst portions feeding each zone of the reaction volume. Thus, in the first stage at $0 < x_i < (2 \div 3)\sqrt{x} \cdot \exp\left(-\frac{x_k}{2}\right)$, catalyst portions must be small: $x_i < \sqrt{x} \cdot \exp\left(-\frac{x_k}{2}\right)$. In the second stage at $x < 2$ or 3 portions of the fed catalyst can be increased up to $x_i \ll 1$. In the third stage, at $x < x_k$, catalyst portions must be raised up to $x_i < \ln x_k$, and then, when the x_k value is reached, all the remaining catalyst must be injected. Calculation results are shown in Tables 3.8 and 3.10.

3.6. Efficiency of external heat elimination under conditions of reacting turbulent flow. The zone model.

Usually, growth of V at the fixed length of the reaction zone 1 leads to a reduction of the contact time of the reacting system with thermostatic controlling surface that must decrease the efficiency of external thermostatic control through the wall. However, growth of V during conduction of fast liquid-phase polymerization processes in PTR leads to a significant growth of D_t, which determines efficiency of heat and mass transfer in the turbulent flow. That is why increase of D_t under conditions of external heat elimination causes a noticeable decrease of temperature in the reaction zone. Hence "washing" of the reaction along the reaction zone length takes place [19]. Growth of V, i.e. intensification of heat and mass exchange in the system, is equal to "deceleration" of the polymerization itself in respect of the transfer process. That is why traditional methods of external heat elimination, usually of very low efficiency in fast standard technologically pre-resulting polymerization processes using mixing reactors, become noticeable and begin to play a significant role at high values of V and D_t, when pipe turbulent reactors (PTR) are used. In this case, the effect of external thermostatic control can be significantly increased at the sacrifice of the zone catalyst feeding.

In the general case, the external heat elimination (at the corresponding D_t) is able to stabilize and regulate the temperature field inside the reactor. This gives additional possibilities for product quality control during conduction of fast polymerization processes.

The conditions determining adequacy of the "zone" reactor model with external heat elimination to the real PTR formed above should be added to by: 1) length of the heat release zone must be greater than that of the reaction zone, i.e. $l_i >> V/k_{ter}$; 2) the heat amount eliminated from the reaction zone through the wall must be sufficiently lower than the heat amount extracted in this zone during polymerization:

$$2\pi \cdot R \cdot \frac{V}{k_{\text{ter}}} \cdot \alpha \cdot \left(T_i - \tilde{T}\right) << \pi \cdot R^2 \cdot V \cdot \rho \cdot q_{II} \cdot \Delta \text{M}_i. \quad (3.54)$$

Here T_i is temperature in the i-th zone; $\tilde{T}$ is temperature of the heat carrier; ΔM_i is the amount of the resulting polymer in the i-th zone.

If PTR with external heat elimination is used, the temperature in each zone is determined not only by the heat balance inside the reaction zone, but also by the amount of the reaction heat eliminated through the wall:

$$T_i = T_0 + \alpha \cdot \sum_{j=1}^{i} \Delta \text{M}_j - \sum_{j=1}^{i} \xi \cdot \left(T_{j-1} - \tilde{T}\right), \quad (3.55)$$

where $\alpha = q_{\text{II}}/c_{\text{p}}$, and numeration of cooling and reaction zones is such that the cooling zone before the reaction zone has the same index. In its turn, the amount of the resulting polymer in the i-th zone is bound to the amount of fed catalyst A* in accordance with the kinetic scheme of polymerization. For example, for isobutylene polymerization it is true that:

$$\Delta \text{M}_i = \text{M}_0 \cdot \exp\left(-\frac{k_{\text{p}}}{k_{\text{ter}}} \cdot \sum_{j=1}^{i-1} \text{A}_j^*\right) \cdot \left[1 - \exp\left(-\frac{k_{\text{p}}}{k_{\text{ter}}} \cdot \text{A}_i^*\right)\right]. \quad (3.56)$$

Knowing the temperature in each zone, it is possible to determine mean MM values:

$$\frac{1}{3} P_z^i = \frac{1}{2} P_w^i = P_n^i = \frac{k_{\text{p}}^0}{k_{\text{M}}^0} \cdot \exp\left(-\frac{E_{\text{p}} - E_{\text{M}}}{RT_i}\right) = \frac{k_{\text{p}}^0}{k_{\text{M}}^0} \cdot \exp\left(-\frac{\varepsilon}{T_i}\right). \quad (3.57)$$

Let us assume that heat transmission conditions are such that the greater part of accumulated heat is eliminated in each cooling zone. This means that polymerization proceeds under a quasi-isothermal regime at the average reaction temperature depending on the amount of the catalyst feeding each zone: $T_i = T_0 + \alpha \cdot \Delta\text{M}$. The temperature field close to the isothermal regime can be obtained by two methods: feeding the catalyst so that $\Delta\text{M}_i = \Delta\text{M}$, and eliminating practically all heat in the cooling zone (Figure 3.33, curve 1); or considering a definite ratio between the amount of catalyst feeding the i-th zone and the length of the i-th cooling zone (Figure 3.33, curve 2). In the latter case, the following conditions must be fulfilled:

$$(1-\xi_i)\cdot(T_1-T_0)+\alpha\cdot\Delta M_i=\alpha\cdot\Delta M_1 \tag{3.58}$$

or

$$\frac{\Delta M_i}{\Delta M_1}=\xi_i=1-\exp(-h\cdot l_i) \tag{3.59}$$

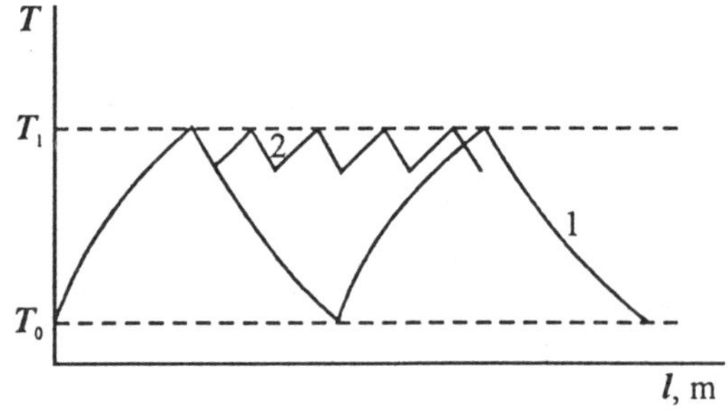

Figure 3.33. Two different quasi-isothermal regimes of operation of the pipe reactor with external heat elimination and multistage catalyst feeding; $hl_{cr} \geq 2$ (1) and << 1 (2).

In this regime, a polymer is obtained possessing an MMD close to the exponential one ($P_w/P_n \cong 2$) and a mean MM corresponding to temperature T_1, which is determined by the ratios of fed catalyst. To put this differently, the higher the mean MM to be obtained, the smaller must be the catalyst portion and greater the cooling zone, the minimal length of cooling zones being coincident with the regime of small portions of the catalyst feeding (Figure 3.33, curve 2). In this case (hl_{cr} << 1), the heat flow from the reactor to the heat carrier is a maximum [19]:

$$l_{\min}=\frac{1}{h}\cdot\left(\frac{\alpha\cdot\Delta M}{T_1-T_0}-1\right)=\frac{R\cdot\rho\cdot c_p\cdot V}{2k}\cdot\left(\frac{\alpha\cdot\Delta M}{T_1-T_0}-1\right). \tag{3.60}$$

Qualitative estimation of the possibility of isobutylene polymerization conduction under this regime is displayed in Table 3.11, which shows calculation results for the minimal length of the cooling zone in respect of the required temperature or the mean MM of the resulting polymer.

Dependence of $l_{\min}$ on other parameters is obvious and expressed by expression (3.60). In particular, $l_{\min}$ grows with V and R.

In the case close to the real one, when the heat elimination is insufficient, it is impossible to keep the temperature constant in the reaction zone, because $l_{\min}$ is significantly higher than real possibilities for reactor construction ($\xi_{cr} = h\cdot l_{cr}$ << 1). In this case, PTR must consist of three zones: two polymerization and one cooling. Then the following correlations are true for the temperature in such zones: $T_1 = T_0 + \alpha\cdot\Delta M_1$ and $T_2 = T_0 + \alpha\cdot\Delta M - \xi\cdot\alpha\cdot\Delta M_1$ [19].

Table 3.11

Values of the minimal length of the cooling zone to obtain isobutylene polymers with assigned MM for different T_0 (ρ = 0.7 g/cm^3, k = 0.05 cal/(cm^2·s·grad), M = 0.33 isobutane–isobutylene fraction, R = 2.5 cm, V = 0.5 m/s) [19].

T_0, K	l_{min}, m	T, K	MM (M_n)
180	17	210	22000
	5.6	250	4700
	3.0	280	2000
	2.0	300	1250
	1.3	320	820
230	24	250	4700
	9.0	280	2000
	6.0	300	1250
	3.7	320	820

Expressions for P_w, P_n and P_z for the two-zone reactor with the heat elimination are as follows:

$$\frac{P_w}{P_w^0} = \frac{x_1}{x} \cdot \exp(-x_1) + \frac{1-x_1}{x} \cdot \exp(-x + \xi \cdot x_1), \tag{3.61}$$

$$\frac{P_n}{P_n^0} = \frac{x_1}{x} \cdot \exp(x_1) + \frac{1-x_1}{x} \cdot \exp(x - \xi \cdot x_1), \tag{3.62}$$

where $x = \dfrac{\varepsilon \cdot \alpha \cdot \Delta M}{T_0^2}$, $x = x_1 + x_2$.

$$P_w^0 = 2\frac{k_p^0}{k_M^0} \cdot \exp\left(\frac{\varepsilon}{T_0}\right) = 2P_n^0. \tag{3.63}$$

The first zone only is sufficient for maximums of P_w and P_z (at $x >> 1$); that is why as in the case without heat elimination $x_1 \cong 1$ and $(P_w)_{max} \cong \dfrac{P_w^0}{e \cdot x}$. At the same time, the second zone is valuable for the maximum (P_n), where the heat elimination plays a significant role decreasing temperature in the reaction zone and raising the mean MM of the resulting polymer. That is why the expression for x_1 also changes [19]:

$$x_1 \cong \frac{x}{1+\xi} - \frac{1}{1+\xi} \cdot \ln \frac{1+\xi+x}{1+\xi+\xi^2 \cdot x}. \tag{3.64}$$

Consequently, higher mean MM values for the polymer, than those without heat elimination, are obtained:

$$\frac{P_w}{P_w^0} \cong \left[\frac{1}{1+\xi}\cdot\left(\frac{1+\xi+\xi^2\cdot x}{1+\xi+x}\right)^{1/1+\xi}+\frac{\xi}{1+\xi}\right]\cdot\left(-\frac{x}{1+\xi}\right)=\varphi(\xi,x)\cdot\exp\left(-\frac{x}{1+\xi}\right), \quad (3.65)$$

where $\varphi(x,\xi)=\begin{cases} x \text{ at } \xi\to 0 \\ 1 \text{ at } \xi\to 1. \end{cases}$

Figure 3.34 displays as an example curves of dependence of $(P_n)_{max}$ of the polymer on the heat elimination quality ξ or the cooling zone length (l). Values of $(P_n)_{max}$ grow, and MMD is narrowed ($P_w/P_n \Rightarrow 2$). In the limit, under good heat transmission ($\xi \Rightarrow 1$), the exponential distribution is obtained. The polymer is produced in two zones in equal amounts, and MM is in respect to temperature, equal to $T_0 + \alpha\Delta M/2$, i.e. significantly higher (at $x >> 1$) than during reaction proceeding without the heat elimination. Analogous is the effect of the heat elimination for three– and more stage catalyst feeding, the mean MM being increased and MMD being narrowed with raising of the heat exchange role with the external medium (heat carrier). The maximum effect is reached at highly effective heat exchange ($\xi \cong 1$) and uniform distribution of the resulting polymer by zones, temperatures in all zones being similar and equal to $T_0 + \alpha\Delta M/3$. Consequently, the polymer with close to exponential MMD and high average MM is obtained.

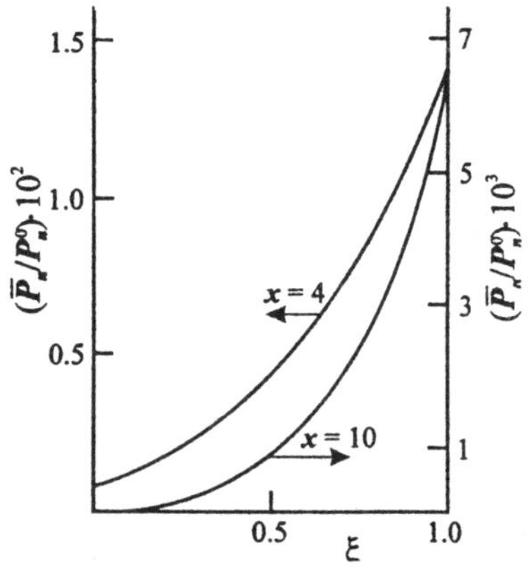

Figure 3.34. Dependence of the maximal average polymerization degree $\overline{P_n}$ on the part of eliminated heat or the cooling zone length ξ.

Introduction of external heat elimination and intensification of the heat elimination, including that at the sacrifice of increase of the heat transmission coefficient, heat capacity of the cooling agent, etc., leads to stabilization of temperature in the reactor and approach of the polymerization regime to isothermal.

The regularities obtained allow quantitative approach to selection of regulating parameters: V, R, α, M_0, T_0, etc. The more so, calculating MMD and mean MM of the resulting polymer with due regard for the final heat capacity of the cooling agent, optimal feeding and volumetric rate of the heat carrier can be selected. This gives additional possibilities to control fast exothermal polymerization processes.

References to Chapter 3

1. Berlin, Al.Al., Minsker, K.S., Sangalov, Yu.A., Oshmyan, V.G., Svinukhov, A.G., Kirillov, A.P., and Enikolopyan N.S., Vysokomol. Soed., 1980, Ser. A, v. 22(3), pp. 566–574. (Rus)
2. Berlin, Al.Al., Minsker, K.S., Sangalov, Yu.A., Novikov, D.D., Proznyak, T.I., Prochukhan, Yu.A., Kirillov, A.P., and Svinukhov, A.G., Vysokomol. Soed., 1979, Ser. B, v. 21(6), pp. 468–471. (Rus)
3. Minsker, K.S. and Berlin, Al.Al., Fast Polymerization Processes, Gordon and Breach Publ. Inc., 1996, 146 p.
4. Minsker, K.S. and Sangalov, Yu.A., Isobutylene and Its Polymers, Moscow, Khimia, 224 p. (Rus)
5. Berlin, Al.Al., Minsker, K.S., Prochukhan, Yu.A., and Enikolopyan, N.S., Polym.–Plast. Technolog. Eng., 1991, v. 30(2–3), pp. 253–297.
6. Berlin, Al.Al., Minsker, K.S., Prochukhan, Yu.A., and Enikolopyan, N.S., Vysokomol. Soed., 1989, Ser. A, v. 31(9), pp. 1779–1798. (Rus)
7. Poznyak, T.I., Lisitsyn, D.M., Novikov, D.D., Berlin, Al.Al., D'yachkovskii, F.S., Prochukhan, Yu.A., Sangalov, Yu.A., and Minsker, K.S., Vysokomol. Soed., 1980, Ser. A, v. 22(6), pp. 1424–1427. (Rus)
8. Berlin, Al.Al., Minsker, K.S., Prochukhan, Yu.A., Karpasas, M.M., and Enikolopyan, N.S., Doklady AN SSSR, 1986, v. 287(1), pp. 145–148. (Rus)
9. Berlin, Al.Al., Minsker, K.S., Prochukhan, Yu.A., Karpasas, M.M., and Enikolopyan, N.S., Internat. Polym. Sci. Technol., 1986, v. 13(11), pp. 95–97.
10. Minsker, K.S., Berlin, Al.Al., Prohukhan, Yu.A., and Enikolopyan, N.S., Vysokomol. Soed., 1986, Ser. B., v. 28(6), pp. 466–469. (Rus)
11. Minsker, K.S., Berlin, Al.Al., Prochukhan, Yu.A., and Enikolopyan, N.S., Internat. Polym. Sci. Technol., 1986, v. 13(12), pp. 90–93.
12. Prochukhan, Yu.A., Berlin, Al.Al., Minsker, K.S., and Enikolopyan, N.S., Doklady AN SSSR, 1986, v. 287(3), pp. 682–685. (Rus)
13. Prochukhan, Yu.A., Minsker, K.S., Berlin, Al.Al., Karpasas, M.M., Kompaniets, V.Z., and Enikolopyan, N.S., Doklady AN SSSR, 1988, v. 298(6), pp. 1428–1430. (Rus)
14. Kompaniets, V.Z., Konoplev, A.A., Berlin, Al.Al., Prochukhan, Yu.A., Minsker, K.S., Karpasas, M.M., and Enikolopyan, N.S., Doklady AN SSSR, 1987, v. 287(5), pp. 1129–1132. (Rus)
15. Prochukhan, Yu.A., Minsker, K.S., Karpasas, M.M., Berlin, Al.Al., Bakhtiyarova, R.Kh., and Enikolopyan, N.S., Vysokomol. Soed., 1988, Ser. A, v. 30(6), pp. 1259–1262. (Rus)
16. Berlin, Al.Al., Karpasas, M.M., Kompaniets, V.Z., Minsker, K.S., Konoplev, A.A., Minsker, S.K., Prochukhan, Yu.A., and Enikolopyan, N.S., Doklady AN SSSR, 1989, v. 305(2), pp. 365–368. (Rus)
17. Berlin, Al.Al., Kompaniets, V.Z., Konoplev, A.A., Minsker, K.S., Minsker, S.K., Prochukhan, Yu.A., Ryabenko, E.A., and Enikolopyan, N.S., Doklady AN SSSR, 1989, v. 305(5), pp. 1143–1146. (Rus)
18. Minsker, K.S., Berlin, Al.Al., Svinukhov, A.G., Prochukhan, Yu.A., and Enikolopyan, N.S., Doklady AN SSSR, 1986, v. 286(5), pp. 1171–1173. (Rus)
19. Berlin, Al.Al., Prochukhan, Yu.A., Minsker, K.S., and Enikolopyan, N.S., Vysokomol. Soed., 1991, Ser. A, v. 33(2), pp. 243–269. (Rus)

20. Berlin, Al.Al., Prochukhan, Yu.A., Minsker, K.S., Tumanyan, E.A., Aleksanyan, G.G., and Enikolopyan, N.S., Vysokomol. Soed., 1988, Ser. A, v. 30(11), pp. 2436–2440. (Rus)
21. Berlin, Al.Al., Minsker, K.S., Prochukhan, Yu.A., Karpasas, M.M., and Enikolopyan, N.S., Vysokomol. Soed., 1986, Ser. B., v. 28(6), pp. 461–465. (Rus)

Chapter 4. Isobutylene copolymerization.

4.1. Specific features of cationic copolymerization.

To determine relative rates of copolymerization reactions between M_1 and M_2, usual correlations, deduced for free-radical copolymerization are appropriate.

In the cationic copolymerization in binary systems, the growing carbocation is able to react with any of the copolymers M_1 and M_2. In this case, two types of carbonium ions occur, able to start four differing chain propagation reactions:

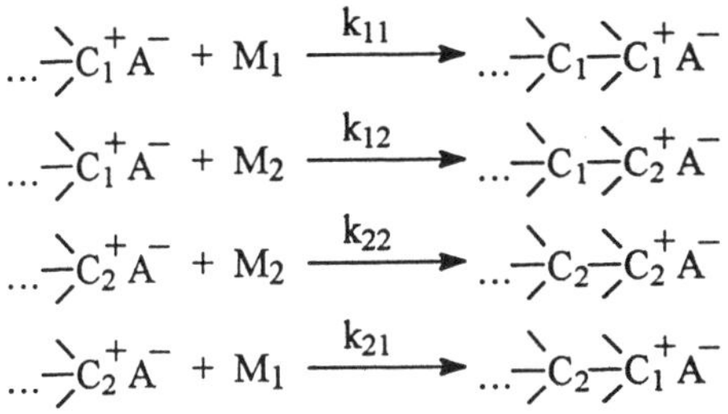

Composition of a copolymer molecule is determined by copolymerization constants r_1 and r_2 ($r_1 = k_{11}/k_{12}$, $r_2 = k_{22}/k_{21}$), which bind rates of carbonium ion interaction between one monomer and another one, existing in the mixture. Composition of the copolymer also depends on appropriate initial concentrations of monomers in the reaction mixture. The composition of resulting copolymer, based on copolymerization constants r_1 and r_2 and concentrations of monomers is described by the Mayo–Lewis equation:

$$\frac{d[m_1]}{d[m_2]} = \frac{[M_1]}{[M_2]} \cdot \frac{r_1[M_1]+[M_2]}{[M_1]+r_2[M_2]}, \tag{4.1}$$

where m_1 and m_2 are monomer units 1 and 2 in the copolymer; $[M_1]$ and $[M_2]$ are their initial concentrations in the initial mixture. In more suitable form, equation (4.1) is written down via molar parts of monomers in the initial mixture f_1 and f_2 and in copolymer F_1 and F_2 (the Feinman–Ross equation):

$$F_1 = 1 - F_2 = \frac{d[m_1]}{d[m_1]+d[m_2]}, \tag{4.2}$$

$$f_1 = 1 - f_2 = \frac{[M_1]}{[M_1]+[M_2]}. \tag{4.3}$$

Substituting these expressions into equation (4.1), we obtain:

$$F_1 = \frac{f_1(r_1 f_1 + f_2)}{r_1 f_1^2 + r f_1 f_2 + r_2 f_2^2}, \tag{4.4}$$

$$\frac{f_1(1-2F_1)}{(1-f_1)F_1} = r_2 + \left[\frac{f_1^2(F_1-1)}{(1-f_1)^2 F_1}\right] \cdot r_1. \tag{4.5}$$

Graphically (the Feinman–Ross method), equation (4.1) is approximated by a straight line with tangent equal to r_1, which cuts r_2 on the ordinates axis[1].

Two important consequences follow from equations (4.1) and (4.4): a) individual rates of homopolymerization are of no importance for predicting the composition of copolymer; b) quantitative data (r_1 and r_2) are tolerable, if they are obtained in experiments during which initial concentrations of monomers can be assumed constant. Usually, this requirement is fulfilled with suitable approximation only at low transformation degrees (3–5%) or with continuous (by a special program) loading of a monomer.

To obtain dependence of the copolymer composition on conversion in the periodic regime of synthesis in accordance with (4.1), it is necessary to fulfill the requirement:

$$\frac{[M_1]}{[M_2]} = X, f_1 = \text{const.}$$

The amount $\Delta[M_1]$ to be added to the system up to the current moment of time τ per specific volume of the reaction mixture, under the condition that copolymer is absorbed by the monomer M_1, and under selected conditions it must be added while the reaction proceeds, is estimated from the expression [1, p. 197]:

$$\Delta[M_1] = \left[M_1^0\right]\left(\frac{r_1 X + 1}{r_2 + X} - 1\right)^{\left(1 - e^{-k_{eff} \cdot \tau}\right)}, \tag{4.6}$$

where $\left[M_1^0\right]$ is the initial concentration of the monomer 1; k_{eff} is the effective constant of the expense rate $[M_2]$ (k_{eff} includes concentrations of both active centers and appropriate propagation constants).

To apply equation (4.6), it is necessary to determine the k_{eff} value. For this purpose, the initial rate of decline of both monomers must be measured by the kinetic curve tangent, and the value obtained must be divided by the initial concentration of the monomers.

The following integral equation is obtained by integrating equation (4.1):

---

[1] At the present time computers are widely used for determination of r_1 and r_2 values. There are special mathematical methods, which enable the difference between experimental and theoretical (calculated) values to be minimized at all experimental points.

$$\lg\frac{[M_2]_P}{[M_1]_P}=\frac{r_2}{1-r_2}\lg\frac{[M_2^0]\cdot[M_1]_t}{[M_1^0]\cdot[M_2]_t}-\frac{1-r_1\cdot r_2}{(1-r_1)(1-r_2)}\cdot\lg\frac{(r_1-1)\cdot\frac{[M_1]_t}{[M_2]_t}-r_2+1}{(r_1-1)\cdot\frac{[M_1^0]}{[M_2^0]}-r_2+1}.$$

Composition for composition, copolymers may differ by the sequence of unit distribution in the chain, that is to say, by the microstructure of the chain [1, p. 198].

The limiting case is a strictly alternating copolymer ($r_1 = r_2 = 0$). If $r_1 = 0$, the monomer M_1 participates in the chain as single units. In all other cases, length of the block of each comonomer changes during the process. Average lengths of chain blocks of M_1 ($\bar{v}_1$) and M_2 ($\bar{v}_2$) are equal to

$$\bar{v}_1 = r_1\cdot\frac{[M_1]}{[M_2]}+1, \quad (4.8)$$

$$\bar{v}_2 = r_2\cdot\frac{[M_2]}{[M_1]}+1. \quad (4.9)$$

The distribution function for blocks is approximated by the equation:

$$M_1=\frac{\left(r_1\cdot\frac{[M_1]}{[M_2]}\right)^{n-1}}{\left(1+r_1\cdot\frac{[M_1]}{[M_2]}\right)^{n}}. \quad (4.10)$$

In some cases, formation of composite-inhomogeneous macromolecules or heterogeneity in the composition during copolymerization of two monomers is observed. This may be indicated by a change of "immediate" composition of the copolymer as the reaction proceeds. However, even the copolymer obtained under conditions of constant composition, for example, in an azeotropic mixture, is inhomogeneous in composition due to accidental chemical reaction. There always exists some probability that the present macromolecule accidentally contains more chains of one monomer than expected from the average composition. Heterogeneity of the composition of this type depends on the mean molecular mass of the polymer and on the size of blocks of comonomers, i.e. on values of r_1 and r_2 and concentrations of monomers. The lower is the mean molecular mass and the higher are values of r_1 and r_2 (the longer are blocks of copolymers), the higher is the composite heterogeneity of the copolymer with respect to composition.

Much higher composite heterogeneity is possible with ionic polymerization due to probable presence of several types of AC and different values of r_1 and r_2.

A specific feature of cationic copolymerization compared with the radical one is the complex character of dependence of the monomers' activity on their structure, nature of solvent, catalyst and temperature. As in the case of any joint polymerization, the process is described by equation (4.1), if the ratio of constants of the cross-chain propagation k_{12}/k_{21} differs from unity by not more than a decimal degree, i.e. activity of the monomers or stability of appropriate growing ions differ little from each other.

To put it differently, $k_{12} \cdot C_1^+ \cdot M_2 = k_{21} \cdot C_2^+ \cdot M_1$ which follows from the principle of stability of the chain propagation. However, modeling of reactions of the cross-chain propagation has shown that even if the stability principle is fulfilled for every comonomer separately, their mutual influence in the cross-chain propagation leads to non-stationary kinetics [2]. Reactivity of growing ions on account of influence of substituents in the α-position changes more than in the case of growing radicals. This supposes noticeable differences in activity of cation-active monomers. If, for example, the reactivity M_1 is much higher than M_2, the reaction rate $k_{12} \cdot C_1^+ \cdot M_2$ will be low, and the condition of the cross-chain propagation is not fulfilled. In this case, full differential equations must take into account partial rates of initiation W_n' and W_n'' of the monomers M_1 and M_2 and partial rates of termination W_0' and W_0'' of cations C_1^+ and C_2^+ [3]:

$$\frac{dC_1^+}{dt} = W_n' + k_{21}C_2^+M_1 - k_{12}C_1^+M_2 - W_0' = 0,$$

$$\frac{dC_2^+}{dt} = W_n'' + k_{12}C_1^+M_2 - k_{21}C_2^+M_1 - W_0'' = 0.$$

At greater differences in activities of monomers, the rates of initiation and termination will affect the ratio of concentrations of propagating chains C_1^+/C_2^+ and, in its turn, composition of the copolymer. If the usual equation of the copolymer composition is used, wrong values of constants of relative activity of monomers are obviously obtained.

In cationic polymerization, the effect of the medium must be taken into account. Dielectric constant ε (polarity) of the medium, which determines the state of an ionic pair together with other factors, must change during the process, excluding cases of azeotropic copolymerization, equality of ε values of both monomers or sufficient prevalence of ε of the solvent over that of the monomers. That is why the rate constants of four reactions of propagation can change in accordance with non-constancy of the mixture composition during copolymerization. In the general case, more than two monomers and more than two types of AC are contained in the systems, which limits application of the equation of copolymerization and the importance of numerical values of copolymerization constants r_1 and r_2.

It is thought that in non-polar solvents an ionic pair is selectively solvated by a more polar monomer, which results in growth of the rate of its interaction with a cation. Polar solvents supersede monomers from the solvate cover; that is why relative activity of monomers is mainly determined by features of their chemical composition. The following sequence of activities of monomers in attachment to carbonium ions was obtained: vinyl esters >> isobutylene > styrene > vinyl acetate > isoprene > butadiene. As is obvious, isobutylene is one of the most active comonomers.

In concordance with the acid–base nature of cationic polymerization, monomers react with an electrophilic center in accord with their nucleophilicity. Addition of a strong nucleophilic monomer to a cation with increased electrophilic characteristics is more probable than attachment to their own cations. That is why the difference between relative reactivities of monomers must be greater than that between appropriate rate constants of homopolymerization of applied monomers k_{11} and k_{22}.

Data on the counterion influence on the copolymer composition are ambiguous. Counterions of different chemical nature can differently "permeate" comonomers to a propagating macrocation on account of steric or electrostatic obstacles. This influence is more typical for non-polar solvents, whereas in some cases in polar solvents it is absent, because ionic pairs are separated in polar solvents, and the solvate cover contains no monomer.

A most important role is usually played by the cocatalyst forming structures of the counterion, which may be determined by comparatively large radii of counterions $[MeX_{n+1}]^{\delta-}$ and changes of their values at the transition from one counterion to another, for example [4, p. 138]:

Counterion	BF_3OH^-	$AlCl_3OH^-$	$TiCl_4OH^-$
Ionic radius, nm	0.28	0.38	0.42

The composition of copolymer is significantly affected by the temperature of the process. Usually, the product $r_1 \cdot r_2$ grows with temperature and often significantly exceeds unity.

The influence of isobutyl-vinyl alcohol with isobutylene on the composition of copolymer, produced under the effect of $AlBr_3$ (hexane, 195 K), with small amounts of an electron-donor (concentration of $AlBr_3$ is $4.8 \cdot 10^{-3}$, isobutylene — 11.0, vinyl ester — 0.8 mol/l, donor agent — 0.67 vol.%), is noted in the work [5]. In particular, content of isobutylene chains in the copolymer in the presence of diethyl ester is 60, dimethylsulfoxide is 52, dimethylformamide is 89 mol%, whereas in the absence of an electron-donor it is only 10 mol%.

Hence it can be stated that the cationic copolymerization is highly sensitive to changing of components of appropriate systems and conditions under which the process occurs. At the present time, there is no general theory that unites, explains and predicts experimentally observed changes of r_1 and r_2, i.e. changes of the reactivity of monomers in cationic polymerization.

4.2. Reactivity of monomers in copolymerization with isobutylene.

For isobutylene $Q = 0.033$, $e = -0.96$. Data on constants of copolymerization — the relative reactivity of isobutylene with other monomers — are generalized in a number of monographs and reviews, for example, [6, pp. 466–494]. Published data on constants of isobutylene copolymerization with some comonomers, correlated with the most important conditions of the experiment, are summarized in Table 4.1. As mentioned above, the constants r_1 and r_2 significantly depend on experimental conditions and, with consideration for features of the cationic polymerization discussed in the previous Section, are determined with greater errors than for free-radical copolymerization of two monomers. The effect of chain alternation in copolymers is practically absent for isobutylene. In most cases, the product $r_1 \cdot r_2 > 1$, and structure of the copolymer is characterized by existence of blocks of homopolymers. In the case of some pairs of monomers, for example, isobutylene–isobutylvinyl ester, no formation of copolymers occurs [3]. Critical consideration of

Table 4.1

Constants r_1 and r_2 in the cationic copolymerization of isobutylene (M_1).

Monomer M_2	r_1	r_2	$r_1 \cdot r_2$	Catalyst	Solvent	Temperature, K	Reference	Note
Styrene	1.10	0.99	1.09	$AlBr_3$	Benzene	273	[4]	
	14.90	0.53	7.90	$AlCl_3$	Nitrobenzene	273	[4]	
	12.20	2.80	34.16	$SnCl_4 \cdot H_2O$	Benzene	273	[4]	
	8.60	1.25	10.75	$SnCl_4 \cdot H_2O$	Nitrobenzene	273	[4]	
	1.60	0.17	0.25	$SnCl_4$	Ethyl chloride	273	[7]	a
	3.1	1.1	3.41	$SnCl_4$	Sulfur dioxide	195	[7]	
	2.2	1.1	2.42	$SnCl_4$	Sulfur dioxide–nitroethane	195	[7]	
	1.5	1.0	1.5	$SnCl_4$	Sulfur dioxide–dichloroethane	195	[7]	
	0.8	0.5	0.4	$SnCl_4$	Sulfur dioxide	195	[7]	
	1.9	0.6	1.14	$SnCl_4$	Sulfur dioxide–benzene	195	[7]	
	3.5	0.33	1.15	γ-radiation	Ethyl chloride	195	[6, 7]	
	9.02±0.77	1.99±0.24	1.79	$AlCl_3$ (0.12%)	Methyl chloride	181	[6, 7]	
	2.51±0.55	1.21±0.06	3.04	$AlCl_3$ (0.5%)	Methyl chloride	243	[6, 7]	
	2.36±0.06	0.76±0.22	1.79	$AlCl_3$ (0.14%)	Methyl chloride	243	[6, 7]	
	1.66±0.02	0.42±0.22	0.70	$AlCl_3$ (0.5%)	Methyl chloride	183	[6, 7]	
	1.79±0.22	0.24±0.22	0.43	$AlCl_3$ (0.14%)	Methyl chloride	183	[6, 7]	
	3±1	0.6±0.3	1.8	$AlCl_3$	Methyl chloride	170	[6]	
	2.8±0.2	1.4±0.2	3.92	$C_2H_5AlCl_2$	Methyl chloride	173	[7]	
	1.78±0.10	1.20±0.10	2.13	$TiCl_4$	Toluene	195	[7]	
	0.37±0.07	2.41±0.12	0.89	$TiCl_4$	Hexane	195	[7, 8]	
	0.54±0.24	1.20±0.11	0.64	$TiCl_4$	Hexane	253	[7]	

	2.63±0.52	5.50±0.55	14.46	$TiCl_4$	Hexane–methylene (75/25)	253	[7, 8]	
	3.25±0.25	2.75±0.25	8.93	$TiCl_4$	Hexane–methylene chloride (50/50)	253	[7, 8]	
	4.11±0.19	1.70±0.07	6.98	$TiCl_4$	Hexane–methylene chloride (25/75)	253	[7, 8]	
	4.48±0.28	1.08±0.07	4.83	$TiCl_4$	Methylene chloride	253	[7, 8]	
	1.51±0.03	2.44±0.15	3.68	$TiCl_4$	Hydrogen sulfide	253	[7]	
	1.53±0.03	2.61±0.4	3.99	$TiCl_4$	Decalin	253	[7]	
	0.75±0.15	1.30±0.15	0.97	$TiCl_4$	Methylcyclohexane	253	[7]	
n-Chloro-styrene	1.01	1.02	1.03	$AlBr_3$	Hexane	273	[6, 7]	
	14.7	0.15	2.20	$AlBr_3$	Nitrobenzene	273	[6, 7]	
	22.5	0.7	15.70	$AlBr_3$	Nitromethane	273	[6, 7]	
	8.6	1.2	10.3	$SnCl_4$	Nitrobenzene	273	[6, 7]	
Butadiene	115±15	0.01±0.01	—	$AlCl_3$	Methyl chloride	170	[6, 7]	b
	43	0	—	$C_2H_5AlCl_2$	Methyl chloride	173	[7, 9]	c
1-Chloro-butadiene	5.5	0.1	0.55	$C_2H_5AlCl_2$– *tert*-C_4H_9OH	Methylene chloride	193	[10]	d
Isoprene	2.5±0.5	0.4±0.1	1	$AlCl_3$	Methyl chloride	176	—	
	0.80	1.28	1.02	$C_2H_5AlCl_2$	Hexane	193	[7]	
	1.17	1.08	1.26	$C_2H_5AlCl_2$	Hexane–methyl chloride (88/12)	193	[7]	
	1.90	1.05	1.99	$C_2H_5AlCl_2$	Hexane–methyl chloride (50/50)	193	[7]	
	2.15	1.03	2.21	$C_2H_5AlCl_2$	Hexane–methyl chloride (12/88)	193	[7]	
	2.5±0.5	—	—	$C_2H_5AlCl_2$-Cl_2	Methyl chloride	238	[7]	
	2.25	0.4	0.9	$C_2H_5AlCl_2 \cdot 0.3H_2O$	Methyl chloride	188	—	

	0.98±0.03	0	—	$C_2H_5AlCl_2 \cdot H_2O$	Isopentane	188	—	
	0.99±0.03	0	—	$(C_2H_5)_{1.5}AlCl_{1.5} \cdot H_2O$	Isopentane	188	—	
	2.4±0.35	4.5±3.0	—	$BCl_3 \cdot C_6H_5C(CH_3)_2OCOCH_3$	Methyl chloride, methylene chloride	243	[11]	e
α-Methyl-styrene	0.27±0.08	1.41±0.25	0.38	$TiCl_4$	Methylene chloride	195	[7]	
	1.2	5.5	6.6	$TiCl_4$	Toluene	195	[7]	
	1.1±0.2	2.2±0.2	2.42	$TiCl_4$	Methylene chloride	201	[7]	
Cis-1,3-pentadiene	5	0	—	$(C_2H_5)_2AlCl$-Cl_2 (*tert*-C_4H_9Cl)	Methyl chloride	218	[5, 12]	f
Trans-1,3-pentadiene	2.3	0	—	$(C_2H_5)_2AlCl$-Cl_2 (*tert*-)C_4H_9Cl	Methyl chloride	218	[5, 12]	f
Cyclo-pentadiene	0.60±0.15	4.5±0.5	2.7	$BF_3 \cdot O(C_2H_5)_5$	Toluene	195	[7]	
	0.73±0.17	1.86±0.20	1.35	$BF_3 \cdot O(C_2H_5)_5$	Methyl chloride	195	[7]	
	0.21±0.02	6.3±0.6	1.32	$SnCl_4$-tetracyanoethylene	Toluene	195	[7]	
	0.80±0.15	1.55±0.20	1.24	$SnCl_4$–tetracyano-ethylene	Methylene chloride	195	[7]	
Cyclo-hexadiene	0.77±0.20	1.05±0.25	0.80	$SnCl_4$–tetracyano-ethylene	Toluene	195	[7, 13]	
	0.98±0.20	1.07±0.45	1.04	$SnCl_4$–tetracyano-ethylene	Methylene chloride	195	[7, 13]	
α-Cyclopen-tadiene-ethyl-ferrocene	1.50	0.07	0.10	$AlCl_3$	Methylene chloride	183	[14]	
Trans,trans-2,4-hexadiene	6.0	0.88	5.28	$C_2H_5AlCl_2$	Methylene chloride–pentane (50/50)	343	[15]	g

Trans,cis-2,4-hexadiene	0.74	1.8	1.3	$C_2H_5AlCl_2$	Methylene chloride–pentane (50/50)	343	[15]	g
Cis,cis-2,4-hexadiene	2.3	1.5	3.45	$C_2H_5AlCl_2$	Methylene chloride–heptane	343	[15]	g
Trans-hexa-triene-1,3,5	0.58	1.57	0.92	$C_2H_5AlCl_2$	Methyl chloride–heptane (50/50)	203	[16]	h
Vinyl acetylene	8	0.13	1.04	BF_3	Methyl chloride	173	[6]	i
Vinylbenzyl chloride	4.5±1	0.7±0.1	3.15	BF_3	Ethyl chloride	170	[6]	
Isoprenylben-zyl chloride	0.4	1	0.4	BF_3	Ethyl chloride	173	[6]	
α-Pinene	11.0	0.09	0.99	$C_2H_5AlCl_2$	Ethyl chloride	168	[17]	
β-Pinene	0.27	3.0	0.81	$C_2H_5AlCl_2$	Ethyl chloride	223	[7, 18]	
	0.52	1.9	0.99	$C_2H_5AlCl_2$	Ethyl chloride	195	[18]	
	0.77	1.5	1.15	$C_2H_5AlCl_2$	Ethyl chloride	173	[18]	
	1.0	0.95	0.95	$C_2H_5AlCl_2$	Ethyl chloride	163	[18]	
	1.0	1.0	1.0	$C_2H_5AlCl_2$	Ethyl chloride	153	[18]	
	1.0	1.0	1.0	$C_2H_5AlCl_2$	Ethyl chloride	143	[18]	
	5.4	0.2	1.08	BF_3	Methyl chloride	173	[19]	
5-Chlorobi-cyclo-[2,2,1]-heptene-2	7.3±1.5	0.09±0.02	0.65	BF_3	Heptene	195	[20]	
Indene	1.1±0.02	2.2±0.02	2.42	$TiCl_4$	Methylene chloride	195	[21]	
Isobutyl-vinyl ester	0.033	28.0	0.92	$AlBr_3 \cdot 0.8(C_2H_5)_2O$	Toluene	195	[22]	
	0.030	12.6	0.37	$AlBr_3 \cdot 1.2(C_2H_5)_2O$	Toluene	195	[22]	
	0.37	22.4	8.28	$AlBr_3 \cdot 1.2(C_2H_5)_2O$	Toluene	195	[22]	
	0.12	12.6	1.51	$AlBr_3 \cdot 2.5(C_2H_5)_2O$	Toluene	195	[22]	

	0.52	13.25	6.89	$AlBr_3 \cdot 1.1HC(O)N(CH_3)_2$	Toluene	195	[22]	
	0.46	10.5	4.83	$AlBr_3 \cdot 2.3HC(O)N(CH_3)_2$	Toluene	195	[22]	
	0.19	55.0	10.45	$AlBr_3 \cdot 0.8(C_2H_5)_2O$	Hexane	195	[22]	
	0.18	50.0	9.0	$AlBr_3 \cdot 1.2(C_2H_5)_2O$	Hexane	195	[22]	
	0.13	12.5	1.62	$AlBr_3 \cdot 1.2(CH_3)_2SO$	Hexane	195	[22]	
	0.033	4.6	0.14	$AlBr_3 \cdot 2.5(CH_3)_2SO$	Hexane	195	[22]	
	0.27	52.0	14.0	$AlBr_3 \cdot 0.6HC(O)N(CH_3)_2$	Hexane	195	[22]	
	0.51	47.0	23.97	$AlBr_3 \cdot 1.2HC(O)N(CH_3)_2$	Hexane	195	[22]	
Trimethyl-sylylvinyl ester	0.37±0.2	5.82±0.86	2.15	Sulfocationite KU-2-8–$C_2H_{55}AlCl_2$ ($SO_3H/Al = 6$)	Toluene	253	[23, 24, 25]	
Dimethyl-(butoxy)-sylylvinyl ester	1.09±0.10	6.1±3.5	6.64	Sulfocationite KU-2-8–$C_2H_{55}AlCl_2$ ($SO_3H/Al=6$)	Toluene	253	[23, 24, 25]	

Notes. a — estimation by the Kelen–Tudos method testifies an approximate value of r_1 and r_2; b — it is supposed that $r_1 \cdot r_2 = 1$; c — polymerization of diene into trans-1,4- and 1,2-positions; d — polymerization of diene into 3,4-position;e — statistic copolymer with 1,4-structure of the diene part under low content of M_2 and a block-copolymer with partially cycled structure under high content of M_2; f — diene reacts into the 1,2-position; g — diene reacts into the 2,5-position; h — diene reacts into the 2,6-position; i — determination by the first experiment; j — approximate values due to heterogeneity of the catalyst.

data on cationic copolymerization and, in particular, isobutylene copolymerization [3], was performed with the help of the Kelen–Tudos linear graphical method — a modification of the well-known Feinman–Ross one [26]. It is based on the approximate type of equation describing cationic copolymerization in the presence of more than two AC in the system ($\sim > C_1^+$ and $\sim > C_2^+$), which leads to a dependence of determining copolymerization constants on the mixture composition and concentration of monomers. If the data on the copolymer composition can be presented by a linear dependence in η—ξ coordinates of the equation

$$\eta = \left(r_1 + \frac{r_2}{a}\right)\xi - \frac{r_2}{a},$$

where a is a constant ($a > 0$), $\eta = \frac{G}{a+F}$, $\xi = \frac{F}{a+F}$, $G = \frac{x(y-1)}{y}$, $F = \frac{x^2}{y}$, $x = \frac{M_1}{M_2}$, and $y = \frac{d[\mathrm{m}_1]}{d[\mathrm{m}_2]}$, the process of copolymerization is characterized by values of r_1 and r_2 (otherwise they are meaningless).

Hence the Kelen–Tudos method enables the suitability of the usual two-parameter model of copolymerization and correctness of determination of constants to be estimated, and also serves to refine their values. Consideration of a number of binary systems, based on the example of isobutylene–styrene, isobutylene–alkylvinyl ester ones in the presence of various catalysts, has shown that in most cases the values of copolymerization constants published are evidently displayed with great errors or are meaningless. Taking into consideration difficulties in application of the copolymerization equation to cationic systems, it would be more correct to compare relative activities of monomers or parameters of copolymerization of monomers as "indexes of relative reactivity" (for a sequence of typical reactions) and to use non-linear mathematical methods to estimate them [27]. These indexes display no connection with the composition of the monomeric mixture and resulting copolymer, but can serve to conduct a high-quality analysis of reactions of the chain cross-propagation, in particular, of selectivity of this reaction (the ratio r_1/r_2 or $\lg(r_1/r_2)$).

Applicability of this approach was demonstrated using the example of isobutylene copolymerization with isobutylvinyl ester (see Table 4.1). When electron-donor compounds are introduced into a hardly polymerizing mixture of these monomers, besides total reduction of the reactivity, convergence of relative reactivities of macroions is observed, because the more electrophilic carrier of the chain must be solvated better than the less electrophilic one. In this case, an increase in the proportion of the less active monomer (isobutylene) in the copolymer (fractionated samples) was observed. In its turn, proportion of the true resulting copolymer reduces with growth of an ester monomer content in separate fractions. Increase of concentration of the donor agent enables selectivity of the reaction of the chain cross propagation to be reduced. An analogous effect is also caused by growth of copolymerization temperature. The connection between selectivity and nature of the catalyst and the solvent is complex.

The processes observed are qualitatively coincident with the behavior of carbocationic centers of both hard and soft acids. The above-mentioned growth of selectivity in selection of monomers, closely related to their activity, on coordination carbonium ions with a soft aromatic base promotes increase in the role of the softness factor. On the contrary, coordination of hard bases (O-, N-, and S-containing compounds) cause the opposite effect: growth of hardness of ionic centers and, as a consequence, reduction of selectivity of the chain cross-propagation during copolymerization take place. The temperature influence is analogous: reduction of dissociation of ionic pairs with the temperature increase correlates with growth of hardiness of polymeric carbonium ions. Table 4.2 shows values of reactivity indexes for the isobutylene–isobutylvinyl ester pair, calculated with the help of linear graphical and non-linear mathematical methods [27].

Even if a refined calculation using a computer is performed, significantly different values of indexes of the relative reactivity of monomers are obtained in relation to the chosen method. Consequently, various structures can be ascribed to isobutylene copolymers (a wide range of changes of r_1/r_2 and $r_1 \cdot r_2$).

Table 4.2

Comparison of indexes of the relative reactivity for isobutylene–isobutylvinyl ester system[2] (n-hexane, $AlCl_3$ ($4.82 \cdot 10^{-3}$ mol/l), dimethylsulfoxide ($2.1 \cdot 10^{-3}$ mol/l)).

Method	r_1	R_2	r_1/r_2	$r_1 \cdot r_2$
Feinman–Ross	0.000±0.007	0.811±0.009	0	0
Kelen–Tudos	0.251±0.025	7.956±5.210	0.031	1.996
Non-linear (numerical integration by the trapezium method)	0.134±0.047	1.726±0.350	0.072	0.231
Non-linear (integral)	0.880±0.296	9.635±1.783	0.092	8.478

The approximate understanding of the reactivity of isobutylene in copolymerization with some monomers (styrene [28], butadiene [29], etc.) is confirmed by non-coincidence of experimentally measured content of dyads and triads of chains in polymers with the ones calculated by copolymerization constants, existing in the literature. All this testifies to the necessity to conduct further deeper studies in the field of cationic copolymerization of isobutylene, for which formation of true copolymers can be thought proved for single cases only. Among them may be systems with low-active cyclic and diene monomers. In this aspect, of interest is the example of isobutylene copolymerization with vinylsilyl esters (see Table 4.1). Although the heterogeneous type of catalyst somehow limits consideration of r_1 and r_2 as the real constants of copolymerization, its application gives the possibility of conducting regulated synthesis of products [30]. Simplicity of separation of homopolymers of isobutylene and trimethylsilylvinyl (dimethylbutoxisilylvinyl) ester by the method of fractional dissolution (the heptane–water system) and the negative result obtained in analysis of products of isobutylene polymerization in the presence of vinylsilyl esters confirm formation of the real copolymers. The solid catalytic surface, irreversibly sorbing more active silyl ester and preventing its homopolymerization, plays the role of regulator of activity of comonomers. The

[2] Constants r_1 and r_2 were calculated using a computer.

presence of isobutylene simplifies desorption of polymeric products and, hence, is the condition of participation of vinylsilyl monomer in the copolymerization.

Data on isobutylene copolymerization can be presented as relative reactivities of monomers in relation to substituted *tert*-butyl cation, expressed by values opposite to r_1 (k_{12}/k_{11}). The relative reactivity of the carbocation with its own monomer equals 1 (Table 4.3).

Approximate data on the relative activity of isobutylene and other monomers in cationic copolymerization are also obtained by application of appropriate values of the parameter *e* from the Alfrei–Price equation, obtained for free-radical polymerization [6, p. 483]. If the effect of end chains is neglected at the stage of polymeric chain propagation, it is assumed that $r_1 = k_{11}/k_{12} = \exp(e_2 - e_1)$ (where e_1 and e_2 are parameters characterizing polar properties of comonomers). Published values of parameters *e* are: –1.77 (isobutylvinyl ester), –1.27 (α-methylstyrene), –1.03 (indene), –0.96 (isobutylene), –0.80 (styrene), –0.33 (*n*-chlorostyrene). They reflect correctly the behavior of isobutylene in copolymerization with the mentioned monomers.

Table 4.3

Relative reactivities of some monomers in respect to the carbonium ion of isobutylene.

Monomer (M_2)	Catalyst	Solvent	Temperature, K	Relative reactivity
Isobutylene	$SnCl_4$	Carbon tetrachloride	273	1.0
Styrene	$SnCl_4$	Ethyl chloride	273	0.62
Isoprene	$AlCl_3$	Methyl chloride	173	0.44
Butadiene	$AlCl_3$	Methyl chloride	173	0.009
β-Pinene	$C_2H_5AlCl_2$	Ethyl chloride	173	1.3
Isopropenyl-benzyl chloride	BF_3	Ethyl chloride	173	2.5

4.3. Specific methods of deriving copolymers of isobutylene.

Progress towards initiation of isobutylene polymerization, in particular, application of complex catalysts, high-energy radiation, or combined methods of influence on monomeric systems, broadens the possibilities of synthesis of copolymeric products. Recently, data have appeared on free-radical and other methods of the synthesis of isobutylene copolymers of various structure, which enable the number of monomers copolymerizing with it to be increased. Contrary to the traditional initiation by cationic catalysts, they lead to isobutylene copolymers of strictly alternate structure or with increased affinity to alternation of various monomeric chains (copolymerization constants smaller than 1).

1:1-Copolymers of isobutylene with chlorotrifluoro- and tetrafluoroethylene were synthesized using free-radical initiation in the mass, solution or emulsion, radiation by γ-rays [31–34]. For isobutylene–vinylidene chloride and isobutylene–

dichlorodifluoroethylene, copolymerization was activated by γ-radiation from ^{60}Co of preliminarily prepared channel complexes of thiocarbamide with appropriate monomers [34]. In the case of vinyl chloride, the copolymer is enriched by a chlorine-containing monomer ($r_{iso\text{-}C_4H_8}$ = 0.34; $r_{C_2H_3Cl}$ = 2.11; 333 K) [35]. 1:1-Copolymers of isobutylene with maleic anhydride, a diethylfumarate, were obtained, which are interesting as emulsifying agents, sorbents of metals from solutions, and modifiers of rubbers [36]. The structure of the polymeric products does not depend on the composition of the initial mixture of monomers and the type of initiator. Copolymers of isobutylene–cyclopentene–SO_2, polyisobutylene–norbornene–SO_2 block-copolymer and isobutylene with butadiene were synthesized by applying the principle of alternation applied [37, 38].

Alternating copolymerization proceeds in isobutylene–acrylic monomer double systems [39, 40] or isobutylene–methacrylate–maleic anhydride triple systems [41], isobutylene–methylacrylate–vinylbenzyl chloride [42] under simultaneous effect of the source of free radicals and the complexing additive HAOS. The generalized mechanism of formation of polymeric products by radical, cationic and other methods has been suggested [43, 44]. A complex of accepting monomer and HAOS is considered as a complex monomer with reduced electronic density on the double bond. Radical initiator activates the main state of the complex and generates stable radicals, which enter the donor–acceptor interaction with isobutylene. Thus, a hexagonal cyclic structure is formed. The donor–acceptor interaction between the monomer existing in the complex with a radical and copolymer (1:1-complex of acrylate with HAOS) enables a double complex to be formed at the propagating end of the chain. As a conjugated system is formed in this case, the chain propagation can proceed according to the adjusted reaction as a simultaneous addition of two monomers at once. An important role is played by aluminum, linked to the propagating end of a macromolecule because, at its detachment, the localized radical becomes quite labile in order to provide inclusion of two monomeric units at once into the chain. Usage of complex formers of the optimum acidity provides the required degree of stabilization of the propagating radical.

Hence the factor of AC complexing is important for both classic cationic copolymerization and alternating copolymerization with participation of isobutylene, which proceeds not necessarily as a cationic process. The mentioned analogy seems pertinent, because it has been shown that the principles of possible realization of continuous cationic polymerization under material initiation and "living" free-radical polymerization are the same.

In the case of alternating cationic polymerization [43], propagation AC are formed at proton detachment from a molecule of a monomer, solvent or complex by a donor–acceptor complex, which includes a molecule of a donor monomer (isobutylene) and a complex of an acceptor monomer with the Lewis acid. In the stage of chain propagation, the monomer complex is introduced into the contact ionic pair of AC. Alternating copolymers of isobutylene with polar and diene monomers are capable of forming various polymer-analogous transformations and vulcanization.

Another group of isobutylene copolymers is represented by products, produced with application of modified methods of cationic initiation. In this manner, properties of known copolymers were improved, and in some cases, new polymeric products were synthesized. For example, application of HAOS combined with various additives, and mixed (polar–non-polar) solvents enables properties (content of gel fractions, molecular mass, degree of unsaturation, ability to be vulcanized, oxidized, etc.) of isobutylene copolymers with diene monomers (butadiene, isoprene, pyperilene, cyclohexadiene, cyclopentadiene, etc.) to be widely regulated [45, 46].

Radiation copolymerization of isobutylene with isoprene in mass gives products with much higher molecular mass, than at the material initiation. In this characteristic they approach the molecular mass of polyisobutylene, synthesized under analogous conditions [47].

Of interest for modification and thermal (chemical) vulcanization of isobutylene polymers is copolymerization with a cyclic diene (1,2-dimethylene cyclobutane, 1-methyl-3-methylene cyclobutane-1) and triene (*trans*-1,3,5-hexatriene, 5-methylheptatriene-1,3,6, 1,6-diphenylhexatriene-1,3,4, octatriene-2,4,6) monomers [48, 49].

Introduction of 2-allyloxytetrahydropyranic units into a polyisobutylene chain during polymerization gives products that can be used directly as additives to oils.

Development of principles of catalysis by Friedel–Crafts halides, in particular, by MeX_n–RHal systems, formed the basis of a method to synthesize block- and grafted copolymers of isobutylene. Similar to reactions of arene alkylation [50] or ionic telomerization of olefins [51], activation of the R–Hal bond with the help of Lewis acids is the effective way to generate cationic particles $R^{\delta+}$, which initiate polymerization and other electrophilic transformations of isobutylene:

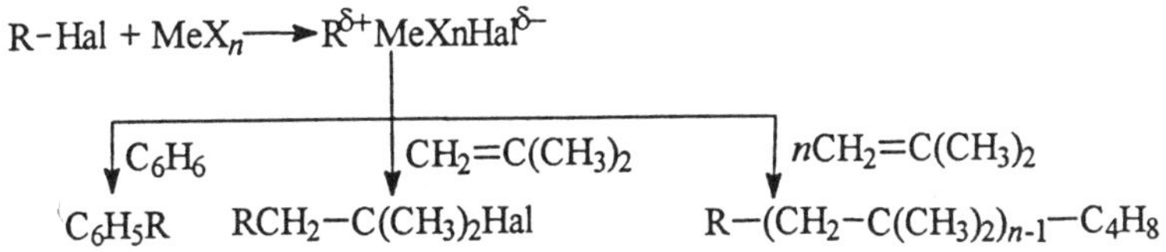

If a polymer activated by MeX_n serves as a carrier of a mobile halogen atom, this leads to formation of block- or graft copolymers. In particular, methods were developed, based on introduction of active atoms of halogens into the polymer at the stage of initiation, transmission and termination of the chain with further excitation of polymerization by the following schemes (A – isobutylene, B and B′ - other monomers, Hal and Hal′ – halogens) [52–54]:

$$\ldots\text{—AAAAAHal} \xrightarrow[R_3Al(R_2AlHal)]{B} \ldots\text{—AAAAABBBBBB—}\ldots$$

$$\ldots\text{—BBBBBBHal} \xrightarrow[R_3Al(R_2AlHal)]{A} \ldots\text{—BBBBBBAAAAA—}\ldots$$

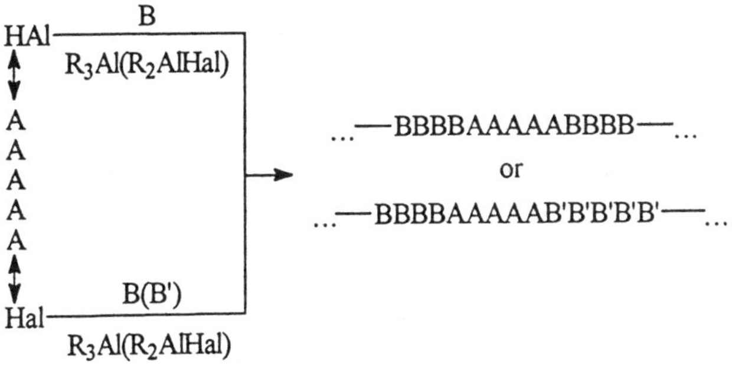

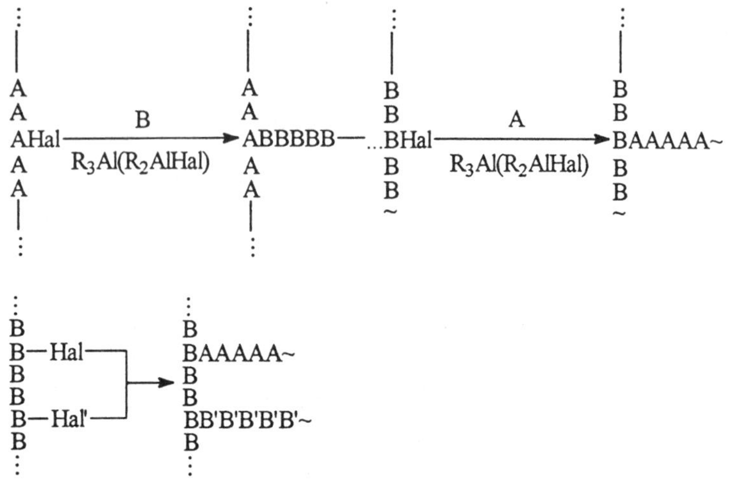

According to the above-mentioned scheme, as variants of synthesis of block- and graft copolymers of isobutylene, the methods are suggested, based on application of a wider range of Lewis acids (BCl_3, $TiCl_4$ and others), other, beside halogens, active functional groups (OH, COOH, acetoxyalkyl, unsaturated, etc.), macromolecular compounds with polyisobutylene chain, polycondensation methods, which include reactions of α,ω-bifunctional oligomers and isobutylene polymers with other functional ones, non-polymerizational methods, via interaction of functional groups of partners. These methods use achievements in the branch of cationic polymerization of isobutylene and other monomers, conduction of polymerization by the principle of "living chains", in particular, in synthesis of telehelic polymers, the effect of inifers, proton traps, etc. [7, 54], as well as polymer-analogous transformations of polymers, in particular, at the synthesis of polyisobutylene with various functional groups [55]. On the other hand, a broad selection of methods of synthetic organic chemistry (polyaddition, condensation, alkylation, etc.) was used for synthesizing block- and grafted copolymers.

The most typical examples of block- and graft copolymers of isobutylene, synthesized according to various schemes, are shown in Table 4.4.

As follows from the data displayed in Table 4.4, methods II and III of production of block- and graft isobutylene copolymers dominate among others. They are directly connected with the achievements in the controlled synthesis of polyisobutylene, mentioned above. Let us note some general and particular positions concerning the theme under consideration.

Partners of polyisobutylene in formation of copolymers represent a wide range of polymeric products: polymers and copolymers of olefins [16, 17, 26–29], polydienes [4–6, 20–23, 25, 36–39], polymers of vinyl monomers [1, 2, 7–12, 14, 17–19, 25, 30–32, 36, 37, 40, 41, 50], heteroatom-containing polymers [13, 42–48], inorganic [15, 33–35] and natural [49] polymers. Practically all possible types of block- and graft copolymers were obtained: double, triple, polyblock, star-like with blocks, approximately equal by in size, and fragmental (with a single shortened block) block-copolymers with blocks of the same and different nature (rigid, elastic), etc. Control of size of blocks, grafting degree and other parameters enables products with various physicomechanical and other properties to be produced, which are of interest from the point of view of both modification of polyisobutylene properties (for example, elimination of the cold flow, stability increase) and its application for improving properties of other polymers (increase of chemical stability of polar products of the polyurethane type, etc.). Additional information about properties of copolymers and their application are discussed in Chapter 6.

Table 4.4

Block- and graft polymers of isobutylene.

No.	Type of copolymer	Production method	Reference
1	Poly(styrene-block-isobutylene)	I	[7, 56-58]
2	Poly(styrene-block-isobutylene-block-styrene)	I	[7]
3	Poly(isobutylene-block-propylene)	II	[7]
4	Poly(isobutylene-block-butadiene)	II	[59]
5	Poly(isobutylene-block-isoprene)	II	[60]
6	Poly(isoprene-block-isobutylene-block-isoprene)	II	[60]
7	Poly(isobutylene-block-styrene)	II	[61-63]
8	Poly(isobutylene-block-α-methylstyrene)	II	[64]
9	Poly(styrene-block-isobutylene-block-styrene)	II	[62, 64]
10	Poly(α-methylstyrene-block-isobutylene-block-styrene)	II	[64, 65]
11	Poly(isobutylene-block-styrene) – three-ray	II	[63]
12	Poly(isobutylene-block-α-methylstyrene) – three-ray	II	[64, 65]
13	Poly(dimethylsiloxane-block-isobutylene-block-dimethylsiloxane)	II	[66]
14	Poly(isobutylene-block-methylmethacrylate)	II	[67]
15	Fragmental poly(isobutylene-block-almoxane-block-isobutylene)	II	[68]
16	Poly(ethylene-graft-isobutylene)	III	[69]
17	Poly(tetrafluoroethylene-graft-isobutylene)	III	[69]
18	Poly(vinyl chloride-graft-isobutylene)	III	[69-73]
19	Poly(vinyl chloride-graft-isobutylene-isoprene)	III	[74]
20	Poly(chloroprene-graft-isobutylene)	III	[75]
21	Poly(chloroprene-graft-isobutylene-graft-α-methylstyrene)	III	[76, 77]
22	Poly(butadiene-graft-isobutylene)	III	[7]
23	Poly(cyclopentadiene-graft-isobutylene)	III	[78]
24	Poly(styrene-graft-isobutylene)	III	[79]
25	Poly(styrene-butadiene-graft-isobutylene)	III	[80]
26	Poly(ethylene-propylene-1,4-hexadiene-graft-isobutylene)	III	[81, 82]
27	Poly(ethylene-propylene-1,4-hexadiene-graft-styrene-graft-isobutylene)	III	[81, 82]
28	Poly(ethylene-propylene-1,4-hexadiene-graft-α-methylstyrene-graft-isobutylene)	III	[83]
29	Poly(ethylene-propylene-dicyclopentadiene-graft-isobutylene)	III	[84]

30	Poly(methylmethacrylate-graft-isobutylene)	III	[85]
31	Poly(methylmethacrylate-2(dimethylamino)-ethylmethacrylate-graft-isobutylene)	III	[86]
32	Poly(butylacrylate-graft-isobutylene)	III	[87]
33	Silica-graft-polyisobutylene	III	[88]
34	Silicagel-graft-polyisobutylene	III	[89]
35	Silicagel-graft-isobutylene-isoprene copolymer	III	[89]
36	Poly(isobutylene-isoprene-graft-styrene)	IV	[61]
37	Poly(isobutylene-graft-α-methylstyrene)	IV	[61]
38	Poly(isobutylene-isoprene-graft-chloroprene)	IV	[61]
39	Poly(isobutylene-isoprene-graft-indene)	IV	[61]
40	Poly(isobutylene-graft-vinyl fluoride)	IV	[89]
41	Poly(isobutylene-block-styrene-block-isobutylene)	V	[74]
42	Poly(isobutylene-block-dimethylsiloxane)	V	[91, 92]
43	Poly(isobutylene-block-dimethylsiloxane-block-isobutylene)	V	[91, 92]
44	Fragmental poly(isobutylene-block-dimethyl-block-siloxane-block-isobutylene)	V	[92]
45	Poly(isobutylene-block-ethylene glycol)	V	[93]
46	Poly(ethylene glycol-block-isobutylene-block-ethylene glycol)	V	[94]
47	Poly(isobutylene-block-ethylene glycol)	V	[95]
48	Fragmental poly(isobutylene-block-urethane)	V	[96, 97]
49	Starch-graft-polyisobutylene	V	[98]
50	Poly(styrene-graft-isobutylene)	V	[99]

Notes. I – block-copolymerization of isobutylene with other polymers; II – block-copolymerization of polyisobutylene with other monomers; III – grafting of polyisobutylene to a polymer of different nature; IV – grafting of other polymers to polyisobutylene; V – interaction of functionalized polyisobutylene with a polymer of different nature.

Among separate representatives, block-copolymers of isobutylene with isoprene should be noted, which give films with unusually high adhesion to metals at chemical and photochemical structuring [60]; fragmental block-copolymer of isobutylene with hydroxyalumoxane [68], as examples of copolymerization of polyisobutylene with a monomer, generated *in situ*; and three-ray copolymer of isobutylene with styrene (α-methylstyrene) [63–65] as demonstration of the possibilities of the fine organic synthesis applied to polymeric objects.

Notice also structural investigations of the three-block copolymer poly-*para*-chlorostyrene-polyisobutylene-poly-*para*-chlorostyrene, in which cylindrical domains of styrene fragments were observed in the polyisobutylene matrix [100].

In conclusion, let us note a number of macromolecular compounds of polyisobutylene, synthesis of which demonstrates progress and possibilities of the "living" cationic polymerization [101] (Table 4.5).

Existence of styrene, olefin and other active groups in polyisobutylene macromolecules gives a possibility of synthesis via copolymerization with other monomers of various comb-like copolymers.

Table 4.5

Characteristics of macromonomers based on polyisobutylene.

End group	Method*	$\overline{M}\cdot 10^{-3}$	$\overline{f_n}$	Reference
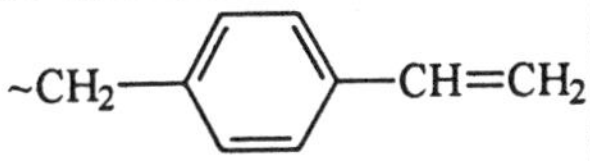	A	10-30	—	102
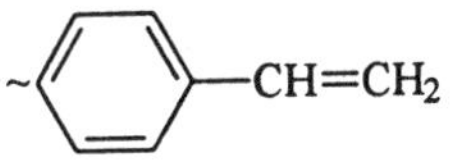	A	4.2-34.2	0.9-1.1	103
$\sim CH_2—C(Me)=CH_2$	B	1.6-3.2	0.95-1.04	104
$\sim CH_2OCOC(Me)=CH_2$	B	1.6-13.9	0.98-1.05	104
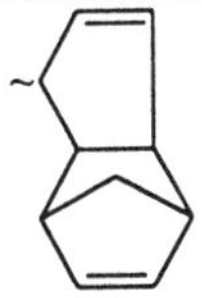	C	27.0	1.0	84
$\sim CH_2OCOCH=CH_2$	B	3.6	1.95**	105

Notes. * Methods of introduction of end groups: A – initiation, B – reactions by end groups, C – the method with application of combined agent-initiator and chain transmitter ("inifer").
** Bifunctional derivative.

Combined with progress in the field of cationic polymerization of isobutylene, the controlled synthesis of block- and graft copolymers of isobutylene represents the visual confirmation of progress in molecular engineering in the field of polyisobutylene, which beside pure scientific advantages, is no doubt interesting from the point of view of obtaining thermal elastoplastics, based on isobutylene, modification of polymers by the method of surface grafting, creation of new composite materials, and solving of other applied tasks.

References to Chapter 4

1. Berlin, Al.Al., Wolfson, S.A., and Enikolopyan, N.S., Kinetics of Polymerization Processes, 1978, 320 p., Moscow, Khimia. (Rus)
2. Heublein, G., Faserforsch. Textitech Z. Polymer Forsch., 1976, Bd. 27(2), S. 57.
3. Kennedy, J.P., Kelen, T., and Tudos, F., J. Polymer Sci., Polymer Chem. Ed., 1975, v. 13, pp. 2277–2289.
4. Erusalimskii, B.L. and Lubetskii, S.G., Processes of Ionic Polymerization, 1974, 256 p., Leningrad, Khimia. (Rus)
5. Heublein, G. and Romer, W.J., J. Plast. Chem., 1973, v. 315(5), pp. 801–809.
6. Cationic Polymerization, Ed. P. Plesha, 1966, 584 p., Moscow, Mir. (Rus)
7. Kennedy, J.P. and Merechal, E., Carbocationic Polymerization, 1982, 510 p., N.Y.
8. Imanishi, Y., Higashimura, T., and Okamura, S., J. Polymer Sci., 1965, Pt. A, v. 3(7), pp. 2455–2474.
9. Corno, C., Priola, A., and Cesca, S., Macromolecules, 1979, v. 12(3), pp. 411–418.
10. Harzallah, F., Tardi, M., Polton, A., and Sigwalt, P., Eur. Polym. J., 1984, v. 20(2), pp. 163–170.
11. Faust, R., Fehervari, A., and Kennedy, J.P., Rubber Chem. and Technol., 1977, v. 60(4), p. 794.
12. Priola, A., Makromol. Chem., 1975, Bd. 176(7), S. 1969–1981.
13. Faust, R., Fehervari, A., and Kennedy, J.P., Brit. Polym. J., 1987, v. 19(3–4), pp. 379–386.
14. Gulimov, V.I., Vishiakov, T.P., and Vlasova, U.D., Izvestiya VUZov, 1988, v. 31(6), pp. 136–138, Khimia i Khim. Tekhnol. (Rus)
15. Priola, A., Corno, C., and Cesca, S., Polymer Bull., 1982, Bd. 7(11–12), S. 599–606.
16. Priola, A., Corno, C., Bruzzone, M., and Cesca, S., Polymer Bull., 1981, Bd. 4(12), S. 735–742.
17. Kennedy, J.P. and Nakao, M., J. Macromol. Sci., 1977, Pt. A, v. 11(9), pp. 1621–1636.
18. Kennedy, J.P., Makromol. Chem., 1974, Bd. 175(4), S. 1101–1115.
19. Kennedy, J.P. and Chou, T., Adv. Polymer Sci., 1976, v. 21(1), pp. 1–30.
20. Sangalov, Yu.A., Nelkenbaum, Yu.Ya., Khusnutdinov, R.I., Jemilev, U.M., and Minsker, K.S., Highmolec. Comp., 1984, Ser. A, v. 26(3), pp. 600–606. (Rus)
21. Marechal, E.-C., R. Acad. Sci., 1968, Ser. C, v. 266(19), pp. 1427–1430.
22. Heublein, G. and Romer, W., Makromol. Chem., 1973, Bd. 163(1), S. 143–153.
23. Sangalov, Yu.A. and Nelkenbaum, Yu.Ya., Physical and Chemical Grounds of Synthesis and Processing of Polymers, 1984, pp. 23–26, Gorky. (Rus)
24. Gladkikh, I.F., Sangalov, Yu.A., and Komarov, N.V., Izvestiya VUZov, 1984, v. 27(6), pp. 707–710, Khimia and Khim. Tekhnol. (Rus)
25. Sangalov, Yu.A., Gladkikh, I.F., Komarov, N.V., and Zisovin, E.G., Intern. Polym. Sci. and Technol., 1984, No. 10, pp. 41–43.
26. Tudos, F., Kelen, T., and Berezhnikh, T.F., J. Polym. Sci., Polymer Symp., 1975, No. 50, pp. 109–132.
27. Heublein, G. and Wondraczek, R., J. Polymer. Sci., Polymer Symp., 1976, No. 56, pp. 359–372.
28. Kennedy, J.P. and Chou, T., J. Macromol. Sci., 1976, Pt. A, v. 10(7), pp. 1357–1369.

29. Corno, C., In: Proc. Europ. Conf. MMR Macromol., Sassari, 1978, S. 1, s.a. 185–191.
30. Sangalov, Yu.A., Gladkikh, I.F., and Minsker, K.S., Highmolec. Comp., 1982, Ser. B, v. 24(5), pp. 356–358. (Rus)
31. Sebenic, A., Guhaniyogi, S., Santee, E.R., and Harwood, H.J., Amer. Chem. Soc. Polym. Prepr., 1986, v. 27(2), pp. 138–139.
32. Kostov, G.K., J. Polymer Sci., Polymer Chem. Ed., 1979, v. 17(12), pp. 3991–4001.
33. Ishigyre, K., Macromolecules, 1976, v. 9(2), pp. 290–293.
34. Schneider, Ch. and Bohlmann, H.P., In: IUPAC Macro, Florence, 1980, Intern. Symp. Macromol, 1980, Prepr., v. 2, Pisa, s.a. 134–137.
35. Takemoto, T., Kunststoffe, 1967, Bd. 57(9), S. 697–701.
36. Hummel, D.O. and Bestgen, J., Angew. Chem., Intern. Ed. Engl., 1974, v. 13(6), p. 414.
37. Raetsch, M. and Bormann, G., Plaste und Kautschuk, 1976, Bd. 23(3), S. 173–176.
38. Priola, A., Ferraris, G., and Cesca, S., Makromol. Chem., 1979, Bd. 180(12), S. 2859–2874.
39. Kuntz, I., Chamberlain, N.F., and Stehling, F.J., Amer. Chem. Soc., Polymer Prepr., 1978, v. 19(1), pp. 216–221.
40. Florjanczyk, Z., Makromol. Chem., 1982, Bd. 183(5), S. 1081–1091.
41. Hirooka, M., Appl. 74-103983, Japan; C.A., 1975, v. 82, 98781 w.
42. Kuntz, I., Pat. 2515381 (FRG), 1975; C.A., 1976, v. 84, 60866 f.
43. Gaylord, N.G., J. Polymer Sci., Polymer Chem. Ed., 1975, v. 13(2), pp. 467–482.
44. Hirooka, M., In: 23rd Intern. Congr. Pure and Appl. Chem., Boston, 1971, Macromol. Prepr., v. 1, S. 1, s.a. 311–318.
45. Thaler, W.A. and Buckley, D.J., Rubb. Chem. Technol., 1976, v. 49(4), pp. 960–966.
46. Pat. 4171414 (USA), 1979.
47. Williams, F.,, Shinkava, A., and Kennedy, J.P., J. Polymer Sci., Polymer Symp., 1976, No. 56, pp. 421–430.
48. Tsao, J.H. and Lenz, R.W., J. Polymer Sci., Polymer Chem. Ed., 1979, v. 17(2), pp. 331–349.
49. Priola, A., Como, C., Bruzzone, M., and Cesca, S., Polymer Bull., 1981, Bd. 4(12), S. 743–750.
50. Olah, G.A., Friedel-Crafts Chemistry, 1973, 581 p., N.Y.
51. Petrov, A.A. and Genusov, M.L., Ionic Telomerization, 1968, 296 p., Leningrad, Khimia. (Rus)
52. Kennedy, J.P., J. Macromol. Sci., 1980, Pt. A, v. 14(1), pp. 1–10.
53. Kennedy, J.P., J. Macromol. Sci., 1979, Pt. A, v. 13(5), pp. 695–714.
54. Kennedy, J.P., Cationic Graft Polymerization (Appl. Polymer Symp., No. 30), 1979, 193 p., N.Y., Wiley.
55. Sangalov, Yu.A. and Yasman, Yu.B., Uspekhi Khimii, 1985, v. 44(7), pp. 1208–1229. (Rus)
56. Kennedy, J.P. and Melby, E.G., Polymer Prepr., 1974, v. 15, p. 180.
57. Kennedy, J.P. and Melby, E.G., J. Polym. Sci., Chem. Ed., 1975, v. 13, p. 29.
58. Kennedy, J.P., Feinberg, S.C., and Huang, S.Y., J. Polym. Sci., Chem. Ed., 1978, v. 16(1), pp. 243–259.
59. Nemes, S. and Kennedy, J.P., J. Macromol. Sci.-Chem., 1971, v. A28(3, 4), pp. 311–328.

60. Puskas, J.E., Kaszas, G., and Kennedy, J.P., J. Macromol. Sci.-Chem., 1991, v. A28(1), pp. 65–80.
61. Kennedy, J.P. and Charles, J.J., J. Appl. Polym. Sci., Appl. Polym. Symp., 1977, v. 30, pp. 119–139.
62. Fodor, Z.S., Kennedy, J.P., Kelen, T., and Tudos, F., J. Macromol. Sci., 1987, v. A24(7), pp. 735–747.
63. Kaszas, G., Puskas, J.E., Kennedy, J.P., and Hager, W.G., J. Polym. Sci., Pt. A, Polym. Chem., 1991, v. 29, pp. 427–435.
64. Kennedy, J.P., Guhaniyogi, S.G., and Ross, L.R., J. Macromol. Sci.-Chem., 1982, v. A18(1), pp. 119–128.
65. Kennedy, J.P. and Smith, R.A., J. Polym. Sci., Polym. Chem. Ed., 1980, v. 18(5), pp. 1539–1546.
66. Wilczek, L., Michka, M.K., and Kennedy, J.P., J. Macromol. Sci.-Chem., 1987, v. A24(9), pp. 1033–1049.
67. Kennedy, J.P. and Chen, F., J. Amer. Chem. Soc., Polymer Prepr., 1979, v. 20(2), pp. 310–315.
68. Tolstikov, G.A., Sangalov, Yu.A., Nelkenbaum, Yu.Ya., and Prokofiev, I.K., Metal-organic Chemistry, 1988, v. 1(3), pp. 676–679. (Rus)
69. Kabanov, V.Ya., Sidorova, L.P., and Spitsyn, V.J., Eur. Polym. J., 1974, No. 10, p. 1153.
70. Chapiro, A., J. Polym. Sci., Polym. Symp., 1976, No. 56, pp. 431–436.
71. Kabanov, V.Ya. and Sidorova, L.P., Highmolec. Comp., 1972, v. 14(10), pp. 2091–2095. (Rus)
72. Kennedy, J.P. and Davidson, D.L., J. Appl. Polym. Sci., Appl. Polym. Symp., 1977, v. 30, pp. 51–72.
73. Gupta, S.N. and Kennedy, J.P., Polym. Bull., 1979, No. 1, pp. 253–259.
74. Minsker, K.S. and Sangalov, Yu.A., Isobutylene and Its Polymers, 1986, 224 p., Moscow, Khimia. (Rus)
75. Kennedy, J.P., Plamthottam, S.S., and Ivan, B., J. Macromol. Sci., 1982, v. A17(4), pp. 637–651.
76. Kennedy, J.P. and Plamthottam, S.S., J. Macromol. Sci.-Chem., 1979, v. A14, p. 729.
77. Kennedy, J.P. and Plamthottam, S.S., Polymer Bull., 1982, v. 7(7), pp. 337–344.
78. Heublein, G. and Freitag, W., J. Prakt. Chem., 1979, Bd. 321(4), S. 544–548.
79. Jiang, Y. And Frechet, J.M.J., 197th ACS Nat. Mect., Dallas, Tex, April 9–14, 1989, Abstr. Pap., 1989, p. 833.
80. Oziomek, J. and Kennedy, J.P., J. Appl. Polym. Sci., Appl. Polym. Symp., 1977, v. 30, pp. 91–104.
81. Kennedy, J.P. and Vidal, A., J. Polym. Sci., Chem. Ed., 1975, v. 13(7), pp. 1765–1773.
82. Kennedy, J.P. and Vidal, A., J. Polym. Sci., Chem. Ed., 1975, v. 13(10), pp. 2269–2276.
83. Vidal, A. and Kennedy, J.P., Polym. Lett., 1976, v. 14, p. 489.
84. Farona, M.F. and Kennedy, J.P., Polym. Bull., 1984, v. 11(4), pp. 359-363.
85. Kennedy, J.P. and Hiza, M., J. Polym. Sci., Polym. Chem. Ed., 1983, v. 21(4), pp. 1033–1044.
86. Chen, D. and Kennedy, J.P., Macromol. Sci., 1988, v. A25(4), pp. 389–401.
87. Kennedy, J.P., Feisch K.C., Ir. Preprints, Macro Florence, IUPAC, 1980, v. 2, p 162.
88. Vidal, A., Guyot, A., and Kennedy, J.P., Polym. Bull., 1980, v. 2(5), pp. 315–320.

89. Vidal, A., Guyot, A., and Kennedy, J.P., Polym. Bull., 1982, v. 6(7), pp. 401–407.
90. Hrublkin, G., Rottmayer, H.H., Stadermann, D., and Vogel, J., Acta Polym., 1987, v. 38(10), S. 559–561.
91. Prokofiev, I.K., Nelkenbaum, Yu.Ya., and Sangalov, Yu.A., Zh. Prikl. Khim., 1989, No. 2, pp. 436–438. (Rus)
92. Nelkenbaum, Yu. and Sangalov, Yu., Macro 87, 31st IUPAC Macromol. Symp., Merseburg, 1987, Abstr. Pap. Microsymp. 2, Microsymp. 3, S. 1, s.a. 45.
93. Macosko, C.W. and Saam, J.C., Polym. Bull., 1987, v. 18(5), pp. 463–471.
94. Kennedy, J.P. and Hongu, Yu., Polym. Bull., 1985, v. 13(2), pp. 115–121.
95. Percec, V., Guhaniyog, S., Kennedy, J.P., and Ivan, B., Polym. Bull., 1982, v. 8(1), pp. 25–32.
96. Kennedy, J.P., Chemtech., 1986, v. 16(1), pp. 694–697.
97. Miyabayashi, T. and Kennedy, J.P., J. Appl. Polym. Sci., 1986, v. 31(8), pp. 2523–2532.
98. Takemoto, K., Kondo, M., and Imoto, M., Memoirs Fac. Eng., 1964, v. 6(2), pp. 211–224, Osaka City Univ.
99. Kennedy, J.P. and Chang, Y.L., J. Polym. Sci., Polym. Chem. Ed., 1971, v. 19(11), pp. 2737–2744.
100. Kennedy, J.P. and Kurian, J., J. Polym. Sci., 1990, Pt. A, v. 28(13), pp. 3725–3738.
101. Kazanskii, K.S., Kubisa, P., and Penchek, S., Uspekhi Khimii, 1987, v. 46(8), pp. 1360–1386. (Rus)
102. Kennedy, J.P. and Frisch, K.S., IUPAC Int. Symp. On Macromolecules, Preprints, Florence, 1980, v. 2, p. 162.
103. Kennedy, J.P. and Lo, C.Y., Amer. Chem. Soc. Polym. Preprints, 1982, v. 23, p. 99.
104. Kennedy, J.P. and Hiza, M., J. Polym. Sci., Polym. Chem. Ed., 1983, v. 21(4), pp. 1033–1044.
105. Liao, T.P. and Kennedy, J.P., Polym. Bull., 1981, v. 6, p. 135.

Part II. Polymers derived from isobutylene

Chapter 5. Properties and production of polyisobutylene

5.1. Physical properties.

A wide spectrum of oligo- and polyisobutylenes with the polymerization degree $\overline{P}$ from 3 to 4000 and higher is obtained during polymerization of isobutylene.

Oligoisobutylenes with molecular mass up to 2,000 are quite mobile liquids; low-molecular products with molecular mass from 5,000 to 50,000 are viscous liquids; high-molecular polyisobutylenes with molecular mass over 70,000 are elastomers, possessing cold flow and ability to crystallize at stretching. Packing density of crystalline and amorphous polyisobutylene is 0.362–0.342; Densities of crystalline and amorphous phases are 937 kg/m^3 and 912 kg/m^3, respectively. T_{melt} of the crystalline phase is 401 K. If crystallized [1, p. 444], polyisobutylene forms a rhomboidal three-dimensional lattice with sizes (nm): a = 0.694, b = 1.196, c = 1.863 (the identity period is 1.85 nm); $C(CH_3)_2$- and CH_2-groups are dislocated helicoidally (the spiral 8_5), which corresponds to the minimum of the potential energy [2] (Figure 5.1); the valence angle is 114°; the spiral step is 0.233 nm. An elementary cell contains two molecular chains with the "head-to-tail" type of attachment. Chains are pseudo-hexagonal packed $\left(b = c\sqrt{3}\right)$. Works [3, 4] note formation of the spiral 8_5 with the valence angle of 122°. A statistic segment of polyisobutylene is composed of ~4 chains. Rotation of the probe molecule of 4-hydroxy-2,2,6,6-tetramethylpyridine-1-oxyl is determined by conformational mobility of small segments of macromolecules (correlation time $\tau = 1.9 \cdot 10^{-16}$ s at the activation energy E = 43 kJ/mol [5]).

As shown with the help of model compounds (2,2,4,4-tetramethylpentane, 2,2,4,4,6,6-hexamethylpentane and 2,2,4,4,6,6,8,8-octamethylnonane), elasticity of polyisobutylene at low temperatures is due to the fact that, despite steric hindrance of macromolecules, a change of angles is possible at the sacrifice of rotation of C–C bonds around each other by up to 30° [4] (Figure 5.1).

Thermodynamic flexibility of polyisobutylene macromolecules (the ratio of unperturbed sizes of the molecule to those of the chain of free rotation) or the factor of deceleration of the internal rotation is 2.2. The inertness z-average radius $\left\langle S_z^2 \right\rangle^{1/2}$ of polyisobutylene is (7.5 ± 0.5) nm in block and (7.7 ± 0.5) nm in a Θ-solvent [6], which indicates the identity of configurations of macromolecules in block and in the Θ-solvent [7].

Beside other characteristics of thermodynamic flexibility smaller than for other non-polar polymers (polyethylene, polypropylene, etc.), for example, length of the segment equal to 1.83 nm, they indicate quite high flexibility of polyisobutylene in solution. On the other hand, concentrated polyisobutylene possesses a relatively

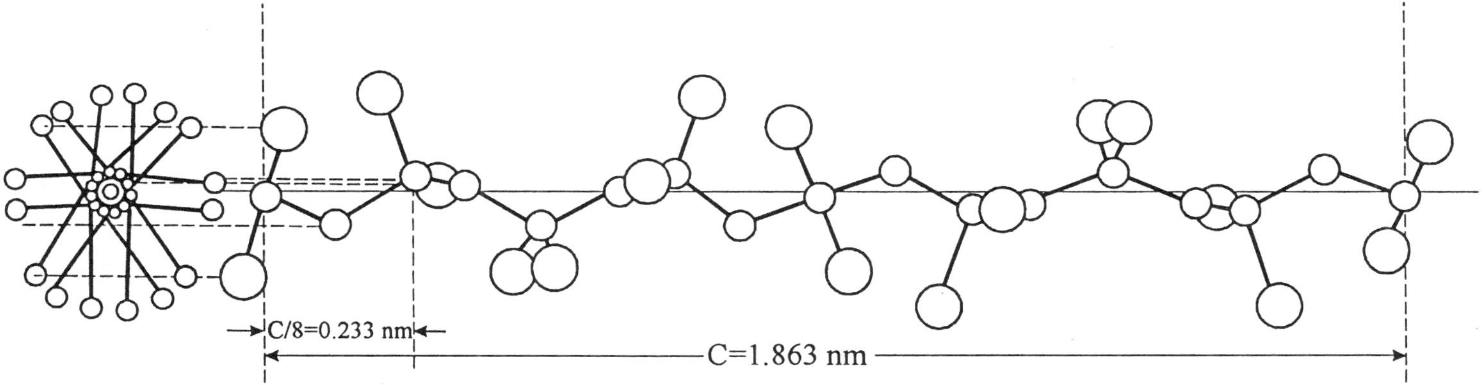

Figure 5.1. Schematic depiction of chains of crystalline polyisobutylene.

rigid chain, the index of which is a low amplitude of vibration implying fast motion of internuclear vector C–H of the methylene group [8]. This marks great steric hindrances of this group, with which, beside classic segmental motion, appearance of a number of secondary relaxation processes (β-process, rotation of CH_3-group, transition of the liquid-liquid type, etc.) in the local dynamics of polyisobutylene is connected [9].

The main physical properties of oligo- and polyisobutylenes are displayed in Table 5.1.

Table 5.1

Physical properties of oligo- and polyisobutylenes.

Index	Oligoisobutylenes, mol. mass < 50,000	Polyisobutylenes, mol. mass = 70,000–225,000
Density, kg/m^3	830–910	910–930
Refraction index, n_d^{298}	1.506	1.5089
Temperature, K, of		
Softening	218–295	373
Glass transition	185–205	199
Heat conductivity coefficient, W/(cm·K)		0.12–0.14
Specific heat capacity, kJ/(kg·K)		1.94
Temperature coefficient of bulk expansion at 298 K, K^{-1}		$6.2 \cdot 10^{-6}$
Strength at elongation, mPa		1.5–6
Relative elongation at break, %		500–1000
Hardiness by Shore		25–35
Elasticity by recoiling, %		10–12
Dielectric permeability at 1kHz	2.0–2.2	2.4–2.9
Specific volumetric electric resistance, TOhm·m	10^{-1}–10^4	10^3–10^4
Dielectric tangent		
at 50 Hz	10^{-3}–10^{-4}	$2 \cdot 10^{-4}$
at 1 kHz	10^{-3}	$5 \cdot 10^{-4}$
Breakdown strength at 353 K, kV/mm	15–20	16–25
Viscosity at 373 K, mm^2/s	75-550	
Dipole moment, D	0	0
Density of cohesion energy, J/cm^3		266
Adhesion, N/cm:		
To copper foil		1.5–3.3
To iron foil		2.45–5.55

The Flory–Huggins parameter (at 293–298 K) of thermodynamic interaction χ_1, which determines changing of the free energy, necessary for transfer of solvent molecule into the polymeric phase, for polyisobutylene–solvent systems equals: polyisobutylene–benzene 0.741; polyisobutylene–toluene — 0.488; polyisobutylene–cyclohexane — 0.390. The upper and the lower critical temperature of mixing for polyisobutylene–benzene systems is 296 K (T_{ucts}) and 433 K (T_{lcts}). Critical temperature of mixing at M $\rightarrow \infty$, Θ-temperature, at which the most of thermodynamic properties do not depend on the solvent concentration for the polyisobutylene–benzene system is equal to 297 K (the upper Θ-temperature) [10].

For polyisobutylene with molecular mass of 13,000–1,600,000, the value of the second virial coefficient (A_2) decreases with growth of the molecular mass according to correlations: $A_2 = 6.88 \cdot 10^{-3} \cdot M_w^{-0.203}$ (cyclohexane, 296 K) and $A_2 = 3.08 \cdot 10^{-3} \cdot M_w^{-0.188}$ (heptane, 296 K). Diffusion coefficients D depend on the molecular mass of polyisobutylene (298 K): $D = 1.78 \cdot 10^{-4} \cdot M_w^{-0.572}$ (cyclohexane), $D = 4.1 \cdot 10^{-4} \cdot M_w^{-0.560}$ (heptane) and $D = 5.94 \cdot 10^{-5} \cdot M_w^{-0.493}$ (isoamyl ester of isovalerianic acid). The inertness radius R of polyisobutylene changes depending on the molecular mass in accordance with equations (in nm): $R = 1.66 \cdot 10^{-2} \cdot M_w^{-0.58}$ (cyclohexanone), $R = 1.34 \cdot 10^{-2} \cdot M_w^{-0.584}$ (heptane) and $R = 3.08 \cdot 10^{-2} \cdot M_w^{-0.5}$ (isoamyl ester of isovaleric acid).

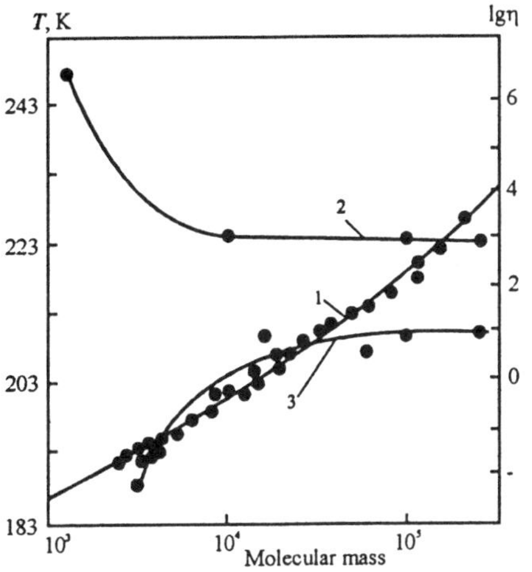

Figure 5.2. Dependence of glass transition temperature T_g (1), and friability T_{fr} (2), and viscosity η (3) on the molecular mass of polyisobutylene.

Changing of many main properties of polyisobutylene is bound to the molecular mass and width of the MMD of polymers based on isobutylene. The glass transition temperature T_g for oligoisobutylenes is greatly increased with the molecular mass, approaching to a limiting value, typical for polyisobutylene in the $\overline{M}_n$ range of 1,000-2,000 (Figure 5.2). Consequently, temperature of friability, determined by the strength material, decreases [1, p. 595]. In these cases, the size of a segment is determined as 30 ± 10 monomeric units. Flow temperature T_f also depends on the molecular mass of polyisobutylene, the following correlation being true in the range of large MM values [1, p. 600]:

$$\lg T_f = 3.3 \lg M + \text{const.} \tag{5.1}$$

Figure 5.2 (curve 3) displays the dependence of viscosity of polyisobutylene melt on the molecular mass at constant temperature [1, p. 601]. The curve possesses clear bending at molecular masses exceeding the limiting values:

$$\lg\eta = 3.4\lg\overline{M}_\eta - \lg K\,. \qquad (5.2)$$

The ratio of viscosities $\eta(T)/\eta(T_0)$ does not depend on the molecular mass of polyisobutylene for values greater than 2,000 [1, p. 602]. Some physical properties of the polymer are much more significantly affected by the structure of the main chain. Contrary to the usual polyisobutylene (attachment of monomeric units by the "head-to-tail" structure), a polymer with different disposition of monomeric units ("head-to-head or "tail-to-tail"), obtained under special conditions, is a partially crystalline product (with crystallinity degree of 50%) with T_{melt} = 460 K and T_g = 360 K [11]. The maximum rate of decomposition of this polymer (about 590 K) is some 70° lower than that for the usual standard polyisobutylene.

Changing of mechanical properties of polyisobutylene in the range of its softening is shown in Figure 5.3 [1, p. 586]. Dynamic modulus of shear G_d abruptly (by more than by 3.5 degrees) decreases with temperature growth at constant frequencies. On moving to higher frequencies, "modulus-temperature" curves shift to the side of lower temperatures.

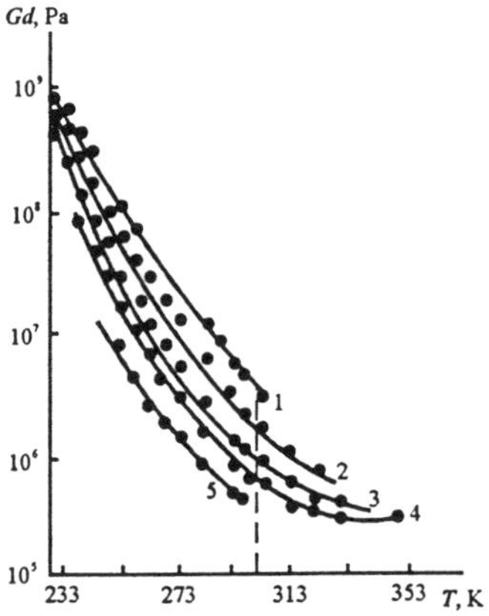

Figure 5.3. Dependence of dynamic modulus G_d of polyisobutylene in the range of softening with temperature rise for constant frequencies: 1 – 3,000 Hz, 2 – 1,000 Hz, 4 – 100 Hz, 5 – 30 Hz.

Modulus–temperature curves, as well as those for modulus–frequency, may be superimposed on each other as a result of their parallel transposition along the abscissa axis, which defines polyisobutylene as a thermodynamically simple polymer. Under these conditions, the general dependence of the modulus on time and temperature is true. This dependence is set with the help of two functions, which themselves depend on a single variable only: a) total "modulus–frequency" curve at the reference temperature T_0 (298 K); and b) the function of reduced variables ("time–temperature").

The shear modulus at frequency ν and temperature T is expressed by a simple dependence relationship [1, p. 589]:

$$G_d(\nu, T) = G_d = (\alpha\nu, T_0), \tag{5.3}$$

where $G_d(\nu, T)$ is the shear modulus at the reference temperature T_0 (298 K). The dependency $\alpha = \alpha(T, T_0)$ represents a function of reduced time–temperature variables and depends specially on T and T_0. To put it differently, all values of relaxation times increase by one and the same value, aliquot to α, at temperature change from T_0 to T:

$$\tau(T)/\tau(T_0) = \alpha(T, T_0). \tag{5.4}$$

In all cases, the function of reduced variables can be represented by the expression [1, p. 590]:

$$\lg\alpha(T, T_0) = -C_1(T - T_0)/(C_2 + (T - T_0)). \tag{5.5}$$

Values of characteristic constants C_1 and C_2 depend on the initial temperature chosen T_0 (Table 5.2).

Table 5.2

Parameters of equation (5.4) for polyisobutylene.

T_0, K	C_1	C_2, K	T_g, K	$f_g \cdot 10^{-2}$	$\alpha_0 \cdot 10^{-2}$	f_g^2/α
298	8.3	193	202	2.6	2.7	2.5
298	7.2	184	202	2.9	3.3	2.5

Here $f_g = a(C_2 - T_g - T_0)$ is the relative free volume at T_0.

Polyisobutylene is able to dissolve in hydrocarbons, chlorinated hydrocarbons, and esters, and partially in *n*-butyl alcohol. It swells in animal and vegetable oils and fats. It is insoluble in ethyl and isopropyl alcohols, acetone, methylethylketone, glacial acetic acid.

Solvents form the following sequence according to entropy of mixing: toluene > carbon tetrachloride > *iso*-octane > cyclohexane. Up to 375 K, polyisobutylene is stable in water and displays low gas permeability $[M^2/(s\cdot N/m^2)]$: $2.0\cdot10^{-18}$ for water; $2.2\cdot10^{-18}$ for nitrogen; $4.9\cdot10^{-18}$ for hydrogen; $9\cdot10^{-18}$ for oxygen.

At 298 K polyisobutylene is non-toxic. The polymer, P-200 trademark, in particular, is permitted for application in the food industry. It is fire hazardous. At air the temperature of 268 K and wind velocity of 5 m/s, the rate of burnout of polyisobutylene (P-118 trademark) is 1.25 kg/(m^2·min); the rate of flame spreading on the surface is 0.25 m/min. When it combusts, the material spreads forming a smoke with fly ash.

5.2. Chemical properties.

The structure of polyisobutylene macromolecules (linear polymer usually containing a single end C=C bond per macromolecule) determines its chemical properties. Polyisobutylene is one of the most chemically inert of the known polymers. Polyisobutylene is resistant to dilute and concentrated acids: H_2SO_4 (98.8%), HNO_3 (50%), HCOOH, HCl (37%), CH_3COOH; ammonia, alkali (NaOH – 40%), solutions of salts, hydrogen peroxide. On heating, it is resistant to the effect of HNO_3 and other acids and is carbonized interacting with H_2SO_4. Already at 290 K, polyisobutylene is unstable to liquid and gaseous Cl_2 and Br_2, their water solutions, ozone, and some energetic influences as well.

Heating above 620 K leads to degradation and decomposition of the polymer. The monomer yield in volatile reaction products is about 20–30 mass%, and yield of C_5 hydrocarbons is higher — 65–70 mass% at depolymerization rate of 2.5–3%/min (623 K) [12]. Among the most important products obtained, besides isobutylene, are: di-, tri- and tetramers of isobutylene, formed at intramolecular chain transmission according to the free-radical mechanism.

Under the effect of ionizing radiation, polyisobutylene decays with random scission of macromolecules by the law of accident and release of volatile products [13–15]. In solution, the process proceeds much faster than in the absence of solvents and is accelerated with diminution of the polymer content. The energy of fission of a single C–C bond in the main chain by electrons or γ-rays is (eV): 77 K — 45, 193 K — 27, 293 K — 20, 343 K — 12, and 363 K — 10 [13, p. 108]. Besides isobutylene, great amounts of H_2 and especially CH_4 are formed (obviously, at the sacrifice of detachment of H and framing CH_3-groups). The rate of formation of CH_4 and H_2 is proportional to the depth of polyisobutylene degradation, and the total yield does not depend on temperature; therewith, formation of isobutylene is increased with the radiation dose (and temperature, as well), which is the consequence of its formation as a result of detachment of end groups of a molecule of the initial polymer, as well as the end groups appearing under the radiation effect. Each single act of chain fission matches formation of 1.87 vinylidene groups $\sim CH_2C(CH_3){=}CH_2$.

The ESR-spectrum of polyisobutylene, radiated by γ-rays at 77–173 K, represents a doublet appropriate to formation of the free radical $\sim C(CH_3)_2–CH–C(CH_3)_2\sim$, which disappears completely at polymer heating up to room temperature [14, p. 225]. Kinetics of termination of free radicals is described by a reaction equation of the second order with effective values $E_a = 77.5 \pm 5.4$ kJ/mol and pre-exponential multiplicand $3.98 \cdot 10^{-5}$ $cm^3/(spin \cdot s)$ [16]. Detachment of a H atom from methyl group does not occur, but even if it does take place, no long living radicals are formed. At 123 K, owing to bipolar widening of signals, the doublet can transform into a singlet in the spectrum.

Generally, transformation of polyisobutylene under the effect of γ-radiation is a complex process. Depending on the conditions, either deep decomposition of the polymer is realized, or processes of degradation and cross-linking of macrochains proceed simultaneously. Conjugated diene structures are formed simultaneously at the sacrifice of transformations of macroradicals [15].

Polyisobutylene is relatively stable in UV-light. However, on long-term photo-radiation (by Hg-lamp for several days at 77 K with further heating up to 195 K) five types of free radicals were observed, which indicated fission of both C–C and C–H bonds [15]. Primary products include formation of isolated double and conjugated double bonds, H_2 and CH_4. Subsequent transformations of polyisobutylene give isobutylene, 2,4,4-trimethyl-1-pentene and 2,4,4-trimethylpentane. Compared with radiolysis, lower yields of products suggest a limited role for triplet excited ionic states.

The main products of oxidative degradation of polyisobutylene in the presence of O-plasma are H_2, H_2O, CO and CO_2 [15]. The presence of even small amounts of ozone (3 mass%) in oxygen promotes formation of CO_2 and H_2O; as a result, C=O and COOH-groups appear in the composition of macromolecules of the polymer, and >C=C< double bonds disappear.

Under various mechanochemical interactions, such as flaring, grinding in a ball mill, vibrogrinding, or ultrasonic radiation, polyisobutylene degrades [17, 18]. The process is accompanied by reduction of the molecular mass, but only down to a definite limiting value, which depends on the initial molecular mass and conditions of the polymer degradation. In this case, the chemical nature of the end groups affects stability of the polymer. During the reaction, not only MM, but also MMD of polyisobutylene changes. Mechanochemical transformations of polyisobutylene are of interest from the point of view of a possibility to obtain block-copolymers or modified (from the end of the chain) isobutylene polymers. The stability index of the polymer to mechanical degradation (by decrease of viscosity) is regulated for all trademarks of polymers, used as thickening additives.

For isobutylene polymers, reactions at both unsaturated C=C bonds and saturated C–C bonds in the chain are typical, including specific processes of degradation by the law of accident and the law of end groups. There is also the possibility of cross-linking of polymeric chains, but this occurs to only a limited extent.

5.2.1. Polyisobutylene reactions at unsaturated bonds.

Existence of end C=C bonds in polyisobutylene raise the possibility of chemical reactions that are typical for olefin hydrocarbons. However, reactivity of end groups decreases with growth of molecular mass of the polymer, which, in turn, usually limits the possibility to make macromolecules functional [19, 20].

The known reactions of chlorination, phosphorylation, sulfonation, sulfochlorination, ozonation, epoxidation of C=C bonds (see Section 1.2) are accompanied by changing of not only the chemical structure of the end

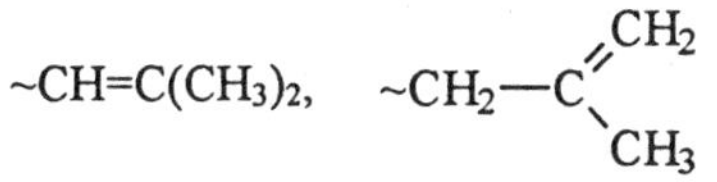

groups, but also often of the chemical nature of the main chain, as well as changing of its length.

Various sulfur-containing derivatives may be obtained by adding sulfur dichloride, 5-sulfenylchloride-4-chloro-3-methyltetrahydrophthalic anhydride or 10-sulfenylchloride-9-chlorostearic acid to polyisobutylene [21, 22]. Reactions with SH_2, S_2Cl_2 and other compounds of this kind are known.

Of interest are hydrosilylation reactions [23, 24], in particular, by trichlorosilane or alkylchlorosilanes in the presence of H_2PtCl_6 as the catalyst:

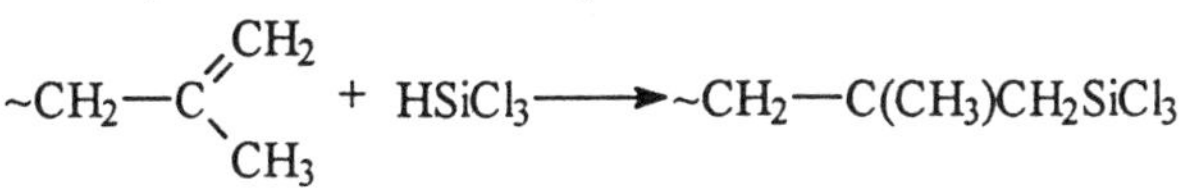

In the presence of electrophilic catalysts, polyisobutylene alkylates arenes up to the highest alkyl-aromatic compounds; as a result, partial depolymerization of polyisobutylene with subsequent alkylation or depolyalkylation of macromolecules can take place [25] with regard to acidity of the catalyst applied.

The main conditions that determine the ratio of reactions of direct (non-degradation) alkylation and partial depolymerization of polymeric chains with subsequent alkylation (depolyalkylation) are temperature, acidity of the catalyst, and ratio of components [26, 27], i.e. the same factors that are important for arene alkylation, conjugated with the chain propagation at isobutylene polymerization.

Processes of non-degradation alkylation and depolyalkylation of polyisobutylene proceed according to Scheme 5.1 [25, 28]:

Scheme 5.1

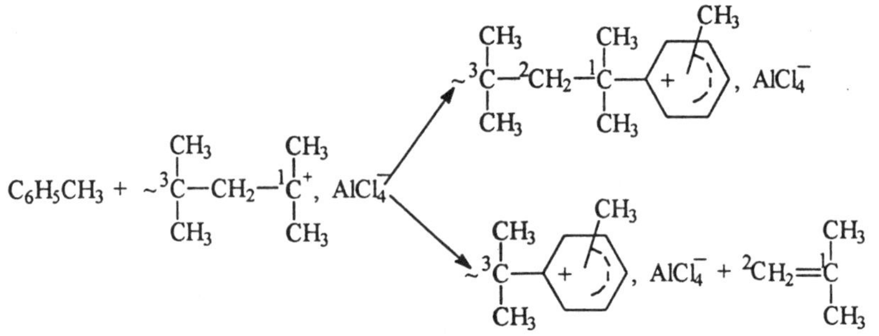

I

II

i.e. by the type of reaction:

$$C_4H_8 + C_6H_5CH_3 \xrightarrow{\text{the Gustavsson complex}} C_4H_9C_6H_4CH_3.$$

The role of reaction I (Scheme 5.1) reduces with growth of temperature and catalyst concentration due to strengthening of interaction between carbocation and counterion. The process of polyisobutylene depolymerization prevails because of interaction of a carbonium center with electrons in the β-position to the C–C bond, which leads to fragmentation of polyisobutylene, i.e. to β-decay of macromolecules by reaction II (Scheme 5.1). The condition for carbonium ion fragmentation in solution is high stability of resulting end carbocations, for example, at the sacrifice of processes of internal stabilization of conjugation or induction. In the case of catalytic degradation of polyisobutylene, arene, probably, plays the role of external stabilizer of carbonium ions, making fragmentation of the polymer by reaction II (Scheme 5.1) easier. In this case, a macromolecular fragment of the initial carbocation, formed during degradation of polyisobutylene, enters the reaction of conjugated alkylation with formation of arenonium structures of polyisobutylene–aromatic compounds possessing molecular mass $M < M_0$. In moving from benzene and toluene to more basic arenes, the depth of polyisobutylene degradation decreases, which is connected to growth of steric obstacles to polymer fragmentation. Simultaneously, yielding isobutylene alkylates a new molecule of arene forming *tert*-butyltoluene by reaction III (Scheme 5.1). Chemical linking of isobutylene by toluene (similar to removal of the monomer from the reaction zone in a different way) decreases the equilibrium concentration of the monomer and causes a decrease of T_{border} at polyisobutylene degradation.

Consequently, arene activates thermal degradation of polyisobutylene, catalyzed by acidic catalysts, assisting alkylation in conjugation with depolymerization or depolyalkylation of polyisobutylene.

Of practical significance are reactions of polyisobutylene with maleic anhydride [29]: hydrogenation (catalyst is mixed Ni–W sulfide, 645 K, 2.4 MPa) and alkylation of phenols in the presence of electrophilic catalysts (300–360 K) [30–36]. Modification of polyisobutylene by phenol in the presence of benzosulfoacid (375–395 K) gives the maximum yield of polyisobutylenyl-phenol of 31 mass% [37]. Significantly better results are obtained at alkylation of phenols by polyisobutylene in the presence of $Me[AlCl_4]{\cdot}H_2O$ (Me — Li, Na, K), differed by lower (than for $AlCl_3$) relative acidity (mg equiv. OH/g of catalyst): $AlCl_3$ — 5.0; $Li[AlCl_4]$ — 2.2; $Na[AlCl_4]$ — 1.6; $K[AlCl_4]$ — 0.9 [38] (Table 5.3).

Table 5.3

Alkylation of phenols by polyisobutylene 3.5 hours, heptane, at phenol: polyisobutylene: catalyst ($Na[AlCl_4]$) molar ratio of 2:1:0.5.

Phenol (385 K)	Yield	Aminophenol (355 K)	Yield
Phenol	98	2-Aminophenol	90
2-Methylphenol	95	4-Aminophenol	95
4-*iso*-Propylphenol	96	2-Benzylidenaminophenol	98
3-Methylphenol	91	4-Benzylidenaminophenol	96
2,6-di-*tert*-Butylphenol	85		
Pyrocatechol	94		
Resorcinol	98		
Hydroquinone	91		

The yield of polyisobutylene-phenols decreases with growth of the molecular mass of polyisobutylene.

Functionalization of polyisobutylene by phenols and aminophenols proceeds easily with high yield (2–4 hours; 373–393 K) in the presence of complex salts of $Me[AlCl_4]$ type, where Me is Li, Na, K, the yield of the product being practically independent on nature of alkaline metal in the catalyst. If formation of products of substitution into the aromatic ring is typical for phenols in alkylation by polyisobutylene by the Friedel–Crafts reaction, obtaining N-substituted compounds is typical for amiophenols [31]:

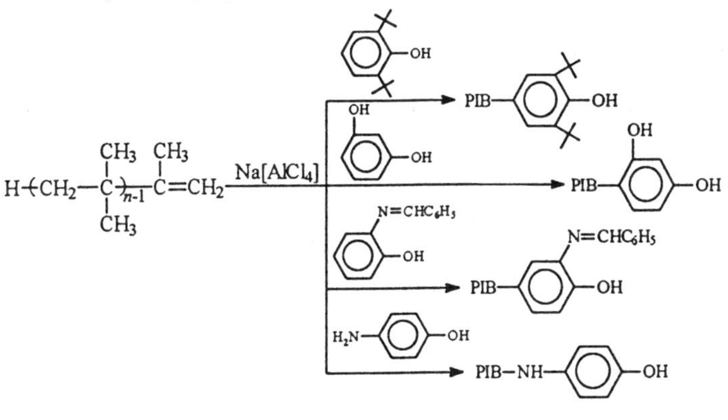

High-molecular phenols and aminophenols are capable of subsequent inter- and intramolecular polymer-analogous transformations, in particular, with 3,4-dibromine-4-methyltetrahydropyrane (dibromide). Synthesis of phenyl esters is performed by interaction of phenols with dibromide in hexane medium in the presence of K_2CO_3 (24 hours).

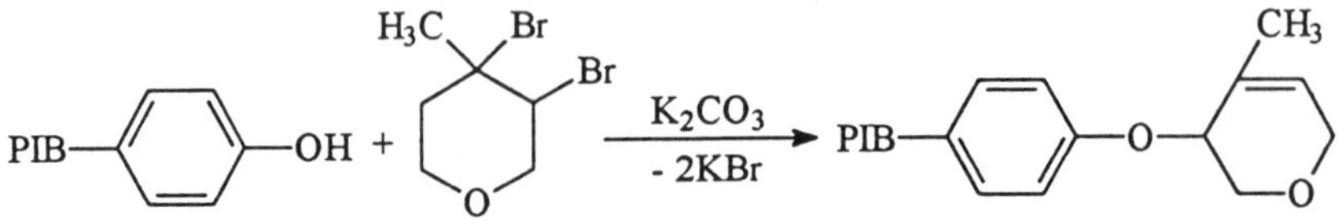

Allyl esters at 423 K (diethylaniline, 5 hours) are regrouped by the Claisen method into appropriate substituted phenols with quantitative yield [32]:

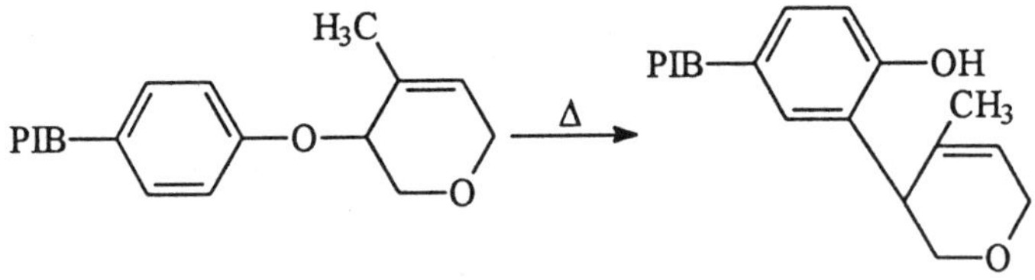

In the case of application of high-molecular Schiff bases, the process proceeds analogously:

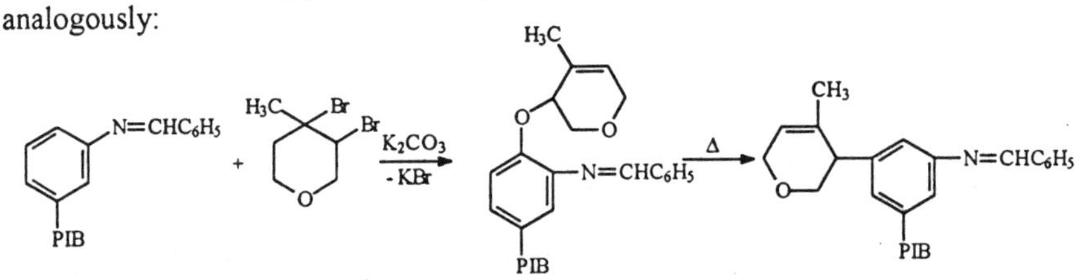

Basing on two-atomic phenols, production of their mono-ethers is possible under the condition of preliminary protection of one of hydroxy groups by the reaction of benzoilation of appropriate phenols [32]:

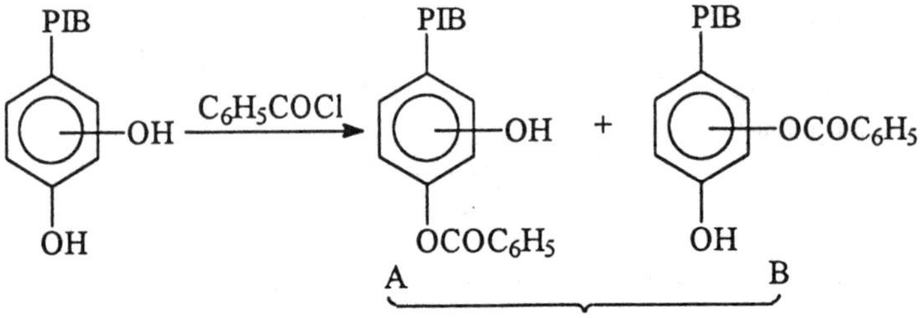

Existence of two hydroxy groups implies occurrence of two isomers in products of benzoylation: A and B. Condensation of two-atomic phenols, substituted by one HO-group, with dibromide proceeds easily (12 hours). After hydrolysis in a water-alcohol solution of NaOH, protection was removed (373 K, 6 hours), and then isomerization of esters by the Claisen reaktiontook place (443 K, 5 hours):

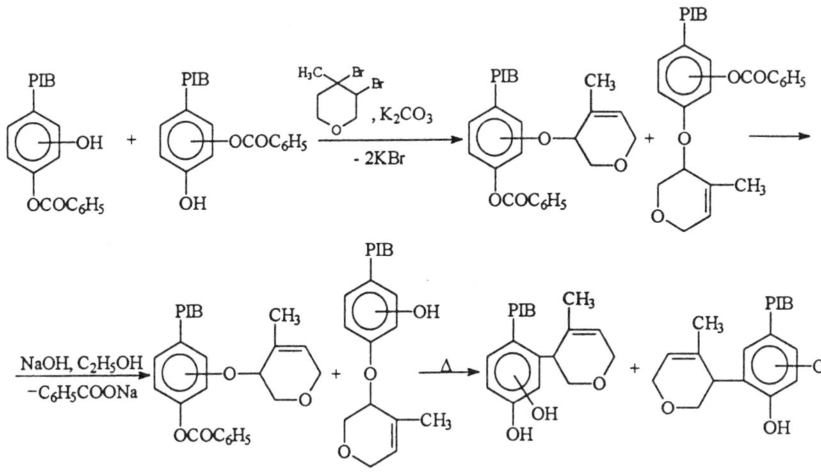

Obtaining full esters from high-molecular two-atomic phenols is also possible, if the ratio of reagents is changed:

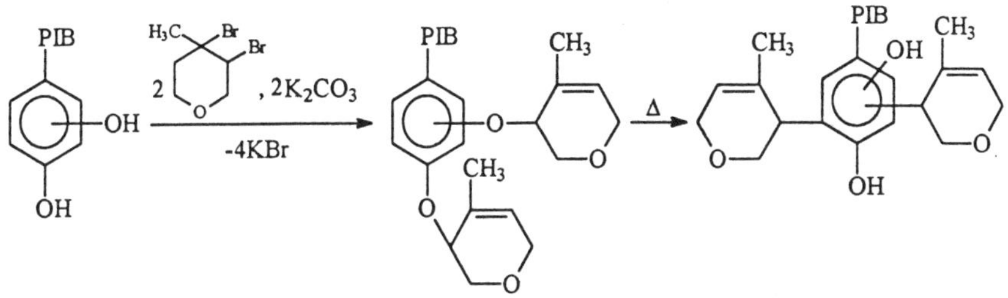

Regrouping by the Claisen reaction in the sequence of allyl–aryl esters based on high-molecular two-atomic phenols (full esters) (443 K, 5 hours) proceeds with lower yields, compared with isomerization of non-full esters of two-atomic phenols (mono-esters), which is stipulated apparently by steric hindrances with regard to introduction of the second pyranyl substituent of non-planar structure into the aromatic ring of the phenol [33].

Intermolecular polymer-analogous transformations of polyisobutylene-phenols with triphenylphosphite enabled the class of high-molecular esters of phosphoric acid to be obtained. Re-etherification of one-atomic polyolephenylphenols under soft conditions (363 K, 1.5–3.0 hours) and with quantitative yield leads to formation of mono-phosphites:

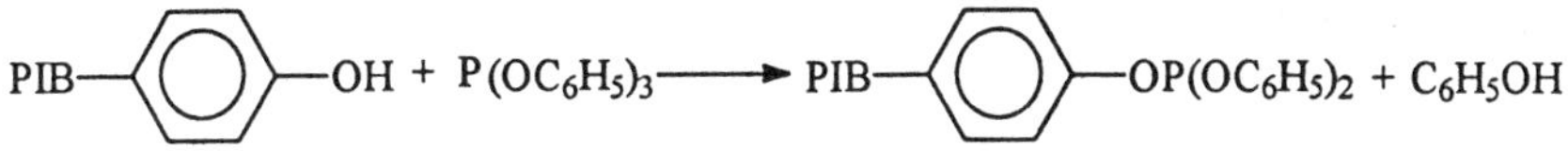

For two-atomic high-molecular phenols, depending on the ratio of initial substances, a sequence of mono- and di-substituted phosphites can be obtained; Condensed cyclic phosphites are formed that are derivatives of polyisobutylene-pyrocatehols [34]:

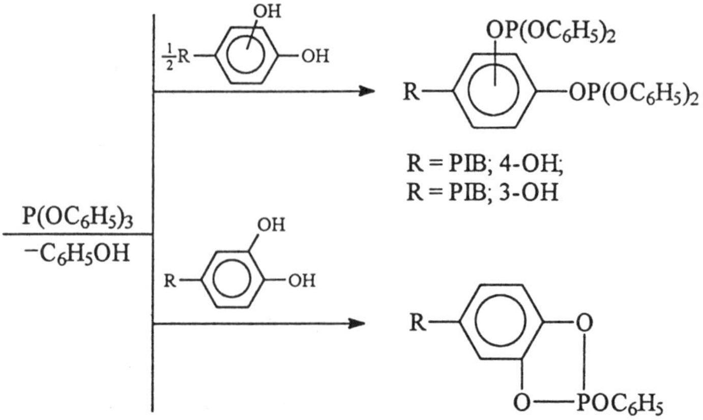

R = PIB

Functionalization of high-molecular phenols by silicon-containing compounds was conducted by re-etherification of framing OC_2H_5-groupings in polyethoxysiloxane molecule (ethylsilicate-40) by polyisobutylene-phenols and polyisobutylenyl-aminophenols, which enables a new class of high-molecular compounds to be formed — polyolephenyl(aminophenoxy)siloxanes [35]:

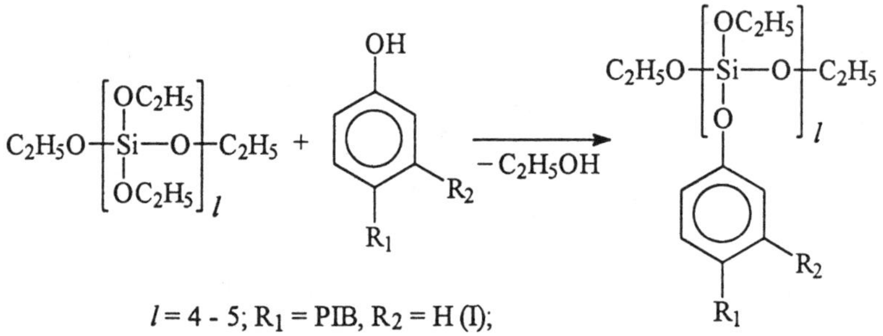

$l = 4 - 5$; R_1 = PIB, R_2 = H (I);
R_1 = PIB, R_2 = OH (II);
R_1 = -NH-PIB, R_2 = H (III);

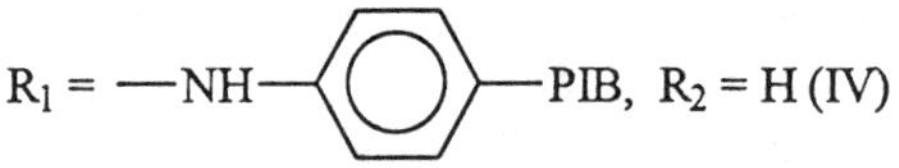

Existence of two reactive centers (HO- and H_2N-groups) in aminophenols gives the possibility of interaction between aminophenols and 3,4-dibromine-4-methyl tetrahydropyrane in two directions, prescribing the mechanism of the subsequent regrouping by the Claisen reactionand chemical structure of resulting compounds (Scheme 5.2):

Scheme 5.2

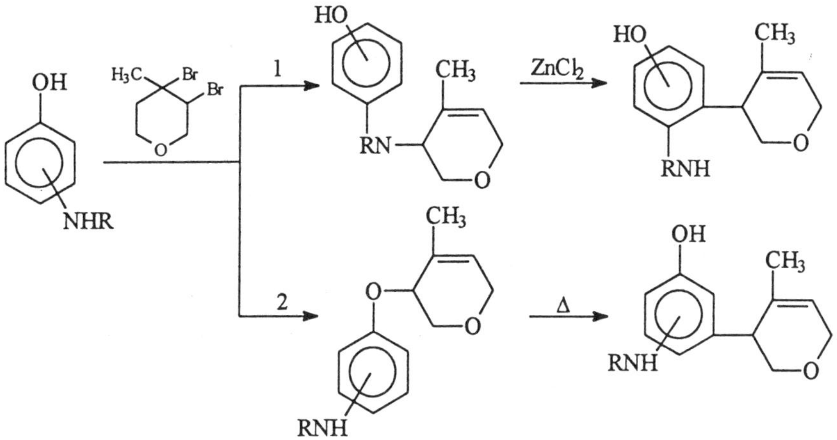

Condensation of aminophenols with dibromide (direction 1 in Scheme 5.2) proceeds in triethylamine at 360 K, which participates first in formation of a bromine-allyl bond. Subsequently, 5-bromine-4-methyl-5,6-dihydro-2H-pyrane enters the reaction of nucleophilic substitution of mobile allylic bromine with formation of products of the interaction by the amino group [36]:

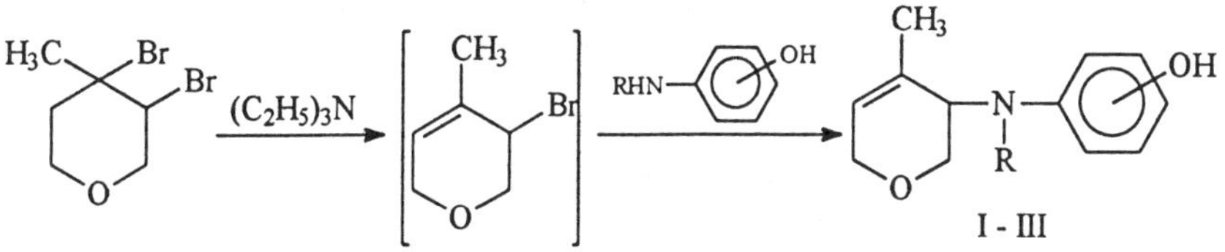

I - III

$R = H$, 2-OH (I); $R = H$, 4-OH (II); $R = C_6H_5$, 4-OH (III).

Alkylation of N-allyl-substituted aminophenols by polyisobutylene (348 K, 3 hours) enables high-molecular analogues to be synthesized, isomerization of which proceeds in nitrobenzene (390 K) only in the presence of catalytic amounts of Lewis acids, $ZnCl_2$, in particular:

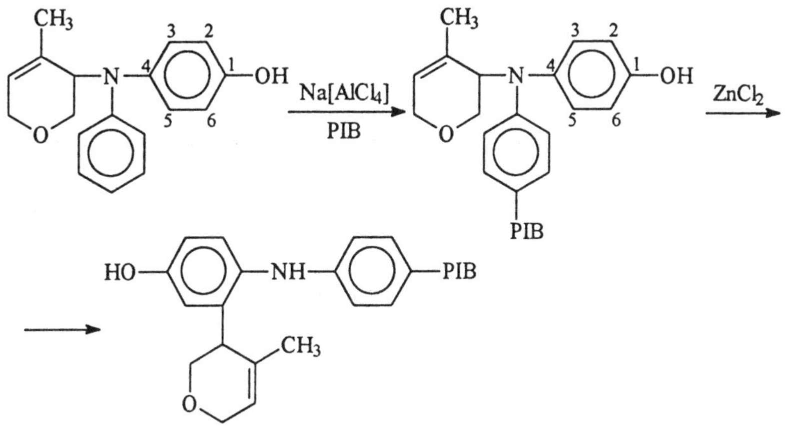

Contrary to esters from the sequence of dihydropyranes, for which thermal regrouping (443 K) into dihydropyranylphenols with quantitative yield in the base medium (N, N-diethylaniline) is typical, the Claisen regrouping of N-allylaminophenols proceeds with a yield of 81-85%.

To realize direction 2 (Scheme 5.2), i.e. condensation of dibromide with aminophenol by hydroxy group

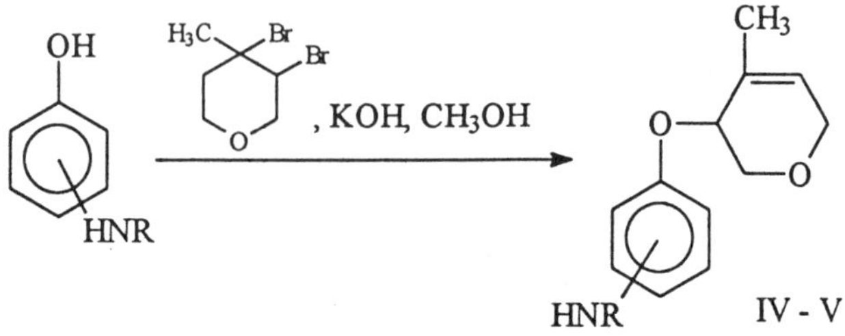

R = H, 2-OH (IV); R = H (V); R = C_6H_5, 4-OH (VI).

it is necessary to use harder bases (for example, KOH in methanol), which perform the same functions as triethylamine at proceeding of condensation by direction 2. Contrary to the reaction of dibromotetrahydropyrane with aminophenols by H_2N-groups, the reaction by HO-group in aminophenol proceeds at lower temperature (353 K, 1.0–1.5 hours). If in the case of dibromide condensation with aminophenols by H_2N-group, which leads to formation of practically a single product – [N-(4-methyl-5,6-dihydro-2H-pyranyl-5)amino]phenols, no N-substituted compounds – [N-(4-methyl-5,6-dihydro-2H-pyranyl-5)amino]phenols in the mol/mol ratio of 2:1 of interaction products by HO- and H_2N-groups are formed during proceeding of the process by the HO-group together with (4-methyl-5,6-dihydro-2H-pyranyl-5-oxy)-aminobenzenes. Separation of isomeric substances is performed with the help of the column chromatography.

High-molecular analogues were obtained by alkylation of compounds from the class of (dihydropyranyloxy)aminobenzenes of polyisobutylenes (383 K, 3 hours), and their regrouping was conducted subsequently:

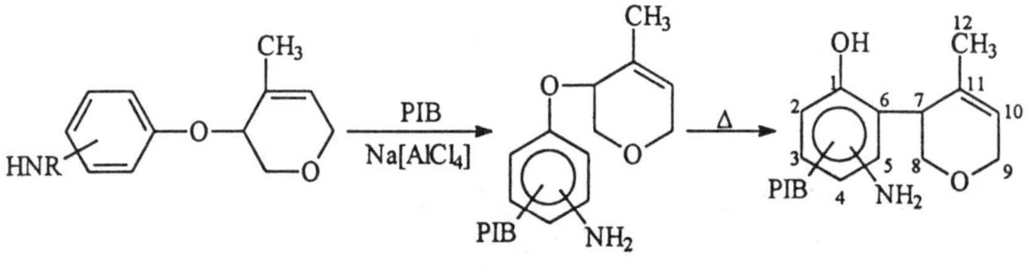

Isomerization of (4-methyl-5,6-dihydro-2H-pyranyl-5-oxy)-polyisobutylenyl-aminobenzenes proceeds by the classic scheme [3,3] of sigmatropic transformations in N,N-diethylalanine at 443 K over 4 hours with practically quantitative yield. By virtue of the fact the that allyl fragment in the molecule of (4-methyl-5,6-dihydro-2H-pyranyl-5-oxy)-aminobenzenes, used in the Claisen isomerization, is allyl-symmetric, the product of regrouping is identical to the initial structure, and inversion cannot be performed.

End functional groups of new high-molecular compounds, synthesized during the past 20 years by modification of polyisobutylene by >C=C< bonds by various low-molecular substances are displayed in Table 5.4.

Table 5.4

Modification of polyisobutylene by end groups [19, 31-38].

Initial groups in PIB[1]	Reagents	Groups in modified PIB
tert-C_4H_9~CH_2—$C(CH_3)$=CH_2	HCl (cat.) C_6H_5R (cat.) H_2 (cat.) 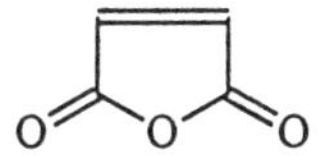 $HSiCl_3$ RSCl	*tert*- C_4H_9~$C(CH_3)_2Cl$ *tert*- C_4H_9~C_6H_4R *tert*- C_4H_9~$CH_2CH(CH_3)_2$ *tert*- C_4H_9~ ~CH=$C(CH_3)CH_2$— 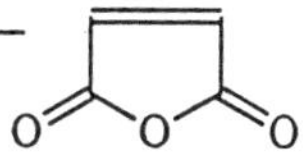 *tert*- C_4H_9~$SiCl_3$ *tert*- C_4H_9~$CH_2C(CH_3)(Cl)CH_2SR$
$ClC(CH_3)_2$~$CH_2C(CH_3)_2Cl$	*tert*- C_4H_9OK	CH_2—$C(CH_3)CH_2$~$CH_2C(CH_3)$=CH_2
$ClCH_2C(CH_3)_2$~CH_2— —$C(CH_3)_2Cl$	MeOH	$ClCH_2C(CH_3)_2$~$CH_2C(CH_3)$=CH_2
CH_2=$C(CH_3)CH_2$~$CH_2C(CH_3)$=CH_2	*m*- Cl—$C_6H_4CO_3H$ BH_3; O_2(NaOH)	$CH_2(O)$—$C(CH_3)HCH_2$~$CH_2C(CH_3)H$—$(O)CH_2$ (epoxide rings) $HOCH_2CH(CH_3)CH_2$~$CH_2CH(CH_3)CH_2OH$
tert-C_4H_9~C_6H_4R	RuO_4	*tert*-C_4H_9~COOH
RC_6H_4~C_6H_4R	RuO_4	HOOC~COOH
HOOC~COOH	$SOCl_2$ $LiAlH_4$ HOROH H_2NRNH_2 CH_2—CH—R (epoxide, O bridge)	ClOC~COCl $HOCH_2$~CH_2OH HOROC(O)~C(O)OROH $H_2NRNHC(O)$~$C(O)NHRNH_2$ $HOCH_2CH(R)OC(O)$~$C(O)OCH(R)CH_2OH$ ~$C(O)OCH(R)CH_2OH$
ClC(O)~C(O)Cl	HS—P(=S)$(OR)_2$	$(RO)_2P(=S)$—C(O)~C(O)—$P(=S)(RO)_2$ ROC(O)~C(O)OR
$HOCH_2$~CH_2OH	HOROH	
	$ClSi(CH_3)_3$ C_6H_5NCO ClC(O)NCO	$(CH_3)_3SiOCH_2$~$CH_2OSi(CH_3)_3$ $C_6H_5NHC(O)O$~$OC(O)NHC_6H_5$ OCNCOO~OCONCO

[1] Symbol ~ marks –$(CH_2$-$C(CH_3)_2$-$)_n$ group.

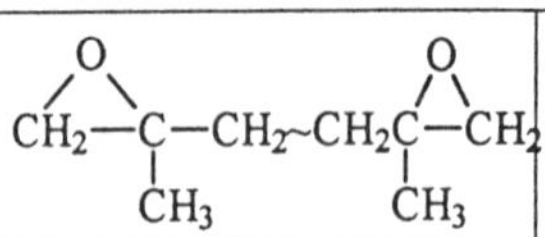	$ZnBr_2$	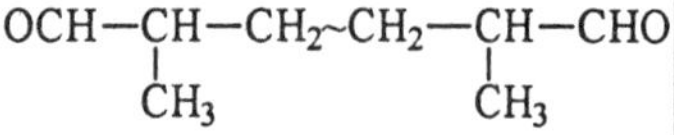
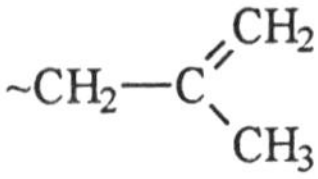	$Me[AlCl_4]$ (cat.), Me = Li, Na, K, 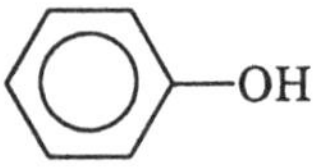	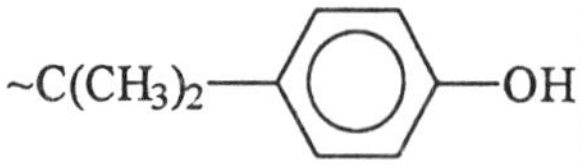
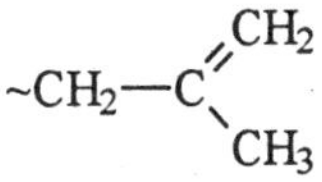	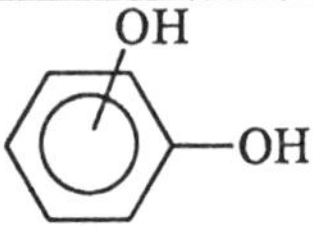	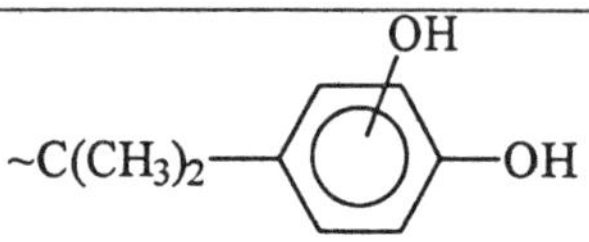
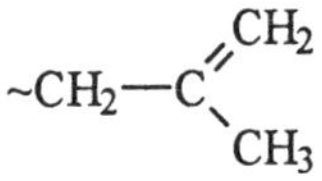	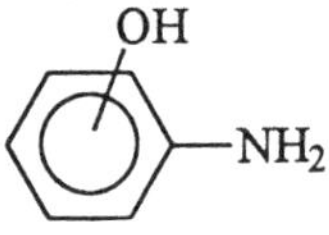	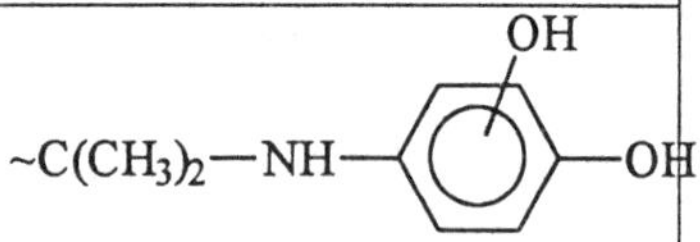
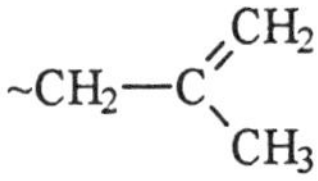	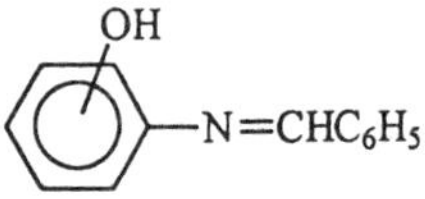	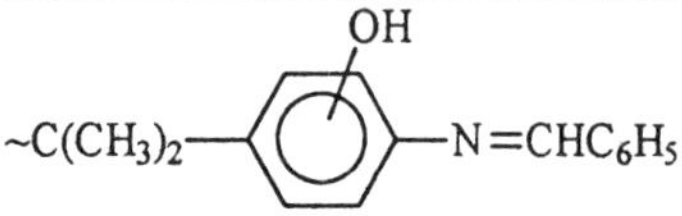
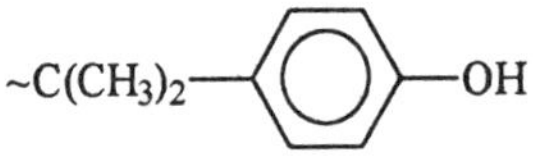	K_2CO_3 (cat.), 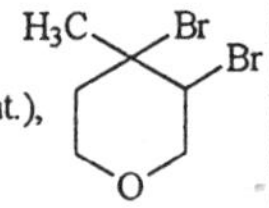	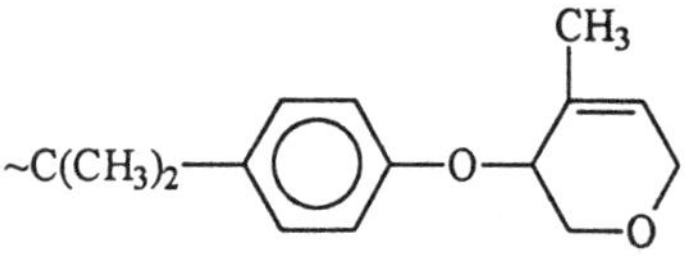
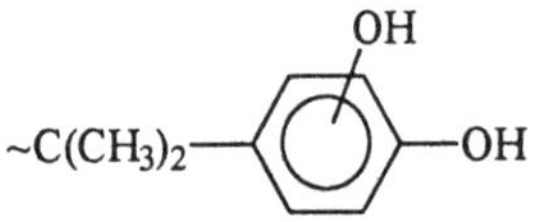	C_6H_5COCl	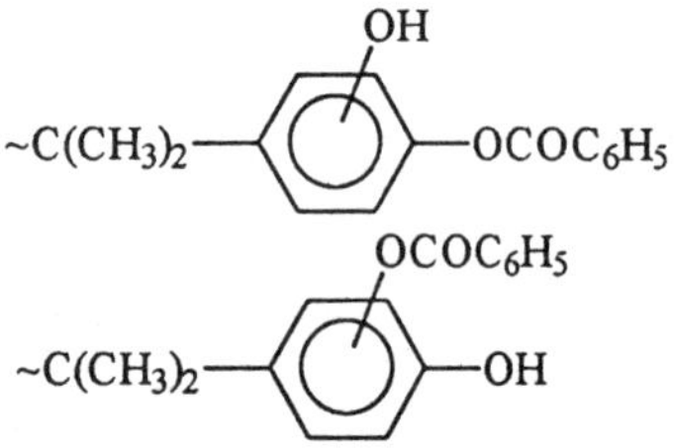
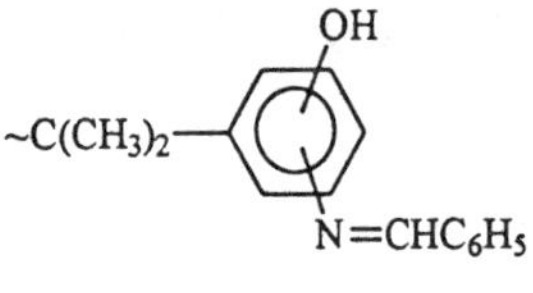	K_2CO_3, 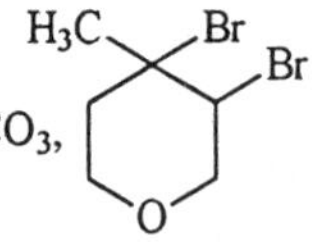	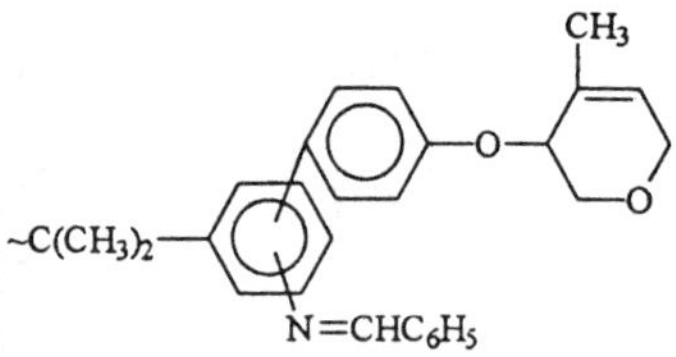

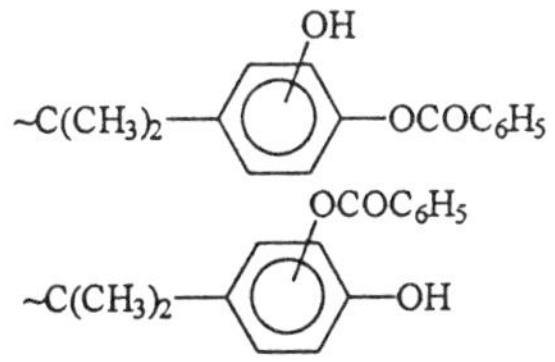	K_2CO_3 (cat.), 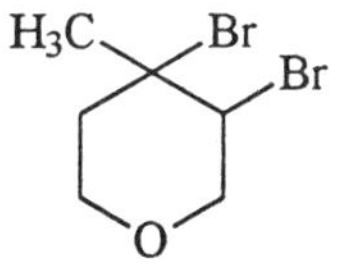	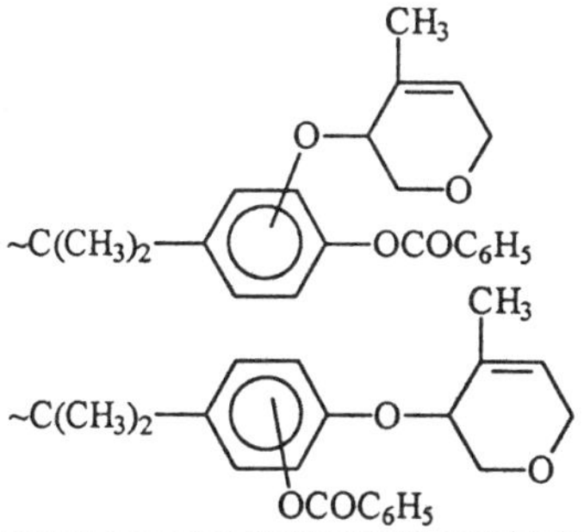
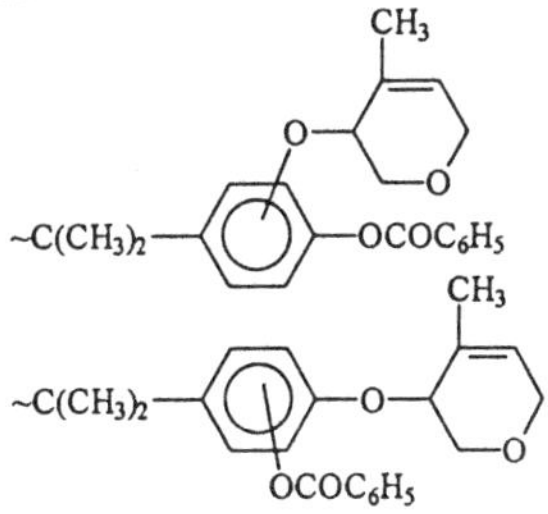	NaOH, C_2H_5OH	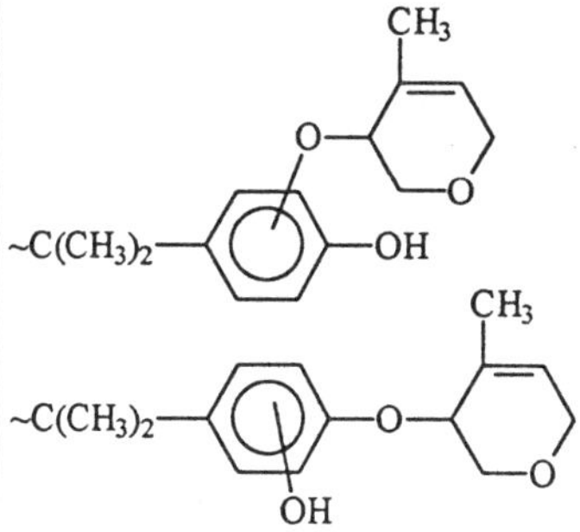
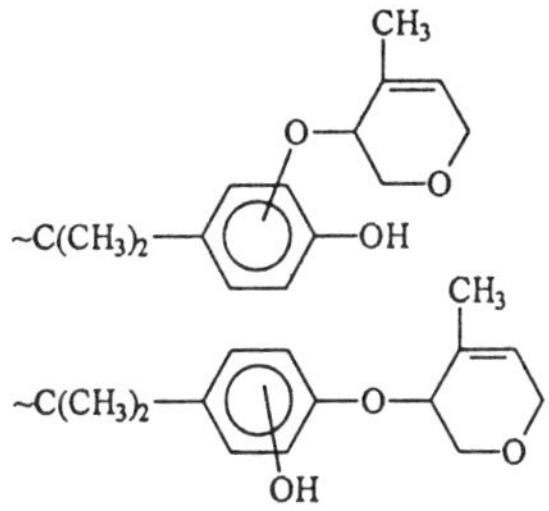	t°	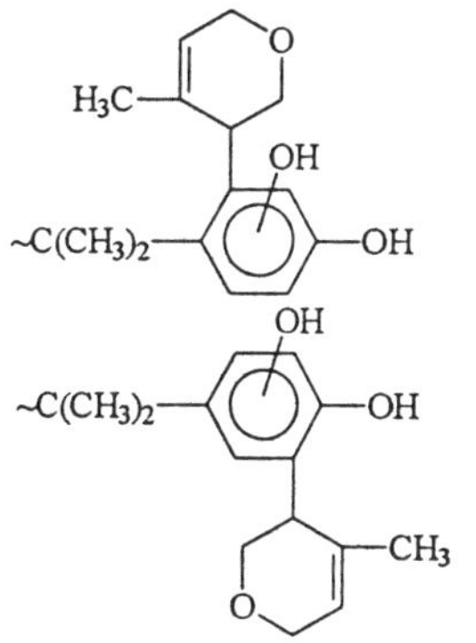
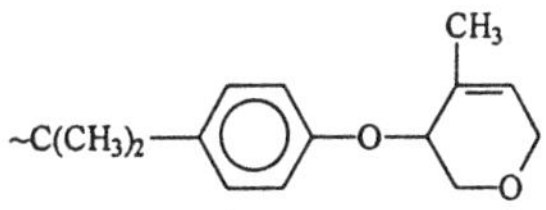	t°	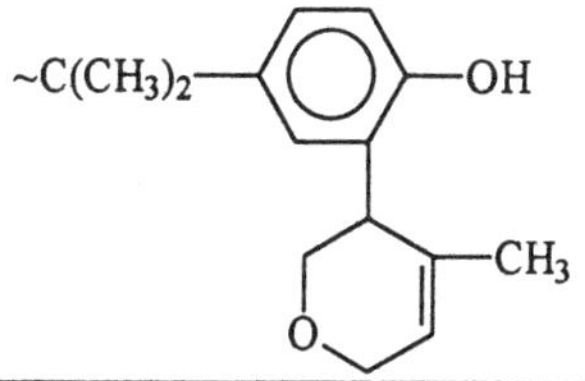
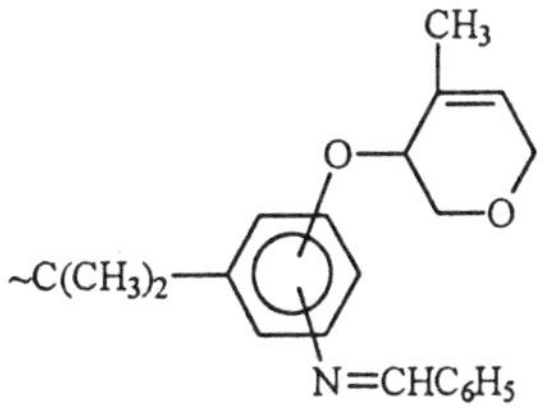	t°	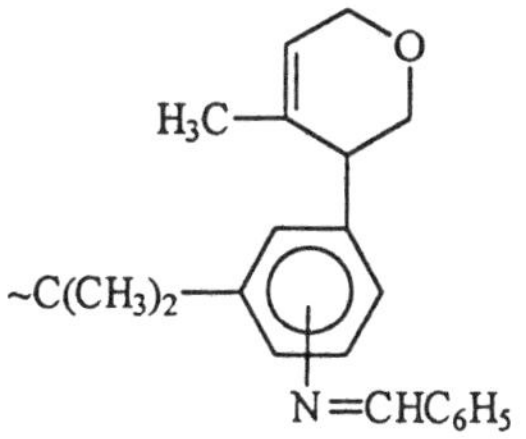

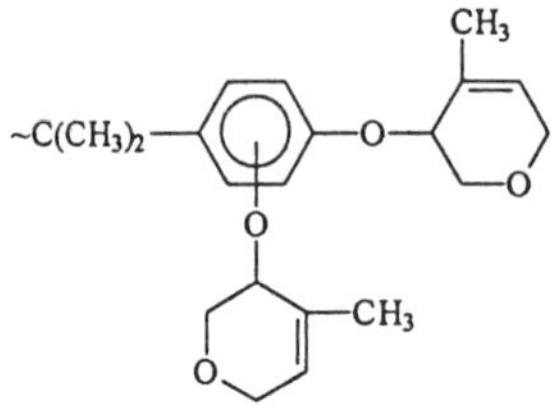	t°	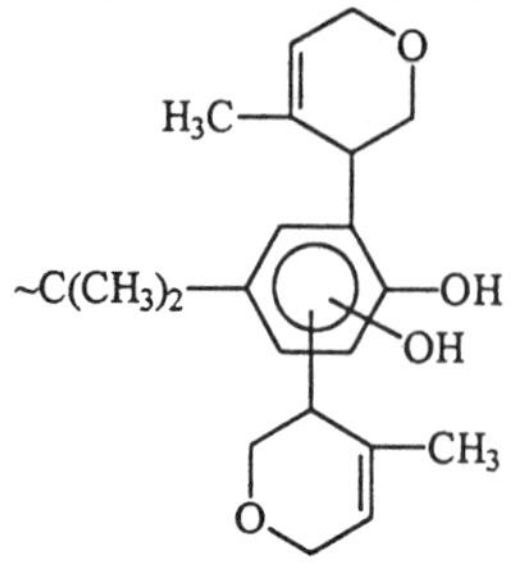
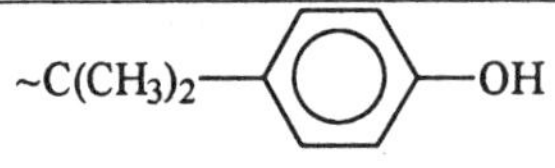	$P(OC_6H_5)_3$	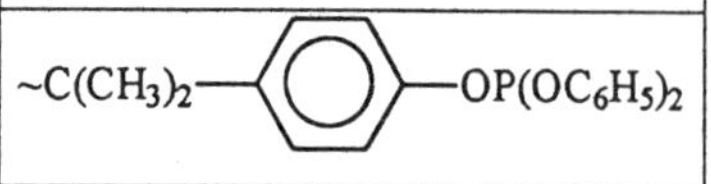
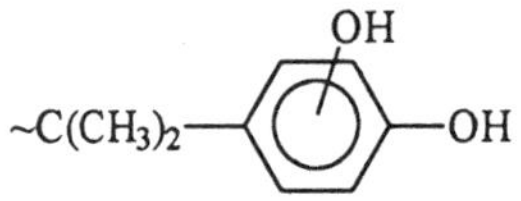	$P(OC_6H_5)_3$	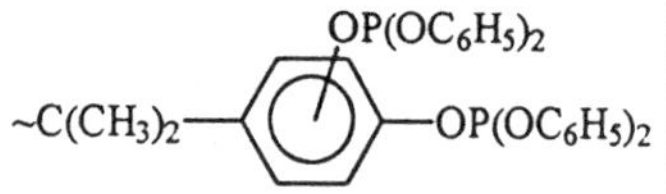
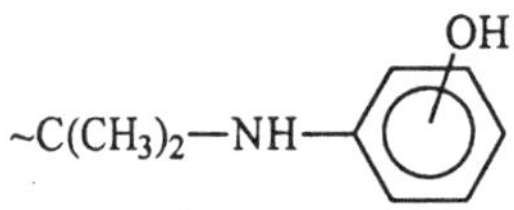	$P(OC_6H_5)_3$	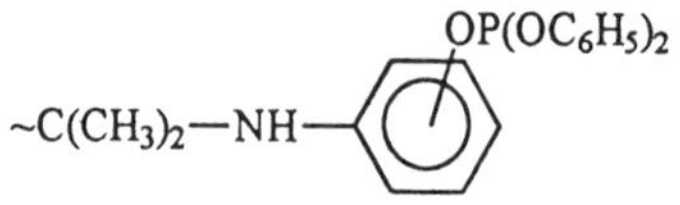
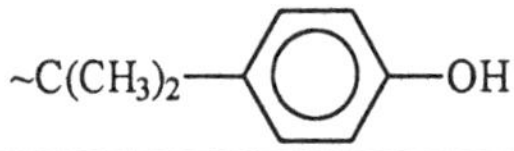	HNO_3	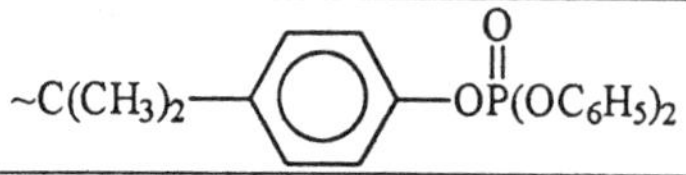
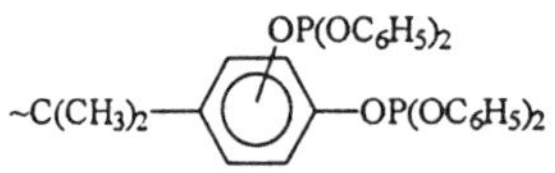	HNO_3	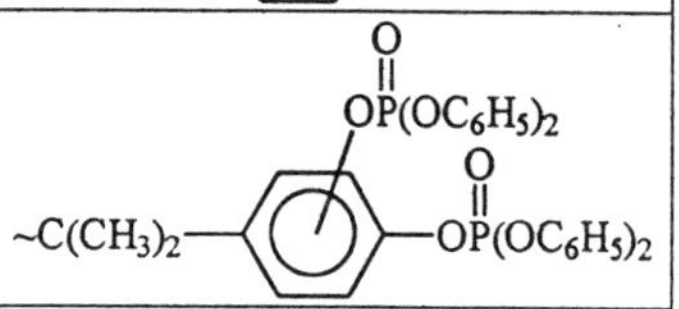
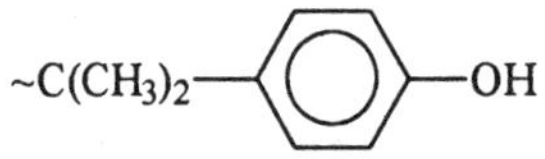	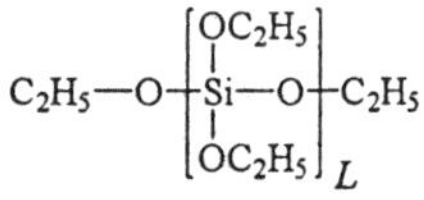 *L* = 4-5	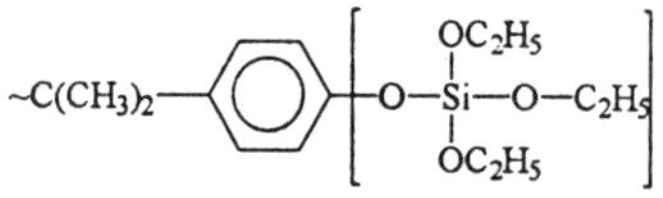
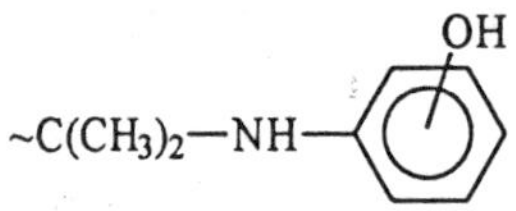	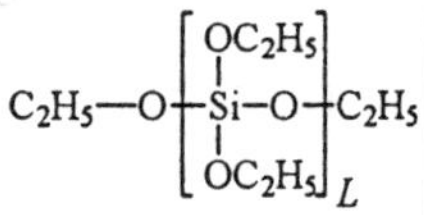	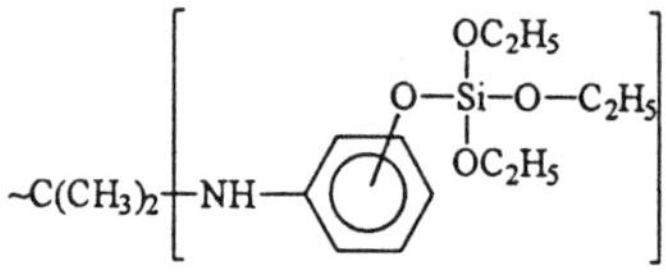
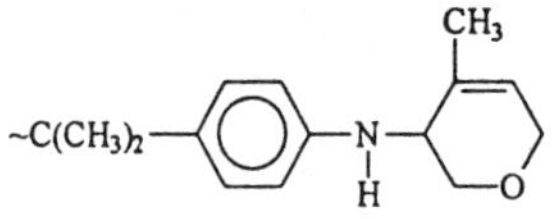	t°, $ZnCl_2$ (cat.)	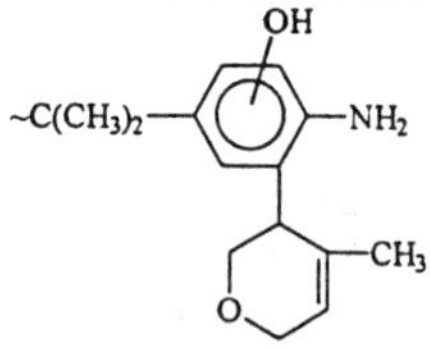

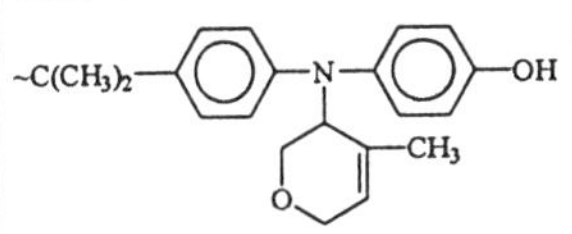	t°, $ZnCl_2$ (cat.)	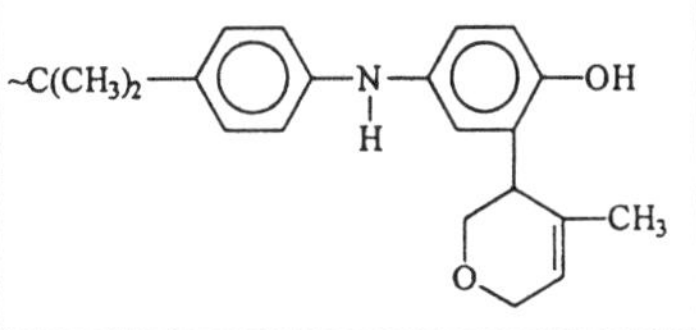
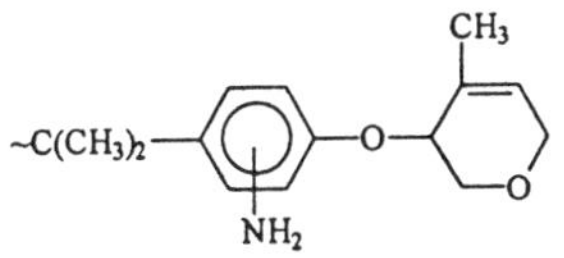	t°	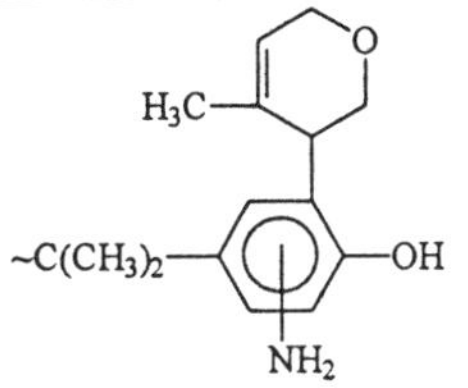
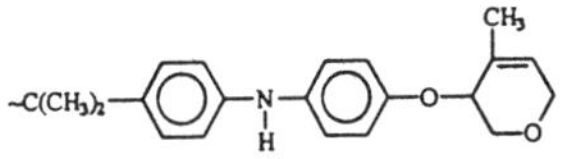	t°	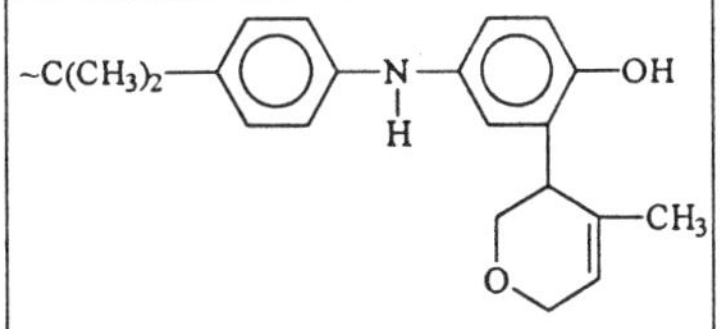

Polyisobutylenylphenols display high activity as stabilizers/antioxidants for many polymers and oils.

Aromatic hydrocarbons with mobile atoms of hydrogen or CH-acid, for example, *iso*-propylenebenzenes, under appropriate conditions cause ionic hydrogenation of polyisobutylenes.

The reaction of ionic (electrophilic) hydrogenation of polyisobutylene, representing consecutive electrophilic (H^+) and nucleophilic (H^-) of the olefin attack [39–43],

$$H^+, A^- + R^= + HR' \rightarrow R^{\delta+}, A^{\delta-} + HR' \rightarrow RH + R'^{\delta+}, A^{\delta-}$$

proceeds, if the cation R'^+ is more stable than R^+, and the acidic anion differs by low nucleophility. Selection of appropriate donors of hydride-ions, catalyst and conditions of the reaction determines the direction and efficiency of carbocation transmission compared with competing conjugated processes – depolymerization or depolyalkylation of polymer, alkylation and other reactions. In particular, dimers and trimers of isobutylene in the presence of 1,4-methylisopropylbenzene and acidic catalyst transform with high conversions into appropriate alkanes (I) and alkyl-derivatives of indane (II) by the reaction

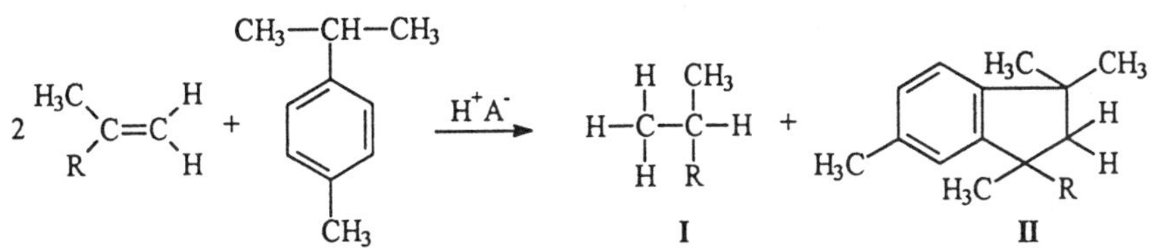

Main products of the reaction are compounds I and II independent of the ratio of vinylidene and three-substituted end C=C bonds in macromolecules. These products are formed in approximately equimolar amounts. The presence of

polyisobutylene with end indane groups is indicated by the data of UV-spectra of the products (absorption with the maximum at 270 nm, lgε = 3.01), identical UV-spectra of 1,1,3,5-tetramethyl-3-neopentylindane.

Table 5.5 displays experimental data on the reaction of ionic hydrogenation of polyisobutylene in the presence of various hydrogenating agents.

Table 5.5

Ionic hydrogenation of isobutylene dimer under the effect of $Al_2Cl_6{\cdot}HCl{\cdot}3C_6H_3(CH_3)_3$ and various isopropylbenzenes (the ratio polyisobutylene:hydrogenating agent = 1:5, 298 K, 1 hour).

Hydrogenating agent	Yield of products of hydrogenation, mass%, for catalyst:substrate ratios		
	0.01	0.03	0.1
Isopropylbenzene	0	—	33
1,4-Methylisopropylbenzene	94	82	62
1,3,5-Triisopropylbenzene	0	92	95
Poly-*n*-isopropylstyrene	0	0	5

Table 5.6

Activity of various catalytic systems in the reaction of ionic hydrogenation of polyisobutylene with MM = 112 in the presence of 1,4-methylisopropylbenzene (polyisobutylene:$C_3H_7C_6H_4CH_3$ = 1:4, 298 K, 1 hour).

Catalytic system	Ratio of catalyst and polyiso-butylene, mol/mol	Conver-sion of polyiso-butylene, mass%	Yield of hydroge-nation products, mass%	Yield of C_8H_{18}, mol/mol	Dispro-portiona-tion degree of $C_3H_7C_6H_4CH_3$, %
$HCl{\cdot}2AlCl_3{\cdot}3C_6H_3(CH_3)_3$	0.01	93	94	25 (50)	0.8
$HCl{\cdot}2AlCl_3{\cdot}3C_6H_5CH_3$	0.01	95	80	20 (40)	1.0
$C_2H_5AlCl_2{\cdot}nH_2O$*	1.0	100	70	0.5 (500)	0.7
$C_2H_5AlCl_2{\cdot}nH_2O$*	0.1	60	—	3.0 (3000)	—
$(C_2H_5)_2AlCl{\cdot}$*tert*-C_4H_9Cl	1.0	100	80	0.3 (—)	—
$C_2H_5AlCl_2$ – KU-2-8**	0.2	100	95	2.5 (0.5)	—
HF***	8.0	33	34	— (0.1)	—

* Contains 10^{-3} mole of H_2O per mole of Al.

** Prepared under conditions shown in Table 3.1; medium — 1,4-methylisopropylbenzene.

*** According to [43].

All studied isopropylbenzenes participate in the reaction of ionic hydrogenation and form a sequence: poly-*n*-isopropylstyrene < isopropylbenzene < 1,4-methylisopropylbenzene ≈ 1,3,5-triisopropylbenzene.

The reaction of ionic hydrogenation of polyisobutylene is often accompanied by secondary processes, first of all, alkylation of isopropylbenzene (polyisobutylene).

Direction and depth of the reaction are also significantly affected by the nature of applied catalyst (Table 5.6).

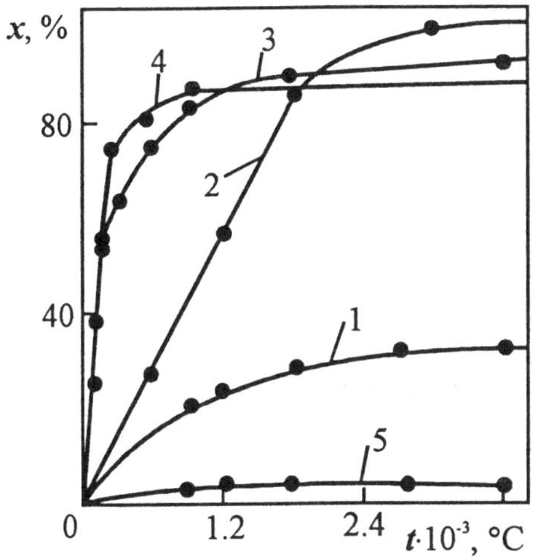

Figure 5.4. Dependence of diisobutylene conversion on hydrogenation time (298 K) in the presence of isopropylbenzene (1), 1,4-methylisopropylbenzene (2, 3), triisopropylbenzene (4, 5) and catalysts $C_2H_5AlCl_2$ — KU-2-8 (1, 2, 5) and $3C_6H_3(CH_3)_3{\cdot}HCl{\cdot}Al_2Cl_6$ (3, 4) under different molar ratio of substrate:hydrogenating agent:catalyst — 1:5:0.2 (1, 2, 5); 1:5:0.1 (3, 4).

Complex catalysts exceed HF in activity and selectivity of action (yield of hydrogenation products). The low efficiency of a heterogeneous catalyst (at the highest selectivity) calculated for SO_3H-group is explained by inertness of an inactivated cation. At the same time, efficiency of the $C_2H_5AlCl_2{\cdot}H_2O$ aquacomplex points out that initiation of the reaction can proceed at low concentration of a proton-donor.

Kinetically, ionic hydrogenation of polyisobutylene under the effect of homogeneous catalysts represents a comparatively fast reactions (Figure 5.4, curves 3 and 4).

General scheme of the process is shown in Scheme 5-3 (see below).

The main direction can be separated in Scheme 5.3 – the reaction of ionic hydrogenation (1, 2, 3), and transformations of carbocations of the first (I) and the second (II) generation connected to it (Ia and IIa — main products; Ib and IIb — secondary products).

A specific feature of the reaction of ionic hydrogenation of polyisobutylene under the effect of the acidic catalyst–isopropylbenzenes system is attachment of initial olefin to carbocation II (Scheme 5.3, reaction 5). Notice that, independent of the type of diisobutylene isomer, 1,1,3,5-tetramethyl-3-neopentylindane is formed only instead of two supposed indane structures Ia. Probably, 2,4,4-trimethylpentene-2 is isomerized into the associated α-isomer. This way of forming indane structures differs from the traditional one, which supposes cyclic dimerization of 1-methyl-4-isopropylbenzene, intermediately formed from cation II.

Formation of 2,2-ditolylpropane during hydrogenation (reaction 6, Scheme 5.3) is the result of conjugated alkylation of toluene by carbocation II of the second generation. Reactions 4-6 in Scheme 5.3 possess regeneration of the catalyst, which predetermines the catalytic character of ionic hydrogenation.

Scheme 5.3

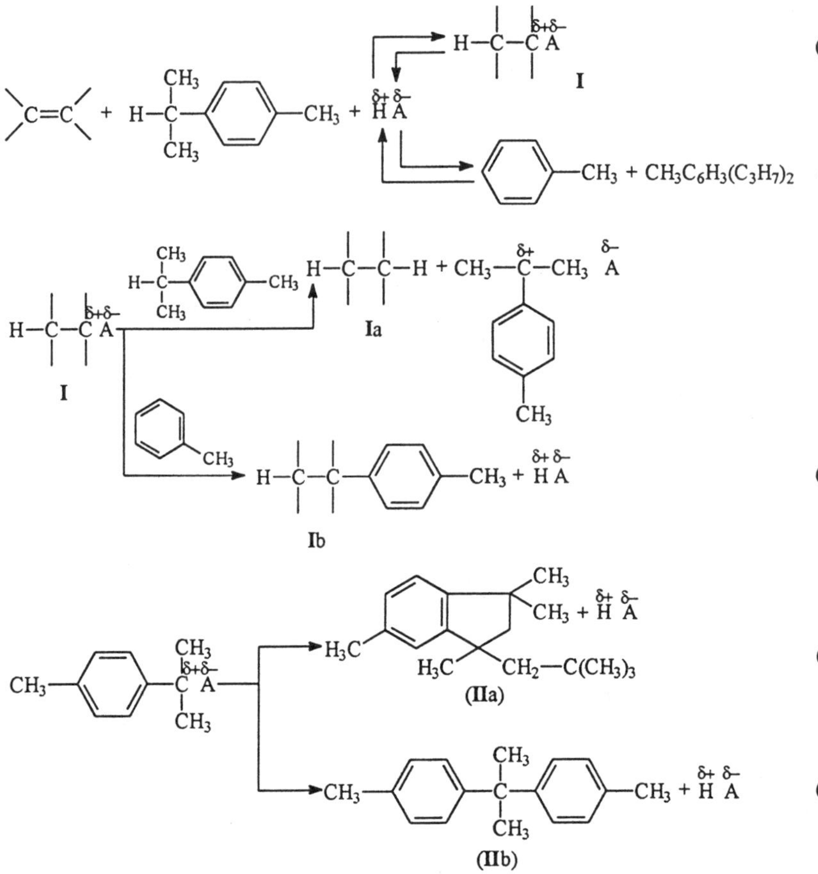

When the molecular mass of polyisobutylene grows, an excess of hydrogenating agent must be taken: approximately 40-fold for polyisobutylene with MM = 450 and 50-fold for polyisobutylene with MM = 900.

Necessity of a great excess of hydrogenating agent is apparently bound to low concentration of C=C double bonds and its decrease with growth of the molecular mass of polyisobutylene. That is why there is a "critical" molecular mass below which samples of polyisobutylene barely participate in the reaction of ionic hydrogenation (1,500 for heterogeneous catalyst).

Rise of the ratio $C_3H_7C_6H_4CH_3$:polyisobutylene is insufficient for total high conversion of iso-olefin, through it increases yield of iso-octane and selectivity of the main reaction 3, Scheme 5.3, at the sacrifice of limiting the secondary reaction 4 (Table 5.7).

Of interest is one more feature of ionic hydrogenation of polyisobutylene in the presence of a solid catalyst — the absence of temperature influence in the range of, at least, 240–335 K on depth and rate of the process. Because of the technological advantages of this system (for example, ease of release of reaction products), the process of non-degradative ionic hydrogenation of polyisobutylene seems to have considerable potential as a route to obtain synthetic oils, resistant to oxidation by O_2 and O_3.

Table 5.7

Influence of the ratio hydrogenating agent:polyisobutylene (C_8H_{16}) on the depth and selectivity of ionic hydrogenation (298 K, 1 hour, C_8H_{16} – $1.6 \cdot 10^{-3}$ kg, catalyst - $3 \cdot 10^{-3}$ kg).

$C_3H_7C_6H_4CH_3$:polyiso-butylene	Reaction products, mass%			Conversion of polyiso-butylene, mass%	Selectivity by yield of C_8H_{18}, %
	C_8H_{18}	Ib**	IIb**		
0:1*	0	100	0	100	0
0.5:1	15.5	40.6	44.1	100	23
1:1	18.7	19	62.3	76	27
2:1	40.7	15.7	43.6	100	73
4:1	40.0	0	62	100	100

* In the absence of *n*-isopropyltoluene, molar ratio of polyisobutylene and toluene is 1:1.

** Products of toluene alkylation by carbocations I and II (reactions 4 and 6).

5.2.2. Polyisobutylene reactions involving the main chain.

Substitution reactions proceeding by the free-radical mechanism are typical for polyisobutylene, and are quite often accompanied by simultaneous degradation of macromolecules.

Chlorine derivatives of polyisobutylene are formed at chlorination in solution of dichloroethane, chloroform or tetrachlorocarbon with catalysis by UV-light, I_2 or $FeCl_2$. Chlorination is favored by the presence of HNO_3 or O_3. The process proceeds faster if liquid chlorine is used [44, p. 238]. Attendant sulfochlorides are formed at simultaneous effect of Cl_2 and SO_2 on polyisobutylene in CCl_4 at UV-radiation.

If polyisobutylene is reacted with oxalyl chloride or phosgene, chloroformyl groups can be introduced into the composition of the molecules [14, p. 239]:

$$\sim C(CH_3)_2 - CH_2 \sim \xrightarrow{Cl\bullet} \sim C(CH_3)_2 - \dot{C}H \sim + HCl$$

$$\sim C(CH_3)_2 - \dot{C}H\sim + ClC(=O)-C(=O)Cl \xrightarrow{-\dot{C}OCl} \sim(CH_3)_2 - C(H)C(=O)Cl$$

Reaction of polyisobutylene phosphorylation in the presence of O_2 hardly proceeds and generally leads to polymer degradation:

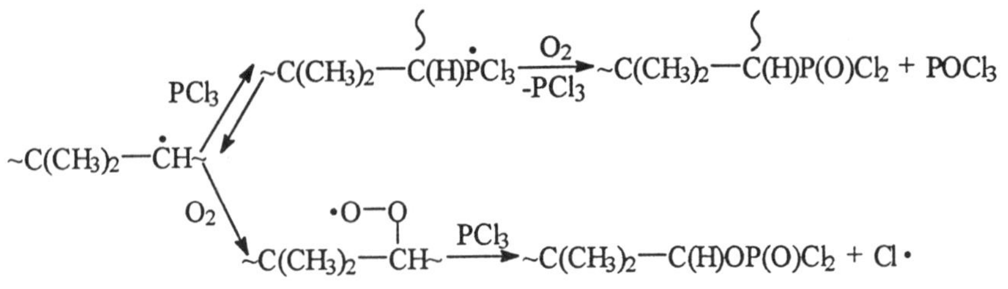

Diminution of the molecular mass of the polymer and accumulation of oxygen-containing functional groups (acids, ketones, peroxides, etc.) is observed at ozone effect on solutions of polyisobutylene, also including temperatures below 273 K.

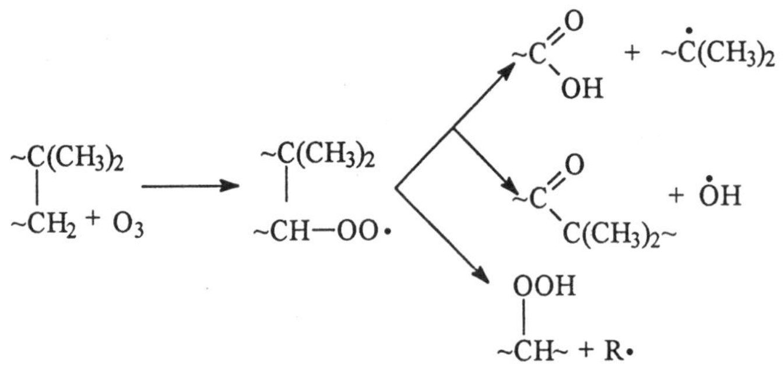

The rate constant for O_3 reaction with polyisobutylene (293 K, CCl_4) equals $K_{8.1} = 0.012$ l/(mol·s) with 0.05 breaks per single act of the reaction [45]. Obviously, in the sequence of polymers with a saturated hydrocarbon chain, polyisobutylene is the most resistant to ozone. The method of ozonation is suitable for application to obtain grafted copolymers, for example, with styrene [46, p. 122]. Hydroperoxide groups, stable under temperatures up to 380–390 K, are formed at oxidation of polyisobutylene (340–365 K).

A noticeable decrease of molecular mass of polyisobutylene is observed in the presence of peroxides; in particular, at application of dicumyl peroxide at 413 K (1 hour) it was 40-fold decreased and no formation of isobutylene was observed [14, p. 223]. Degradation of polyisobutylene also proceeded in the presence of triazene (378 K, $RN_2NHC_6H_5$). The reactivities of free radicals $R^{\bullet}$ form a sequence: $\dot{C}H_3 > \dot{C}_2H_5 > \dot{C}_3H_7 > tert-\dot{C}_4H_9$.

Of importance from scientific and practical point of view are intramolecular transformations of polyisobutylene, proceeding by the law of accident and end groups, as methods for regulating decrease of the molecular mass of polyisobutylene and MMD width (degradation of macromolecules) and obtaining pure monomer (depolymerization) [12, 47–62].

The thermodynamic parameters of polyisobutylene make electrophilic depolymerization easy (Q_{polym} = 53 kJ/mol; T_{border} = 323 K, $E_{depolym}$ = 217 kJ/mol). However, the temperature at which significant polymer decay begins is about 610 K, and thereafter the monomer yield does not exceed 20 mass%.

As can be seen, temperature range of intensive decay of polyisobutylene independent of the molecular mass lies much higher than the limiting temperature of isobutylene polymerization, which is determined by decay of polyisobutylene by

Table 5.8

Yield and composition of gas products at thermal and catalytic degradation of polyisobutylene ($C_{cat} = 5 \cdot 10^{-4}$ mol/g of polyisobutylene).

Molecular mass of polyiso-butylene, M_n^0	Catalyst	Reaction temperature, K	Yield of gas products, mass%	Composition of gas products of the reaction, mass %						
				C_1-C_3	*iso*-C_4H_{10}	*n*-C_4H_{10}	α-C_4H_8	β-C_4H_8	*iso*-C_4H_8	*iso*-C_4H_8 + *iso*-C_4H_{10}
1100	Without catalyst	653	75.9	1.9	2.1	—	0.1	0.9	16	18.1
1100	$AlCl_3$	643	62.7	1.5	2.1	—	Traces	1.0	19.1	21.2
1100	Li[$AlCl_4$]	583	91.8	1.1	23.1	Traces	0.3	3.1	72.3	95.5
1100	Li[$AlCl_4$]	543	61.1	1.1	40.2	0.1	0.2	3.0	55.4	95.6
1100	Li[$AlCl_4$]	523	52.4	1.4	45.6	0.2	0.2	3.7	49.5	95.1
1100	Li[$AlCl_4$]	503	38.2	1.5	53.1	0.1	0.2	3.0	42.4	95.5
1100	Li[$AlCl_4$]	473	21.5	0.4	63.3	0.2	Traces	1.8	31.4	97.7
575000	Li[$AlCl_4$]	583	10.6	9.5	38.7	1.4	4.7	5.3	40.4	79.1
1100	Na[$AlCl_4$]	623	62.8	1.3	3.9	Traces	0.3	1.7	92.5	96.4
1100	Na[$AlCl_4$]	603	55.0	1.2	6.4	0.1	0.3	2.2	89.8	96.2
1100	Na[$AlCl_4$]	583	47.9	1.1	7.8	Traces	0.3	2.0	88.8	96.6
1100	Na[$AlCl_4$]	553	35.4	0.6	13.5	0.1	0.3	6.6	78.8	92.3
1100	Na[$AlCl_4$]	523	21.7	0.8	32.7	0.2	0.2	4.5	61.5	94.2
575000	Na[$AlCl_4$]	623	53.3	6.2	22.4	1.4	9.4	2.5	58.0	80.4
575000	Na[$AlCl_4$]	603	19.6	16.1	34.3	2.9	0.8	11.0	33.9	68.3
575000	Na[$AlCl_4$]	583	10.8	10.9	48.3	1.7	0.6	10.5	26.7	75.0
1100	K[$AlCl_4$]	603	16.3	2.3	5.0	0.1	4.0	1.6	87.1	92.1
1100	K[$AlCl_4$]	583	6.3	2.1	5.6	Traces	0.5	6.2	85.7	91.2
1100	$MgCl_2AlCl_3$	573	93.8	1.4	24.9	0.1	0.1	0.1	73.4	98.3

Note: Hydrogen traces were identified in gas products.

a free-radical, but not carbocationic mechanism. The yield and distribution of low-molecular products of free-radical degradation of polyisobutylene are determined by two typical reactions: degradation of macromolecules by free-radical mechanism and radical substitution reactions (chain transmission to the polymer), which generally leads to formation of hydrocarbons of isomeric structure (Table 5.8) from C_1 to C_{30} [52]. In this case, the idea of selectivity of the process becomes meaningless.

Selectivity and efficiency of the process grows sharply in catalytic degradation of polyisobutylene in the presence of electrophilic catalysts, especially with lower acidity compared with $AlCl_2$ (Table 5.8) [12, 53, 56, 58, 60–62].

The mechanisms of thermal and thermocatalytic degradation of polyisobutylene are different. Thermal degradation of polymeric products proceeds predominantly by the free-radical mechanism [57] and, at polyisobutylene degradation in the presence of electrophilic catalysts (H_2O as cocatalyst), by the ionic mechanism [58–60], possibly, with participation of free-radical processes [63].

In the general case, depolymerization of polyisobutylene, dependent on the nature of the catalyst, is initiated by both the law of accident (so that the possibility of non-catalytic rupture of internal C–C bonds in macromolecules must be taken into account) and the law of end groups (Scheme 5.4):

Scheme 5.4

a) Polyisobutylene degradation by the law of accident:

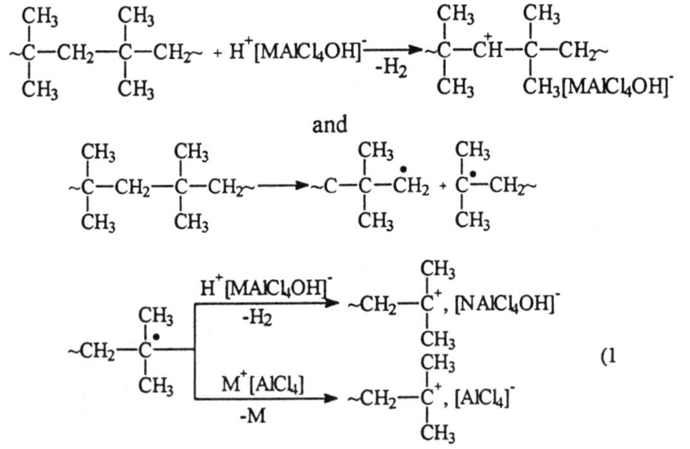

b) Polyisobutylene decomposition by the law of end groups:

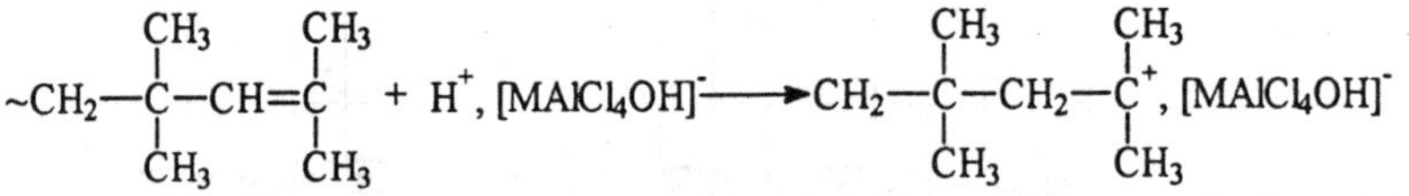
(2

Macrocations formed at temperature higher than T_{border} are β–degraded with isobutylene elimination.

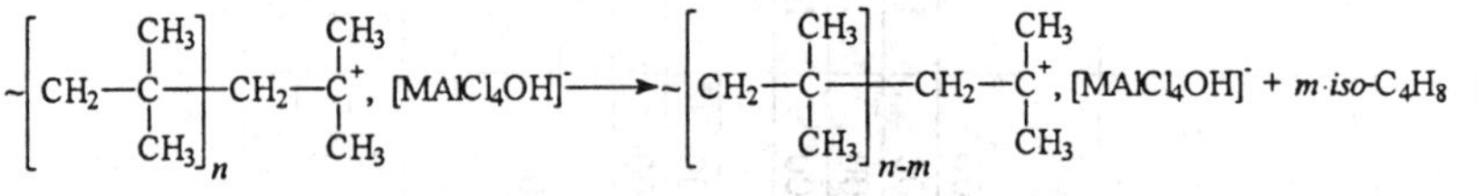
(3

Diminution of E_a of the process from 217 ± 9 kJ/mol at thermal degradation of polyisobutylene down to 93 ± 4 kJ/mol (reaction 1, Scheme 5.4) and 70 ± 9 kJ/mol (reaction 2, Scheme 5.4), and simultaneous elongation of kinetic Z (zip) degradation of the chain from 4 to 20-35 monomeric units at introduction of electrophilic catalysts indicate substitution of the free-radical mechanism of polyisobutylene degradation by a predominantly cationic one. Consequently, the monomer yield grows noticeably. According to these data, seeming noncorrespondence, displayed in the literature, between low yield of the monomer at thermal degradation of polyisobutylene and low values of polymerization heats, and, as a consequence, the upper limit of formation temperature of the polymer, is explained by difference in the mechanism of polyisobutylene degradation (cationic or free-radical mechanism at thermal effect). The possibility of providing degradation of the polymer according to the scheme of cationic depolymerization eliminates this noncorrespondence.

The chemical nature of the catalyst determines the rate and sequence of the reactions of polyisobutylene degradation.

In the general case [53, 58, 60], the following kinetic scheme of the process proceeding can be assumed:

$$\left.\begin{aligned} &\mathrm{P} \xrightarrow{W_\mathrm{P}} 2\mathrm{R}^\bullet \\ &\mathrm{P} + \mathrm{cat} \xrightarrow{k} 2\mathrm{R}^\bullet \\ &\mathrm{R}^\bullet + \mathrm{cat} \xrightarrow{k'_\mathrm{in}} \mathrm{R}^+ \\ &\mathrm{P} + \mathrm{cat} \xrightarrow{k'_\mathrm{in}} \mathrm{R}^+ \end{aligned}\right\} \quad \text{(the law of accident)} \tag{5.1}$$

$$\begin{aligned} &\mathrm{P} + \mathrm{cat} \xrightarrow{k'_\mathrm{in}} \mathrm{R}^+ \quad \text{(the law of end groups)} \\ &\mathrm{R}_n^+ \xrightarrow{k'_\mathrm{depol}} \mathrm{R}_{n-z}^+ + {}_z\mathrm{M}; \quad \mathrm{R}_n^\bullet \xrightarrow{k_\mathrm{depol}} \mathrm{R}_{n-z}^\bullet + {}_z\mathrm{M} \\ &R^\bullet \xrightarrow{k_{\mathrm{deg}r}} R_n^\bullet + R_m^\bullet; \quad \mathrm{R}^+ \xrightarrow{k'_{\mathrm{deg}r}} \mathrm{R}_n^+ + \mathrm{R}_m^+ \\ &\mathrm{R}^\bullet + \mathrm{P} \xrightarrow{k_\mathrm{pol}} \mathrm{RH} + \mathrm{P}^\bullet; \quad R^+ + P \xrightarrow{k_{pol}} RH + P^+ \\ &\qquad\qquad \mathrm{R}^+ \xrightarrow{k_{ter}} \mathrm{P} \end{aligned} \tag{5.2}$$

Here k'_in, k_in, k'_depol, k_depol, k'_degr, k_degr, k'_pol, k_pol, k_ter are rate constants of initiation, depolymerization and degradation of macromolecules under the effect of ionic and radical centers and chain termination, respectively.

For degradation of polyisobutylene by the law of accident, the change of concentration of macrocations $[\mathrm{R}]^+$ is described by the equation:

$$\frac{\mathrm{d}\left[\mathrm{R}^+\right]}{\mathrm{d}t} = k_\mathrm{in}[\mathrm{Cat}]\left[\mathrm{R}^\bullet\right] - k_\mathrm{ter}\left[\mathrm{R}^+\right]. \tag{5.3}$$

In the stationary state

$$\left[\mathrm{R}^+\right] = \frac{k_\mathrm{in}[\mathrm{Cat}]\left[\mathrm{R}^\bullet\right]}{k_\mathrm{ter}},$$

where

$$\left[\mathrm{R}^{\bullet}\right]=-\frac{k_{\mathrm{in}}[\mathrm{Cat}]}{2k_1}+\sqrt{\left(\frac{k_{\mathrm{in}}[\mathrm{Cat}]}{2k_1}\right)^2+\frac{W_{\mathrm{p}}}{k_1}}=\frac{k_{\mathrm{in}}[\mathrm{Cat}]}{2k_1}\sqrt{\frac{4k_1\left(W_{\mathrm{p}}+k_1[\mathrm{Cat}]\right)}{\left(k_{\mathrm{in}}[\mathrm{Cat}]\right)^2}+1}-1, \quad (5.4)$$

with $W_{\mathrm{p}} = W_T + k_1[\mathrm{Cat}]$, the rate of formation of free radicals as a result of thermal termination of the polymeric chain under the effect of catalyst.

In the case of application of small amounts of catalyst, when only a small part of the free radicals formed are transformed into polymeric carbonium ions, the expression for the rate of statistical decay of the polymer by the law of accident is the following:

$$V_1=\left(k'_{\mathrm{depol}}+k'_{\mathrm{degr}}\right)\frac{k_{\mathrm{in}}[\mathrm{Cat}]}{k_{\mathrm{ter}}}\sqrt{\frac{W_T+k_1[\mathrm{Cat}]}{k_1}}+\left(k_{\mathrm{depol}}+k_{\mathrm{degr}}\right)\sqrt{\frac{W_T+k_1[\mathrm{Cat}]}{k_1}}. \quad (5.5)$$

In this case, the limiting stage of the process is the reaction of catalyst interaction with free radical.

If small amounts of catalyst are used (all resulting radicals transform into carbocations), the rate of the process is described by the equation:

$$V_1=\left(k'_{\mathrm{depol}}+k'_{\mathrm{degr}}\right)\frac{W_T+2k_1[\mathrm{Cat}]}{k_{\mathrm{ter}}}. \quad (5.6)$$

For degradation of polyisobutylene by the law of end groups, the equation for the rate of depolymerization becomes:

$$V_2=\frac{k_{\mathrm{p}}}{\left[\mathrm{R}^+\right]}=\frac{\left(k'_{\mathrm{depol}}+k'_{\mathrm{degr}}\right)k'_{\mathrm{in}}[\mathrm{Cat}]C_{\mathrm{p}}}{k_{\mathrm{ter}}}=\frac{\left(k'_{\mathrm{depol}}+k'_{\mathrm{degr}}\right)k'_{\mathrm{in}}\mathrm{A}_0[\mathrm{Cat}]}{k_{\mathrm{ter}}\overline{P}_n}. \quad (5.7)$$

Here, $C_{\mathrm{p}}=\mathrm{A}_0/\overline{P}_n$ is the content of double bonds in polyisobutylene; A_0 is the polymer concentration; $\overline{P}_n$ is the degree of polymerization of polyisobutylene. Taking into account that concentration of active centers $[\mathrm{R}^+]$ is determined by the correlation: $\alpha\left[\mathrm{R}^+\right]/\mathrm{d}t=k'_{\mathrm{in}}[\mathrm{Cat}]C_n-k'_{\mathrm{ter}}\left[\mathrm{R}^+\right]$, in the general case, the rate of polyisobutylene depolymerization is described by the equation:

$$\begin{aligned}V=V_1+V_2&=\left(k'_{\mathrm{depol}}+k'_{\mathrm{degr}}\right)\frac{k_{\mathrm{in}}[\mathrm{Cat}]}{k_{\mathrm{ter}}}\sqrt{\frac{W_T+k_1[\mathrm{Cat}]}{k_1}}+\\&+\left(k_{\mathrm{depol}}+k_{\mathrm{degr}}\right)\sqrt{\frac{W_T+k_1[\mathrm{Cat}]}{k_1}}+\frac{\left(k'_{\mathrm{depol}}+k'_{\mathrm{degr}}\right)k'_{\mathrm{in}}\mathrm{A}_0[\mathrm{Cat}]}{k'_{\mathrm{ter}}\overline{P}_n}\end{aligned} \quad (5.8)$$

The equation (5.8) allows for reactions of degradation and depolymerization of polyisobutylene on ionic and radical active centers occurring simultaneously, but does not allow for any judgements about the reasons for the difference in selectivity of the process by the monomer yield in the presence of a catalyst. Selectivity of action of catalysts can be quantitatively estimated by value Z – the length of the kinetic chain, which characterizes the amount of monomer yielded on average for each single scission of the macromolecule:

$$Z = \frac{k'_{\text{depol}}\left[\text{R}^+\right] + k_{\text{degr}}\left[\text{R}^+\right] + k_{\text{degr}}\left[\text{R}^\bullet\right]}{W_T + k'[\text{Cat}] + k_{\text{p}}\left[\text{R}^+\right] + k'_{\text{p}}\left[\text{R}^+\right]}. \quad (5.9)$$

The value of Z and the gross rate of degradation of polymeric products grow with the catalyst activity in the sequence $LiAlCl_4 > NaAlCl_4 > KAlCl_4$ (Table 5.9).

Table 5.9

Gross parameters of thermocatalytic degradation of polyisobutylene ($C_{cat} = 1 \cdot 10^{-3}$ mol/g, reaction time is 1 hour).

Catalyst	Reaction temperature, K	$Z \pm 10\%$	$k'_{in} \cdot 10^7 + 5\%$, s^{-1}	$E_a \pm 5\%$, kJ/mol
$LiAlCl_4$	583	10	0.6	100
	603	8	1.7	
	623	6	4.3	
$NaAlCl_4$	583	10	0.3	105
	603	7	0.8	
	623	6	2.4	
$KAlCl_4$	583	4	0.4	140
	603	3	1.1	
	623	3	3.7	

Length of the kinetic chain Z of polyisobutylene depolymerization decreases with growth of the reaction temperature, whereas the rate constant of polyisobutylene polymerization increases. This is connected with growth of contribution of the thermal component, which increases probability of statistic (by the law of accident) degradation of macromolecules, as well as with reactions of chain transmission and termination.

The process of electrophilic catalytic degradation of polyolefins can be described by a general equation taking into account contribution of the reactions by both the law of accident and the law of end groups [56]:

$$-\frac{\mathrm{d}P}{\mathrm{d}t} = z_{\text{eg}} k_{\text{in-eg}} C_{\text{eb}} + k_{\text{in-acc}} z_{\text{acc}} P. \quad (5.10)$$

Here P is diminution of the polymer (in parts of initial content); $k_{in\text{-}acc}$ and $k_{in\text{-}eg}$ are rate constants of initiation of polymer degradation by the law of accident and the law of end groups, respectively; z_{acc} and z_{eg} are the length of kinetic chain at degradation by the law of accident and the law of end groups, respectively; C_{eb} is the concentration of end double bonds. Supposing that end groups able to initiate depolymerization are consumed during degradation by the exponential law $C_{eb} = C_{eb0}\exp(-k_{in\text{-}eg}t)$ (C_{eb0} is the initial content of end double bonds), which is typical for polyisobutylene, in particular, we obtain that

$$P = \exp(-k_{in-acc}z_{acc}t) + \frac{k_{in-eg}z_{eg}C_{eb0}}{k_{in-eg} - k_{in-acc}z_{acc}}\left[\exp(-k_{in-eg}t) - \exp(-k_{in-acc}z_{acc}t)\right]. \quad (5.11)$$

Numerical values of $k_{in\text{-}eg}$, z_{eg}, $k_{in\text{-}acc}$ and z_{acc} can be estimated for the initial stages of the process, when $P << 1$, $t << 1/k_{in\text{-}eg}$ and $t << 1/k_{in\text{-}acc}z_{acc}$ (experimentally, this is possible, if the process is carried out under low-temperature conditions — 200–300°C, when the rate of burnout of chains is rather low). Then diminution of the polymer (in parts of initial content) is described by the equation:

$$P = 1 - k_{in\text{-}acc}z_{acc}t - k_{in\text{-}eg}z_{eg}tC_{eb0}. \quad (5.12)$$

Changes in the molecular mass of polyisobutylene can be determined with regard to accumulation of chain breaks. The amount of macromolecules is equal to $n = n_0 + k_{in\text{-}acc}t$, and their molecular mass is determined by the expression:

$$\overline{M}_n = \frac{P}{n} \cong \frac{1 - k_{in-acc}z_{acc}t - k_{in-eg}z_{eg}C_{eb0}t}{n_0(1 + k_{in-acc}T/n_0)} \cong$$
$$\cong \overline{M}_n^0\left(1 - k_{in-acc}z_{acc}t - k_{in-eg}z_{eg}C_{eb0}t - \frac{k_{in-acc}t}{n_0}\right) \quad (5.13)$$

Here $\overline{M}_n^0 = 1/n_0 = 1/C_{eb0}$, which corresponds to the presence of one end double bond in each macromolecule.

The expression for relative degree of polymerization of polyisobutylene during degradation is [53]:

$$\frac{d\left(\overline{P}_n/d\overline{P}_n^0\right)}{d\alpha} = -\frac{k_{in-acc}\overline{P}_n^0 + k_{in-eg}z_{eg}/\overline{P}_n^0}{k_{in-acc}z_{acc} + k_{in-eg}z_{eg}/\overline{P}_n^0}. \quad (5.14)$$

To calculate kinetic parameters, it is necessary to consider a number of limiting cases for various ranges of the molecular mass of the polymer:

1. At low contribution of reactions proceeding by the law of end groups, i.e. at $k_{in}z_{eg}/\overline{P}_n^0 << k_{in-acc}z_{acc} << k_{in-acc}\overline{P}_n^0$, the expression (5.14) for reduced relative degree of polymerization will be the following:

$$\frac{d\left(\overline{P}_n/d\overline{P}_n^0\right)}{d\alpha} = -\frac{\overline{P}_n^0}{z_{acc}}. \quad (5.15)$$

2. At small contribution of reactions proceeding by the law of accident, when degradation of polyisobutylene is determined by the law of end groups (low values of), it is obtained:

$$\frac{d\alpha}{dt} \approx \frac{k_{\text{in-acc}} z_{\text{eg}}}{\overline{P}_n^0}. \tag{5.16}$$

In this process, the value of $\overline{P}_n^0$ for initial, most probable MMD $\left(\overline{P}_w/\overline{P}_n \approx 2\right)$ is practically unchanged, i.e. $\mathrm{d}\left(\overline{P}_n/\overline{P}_n^0\right)/\mathrm{d}\alpha \approx 0$. This conforms to the experimental data (Figure 5.5, curves 1–3). In this case, application of equation (5.14) is meaningless, because "burnout" of macromolecules at the sacrifice of their carbocationic depolymerization was not taken into account in its deduction, which was taken into consideration in [53].

3. When the main role in changing of the number of chains is played by the law of accident, and in depolymerization then the law of end groups, i.e. at $k_{\text{in-acc}}\overline{P}_n^0 >> k_{\text{in}} z_{\text{eg}}/\overline{P}_n^0 >> k_{\text{in-acc}} z_{\text{acc}}$:

$$\frac{\mathrm{d}\left(\overline{P}_n/\overline{P}_n^0\right)}{\mathrm{d}t} = -\frac{k_{\text{in-acc}}\left(\overline{P}_n^0\right)^2}{k_{\text{in-eg}} z_{\text{eg}}}. \tag{5.17}$$

The three above-mentioned variants correspond to three parts on curves of the dependence of calculated values $\mathrm{d}\left(\overline{P}_n/\overline{P}_n^0\right)/\mathrm{d}\alpha$ on $\overline{P}_n^0$, obtained by tangents of appropriate experimental curves (Figure 5.5). Each of these three possible variants – b, c, or a, respectively, is consequently realized with increase in the degree of polymerization (Figure 5.6).

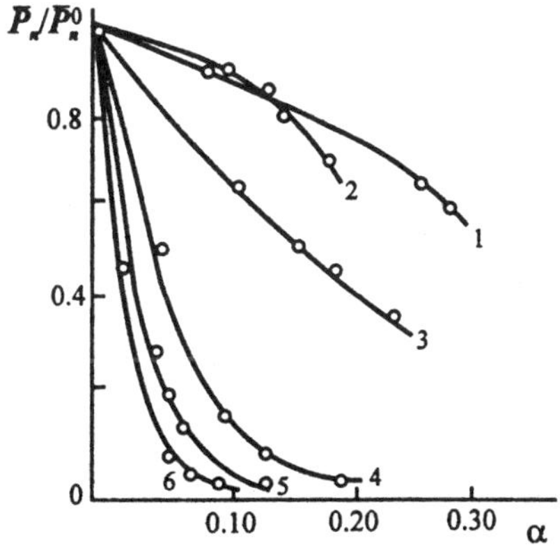

Figure 5.5. Dependence of the relative degree of polymerization on the depth of transformation (α) during thermocatalytic degradation of polyisobutylene (PIB) at 573 K in the presence of H[NaAlCl$_4$OH]. Catalyst concentration is $1 \cdot 10^{-3}$ mol/g of polyisobutylene. The initial degree of polyisobutylene polymerization: 1 – 17; 2 – 20; 3 – 49; 4 – 360; 5 – 1790; 6 – 3570.

Values of z_{acc} and $k_{\text{in-acc}}/k_{\text{in-eg}} z_{\text{eg}}$, respectively, can be calculated by tangents for parts a) and c) of the curve of $\mathrm{d}\left(\overline{P}_n/\overline{P}_n^0\right)/\mathrm{d}\alpha$ against $\overline{P}_n^0$ (Figure 5.6). The value of

$k_{\text{in-acc}}$ can be calculated taking into account that the main contribution to reduction of molecular mass of the polymer is made by the law of accident. This correlates with the dependence $d\left(\bar{P}_n/\bar{P}_n^0\right)/dt = -k_{\text{in-acc}}\bar{P}_n^0$, which allows estimation of the $k_{\text{in-acc}}$ value.

Values of z_{acc} and $k_{\text{in-acc}}$, obtained at treatment of experimental data, can be used for estimation of numerical values of $k_{\text{in-eg}}$ and z_{eg} by equation (5.11). The values of kinetic parameters obtained — the rate constant of initiation and length of kinetic chain of depolymerization of polymers (for polyisobutylene degradation in the presence of catalysts with moderate relative acidity) — are shown in Table 5.10. They indicate the prevalence of reactions of initiation of catalytic degradation by the law of end groups under low-temperature conditions. The ratio of $k_{\text{in-eg}}$ and $k_{\text{in-acc}}$ decreases with rise of temperature of degradation of polyolefins, and at 260°C their values become commensurable.

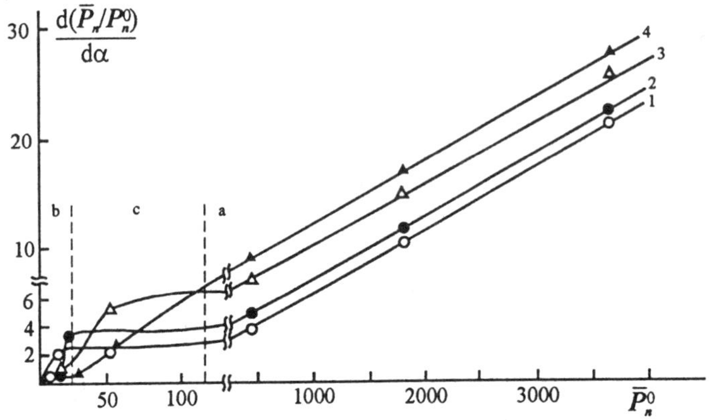

Figure 5.6. Change of $d\left(\bar{P}_n/\bar{P}_n^0\right)/d\alpha$ in degradation of polyisobutylene on the catalyst $NaAlCl_4$ under various temperatures. Temperature, K: 1 – 473, 2 – 493, 3 – 513, 4 – 533.

Independent of chemical structure of metal-complex catalysts, numerical values of both $k_{\text{in-eg}}$ and $k_{\text{in-acc}}$ hardly changes (within the range of the experimental error). This confirms (with regard to the total yield of gas products, as well as to content of *iso*-C_4-hydrocarbons) the conclusion about connection between activity and selectivity of electrophilic catalysts, in particular, of complexes based on metal chlorides with their indexes of relative acidity (Table 5.11). The ratio of $k_{\text{in-eg}}$ and $k_{\text{in-acc}}$ (Table 5.10) indicates that initiation of depolymerization of macromolecules at relatively low temperatures of degradation of polymeric products by the law of end groups is preferred. The contribution of reactions proceeding by the law of accident grows with temperature of the process conduction. However, in this case, shortening of the length of the kinetic chain of decomposition of polyolefins at the initiation of the process by the law of accident (z_{acc}) comparing with the length of kinetic chain of the process initiated by the law of end groups (z_{eg}) are, probably, only apparent. The real growth of amount of monomeric chains, formed at the process initiation by the law of accident, also takes place because in this case the energy of β-decomposition of the carbonium ion decreases but, on the background of sharp increase of the number of breaks in macrochains, no degradation of macromolecules is displayed.

Table 5.10

Main kinetic parameters of catalytic degradation of polyisobutylene.

Catalyst	T, K	$K_{in\text{-}acc}\cdot 10^6$, s^{-1}	$k_{in\text{-}eg}\cdot 10^6$, s^{-1}	Z_{acc}	z_{eg}	$k_{in\text{-}eg}/k_{in\text{-}acc}$
$H[Na(C_2H_5)AlCl_3OH]$	473	0.45	3.50	8	8	7.8
	493	0.48	2.50	7	16	5.2
	513	1.12	2.10	4	26	1.9
	533	1.50	1.90	3	37	1.3
$H[NaAlCl_4OH]$	473	0.53	3.50	7	4	6.6
	493	0.62	3.30	5	9	5.3
	513	0.80	2.90	5	15	3.6
	533	1.20	2.50	4	20	2.1
$H[Mg(C_2H_5)AlCl_4OH]$	473	0.36	4.90	12	10	13.6
	493	0.45	3.10	11	21	6.9
	513	0.69	2.40	9	33	3.5
	533	0.89	2.20	8	42	2.5
$2MgCl_2\cdot AlCl_3\cdot H_2O$	473	0.54	4.60	9	7	8.5
	493	0.77	2.70	7	17	3.5
	513	1.07	2.20	6	28	2.1
	533	1.17	2.00	6	38	1.7

Note. Conditions of the reaction conduction: [Cat] = $1\cdot 10^{-3}$ mol/g of polyisobutylene. Error of k determination is ±3%, z – ±5%.

Table 5.11

Acidity indexes for electrophilic catalysts based on metal chlorides.

Catalyst	RAI, mg-eq/g of catalyst (±0.1)	pK_a (±0.1)	q_{H^+} (±0.02)
$H[AlCl_3OH]$	5.0	–22	0.45
$H[AlCl_4]$	4.4	–16	—
$H[FeCl_2OH]$	3.5	–10	0.34
$H[BiCl_3OH]$	2.8	–4.2	0.29
$H[InCl_3OH]$	2.6	–2.5	0.27
$H[MgInCl_5OH]$	1.9	+3.5	0.21
$H[CaInCl_5OH]$	1.6	+4.5	0.22
$H[MgCl_2OH]$	1.5	+7.1	0.19
$H[MgCl_2OH]\cdot H_2O$	1.5	+7.7	0.18
$H[Mg(C_4H_9)ClOC_3H_7]$	—	+7.7	0.19
$H[AlCl_3OC_2H_5]$	1.5	+7.7	0.19
$H[Mg(C_2H_5)ClOH]$	—	+8.1	0.17
$H[Mg(C_3H_7)ClOC_3H_7]$	—	+8.5	0.18
$H[MgCl_2OH]\cdot 3H_2O$	1.0	+10.5	0.16
$H[CaCl_2OH]$	1.0	+10.5	0.16
$H[MnCl_2OH]$	1.0	+10.5	0.16
$H[BaCl_2OH]$	0.9	+11.6	0.15

$H[SnCl_2OH]$	0.6	+13.8	0.13
$H[NiCl_2OH]$	0.5	+14.5	0.12
$H[ZnCl_2OH]$	0.2	+17.0	0.10
$H[CdCl_2OH]$	0.1	+17.8	0.10
$H[Al(C_2H_5)Cl_3]$	4.3	–16	—
$H[MgAlCl_5OH]$	2.3	0	0.24
$H[MgAl(C_2H_5)Cl_4OH]$	2.2	+0.9	0.25
$H[CaAlCl_5OH]$	2.2	+0.9	0.25
$H[CaAl(C_2H_5)Cl_4OH]$	1.8	+4.0	0.21
$H[LiAl(C_2H_5)Cl_3OH]$	1.8	+4.1	0.18
$H[Al_2Cl_4(O)OH]$	—	+4.4	0.21
$H[NaAlCl_4OH]$	1.6	+4.5	0.23
$H[SrAlCl_5OH]$	1.6	+5.0	0.21
$H[BaAl(C_2H_5)Cl_4OH]$	1.4	+7.3	0.18
$H[NaAl(C_2H_5)Cl_3OH]$	1.4	+7.2	0.19
$H[Al(C_2H_5)Cl_2OH]$	—	+7.5	0.20
$H[Al(C_2H_5)_2ClOH]$	—	+8.5	0.18
$H[Al(OH)Cl_2OH]$	—	+8.5	0.18
$H[Al_2(CH_3)_3Cl(O)OH]$	—	+8.5	0.18
$H[ZnAlCl_5OH]$	1.1	+9.8	0.16
$H[CdAlCl_5OH]$	1.1	+9.8	0.16
$H[KAlCl_4OH]$	0.9	+11.6	0.13
$H[KAl(C_2H_5)Cl_3OH]$	0.8	+12.0	0.13

Note. RAI is the relative acidity index (experiment) [1]; pK_a is the universal acidity index [2, 3]; q_{H^+} is the charge on hydrogen atom of the catalyst (quantum-mechanical calculation) [2–5].

At low degree of polymerization $\left(\overline{P}_n^0\right)$, values of z_{eg} are equal to the length of the chain of depolymerization of macromolecules. At high values of $\overline{P}_n^0$, the value of z_{eg} becomes constant, which, obviously, is determined by the ratio of reactions of chain propagation and termination on ionic carbonium active centers, resulting from end groups [8, 9].

Existence of a tertiary carbon atom and especially aliquot C=C bonds (both end and internal ones) in the composition of macromolecules significantly affects the thermocatalytic degradation of polyolefins. In particular, initiation of the process preferably by the law of end groups is typical for low-molecular samples of polyisobutylene with high specific weight of unsaturated end C=C-bonds (Figure 5.5), because little change of the molecular mass of polymeric products (in the initial stages of the process $\overline{P}_n/\overline{P}_n^0 \approx 1$) is observed during thermocatalytic degradation of the polymer. The ratio $\overline{P}_n/\overline{P}_n^0$ decreases only at greater degrees of transformation. For high-molecular samples of polyisobutylene, isobutylene copolymers with isoprene, containing labile bonds in connection knots of isobutylene clusters and isoprenyl chain with C=C-bond, as well as for polyethylene (independent of the molecular mass) containing internal vinylidene and *trans*-vinylidene C=C-bonds, initiation of the gross-process of degradation of macromolecules proceeds simultaneously by both the law of accident and the law of end groups, but with different contribution of these mechanisms to the total process (Figure 5.7). The possibility of initiation of degradation of macromolecules by end vinyl groups also cannot be excluded.

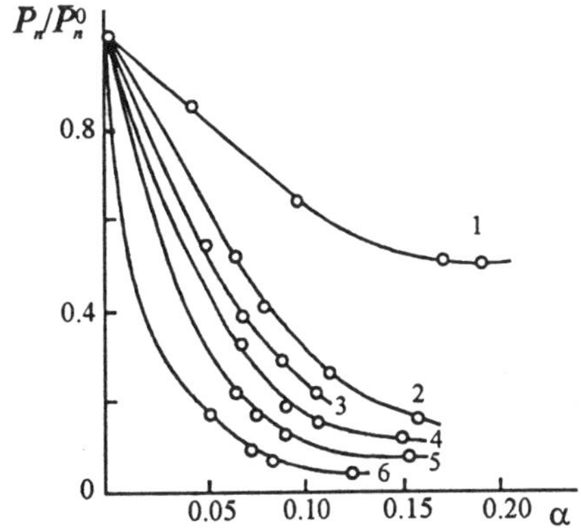

Figure 5.7. Dependence of the relative degree of polymerization on the transformation depth (α) of low density polyethylene (LDPE) in thermocatalytic degradation at 633 K in the presence of $H[NaAlCl_4OH]$. Catalyst concentration is $1 \cdot 10^{-3}$ mol/g of LDPE. The initial degree of polymerization of LDPE: 1 - $2 \cdot 10^4$; 2 - $6 \cdot 10^4$; 3 – $1.45 \cdot 10^5$; 4 - $2.5 \cdot 10^5$; 5 - $2.7 \cdot 10^5$; 6 - $8.7 \cdot 10^5$.

$MgCl_2 \cdot AlCl_3$ is characterized by somewhat higher acidity compared with $NaAlCl_4$. This results in higher activity and selectivity of $MgCl_2 \cdot AlCl_3$ compared with $NaAlCl_4$ in the process of polyisobutylene degradation: total yield of gas products and yield of isobutylene itself grow simultaneously. Rise of selectivity of the process is bound to growth of the role of polyisobutylene depolymerization (degradation of macromolecules by the law of end groups). The index of the ratio of rate constants of initiation by the law of accident and the law of end groups in polyisobutylene degradation in the presence of $MgCl_2 \cdot AlCl_3$ is higher than in the presence of $NaAlCl_4$ (Table 5.10). There is a tendency to reduction of the ratio with temperature rise.

Cationic selective degradation of polyisobutylene is practically interesting for processing of unusual and other polymers down to a monomer [64].

Let us point out one more case of decomposition of polyisobutylene polymeric molecules, accompanied by modification of polymer properties — ultrasonic degradation of polyisobutylene in aromatic solvents. In this case, two processes proceed in parallel: degradation of macromolecules and polymerization of arene [65, 66].

The former is characterized by regularities of ultrasonic decomposition of polymers in solvents and proceeds up to reaching some limiting molecular mass, which depends on conditions and method of polymer processing (solution viscosity, thermodynamic quality of the solvent, etc.) [67].

As follows from Table 5.12, the above factors influence polyisobutylene degradation, but they are not of major significance.

Table 5.12

Rate constants of polyisobutylene degradation in various solvents.

Solvent	ΔH_i^{297}, kJ/mol	Higgins constant χ_1	Rate constant of degradation and viscosity of solutions for times τ			
			$0–1.2\cdot10^4$, s		$1.2\cdot10^4–3.6\cdot10^4$, s	
			$k\cdot10^4$, g/l·min	ν_{30}, cSt	$k\cdot10^4$, g/l·min	ν_{30}, cSt
n-Heptane	36.7	0.468	5.81–1.64	21.8–1.00	1.64–0.64	1.00–0.85
Toluene	38.2	0.488	3.45–1.74	17.9–1.15	1.74–0.76	1.15–0.97
Isopropylbenzene	45.0	—	4.02–1.99	28.5–1.70	1.99–0.69	1.70–1.36

Note. Polyisobutylene – $\overline{M}_n$ = $5.4\cdot10^5$, $\overline{M}_w/\overline{M}_n$ = 3.5; radiation conditions – 22 kHz, 294 K, 100 W power output, exponential radiator; rate constants of ultrasonic degradation, calculated by the Schmidt equation in the integral form [68]:

$$\frac{M_e}{M_\tau}+\ln\left(1-\frac{M_e}{M_\tau}\right)=-\frac{K}{C(M_e/M_0)^2\tau}+\frac{M_e}{M_i}+\ln\left(1-\frac{M_e}{M_i}\right),$$

where M_e, M_i, M_τ are values of the molecular mass of polyisobutylene: the limit ($5\cdot10^3$), initial and current, respectively.

In all cases, processes are characterized by a non-stationary constant (during the initial period), and then proceed at a constant rate (k = $(0.7\pm0.06)\cdot10^{-9}$ g/l·min). However, physicochemical properties of polyisobutylene after sonication in aromatic solvents differ from those of polyisobutylene, degraded in an inert solvent (heptane), and properties of the initial polymer. First of all, there is a stable yellow-brown color and increased thermostability of products. They may be bound to a change of end aromatic fragments of polymer chains (wide absorption in the range of 220–270 nm in UV-spectra, an exothermic peak on DTA curves in the range of 673–677 K, typical for linear polyphenylene structures).

Transformations of aromatic hydrocarbons under conditions of cavitational influence on polyisobutylene are accompanied by formation of low-molecular polymers, which proceeds via intermediate cation-radical centers (this was proved by the method of selective inhibition of free-radical and cationic centers), the latter being presented as products with polyquinoid structure (ozone absorption) (Table 5.13).

At heating or long-term exposure in air, the reaction products transform into insoluble powders, approaching polyolefins in properties (thermal stability above 700 K, concentration of unpaired electrons of $150–200\cdot10^{18}$ spin/mol). Obviously, polymerization of arenes under ultrasonic effects proceeds similar to catalytic polymerization in systems of the Lewis acid ($AlCl_3$)–oxidant ($CuCl_2$) type [69], but represents a more complex process.

Interaction (recombination) of intermediate products of polyisobutylene decomposition and polymerization of arenes at sonication of solutions causes formation of fragmentary block-copolymers of AB type with end polyphenylene grouping (3–5 units):

Table 5.13

Characteristics of transformation products of aromatic compounds under ultrasonic effect.

Initial compound	Element composition, mass%			$\overline{M}_\eta$	UV-spectrum λ, nm (ε, $m^{-3}kg^{-1}m^{-1}$)	Temperature of decay initiation, K		Absorption of $O_3 \cdot 10^3$, kmol/kg
	C	H	X(Cl, NO_2)			Liquid	Powder	
C_6H_6	92	8	—	—	250	—	—	1.40
$C_6H_5CH_3$*	88.3	11.7	—	340	223 (1860)	—	—	1.51
$C_6H_5C_3H_7$**	92.2	7.8	—	410	246 (2140)	365	—	1.31
$C_6H_5C_3H_7$	94.6	5.4	—	—	—	—	710	—
C_6H_5Cl	75.9	4.2	19.9	—	254 (5640)	365	—	1.10
$C_6H_5NO_2$*	84.0	5.6	10.4	296	246 (4210)	361	—	1.50
$C_6H_5NO_2$**	81.6	5.3	13.1	—	—	—	730	—

* - relates to the soluble part of the product;

** - relates to insoluble product.

$$\sim CH_2\text{-}C(CH_3)_2\text{-}CH_2\text{-}C(CH_3)_2\sim \rightarrow CH_2C^{\bullet}(CH_3)_2 + CH_2\text{-}C^{\bullet}(CH_3)_2\sim$$

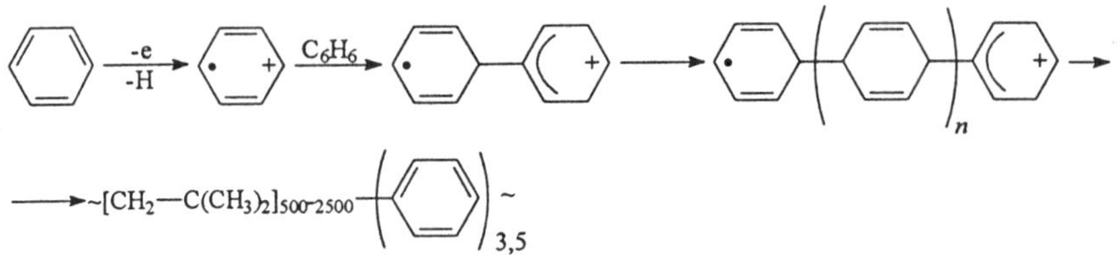

Owing to suppression of polyisobutylene decomposition by aromatic fragments from the end of the chain and decrease of the total rate of decomposition, the products obtained are of interest as a basis for synthetic lubricants and additives with improved performance [46].

5.3. Analytical characteristics of polyisobutylene.

The IR-spectrum of polyisobutylene is comparatively simple [70]. The band at 1485 cm^{-1} is appropriate for a deformation oscillation of the methyl group. Compared with IR-spectra of paraffin hydrocarbons, this band is shifted by 25 cm^{-1} towards higher frequencies. The doublet at 1370 and 1400 cm^{-1} is appropriate for symmetric deformational oscillations of both methyl groups. In the range of valence oscillations, absorption of methyl groups is displayed at 2880 and 2960 cm^{-1}. Absorption in the area of 1250 cm^{-1} is due to the presence of dimethyl groups in macromolecules. The doublet of absorption bands of equal intensity at 920 and 950 cm^{-1} is also typical. End groups, specially displayed in IR-spectra of low-molecular products, are represented by three-substituted ethylene bonds $\sim CH_2\text{–}C(CH_3)_2\text{–}CH{=}C(CH_3)_2$ (absorption at 830 and 1660 cm^{-1}) and vinylidene double bonds $\sim CH_2\text{–}C(CH_3)_2\text{–}CH_2\text{–}C(CH_3)_2{=}CH_2$ (absorption at 890 and 1640 cm^{-1}). The relation between exo- and endoforms of olefins depend on the type of catalyst, used in the polymer synthesis, and is, for example, 30:70 for application of organoaluminum haloid compounds and 20:80 for Brönsted acids. Vinyl double bonds are absent [71, 72].

IR-spectra of industrial polyisobutylenes, especially of low-molecular samples (octoles), are much more complex, which is associated with influence of additives, inconstancy of the temperature regime and other factors in the polymerization stage. It specially affects the type of double bonds. Beside end vinyliden and three-substituted double bonds, the following ones were observed in IR- and ESR-spectra [73]: internal vinyliden double $\sim CH_2C({=}CH_2)\text{–}CH_2\text{–}C(CH_3)_3$ (1), three-substituted double $\sim C(CH_3)_2\text{–}C(CH_3){=}CH\text{–}CH_3$ (2) and tetra-substituted ethylene $\sim C(CH_3){=}C(CH_3)\text{–}CH(CH_3)_2$ (3) bonds. Their formation is ascribed to processes of isomerization, skeleton regroupings, in particular, 1,3-methide shear (1), concert 1,2-hydride-2,3-methide shear (3), attachment by the "head-to-head" type with 1,2-methide shear (2). The relation of groupings depends on the nature of the applied catalyst and conditions of synthesis. For example, in the case of BF_3, concentration of vinylidene groupings in macromolecules decreases with time due to their protonation with simultaneous rise of content of three-substituted ethylene bonds. Formation of tetra-substituted ethylene bonds, which demands an unlikely attachment of monomers by the "head-to-head" type, is doubtful [74]. Admittedly, the argument proposed for confirmation — observance of analogously composed fragments in products of thermal decomposition of a high-molecular polymer [74] — is not convincing, because thermal decomposition proceeds according to the free-radical mechanism of polymer degradation, and occurrence of various products could be the consequence of secondary reactions of polymeric radicals.

More complicated spectra are observed for copolymers of isobutylene with small amounts of butenes or diene monomers. Besides isobutylene blocks, the following fragments are identified in them [75]: end and methyl groups $CH_3–CH_2–$, $CH_3–CH_2–CH_2–$, $CH_3–CH(CH_3)–CH_2–$, isopropyl groups $–CH(CH_3)_2$, di- and polymethyl groups $–(CH_3)_2C–(CH_2)_n–C(CH_3)_2$, internal double bonds, appropriate to 1,4-polymerization of diene hydrocarbons at $n \geq 2$

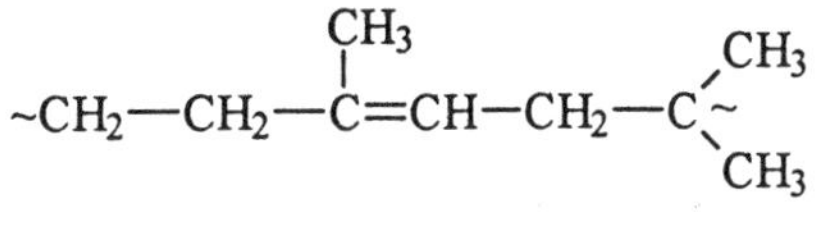

Existence of double bonds, different in reactivity, must be taken into account in the study and application of oligoisobutylenes, for example, as additives to oils, lacquers, etc., which supposes their functionalization by places of double bonds.

NMR-spectrum of H (*o*-dichlorobenzene) is: δ_{CH_2} = 1.4 m.p., δ_{CH_3} = 1.2 m.p. NMR-spectrum of ^{13}C (*o*-dichlorobenzene, 353 K) is: δ_{CH_3} = 31.6 m.p. (quadruplet), δ_{CH_2} = 60.0 m.p. (triplet), $\delta_{>C<}$ = 38.5 m.p. (singlet).

Molecular masses of moderate- and high-molecular polyisobutylenes are convenient to estimate by the data of characteristic viscosity of samples by the Mark–Hauvink equation $[\eta] = K_\eta M^\alpha$. Constants K_η and α for polyisobutylene–solvent systems are shown in Table 5.14 (concentration is expressed in g/100 cm^3) [76].

Table 5.14

Constants K_η and α in the equation $[\eta] = K_\eta$ and M^α.

Solvent	$K_\eta \times 10^4$	α	Temperature, K	Method*	$M \times 10^{-4}$
Polyisobutylene					
Isoamylisovalerate	11.4	0.5	295.1	L	16–470
n-Heptane	1.58	0.69	298	L	16–470
Iso-octane	6.75	0.576	293	L	0.8–3
Benzene	11.0	0.5	298	—	—
Cyclohexane	1.35	0.74	298	—	—
Cyclohexanone	2.63	0.69	303	—	—
Isobutylene copolymer with isoprene (BC-2045)					
Carbon tetrachloride	1.07	0.78	298	O	11–40
Toluene	6.6	0.60	298	O	11–40

* O – osmometry; L – light scattering.

Typically, $[\eta]$ values in good solvents do not practically depend on temperature (Table 5.15) [76].

Table 5.15

Characteristic viscosity $[\eta]$ of polyisobutylene with $\overline{M}_\eta$ = 1,300,000 in good and bad solvents.

Solvent	Characteristic viscosity $[\eta]$ of polyisobutylene at temperatures			
	298 K	303 K	313 K	333 K
Diisobutylene	2.98	2.90	3.03	2.95
Carbon tetrachloride	4.25	4.30	4.30	4.38
Cyclohexane	4.78	4.75	4.81	4.83
Benzene	1.45	1.77	2.20	2.87

Table 5.16

Comparative analysis of polyisobutylene samples.

$\overline{M}_n$			Content of C=C bonds, mass%	
Ebullioscopy	MCHE	ADC	Iodometry	ADC
500	510	500	7.1	7.0
540	560	550	6.8	6.9
630	700	690	5.2	5.1
800	820	820	4.5	4.4
820	840	840	4.2	4.2
920	940	930	3.9	3.8
940	950	950	3.9	3.6
980	1000	1000	3.8	3.5
1150	1210	1200	3.0	2.9
1250	1250	1210	2.9	2.5
1450	1500	1480	2.5	2.3

For oligomers of isobutylene and low-molecular polymeric products, in determination of average molecular masses $\overline{M}_n$, it is expedient to use the method of measurement of heat condensation effects (MHCE) [77], as well as the ozonolysis method, more accurate and fasten with high selectivity of unsaturated >C=C<-bonds, using ADS device of the ICPh RAS system (selectivity of the method is 10^{-8} g/l against 10^{-4} g/l for MCHE) [78, 79]. The method is based on the quantitative interaction of ozone with double C=C-bonds in macromolecules of polyisobutylene. Values of $\overline{M}_n$ are calculated from the equation:

$$\overline{M}_n = \frac{1}{D}, \quad \text{where } D = \frac{C_{st} V^{st}_{border} S_{samp} V_{solv}}{S_{st} V^{samp}_{border} q_{samp}}.$$

Here D is the non-saturation degree; C_{st} is the stilbene (the standard substance) concentration; V^{st}_{border}, V_{solv} and V^{samp}_{pr} are volumes of probes of stilbene, solvent and sample, respectively; S_{samp} and S_{st} are squares of peaks on ozonograms of samples of polyisobutylene and stilbene; q_{samp} is the length of pendant probe.

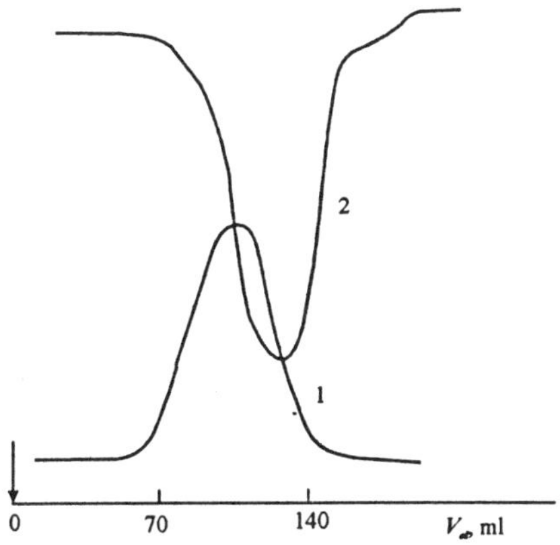

Figure 5.8. Gel- (1) and ozonochromatograms (2) of polyisobutylene with $\overline{M}_n$ = 970. The arrow marks the probe injection.

Comparative analysis in estimation of molecular masses of low-molecular polyisobutylene can also be used as the method of fine-layer adsorption chromatography (standard "Silufol" plates, the probe of 1–2 mg, elutriator cyclohexane-chloroform in relation 7:1, developer – iodine, determination time 5–10 min). The method is based on measuring the path length of a polymer probe patch and determining values of the molecular mass of polyisobutylene by the calibration curve, and enables molecular masses of polyisobutylene in the composition of polymerizates to be determined, which is important for analytical control in modern high yield processes of polyisobutylene industrial production.

Traditional methods of determination of polyisobutylene MMD: fractionation of polymeric products, gel-chromatography and the universal method based on combination of liquid chromatograph with a specific detector of unsaturated compounds with the ADC method.

For polyisobutylene, ADC-chromatograms differ from chromatograms, because the part of nonsaturation in the low-molecular range is higher, than in the high-molecular one (Figure 5.8). As practically every macromolecule contains a single end bond, application of a specific detector enables a calibration curve to be composed for isobutylene oligomers, which connects values of average molecular masses to elutriator volume, by combined treating of two chromatograms. Calibration is performed on fractions of isobutylene oligomers with $\overline{M}_n$ = 600-1200 by dividing mass parts of the polymer by the appropriate number of double bonds and determination of values of molecular masses, related to the elutriator volume. Semi-logarithmic anamorphose of $\overline{M}_n$ values on the elutriator volume V_{el} is depicted in Figure 5.9.

Comparison of MMD of polyisobutylene samples, obtained by ozono- and gel-chromatograms, based on the calibration dependence and calibration on narrow fractions of the polymer, verifies the high accuracy of the universal method. Application of the specific method to decode MMD curves allows fast and reliable determination of $\overline{M}_n$ value of polyisobutylene in every point of the chromatogram. Consequently, the universal method gives information about MMD in connection with the nonsaturation degree of the sample dependent on conditions of synthesis.

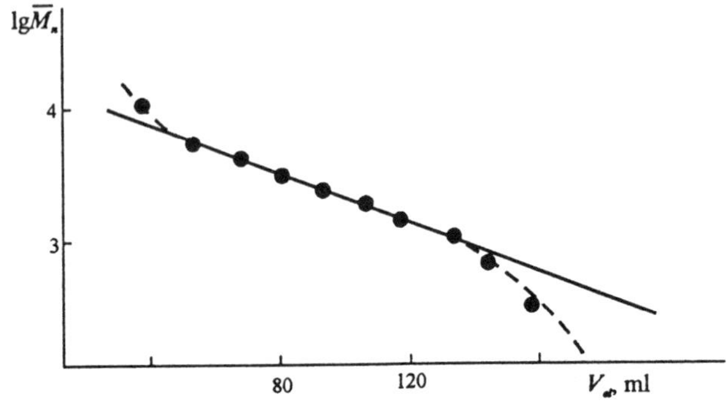

Figure 5.9. Dependence of logarithm of the average molecular mass of polyisobutylene on the elutriant volume.

Values of molecular mass and kinematic viscosity of oligoisobutylene are in clear dependence (Table 5.17).

It can be easily predicted that widening of MMD, keeping $\overline{M}_n$ and the iodine number of oligoisobutylenes constant, will lead to a significant growth of viscosity.

Table 5.17

Relation between some molecular characteristics of oligoisobutylenes*.

Molecular mass $\overline{M}_n$	Content of C=C bonds, mass%	Viscosity (373 K), m^2/s
500	7.1	0.59
600	5.9	1.00
660	5.5	1.12
700	5.2	1.50
740	5.0	1.70
800	4.6	2.00
830	4.3	2.08
875	4.1	2.68
900	4.0	3.00
960	3.9	4.80
1000	3.8	4.90
1050	3.6	5.10
1200	3.0	9.00
1420	2.6	12.20

* MMD of polymer fractions is not wider than 1.5.

References to Chapter 5

1. Huivink, R. and Staverman, A., Chemistry and Technology of Polymers, (Translation from German), 1965, v. 1, 675 p., Moscow–Leningrad, Khimia. (Rus)
2. Suter, U.W., Saiz, E., and Flori, P.J., Macromolecules, 1983, v. 16(8), pp. 1317–1328.
3. Allegra, G., Benedetti, E., and Pedone, C., Macromolecules, 1970, v. 3(6), pp. 727–785.
4. Boyd, R.H. and Breitling, S.M., Macromolecules, 1972, v. 5(1), pp. 1–7.
5. Bartos, J. and Hlonskova, Z., Eur. Polym. J., 1989, v. 25(1), pp. 21–24.
6. Hayashi, H., Flori, P.J., and Wignall, G.D., Macromolecules, 1983, v. 16(8), pp. 1318–1325.
7. De Boll, L.C. and Suter, W.W., Macromolecules, 1987, v. 20(6), pp. 1424–1425.
8. Dejean de Ca Batle, R., Zauprete, F., and Monnerie, L., Macromolecules, 1989, v. 22(6), pp. 2617–2622.
9. Boyer, Raymond F. and Enns, John B., J. Appl. Polym. Sci., 1986, v. 32(3), pp. 4075–4107.
10. Raficov, S.R., Budtov, V.P., and Monakov, Yu.B., Introduction into Physicochemistry of Polymers, 1978, 328 p., Moscow, Nauka. (Rus)
11. Malanga, M. and Vogl, O., Polym. Eng. Sci., 1983, v. 23(10), pp. 597–600.
12. Sangalov, Yu.A., Minsker, K.S., and Berlin, Al.Al., Highmolec. Comp., 1983, Ser. A, v. 25(7), pp. 1451–1457. (Rus)
13. Chemical Reactions of Polymers, Ed. By E.M. Fettes, (Translation from English), 1967, v. 2, 536 p., Moscow, Mir. (Rus)
14. Chemical Reactions of Polymers, Ed. By E.M. Fettes, (Translation from English), 1967, v. 1, 503 p., Moscow, Mir. (Rus)
15. Chandra, R., Progr. Polym. Sci., 1985, v. 11(1–2), pp. 1–22.
16. Bartos, J., Collect. Czech. Chem. Commun., 1985, v. 50(8), pp. 1699–1713.
17. Baramboim, N.K., Mechanochemistry Of High–molecular Compounds, 1978, 383 p., Moscow, Khimia. (Rus)
18. Cazale, A. And Porter, R., Reactions of Polymers Under Tensions, 1983, 441 p., Leningrad, Khimia. (Rus)
19. Sangalov, Yu.A. and Yasman, Yu.B., Uspekhi Khimii, 1985, v. 54(7), pp. 1206–1229. (Rus)
20. Sangalov, Yu.A., Yasman, Yu.B., Minsker, K.S., and Prokofiev, K.V., Chemistry and Technology of Petroleum and Oil, 1984, No. 6, pp. 30–32. (Rus)
21. Shamaeva, Z.G., Popova, V.V., Monakov, Yu.B., and Tolstikov, G.A., In: Thes. of VI Conf. on "Ageing and Stabilization of Polymers", Ufa, 1983, p. 166. (Rus)
22. Shamaeva, Z.G., Monakov, Yu.B., Galin, F.Z., Kovardakova, V.V., and Tolstikov, G.A., Acta Polym., 1984, v. 35(8), pp. 561–564.
23. Chang, V.S.C. and Kennedy, J.P., Polymer Bull., 1981, v. 5(7), pp. 379–384.
24. Macosko, C.W. and Saam, J.G., Polymer Bull., 1987, v. 18(5), pp. 463–471.
25. Sangalov, Yu.A., Yasman, Yu.B., Khudaiberdina, Z.I., and Minsker, K.S., Highmolec. Comp., 1981, Ser. A, v. 23(12), pp. 2652–2656. (Rus)
26. Kasianov, V.V. and Naverro, E.A., Izv. VUZov. Khimia i Khim. Tekhnol., 1977, v. 20(11), pp. 1623–1626. (Rus)
27. Kolyavina, S.I., Neftepererabotka i Neftekhimia, 1981, No. 9, pp. 16–18. (Rus)

28. Berkly, E.Yu., Novikov, D.D., and Entelis, S.G., Highmolec. Comp., 1967, v. A9(12), pp. 2754–2763. (Rus)
29. Kuliev, A.M., Chemistry and Technology of Additives to Oil and Petroleum, 1972, 358 p., Moscow, Khimia. (Rus)
30. Belov, P.S., Netepererabotka i Neftekhimia, 1991, No. 9, pp. 14–16. (Rus)
31. Biglova, R.Z., Malinskaya, V.P., Sagitdinova, Kh.F., and Minsker, K.S., Highmolec. Comp., 1994, v. A36(8), pp. 1276–1280. (Rus)
32. Biglova, R.Z., Malinskaya, V.P., Sagitdinova, Kh.F., and Minsker, K.S., Highmolec. Comp., 1995, v. A37(9), pp. 1474–1481. (Rus)
33. Biglova, R.Z., Zaikov, G.E., and Minsker, K.S., Intern. J. Polymeric Mater., 1998, v. 42, pp. 219–248.
34. Mukmeneva, N.A., Cherezova, E.N., Biglova, R.Z., Malinskaya, V.P., and Minsker, K.S., Highmolec. Comp., 1997, v. A39(6), pp. 953–959. (Rus)
35. Sangalov, Yu.A., Biglova, R.Z., Kamalova, N.G., and Minsker, K.S., Bashkir. Khim. Zh., 1998, v. 5(4), pp. 37–41. (Rus)
36. Minsker, K.S., Biglova, R.Z., Kamalova, N.G., and Zaikov, G.E., Russian Polymer News, 1997, v. 2(3), pp. 4–10.
37. General Organic Chemistry, Ed. by N.K. Kochetkov, 1982, v. 2, p. 239, Moscow, Khimia. (Rus)
38. Gumerova, E.F., Ivanova, S.R., Ponomareva, E.D., Budtov, V.P., Berlin. Al.Al., and Minsker, K.S., Highmolec. Comp., 1989, v. 31(8), pp. 607–611. (Rus)
39. Boone, D.E., Eisenbraum, P.W., Flanagan, P.W., and Grigsby, R.D., J. Org. Chem., 1971, v. 36(15), pp. 2042–2048.
40. Yasman, Yu.B., Sangalov, Yu.A., Prokofiev, K.V., Khudaiberdina, Z.I., and Nelkenbaum, E.M., Neftekhimia, 1983, v. 23(4), pp. 500–507. (Rus)
41. Kursanov, D.N., Parnes, Z.N., Kalinkin, M.I., and Loim, N.M., Ionic Hydrogenation, 1979, 192 p., Moscow, Khimia. (Rus)
42. Sangalov, Yu.A., Yasman, Yu.B., Gladkikh, I.F., and Minsker, K.S., Doklady AN SSSR, 1981, v. 261(2), pp. 428–432. (Rus)
43. Ipatief, V.N., Pines, H., and Olberg, R.L., J. Amer. Chem. Soc., 1978, v. 70(6), pp. 2123–2128.
44. Guterbock, G., Polyisobutylene and Isobutylene Copolymers, 1962, 363 p., Leningrad, Gostoptekhizdat. (Rus)
45. Razumovsky, S.D. and Zaikov, G.E., Ozone and Its Reactions with Organic Compounds, 1974, 322 p., Moscow, Nauka. (Rus)
46. Minsker, K.S. and Sangalov, Yu.A., Isobutylene and Its Polymers, 1986, 224 p., Moscow, Khimia. (Rus)
47. Grassi, N., Chemistry of Degradation of Polymers, 1959, 252 p., Moscow, Izdinlit. (Rus)
48. Jamaguchi, T., Kamiguchi, T., and Ito, T., J. Chem. Soc. Jap. Chem. Ind. Chem., 1976, pp. 1171–1174.
49. Kodaira, J., Osawa, Z., and Ando, H., J. Chem. Soc. Jap. Chem. Ind. Chem., 1977, pp. 1892–1898.
50. Sangalov, Yu.A., Prochukhan, Yu.A., and Minsker, K.S., Highmolec. Comp., 1978, v. B20(3), pp. 686–688. (Rus)
51. Seger, M. and Cantow, H.I., Polym. Bull., 1979, v. 5, pp. 347–354.
52. Minsker, K.S., Ivanova, S.R., and Biglova, R.Z., Russian Chemical Reviews, 1995, v. 64(5), pp. 462–479.
53. Berlin, Al.Al., Minsker, K.S., Sangalov, Yu.A., and Prochukhan, Yu.A., Highmolec. Comp., 1983, v. A25(7), pp. 1458–1466. (Rus)

54. Vasile, C. and Onu, P., Acta Polym., 1985, v. 36, pp. 543–550.
55. Ivanova, S.R., Ponedelkina, I.Yu., Berlin, Al.Al., and Minsker, K.S., Highmolec. Comp., 1986, v. A28(2), pp. 266–271.
56. Berlin, Al.Al., Gumerova, E.F., Ivanova, S.R., Minsker, K.S., and Karpasas, M.M., Highmolec. Comp., 1987, v. B29(8), pp. 604–607. (Rus)
57. Grassi, N. and Scott, J., Degradation and Stabilization of Polymers, 1988, 248 p., Moscow, Mir. (Rus)
58. Ivanova, S.R., Gumerova, E.F., Minsker, K.S., Zaikov, G.E., and Berlin, Al.Al., Prog. Polym. Sci., 1990, v. 15, pp. 193–215.
59. Ivanova, S.R., Gumerova, E.F., Berlin, Al.Al., and Minsker, K.S., Highmolec. Comp., 1991, v. A33(2), pp. 342–349. (Rus)
60. Ivanova, S.R., Gumerova, E.F., Berlin, Al.Al., and Minsker, K.S., Highmolec. Comp., 1991, v. A33(2), pp. 430–448. (Rus)
61. Ivanova, S.R., Ponedelkina, I.Yu., Romanenko, T.V., Minsker, K.S., Karpasas, M.M., and Minsker, K.S., Highm Sangalov, olec. Comp., 1986, v. A28(6), pp. 1217–1221. (Rus)
62. Nanbu, H., Yumiko, Y., and Ikemura, T., Polym. J., 1986, v. 18(11), pp. 871–875.
63. Khalafov, F.R., Novrusova, F.A., Ismailov, E.G., and Krentsel, B.A., Acta Polym., 1990, v. 41(4), pp. 226–229.
64. Sangalov, Yu.A. and Prokofiev, K.V., Chemistry and Technology of Petroleum and Oils, 1985, No. 3, pp. 2–4. (Rus)
65. Nelkenbaum, Yu.Ya., Prokofiev, I.K., and Sangalov, Yu.A., Highmolec. Comp., 1986, v. A28(5), pp. 1058–1063. (Rus)
66. Sangalov, Yu.A., Prokofiev, I.K., Polovinkina, G.M., and Zorin, Bashkir. Khim. Zh., 1995, v. 2(2), pp. 34–38. (Rus)
67. Baramboin, N.K., Mechanochemistry of High–molecular Compounds, 1978, 384 p., Moscow, Khimia. (Rus)
68. Basedow, A.M., Ebert, K., Adv. Polym. Sci., 1977, v. 22, pp. 83–148.
69. Engstrom, G.G. and Kovacic, P., J. Polym. Sci., Polymer Chem. Ed., 1977, v. 15(10), pp. 2453–2468.
70. Thompson, H.W. and Torkington, P., Trans. Faraday Soc., 1945, v. 41(3), pp. 246–260.
71. Siriuk, A.G., Chemistry and Technology of Petroleum and Oils, 1972, No. 6, pp. 52–57. (Rus)
72. Puskas, I., Banas, E.M., and Nerheim, A.G., J. Polym. Sci., Symp., 1976, No. 56, pp. 191–202.
73. Sangalov, Yu.A., Khudaiberdina, Z.I., and Yasman, Yu.B., In: Thes. VI Conf. on Aging and Stabilization of Polymers, Ufa, 1983, p. 8, UNI. (Rus)
74. Slobodin, V.M. and Matusevitch, N.I., Highmolec. Comp., 1963, v. 5, pp. 774–776. (Rus)
75. Kichkin, G.I., Kupreev, A.I., and Lakhshi, V.A., Chemistry and Technology of Petroleum and Oils, 1973, No. 7, pp. 47–49. (Rus)
76. Fetters, L.J., Hadjichristidis, N., J.S. Linder, Mays, J.W., and Wilson, W.W., Macromolecules, 1991, v. 24(11), pp. 3127–3135.
77. Bekli, E.Yu., Novikov, D.D., and Entelis, S.G., Highmolec. Comp., 1967, Ser. A, v. 9(12), pp. 2754–2763. (Rus)
78. Pozniak, T.I., Lisitsin, D.M., Novikov, D.D., Berlin, Al.Al., Diachkovsky, F.S., Prochukhan, Yu.A., Sangalov, Yu.A., and Minsker, K.S., Highmolec. Comp., 1980, Ser. A, v. 22(6), pp. 1424–1427. (Rus)

79. Pozniak, T.I., Lisitsin, D.N., Novikov, D.D., and Diachkovsky, F.S., Highmolec. Comp., 1977, Ser. A, v. 19(5), pp. 1168–1170. (Rus)

Chapter 6. Butyl rubber and its derivatives

6.1. Butyl rubber.

Among isobutylene copolymers, of special interest are products with a low amount of diene hydrocarbon — butyl rubbers (BR).

To obtain vulcanizing rubbers, various diene monomers (butadiene, isoprene, pyperilene, 2,3-dimethylbutadiene, cyclopentadiene, etc.) can be used, but high quality butyl rubber is obtained only by copolymerization of isobutylene with isoprene.

Initially, butyl rubber was synthesized by Thomas and Sparks in 1937. Industrial production of isobutylene copolymers with 1-5% of isoprene under the effect of $AlCl_3$ in a methyl chloride (or ethyl chloride) medium at 170-185 K was started in 1941.

First producers of butyl rubber were the companies Standard Oil of New Jersey (USA) and Polymer Corporation Ltd. of Sarnia (Canada). A little later, production of butyl rubber was extended to companies Socabu (France), Enjay Co., Inc. (USA), Precision Rubber Products Corporation (USA), U.S. Rubber Co. (USA), and others. Production of butyl rubber in Russia was developed in 1956. Productive capacity of production of butyl rubber and its derivatives exceeds 600 thousand tons a year.

6.1.1. Physical properties.

In butyl rubber macromolecules, 99% of the isoprene units are attached in *trans*-1,4-position, and less than 1% in 1,2- and 3,4-positions. Isoprene units are effectively not connected to each other and are statistically distributed along the macromolecule. Moreover, isoprene units are main end groupings of macromolecules. Butyl rubber obtained from $AlCl_3$ in methyl chloride medium differs by high linearity of macromolecules and relatively narrow MMD ($U = \overline{M}_w / \overline{M}_n = 2-3$). Molar concentration of unsaturated bonds in macromolecules is 0.6-3%. Molecular mass $\overline{M}_\eta$ of industrial butyl rubber usually varies from 30,000 to 700,000. The value of $\overline{M}_\eta$ is determined from the equation $[\eta] = KM_\eta^\alpha$, where $K = 6.6 \cdot 10^{-4}$, $\alpha = 0.60$ (298 K, toluene), $K = 1.07 \cdot 10^{-4}$, $\alpha = 0.78$ (298 K, CCl_4), or $K = 2.76 \cdot 10^{-4}$, $\alpha = 0.69$ (298 K, diisobutylene). At $\overline{M}_\eta$ below 30,000, vulcanizates of butyl rubbers possess unsatisfactory strength and increased stickiness, which complicates production of the polymer by the continuous scheme because of its adhering to cooling surfaces.

Butyl rubber is a product that is amorphous in a wide range of temperatures; crystallization of the polymer proceeds at large (over 500%) elongation. Ability of butyl rubber to crystallize is appreciably determined by the content of isoprene units in the chain. Crystallization of butyl rubber without elongation is possible at double bond content up to 1 mol%. Propensity for crystallization decreases with growth of nonsaturation.

Butyl rubber dissolves in saturated hydrocarbons and a little worse in arenes. It does not dissolve in alcohols, esters, ketones, acetals, or in solvents containing amino-, nitro- and other high-polar groups. Specific features of butyl rubber are its low gas permeability, high dielectric characteristics and resistance to ozone. In gas permeability, it exceeds all known rubbers, excluding thiocol and ethylene–propylene rubber due to high density of packing of macromolecules.

The main physical properties of butyl rubber are shown below [1, p. 352; 2]:

Density at 300 K, g/cm^3	0.92
Refraction index n_D^{278}	1.5078–1.5081
Glass transition temperature, K	~204
Density of cohesion energy, mJ/m^3	271
Specific heat capacity, kJ/(kg·K)	1.94
Heat conductivity coefficient, W/(m·K)	0.091
Gas permeability coefficient $P\cdot 10^{18}$, $m^2/(s\cdot N/m^2)$	
H_2	54.3
O_2	9.77
N_2	2.44
Diffusion coefficient for O_2 ($D\cdot 10^6$), cm^2/s	0.08
Coefficient of solubility of gases, m^2/N	
H_2	$0.36\cdot 10^{-6}$
O_2	$1.20\cdot 10^{-6}$
N_2	$0.54\cdot 10^{-6}$
Dielectric permeability	
at 1 kHz	2.30–2.35
at 50 MHz	2.2–2.3
at 1.3 GHz	2.12
Tangent of dielectric losses	
at 1 kHz	0.0005–0.0009
at 50 MHz	0.0003–0.0009
at 1.3 GHz	0.0004

Initial rubbery modulus G_0 of solutions for butyl rubbers does not depend on solvent or molecular mass of the polymer, but weakly depends on temperature and strictly depends on concentration of the polymer and its MMD. In accord with the Vinogradov equation [3, p. 124]

$$\frac{\eta_0}{\eta} = 1 + b\left(\dot{\gamma}\,\theta_{\text{det}}\right)^n = 1 + b_0\left[\dot{\gamma}\,\frac{\eta_0}{G_{\text{det}}}\cdot\left(\frac{M_w}{M_n}\right)^{a/n}\right]^n, \tag{6.1}$$

where η_0, η is the initial greatest Newtonian and current viscosity, $N\cdot s/m^2$; $\dot{\gamma}$ is the shift rate of butyl rubber solutions (in the mixture of isopentane with isobutylene), s^{-1};

$\theta_0 = \frac{\eta_0}{G_0}$ is the initial relaxation time of the solution, s; G_{det} are determined from $\lg G_0$ dependency on butyl rubber concentration in hexadecane at 298 K.

Dependency of $\lg\eta/\eta_0$ on $\lg\left(\dot{\gamma}\frac{\eta_0}{G_0}\right)$, where $G_0 = G_{det}\left(\frac{M_w}{M_n}\right)^{1.015}$ with regard to polydispersion is displayed in Figure 6.1. Therewith it should be taken into account that η_0 depends only on the mean viscous molecular mass of butyl rubber, but does not depend on MMD.

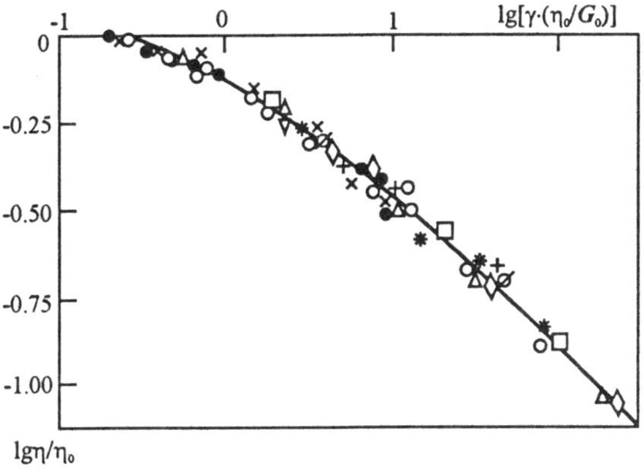

Figure 6.1. Dependence of $\lg\eta/\eta_0$ on $\lg\left(\dot{\gamma}\frac{\eta_0}{G_0}\right)$ for solutions of various butyl rubbers; polydispersion $\overline{M}_w/\overline{M}_n$ is:
• – 2.9; ○ – 3.5; + – 3.3; X – 3.1; * – 5.7; Δ – 4.0; □ – 6.7; ◊ – 4.1, unsaturation; mol%;
•, ○, + – 2.2–2.0; X, *, Δ, □, ◊ – 1.6–1.4, moderate-viscous molecular mass (toluene, 298 K), $M_\eta \cdot 10^{-3}$:
•, ○, + – 360–390; X – 460; * – 510; Δ, □ – 552, ◊ – 610.

Butyl rubber is non-toxic. Some trademarks of vulcanizates can be used in food and medical industries. The polymer is fire hazardous, practically because of its convex form. Usually, ignition temperature of rubbers in the form of granules is some 50–100° lower than for bricks, rolls, and belts. Extremely low values of flash temperature (T_{flash}) and ignition (T_{ign}) are possessed by liquid butyl rubbers and these spreading below 570 K, which is explained by affinity to depolymerization with elimination of isobutylene (about 470 K). Values of T_{flash}/T_{ign} are: 491/524 for BR-1675T, 462/462 for BR at standard temperatures of self-ignition of 684 K and 675 K, respectively [3, p. 149].

6.1.2. Chemical properties.

Butyl rubber is resistant to the influence of oxygen, and is not appreciably affected by structural changes under the effect of metal salts of variable valence. It is noticeably degraded by UV-light, ionizing radiation and mechanical loads. Butyl rubber is resistant to the effects of water, acids, several vegetable oils, etc. It easily reacts with halogens in light. The reaction is accompanied by abrupt decrease of the molecular mass. At partial halogenation under mild conditions in solution (290 K), halogen reacts with macromolecules, but causing no deep degradation of butyl rubber. Industrial synthesis of halogenated butyl rubber is based on this reaction.

Reactions proceeding by double bonds of isoprenyl units are typical for butyl rubber. Epoxidation of butyl rubber in solution proceeds under the effect of peracids (peracetic, perbenzoic, and others), added in excess in relation to double bonds of the polymer, with conversion up to 95–97% (318 K). Epoxided butyl rubber displays typical properties of oxirane compounds, in particular, it reacts with water, acohols, amines, etc.

Chain decomposition at the double bonds proceeds under the effect of RuO_4 in combination with an oxidant [4]:

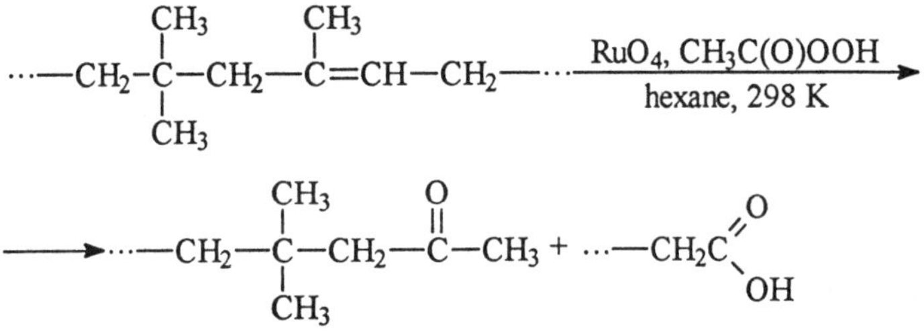

The reaction proceeds practically up to complete exhaustion of double bonds. It can be used for analytical purposes (determination of degree of unsaturation of butyl rubber) and for synthesis of oligoisobutylenes with end ketone and carboxy groups.

Degradation of butyl rubber under exhausting ozonation in combination with appropriate chemical transformations of decomposition products forms the basis of synthesis of α,ω-bifunctional oligomers of isobutylene with various end groups [5, 6]:

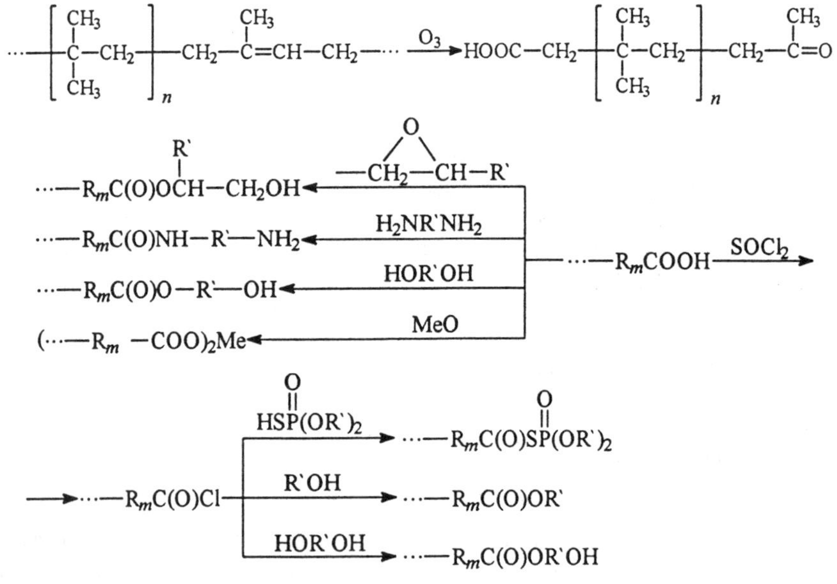

The products obtained are characterized by a functionality, close to the theoretical one, and a narrower molecular-mass distribution than that of the original copolymer (the ratio M_w/M_n varies in the range from 1.15 to 1.4).

High selectivity of butyl rubber ozonation (rate constants of ozone interaction with isoprene and isobutylene fragments are $1{\cdot}10^4$ and $1.2{\cdot}10^{-2}$ l/(mol·s), respectively) enables methods of determination of the molecular-mass distribution and structure of polyisobutylene to be developed [7].

Existence of double bonds in butyl rubber promotes cross-linking of polymeric chains under the effect of various chemical agents (sulfur, organic polysulfides, dinitroso-compounds, alkylphenoloformaldehyde resins, etc.), used individually or combined with activators. These reactions do not necessarily involve the double bonds, but obviously are activated by them.

A typical reaction for butyl rubber is degradation of macromolecules under the effect of electrophilic catalysts, i.e. fission of polymeric chains by the law of accident with no formation of significant amounts of the monomer [3, p. 130; 8].

Suitable catalysts are donor–acceptor complexes of $(C_2H_5)AlCl_2{\cdot}H_2O$ type. The number of breaks of macrochains may reach 60 (concentration of BR-2045 is 20 kg/m^3, concentration of catalyst is 2–20 mass%, 293 K, 600 s).

According to their efficiency, electrophilic catalysts form the following sequence:

$(C_2H_5AlCl)_2O{\cdot}H_2O < (iso\text{-}C_4H_9AlCl)_2{\cdot}H_2O < C_2H_5AlCl_2{\cdot}H_2O < iso\text{-}C_4H_9AlCl_2{\cdot}H_2O$.

No depolymerization of butyl rubber at a noticeable rate was observed below 425 K. Unsaturation of butyl rubber at degradation hardly changes. Degradation of butyl rubber starts immediately after introduction of a catalyst and proceeds at significantly lower temperature (170 K). The reaction forms the basis of the method of industrial reactor cleaning from sediments of linked butyl rubber [3, p. 130].

Thermal degradation of butyl rubber, which causes high yield of isobutylene [9] in the presence of complex catalysts based on HAOS and $AlCl_3$, proceeds most probably via an electrophilic attack on internal C=C-bonds of the chain, formation of polymeric carbocations and their transformation (depolymerization):

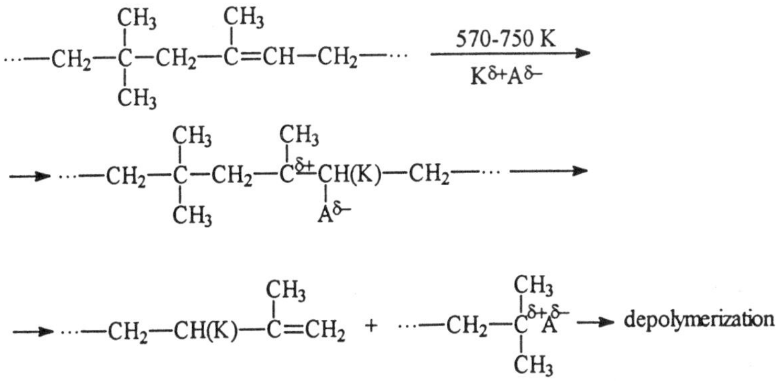

The tertiary carbonium ion, formed during decomposition of butyl rubber macromolecules, may participate in secondary processes.

A feature of catalyzed degradation of butyl rubber is formation, beside isobutylene, of noticeable amounts of isobutane with simultaneous rise of unsaturation of macromolecules and parallel formation of high-molecular products, including some that are highly structured. It may be concluded that the presence of isobutane in the products is due to the reaction of ionic hydrogenation of isobutylene, formed according to the above-mentioned scheme, in which isomerized chains of butyl rubber play the role of donor of hydride ions. Cationic mechanism of the process is confirmed by low activation energies (50–55 kJ/mol) and significant elongation of the kinetic chain of the polymer degradation from 18 monomeric units (thermal process) to 50–60 (catalyzed degradation), which is close to the average length of polyisobutylene blocks between isoprene chains of butyl rubber. Contrary to degradation of butyl rubber under the effect of usual acidic catalysts (such as $AlCl_3$), thermocatalytic degradation of the polymer in the presence of complex catalysts represents a higher selective process — depolymerization with high yield of isobutylene (80% and higher).

This reaction is interesting from the point of view of processing of substandard butyl rubbers.

6.1.3. Main areas of butyl rubber application

The utilization of butyl rubber is determined by its molecular mass and content of isoprene chains (nonsaturation), as well as by type of antioxidant, which, by the way, determines the color of the product (from light- to dark-yellow) (Table 6.1).

The first two digits in USSR trademarks of butyl rubbers characterize the average value of unsaturation in decimal parts of the molar percent; two other digits mark the Muni viscosity, which is approximately appropriate to $\overline{M}_\eta$ in thousand units in this interval.

Table 6.1

Equivalent trademarks of butyl rubber, produced in different countries.

Butyl rubber (USSR)	Polysar-butyl (Canada)	Exolbutyl (USA), Essobutyl (Great Britain)	Bucarbutyl (USA)	Viscosity by Muni at 373 K (8 min)	Unsaturation, mol%
BR-0845T	100	0.35	1000S	45 ± 4	0.8 ± 0.2
BR-0845H			1000NS	45 ± 4	0.8 ± 0.2
BR-1675T		218	5000S	45 ± 5	1.6 ± 0.2
BR-1675H	301	268	5000NS	45 ± 5	1.6 ± 0.2
BR-2045T	400	325	6000S	45 ± 4	2.0 ± 0.2
BR-2045H	402	365	6000NS	45 ± 4	2.0 ± 0.2
BR-2545H	600			45 ± 4	2.5 ± 0.2

Note. Letters T and S in the trademark indicate rubber containing coloring antioxidant; letters H, N, NS indicate rubber containing non-coloring antioxidant.

Characteristics of native trademarks of butyl rubber are shown in Table 6.2.

Physicomechanical and other exploitation properties of vulcanizates of non-filled and filled butyl rubber (on the example of BR-1675) are shown in Tables 6.3–6.5 [1, 2].

Table 6.2

Characteristics of butyl rubber trademarks, produced in Russia.

No.	Trademark	Muni viscosity at 373 K (8 min)	Unsaturation, Mol%	Conditional strength at elongation, MPa, not less	Conditional tension at elongation by 400%, MPa, not less	Relative elongation at break, %, not less	Residual elongation, %, not less	Ash content, mass%	Neozone D content, mass%	Area of application
1	BR-0845T (H)	45 ± 4	0.8 ± 0.2	17.5	—	700	—	0.35	0.1–0.3	Ozone-resistant cable rubbers
2	BR-1040T	42 ± 4	1 ± 0.2	17.5	—	700	—	0.40	0.1–0.3	Rubber articles for general purposes
3	BR-1050T	≥ 47	1 ± 0.2	17.5	—	700	—	0.40	0.1–0.3	Rubber articles for general purposes
4	BR-1675T	45 ± 4	1.6 ± 0.2	17.5	—	600	—	0.35	0.1–0.3	Latex
5	BR-1675T (H)	75 ± 5	1.6 ± 0.2	19	7.8	600	—	0.40	0.1–0.3	Car balloons, diaphragms of formers–vucanizers
6	BR-2045T (H) (M)	45 ± 4	2 ± 0.2	17.5	7.3	600	—	0.40	0.1–0.3 (0.3-0.5 of antioxidant NG-2246 for brand H)	Articles for medical and food purposes, heat-resistant articles, light colored articles, low-voltage cable rubbers
7	BR-2055T	55 ± 5	2 ± 0.2	17.5	7.3	600	—	0.35	0.1–0.3	—

8	BR brand A	40000	1.0 ± 0.2	—	—	700	20	0.25	0.1–0.5	Rubber articles for general purposes
	BR brand B	35000-39000	1.0 ± 0.2	—	—	700	20	0.25	0.1–0.5	
	BR brand C	30000-34000	1.0 ± 0.2	—	—	700	25	0.25	0.1–0.5	

Notes. **1.** Brands of butyl rubber No. 1-7 – TU 38.003 169-79, No. 8 – TU 38.10320-77. **2.** The following indexes are regulated for all butyl rubber brands: mass loss at drying – not less than 0.2%; Ca stearate content – not less than 1.2 mass%, Cu stearate – $(2–3)\cdot 10^{-4}$ mass% (excluding butyl rubbers of brands A, B and C). **3.** Instead of Muni viscosity, molecular mass by Staudinger is determined for butyl rubber of brands A, B and C; appearance of the product is regulated – from light brown to dark brown color.

Table 6.3

Some properties of butyl rubber vulcanizates.

Index	Vulcanizate	
	Non-filled*	Filled**
Tension at elongation by 300%, MPa	4–6	3–6
Tension at elongation by 500%, MPa	1.2	11
Resistance to break, MPa		
at 293 K	19–23	19–24
at 373 K	10–15	10–16
after aging during 72 h at 373 K	14–16	14–16
Relative elongation at break, %	650–950	650–850
Residual elongation, %	15–40	30–50
Resistance to tearing, kN/m	8.8	83
Hardness by TM-2[1]	30–65	60–70
Elasticity by recoiling, %		
at 293 K	8–11	8–11
at 373 K	34–40	30–40
Friability temperature, K	~218	~225
Resistance to intergrowth of cutting (expansion of cracks), thousand cycles	600-900	600
Abradability, mm^3/J	0.090	0.084
Resistance to cold at 232 K	—	0.40
Resistance to temperature at 373 K		
by strength at elongation	—	0.6
by relative elongation	—	1.1
Heat aging coefficient (724 h at 403 K)		
by strength at elongation	—	0.7
by relative elongation	—	0.7
Residual deformation of compression after aging (724 h at 403 K), %	—	90
Specific volumetric electric resistance, TOhm·m	100	—
Electric strength, mV/m	16-24	—

* Typical composition of the mixture (mass parts): rubber – 100, sulfur – 2, thiuram – 1, ZnO – 5, stearin – 2; vulcanization during 20-30 min, 416 K.

** Typical composition of the mixture (mass parts): rubber – 100, sulfur – 2, captax – 0.65, thiuram – 1.3, ZnO – 5, stearin – 2, technical carbon DG-100[2] - 50; vulcanization during 30-40 min, 416 K.

Mainly, butyl rubber is used to produce rubber for the tire industry. For example, in the USA more than 2/3 of the total volume of butyl rubber is consumed by the tire industry [1, 2, 10, 11]. Car air-bags and the internal layer of tubeless tires from butyl rubber exceeds natural rubber by air impermeability by 10 times. Digesting chambers, used for tire production, more than twice exceed analogous articles from natural rubber by a complete lifetime.

[1] Russian specification.

[2] Russian trademark.

Table 6.4

Chemical resistance of butyl rubber vulcanizates to some aggressive media.
Designations: + - stable for several years; – - unstable; ± - stable at short-time (up to several months) or periodic influence.

Medium	Medium temperature, K	Estimation of stability
40% formaldehyde	335	+
Water (fresh, sea, overheated)	400 450	+ –
40% hydrogen bromide	295	–
Iodine (dissolved)	305	+
Fluorine (dry and liquid)	295	±
15% hydrogen fluoride	345	+
Acetone	295	+
Methylethylketone	295	+
Fish oil	295	–
Castor oil	340	+
Corn, linseed oil	340	–
Gasoline, kerosene, black oil	295	–
Phenol, butylphenol	295	+
Hydraulic oils	295	±
Nitrogen oxides	375	±
Sulfur oxides	295	±
30% hydrogen peroxide	295	+
100% hydrogen sulfide	355	+
70% sulfuric acid	345	+

Thermal resistance of butyl rubber allows its broad application in production of steam-conducting hoses or pipes and transportation belts, used at high temperature. Chemical stability of butyl rubber vulcanizates to influence of many aggressive media (acids, alkali, solutions of salts, ketones, alcohols, H_2O_2, nitrogen-containing solvents, river and sea water, many of vegetable oils, etc.) suggests application of butyl rubber for rubberizing of chemical devices, exploited at temperatures up to 375-400 K, facings of rolls, production of acid-resistant gloves, hoses for delivering of aggressive agents, etc. [1; 12, p. 40; 13].

Combination of chemical stability, gas impermeability, atmosphere- and water-resistance generates application of butyl rubber for production of rubberized fabrics of various purposes, gas masks, anticorrosion facing rubbers and articles for pharmacological industry.

Butyl rubber resistance to swelling in milk and food fats enables the polymer to be used for production of milking devices and other rubber articles, contacting foodstuffs. For these purposes, kinds of butyl rubber are applied that are stabilized by non-toxic antioxidants. Butyl rubbers are widely applied in production of spongy rubbers, various sealing composites and molded articles, used under conditions of intensive dynamic influences. Combination of high dielectric characteristics, water and ozone resistance generates wide application of butyl rubber for production of insulation for high and low voltage cables [1].

Butyl rubber can be used as waterproof and roofing material. Film, sheet, roll and mastic articles from it can be used in northern regions, because they retain their useful properties in broad ranges of temperature. For example, resistance to cold of hydrobutyl rubber is down to 208 K.

Besides traditional application, in the vulcanized form, there are non-hardening (non-drying) sealing materials based on butyl rubber or its mixtures with polyisobutylene, ethylene–propylene copolymer, as well as with bitumen, natural or artificial resins, etc. [13]. Besides powder-like or fibrous fillers, such compositions usually contain solvents (mineral or vegetable oils), low-molecular rubbers, non-migrating plasticizers and other low-boiling liquids. Adhesives, fireproof substances and other gel additives are introduced into sealing materials depending on application purposes. Sealing composites with butyl rubber, produced in Russia (buteprol-1 or -2), USA (butyl-extrasyl), Germany (terostats, vidolastic) and other countries, designed for butt joining thickening of external panels of buildings, joint sealing in stained glass compositions and air conditioning systems, etc., can be used at 220–360 K. Covered by thin film of oxidized oil in the air, sealing material or mastic does not change its properties over many years.

In some cases, butyl rubber is used for production of latex, which is applied in various fields of technology. One of the main methods of preparation of water dispersions consists of dispersing of concentrated butyl rubber solution with consecutive elimination of the solvent.

In some spheres substandard kinds of butyl rubber (rejects) or products of its deep chemical transformations, which simultaneously are efficient methods of secondary processing of the polymer, are used. For example, saturated oligoisobutylenes of narrow fractional composition with end aldehyde, ketone, carboxy- and other groups were obtained in selective oxidative splitting of butyl rubber at the double bonds with further thermal (chemical) processing of degradation products [4, 6]. Owing to the saturated nature of the main chain, they may form the basis for highly efficient lubricants, resistant to thermal, thermooxidative and mechanical degradation. Products of further transformations of oligobutylenes at the end groups found uses as multifunctional additives to lubricants (thickening, antioxidative, antiscratching, antiabrasion, etc.), which take on many of the most desirable properties of the polymer [14, 15], as well as stabilizers for many industrial polymers. They are also stabilizers of many industrial polymers. Good adhesive properties and compatibility with rubbers enable functional oligoisobutylenes to be applied to rubber composites (with butadiene-nitrile, chloroprene rubbers) to increase stickiness, resistance to cold, chemical stability, and resistance to ozone aging [6, 16].

Butyl rubber is highly compatible with many polyolefins, in particular, with polyethylene, polyisobutylene, copolymers of ethylene, styrene, isobutylene, etc. [1, 2, 17]. To increase resistance to cold of butyl rubber vulcanizates, it is combined with ethylene–propylene rubber, and to increase chemical stability, with polyethylene. Butyl rubber mixtures with natural rubber, natural and butadiene–styrene or natural, ethylene–propylene triple and butadiene–styrene rubbers are applied in production of details of tires for cars and vehicles.

6.2. Modified butyl rubber

To obtain butyl rubber with new properties, it is chemically modified by halogenation, sulfonation, cross-linking (vulcanization), and so on. Of special interest for industry are chlorine butyl rubber and bromine butyl rubber, the price of which is no more than 20–25% higher than usual butyl rubber. That is why halogenated butyl rubbers may be considered as polymers for a wide range of purposes.

For the first time bromine modification of butyl rubber was performed in the laboratory of the company B.F. Goodrich Chemical Co. (USA) in 1952, and chlorine modification, in 1955 by Esso Industrial and Engineering Co. Linden New Jersey [18, p. 4].

6.2.1. Chlorobutyl rubber.

Chlorobutyl rubber represents the product of butyl rubber chlorination, which contains 1.1–1.3% of chlorine, mainly attached in the α-position to the double bond of isoprene chains of butyl rubber macromolecule. Since the 1950s, chlorobutyl rubber has been produced by the company Enjay Co. (USA). In 1971–1972 a reactor for production of chlorobutyl rubber was built in England on the butyl rubber plant in Foley, and in 1973–1974, in Japan. In 1979, production of chlorobutyl rubber was started in Canada, and in 1980, in Belgium.

Growth of chlorobutyl rubber production [5, 19, 20] is follows: 1960 – 600 tons, 1965 – 13,000 tons, 1972 – over 55,000 tons, 1982 – 75,000 tons. Hence, the tendency of an increasing proportion of chlorobutyl rubber (as well as of other halogenated butyl rubbers) is clearly seen in the structure of production of butyl rubbers, which has grown up from 24.8% in 1975 to 32.1% in 1982.

Three brands of chlorobutyl rubber are produced in Russia (CBR-155, CBR-165 and CBR-175), corresponding in properties to foreign brands HT-1066, HT-1067, and HT-1068.

6.2.1.1. Structure and physical properties.

Macromolecules of chlorobutyl rubber consist of polyisobutylene fragments, representing chains of a compact helicoidal form, statistically separated by chlorine-containing isoprenyl chains of various structure. Chlorobutyl rubber may also include non-chlorinated isoprenyl chains. Chlorobutyl rubber preserves about 75% of unsaturation of the initial butyl rubber. At chlorination of butyl rubber, several chlorine-containing structural units are formed in its macromolecules [18, 21–25], establishing bonds between them is an important task from both points of view of science and application. However, in spite of application of chemical and spectral methods of analysis, no general opinion on the presice structure of chlorobutyl rubber exists yet. One can state only that isobutylene chains remain unchanged. Generally (by some 95%), the substitution reaction proceeds according to the ionic mechanism at *cis*- and *trans*- (40–45):(60–55) isoprenyl fragments, statistically distributed along macromolecules [21]; therewith, up to 5 mass% of products of H atoms substitution by Cl ones are formed.

Table 6.5

Physicochemical properties of butyl rubber vulcanizates before and after tests in aggressive media (373 K, duration of tests is 25 days).

Medium	Estimation of resistance	Swelling degree, %	Strength at elongation, MPa		Relative elongation at break, %	
			Before tests	After tests	Before tests	After tests
Non-filled vulcanizate						
Acetic anhydride (concentrated)	—	6	18.5	8.5	785	775
Acetic acid (concentrated)	—	3.7	19.5	5.5	790	555
Formic acid (20%)	—	41	20.5	9.5	920	800
Nitric acid (30%, 323 K)	±	5.5	21.5	15	865	890
Sulfuric acid (70%)	±	1	21.5	12.5	910	765
Hydrochloric acid (20%)	—	8.5	21.5	8.5	860	735
Filled vulcanizate						
Acetic anhydride (concentrated)	±	13	18	12.5	805	670
Acetic acid (concentrated)	±	11.5	18	12.5	805	730
Formic acid (20%)	+	13	19	16	810	725
Nitric acid (30%, 323 K)	+	6	19.5	17.5	810	800
Sulfuric acid (70%)	+	1	19.5	20.0	810	685
Hydrochloric acid (20%)	+	11.5	19.5	16.0	810	625

Note. Designations as in Table 6.4.

Chlorination of butyl rubber was identified by interaction of the low-molecular model compound 2,2,4,8,8-pentamethylnonene-4 with Cl_2 [2, 21]:

$(CH_3)_3C \vdots CH_2{-}C(CH_3){=}CH{-}CH_2 \vdots CH_2C(CH_3)_3$ - model compound

$\cdots{-}CH_2C(CH_3)_2 \vdots CH_2{-}C(CH_3){=}CH{-}CH_2 \vdots CH_2C(CH_3)_2{-}\cdots$ - butyl rubbe

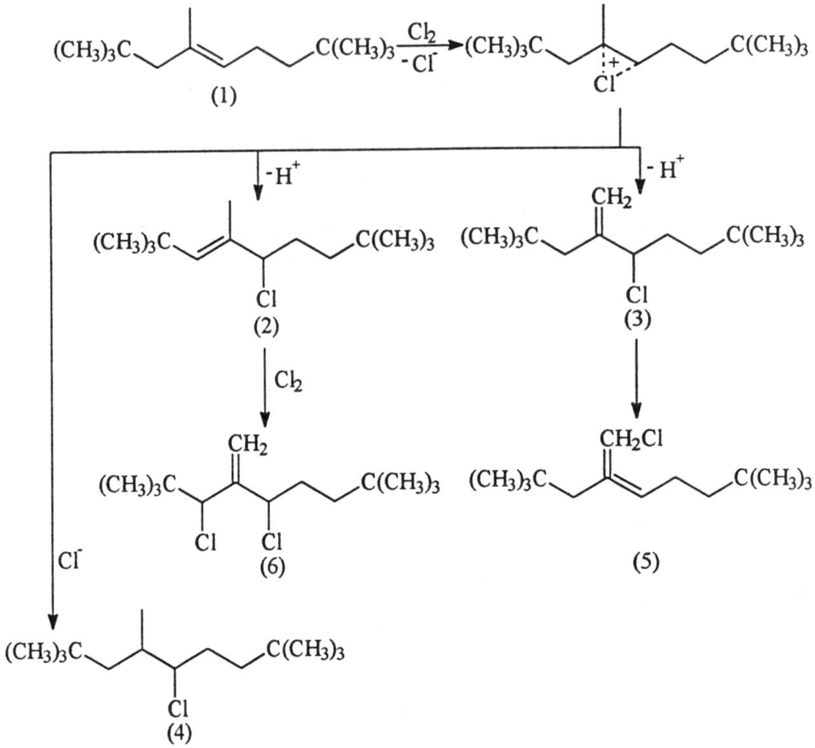

Table 6.6

Data of ESR-spectra (ppm, δ) for chlorinated 2,2,4,8,8-pentamethylnonene-4.

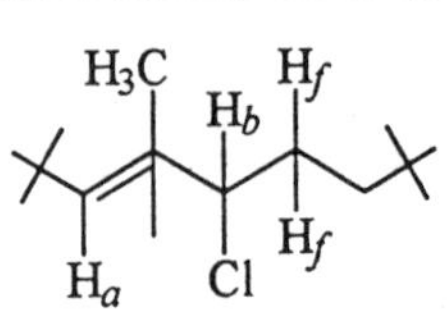 (2)		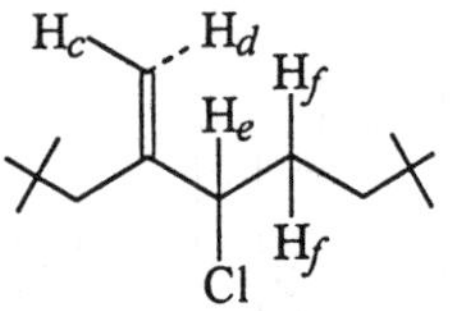 (3)			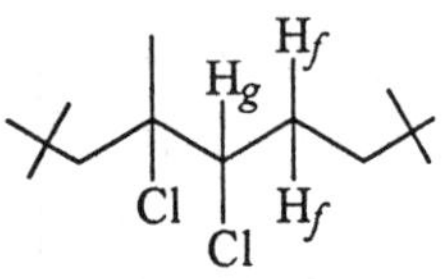 (4)	
H_a	H_b	H_c	H_d	H_e	H_f	H_g
5.4	4.2	5.0	5.3	4.2	~2.0	3.7-3.9
s, wide	*t* (J = 6 Hz)	s	s	*t* (J = 6 Hz)	*q* (J = 14 Hz)	*m*

According to data of ESR-spectra (Table 6.6.), including ESR of high resolution [22], main products are represented by structures (2) and (3). Product (4) is mainly formed at low temperatures (Table 6.7).

Table 6.7

Chlorination of 2,2,4,8,8-pentamethylnonene-4.

Reaction tempera-ture, K	Amount of chlorine, introduced into the reaction medium, mol/mol	Reac-tion time, s	Yield of products by ESR-spectrum data, mass%			Chlorine content in products, mol/mol
			2 (± 2%)	3 (± 2%)	4(± 2%)	
298	2.6	300	24	76	0	1.05
298	2.6	300*	24	76	0	1.05
298	2.4	1800	—	**	—	1.05
323	2.4	300	24	76	0	—
195	2.4	300		50	50	1.20

* In light (other reactions – in darkness).

** Identified as components of the mixtures of compounds (5) and (6).

The presence of a noticeable amount of exomethylene $=CH_2$-groups in chlorobutyl rubber is indicated by the occurrence of bands at 909, 1645 and 3095 cm^{-1} in IR-spectra [18, p. 16]. As butyl rubber is chlorinated, the total unsaturation of macromolecules and, especially, the content of C=C-bonds decreases, which can be easily determined by the method of splitting ozonolysis.

Analysis of industrial samples of chlorobutyl rubbers confirms the conclusion about the preferable content of exomethylene groups in the product (structure (3)). In particular, the following structure of macrochains is prescribed for chlorobutyl rubber of HT-1068 brand [21]:

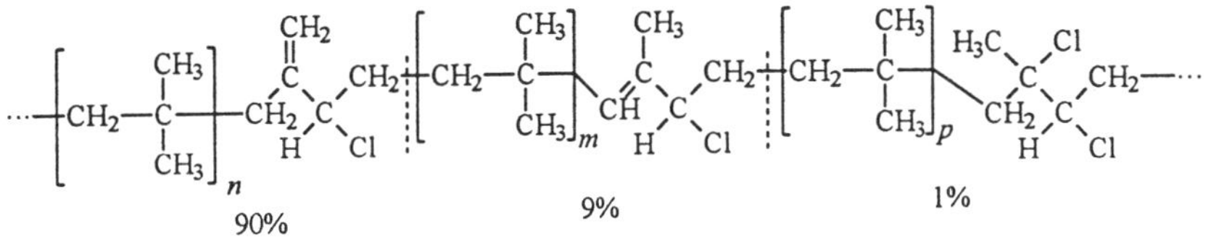

Table 6.8

Adhesive properties of halogenated butyl rubbers.

Indexes	Butyl rubber	Chlorobutyl rubber	Bromobutyl rubber
Adhesion to rubbers from natural caoutchouc, kN/m			
at 293 K	0.84	2.63	8.75
at 373 K	0	0.52	2.63
Adhesion to brass, kN/m			
at 293 K	0	5.25	17.4
at 373 K	0	2.97	15.0

Participation of ...—CH_2—$C(CH_3)$=CCl—CH_2—... in the composition of macromolecules is also probable [24, 25].

Chlorobutyl rubber is soluble in aliphatic and aromatic hydrocarbons, is stable to the influence of polar organic and inorganic compounds, alcohols, ketones, animal and vegetable fats.

Chlorobutyl rubber possesses high gas and water impermeability, relatively low elongation, high modulus and strength at elongation. Introduction of a halogen into butyl rubber improves its adhesive properties (Table 6.8).

It is suitable to perform sedimentation of chlorobutyl rubber from *n*-pentane solutions into acetone (*n*-hexane–ethyl alcohol or toluene–acetone systems can be used [25]).

6.2.1.2. Chemical properties.

Chlorobutyl rubber is quite stable under the effect of O_2 and O_3, which is due to the lower reactivity of its molecules to these agents compared with butyl rubber. The rate constant of O_3 interaction with C=C-bonds in isoprenyl chains of butyl rubber is k_{BR} = (1.1 ± 0.2)·10^5 l/(mol·s), and in chloropentene, preferably exomethylene groupings ...—CH_2—C(=CH_2)—CHCl—CH_2—..., by one and half or more degrees lower: k_{CBR} = (0.3 ± 0.1)·10^4 l/(mol·s). Ozonation of the grouping of ...—CH_2—C(CH_3)=C(Cl)—CH_2—... structure displays greater difficulties [26]. Ozonolysis of samples of chlorobutyl rubber in tetrahydrofurane solution with consequent oxidative hydrolysis of polymeric products by H_2O_2 is accompanied by reduction of characteristic viscosity $[\eta]_0$ down to a definite limiting value, determined by concentration $\bar{\gamma}$ of internal isolated C=C-bonds in macromolecules (Figure 6.2). Numerical value of $\bar{\gamma}$ can be determined easily, based on changes in values of $\overline{M}_\eta$ for initial and ozone-treated samples. According to works [27, 28]

$$\bar{\gamma} = m\left(\frac{Z}{\overline{M}_\eta} - \frac{Z_0}{\overline{M}_\eta^0}\right), \qquad (6.1)$$

where *m* is the molecular mass of the monomeric unit.

The value of *Z* is determined from the relation (at $\overline{M}_w/\overline{M}_n \approx 2$)

$$Z = \overline{M}_\eta / M_n = [(1+\alpha)\Gamma(1+\alpha)]^{1/\alpha},$$

where Γ is the gamma-function; α is the coefficient in the Kun–Mark–Hauvick equation.

As for narrow fractions of butyl rubber (293 K, *n*-heptane) [29]

$$[\eta] = K\overline{M}_\eta^\alpha = 3.6\cdot 10^{-4}\cdot \overline{M}_\eta^{0.62} \quad \text{and} \quad Z = 1.82, \qquad (6.2)$$

then

$$\bar{\gamma} = 56\cdot 1.82\left(\frac{1}{\overline{M}_\eta} - \frac{1}{\overline{M}_\eta^0}\right) = 102\left(\frac{1}{\overline{M}_\eta} - \frac{1}{\overline{M}_\eta^0}\right). \qquad (6.3)$$

Chlorobutyl rubber is relatively unstable. It releases HCl in various energetic and chemical interactions [24, 25, 30]. The reaction is the result of the formation of a system of conjugated double bonds, and typical color of the polymer occur (mass loss is about 1%). Detachment of a sufficient amount of HCl causes structuring of macromolecules rather than forming condensed aromatic fragments [31, 32]. Cross-linking of chlorobutyl rubber can also proceed at intermolecular detachment of HCl.

Typical dependencies of HCl elimination with time for various samples of chlorobutyl rubber are displayed in Figure 6.3 [30].

In the general case, with the presence of several types of chlorine-containing structures, the amount of which decreases during thermal degradation of chlorobutyl rubber, dehydrochlorination of chlorobutyl rubber is described by the expression:

$$a_t = a_0 - \sum_i a_i \exp(-k_i t), \qquad (6.4)$$

where a_t is the amount of HCl, released during time t; $a_0 = \sum a_i$ is the total content of chlorine in the sample; k_i is the rate constant of HCl elimination from the current structure of chlorobutyl rubber.

As the stability of the Cl atom is significantly higher in saturated structures than in β-chloroallyl groupings [33], and the contribution of such structures to the total rate of HCl elimination at moderate temperatures (400-500 K) is low, equation (6.4) with regard to the presence of two types of structures with labile chlorine atom, 2 and 3, in the composition of macromolecules transforms into the expression:

$$a_t = a_0 - a_1\exp(-k_1 t) - a_2\exp(-k_2 t), \qquad (6.5)$$

where a_1 and a_2 are contents of labile chlorine atoms in structures 2 and 3, respectively.

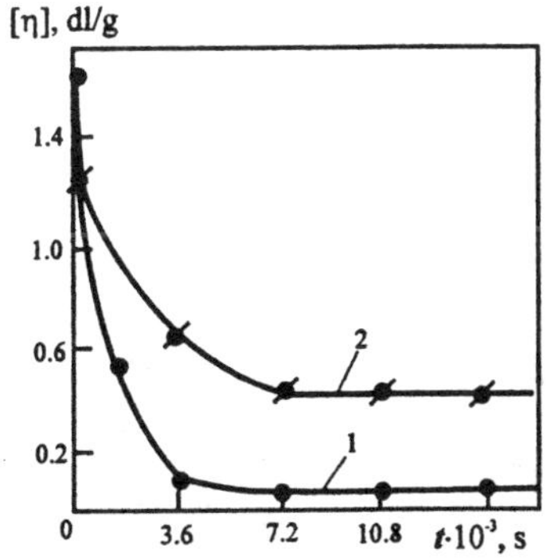

Figure 6.2. Change of characteristic viscosity of 5% solutions of rubbers during ozonation:
1 – butyl rubber; 2 – chlorobutyl rubber (tetrahydrofurane, 298 K).

The content of framing (side) unsaturated groupings (structure 3, Table 6.6) is determined experimentally by difference between total unsaturation of chlorobutyl rubber and concentration of internal C=C-bonds ($\bar{\gamma}$). Therewith the content of structure (2) (Table 6.6) exists in the range of $0 \le a_2 \le \bar{\gamma}_0$ ($a_2 < \bar{\gamma}_0$ at the relation of introduced chlorine to initial C=C-bonds of butyl rubber below 1). Unknown values in equation (6.5) are a_2, k_1 and k_2. For two values a_t and a_{2t} with multiple time intervals, it can be deduced that

$$a_t = a - a_1\exp(-k_1t) - a_2\exp(-k_2t), \qquad (6.6)$$
$$a_{2t} = a - a_1\exp(-2k_1t) - a_2\exp(-2k_2t).$$

Solving the system of equation (6.6) for k_1 and k_2, we obtain:

$$k_1 = -\ln x/t, \qquad (6.7)$$
$$k_2 = -\ln[(a - a_t - a_1x)/a_2]/t, \qquad (6.8)$$
$$x = \frac{a - a_t \pm [a_1a_2(a_1 + a_2)(a - a_{2t})]^{1/2}}{a}. \qquad (6.9)$$

Calculations using formulae (6.7) and (6.8) were conducted at variation of a_2 values in the range $0 \le a_2 \le \bar{\gamma}_0$. The value of a_2 and k_1 and k_2, appropriate to it, were calculated by a computer at better coincidence of experimental and calculated values a_{nt} at $n > 3$, i.e. at the minimal value of the sum:

$$\sum_n [a_{\text{exper}} - a_{\text{theor}}(nt \mid k_1, k_2, a_2)]^2. \qquad (6.10)$$

As seen from Figure 6.3, experimental curves of dehydrochlorination of chlorobutyl rubber successfully coincide with theoretical ones. Calculated values of the rate constant of HCl elimination from structures (2) and (3) (see Table 6.9) coincide within the experimental error for chlorobutyl rubber samples with various content of chlorine.

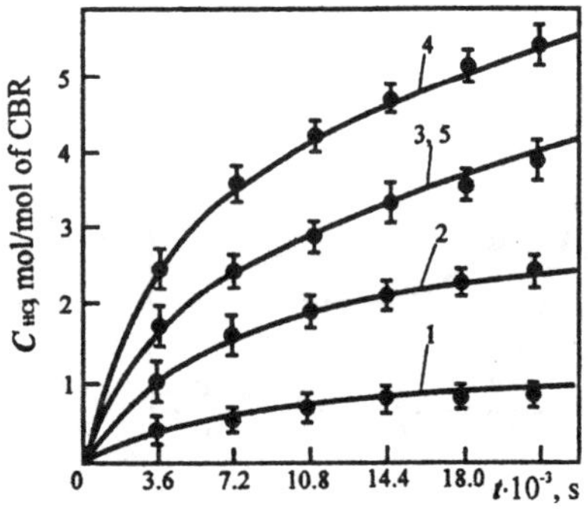

Figure 6.3. Kinetic curves for dehydrochlorination of chlorobutyl rubber (448 K, 10^{-2} Pa, films 20–60 μm thick), chlorine content in the sample of chlorobutyl rubber (mol%):
1 – 0.27; 2 – 0.62; 3 – 1.05; 4 – 1.55; 5 – 1.90.

According to quantum-chemical calculations (PPDP/2 method, the reactivity index – C-Cl bond degree [34, p. 54]), relative stability of structures (2) and (3), Table 6.6, coincide with found constants of HCl elimination from structures (2) and (3) of chlorobutyl rubber (Table 6.9).

Of importance is that stability of the structure (2) ($CH{=}C(CH_3){-}CHCl{-}CH_2{-}$), in which the chlorine atom lies in the β-position to the internal C=C-bond, was found more than a degree lower than that of the structure (3) ($-CH_2{-}C({=}CH_2){-}CHCl{-}CH_2$-). A significant difference in mobility of chlorine atoms in structures (2) and (3) is due to the influence of electronic density of double C=C-bonds (which is higher for the structure (2)).

Table 6.9

Kinetic parameters of dehydrochlorination and parameters of stability of model structures of chlorobutyl rubber.

Structure	$k\cdot10^5$ s^{-1} (±30%)					A	E,	Bond degree	
	448 K	458 K	463 K	473 K	483 K	(±30%)	kJ/mol (±5%)	C-Cl	C=C
(3)	0.09	0.27	0.42	0.91	1.4	$2.0\cdot10^{11}$	140	1.0452	1.8300
(2)	10	22	27	48	66	$4.7\cdot10^{7}$	100	1.0397	1.7392

Calculations have shown (Table 6.10) that content of the structure (2) (a_2) in chlorobutyl rubber is 7–9-fold lower than that of the structure (3) (a_1). Nevertheless, the high rate of chlorobutyl rubber dehydrochlorination in the initial stage of thermal degradation is determined just by this most unstable structure (2). It is appropriate to use this circumstance in industrial production of chlorobutyl rubber.

At introduction of less than one atom per one double bond of the initial butyl rubber ($a_0/b < 1$), the relative content of saturated chlorinated structures in chlorobutyl rubber does not exceed 10%, which is by a degree higher than their content at chlorination of butyl rubber model [21]. If $a_0/b > 1$, the relative content of saturated chlorine-containing structures a_{sat}/a_0 grows sharply. However, in this case, thermal stability of chlorobutyl rubber hardly changes. Samples 3 and 5 (see Table 6.5) display close rates of dehydrochlorination, although they differ by more than 1.5 times in chlorine content. This indicates once again the relative stability of saturated chlorine-containing groupings.

Table 6.10

Concentration of chlorinated structures in chlorobutyl rubber (calculated by equation (6.5)).

Sample	Content of chlorinated structures $C\cdot10^2$, mol/mol of chlorobutyl rubber				a_0/b*	a_1/a_2	a_{sat}/a_0
	a_0	a_1	a_2	a_{sat}			
1	0.52	0.50	0.06	—	0.25	9.2	—
2	1.36	1.1	0.15	0.11	0.65	7.3	0.11
3	1.80	1.5	0.21	0.19	0.86	7.0	0.105
4	2.45	1.6	0.23	0.61	1.17	7.0	0.25
5	2.80	1.4	0.15	1.25	—	8.8	0.45

* Concentration of isoprenyl units in the initial butyl rubber.

Controlled dehydrochlorination of chlorobutyl rubber under the effect of metal carboxylates (IB, IIB, IVA and VIII groups), copper and niobium oxides, applied to activated Al_2O_3, are used for obtaining diene–butyl rubbers and vulcanizates with high cross-link degree. To prevent secondary reactions during HCl elimination, oxides, hydroxides or metal carboxylates, whose halides are insoluble in the reaction medium, CaO, for example, are added.

The allyl Cl atom in chlorobutyl rubber molecule differs by high mobility and can participate in substitution reactions. At the interaction of chlorobutyl rubber with $(CH_3)_3Al$ (298 K, 120 min, 0.8–1 mass% chlorobutyl rubber solution) in solution of methylene chloride, up to 70–75 mass% of labile Cl atoms are substituted by methyl groups [8, 24].

The presence of the Cl atom in the allyl position significantly exceeds the ability of chlorobutyl rubber to vulcanization compared, for example, with butyl rubber; as a result, the choice of vulcanized systems broadens significantly. The latter circumstance is brought about by the presence of two reaction centers of vulcanization (1) and (2):

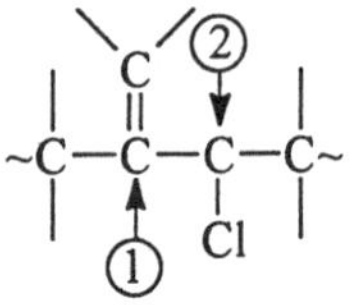

On the level of centers (1), the usual vulcanization systems can be used: sulfur + catalyst, sulfur donors, quinone, dioxime, rubber halides. Vulcanization in the participation of Cl atoms and on the level of carbon (2) can proceed under the effect of zinc oxide, diamines, thiocarbamide, dimercaptanes, dihydroxy-aromatic compounds, etc. [23].

Exchange of mobile Cl atoms for an allyl group promotes decrease of the rate of chlorobutyl rubber dehydrochlorination. Synergic mixtures of alkaline or alkaline-earth metal carboxylates and polyesyers (polyalkylene glycoles) are most efficient. They significantly decrease the rate of HCl elimination from chlorobutyl rubber and prevent gel-formation and color change of the composition.

6.2.1.3. Areas of application.

Usually, strength characteristics of chlorobutyl rubber vulcanizates are lower than those of butyl rubber; in its turn, covulcanization ability of chlorobutyl rubber with natural rubber is much better. Information on chlorobutyl rubbers produced in industry is displayed in Table 6.11.

The main customer of chlorobutyl rubbers is the tire industry [2, 18].

Most rubber is applied to production of the sealing layer of tubeless tires. To provide the maximal strength of connection of this layer with the body of the tire, a combination of chlorobutyl rubber with highly unsaturated elastomer is used. Low gas permeability of the sealing layer from chlorobutyl rubber enables air pressure in the tire body to be reduced, which lowers tire destruction due to stratification of the body. High dynamic strength of the sealing layer under hard exploitation conditions also promotes increase of durability of tires. In the general case, durability of tires, depending on size and type, increases by 50–200%; thereby, it seems possible to use a thinner sealing layer from chlorobutyl rubber.

Table 6.11

Properties of chlorinated butyl rubbers.

Trademark	Company–producer, country	Viscosity by Muni			Unsaturation, mol%	Chlorine content, mass%	Ash content, max., mass %	Mass loss, max., mass%	Stabilizer content, mass%
		373 K (1–8 min)	398 K (1–8 min)	400 K (3 min)					
Esso butyl HT-1066	Ecson (Esso) Chemical Co. (USA, England)	49–58	—	—	1.4	1.1–1.3	0.5	0.7	0.05–0.2
Esso butyl HT-1067		61–70	—	—	1.4	1.1–1.3	0.5	0.7	0.05–0.2
Esso butyl HT-1068		—	45–55	50-60	1.4	1.1–1.3	0.5	0.7	0.05–0.2
Polysar Chlorobutyl 1240	Polysar (Canada)	—	34–42	—	—	1.1–1.3	0.5	0.4	0.05–0.2
Polycar Chlorobutyl 1255		—	45–55	—	—	1.1–1.3	0.5	0.4	0.05–0.2
CBR-155N	USSR	51–59	—	—	1–1.16	1.1–1.3	0.7	0.8	0.2–0.8
CBR-165N	USSR	60–68	—	—	1–1.16	1.1–1.3	0.7	0.8	0.2–0.8
CBR-175N	TU 3840-354-80	—	47–57	—	1–1.16	1.1–1.3	0.7	0.8	0.2–0.8

Notes. 1. As a stabilizer, 2,6-di-*tert*-butyl-4-methylphenol is usually used. 2. To estimate quality of chlorinated butyl rubber, prescription contains (in mass parts) [29]: rubber – 100, stearin – 3, captax – 0.65, thiuram – 1.3, zinc oxide – 5, technical carbon DG-100 – 50, sulfur – 2.0. Reactivity of halogenated isoprene unit is estimated by ability to be covulcanized with natural rubber. For this purpose, 5 mass parts of natural rubber are added to 100 mass parts of the standard mixture (the lower the decrease of strength of vulcanizates with this additive, the better are the properties of chlorinated butyl rubbers).

Addition of 5–10 mass parts of chlorobutyl rubber to facing rubbers based on caoutchoucs of general purpose also promotes decrease of pressure within the body and growth of strength connection with polyamide cord.

Sidewalls of radial tires are deformed more than in usual tires. Consequently, the probability of cracking of sidewalls because of influence of O_3 and fatigue increases. Excellent resistance to ozone and resistance to cracking at numerous deformations of chlorobutyl rubber allow obtaining of high-quality rubbers. Combinations of polymers, in particular, chlorobutyl rubber (35 mass%), ethylene–propylene copolymer (15 mass%) and highly unsaturated rubbers of general purpose (50 mass%), are also very good. High durability under hard exploitation conditions is the definitive property of car tires made from chlorobutyl rubber (heavy vehicles, buses, etc.). In this case, ZnO with low additives of tetrachlorobenzoquinone is used for vulcanization. Chlorobutyl rubber is used in production of digestion chambers, as well as many technical rubber articles. Application of chlorobutyl rubber is specially effective for production of molded articles (high rate of vulcanization, the absence of reversion, heat resistance); heat resistant transportation belts (heat resistance, high resistance to evaporation, high stability of connection with synthetic cords); chemically stable facings of vessels (chemical stability, possibility of vulcanization at low temperature; good stability of connection to various materials); taps for corking pharmacological compounds (simple non-toxic vulcanizing systems); as well as in the food industry, construction and medicine [23].

Contrary to usual butyl rubbers, non-filled vulcanizates of butyl rubber halides do not crystallize at elongation; that is why highly stable vulcanizates cannot be obtained without application of active fillers, the main of which is technical carbon. Other fillers are also applied: silicon dioxide, kaolin, talc, chalk. Resistance to ozone of vulcanizates decreases with growth of filler content (obviously, as a consequence of tension growth in the samples).

Chlorobutyl rubber is also applied to production of intermediate or gluing interlayers of multilayer rubber articles, promoting adhesion increase between layers from butyl rubber and (or) highly unsaturated rubbers, preparation of glues, intended for rubber linkage to metal, etc.

6.2.2. Bromobutyl rubber.

Bromobutyl rubber is the product of butyl rubber bromination, containing 2–3% of bromine, attached in the α–position and by double bonds of isoprenyl units of the macromolecule.

First, bromobutyl rubber of Haikar 2202 brand was manufactored by B.F. Goodrich Chemical Co. (USA). However, this product was not widely applied because of its low stability during storage and high price.

In 1971 a new process of bromobutyl rubber production was developed by the firm Polymer Corporation Ltd., of Sarnia (Canada). This compound is superior to chlorobutyl rubber Polycar Bromobutyl X-2 in some of its characteristics [35]. By 1973–1975 productivity of the plant was 25–35 thousand tons per year [36], i.e. about 75% of the butyl rubber produced in Canada is bromobutyl rubber. At the present time bromobutyl rubber is produced in Belgium, Great Britain and USA. Total productivity of processes producing bromobutyl rubber is about 50,000 tons a year.

6.2.2.1. Structure and physical properties.

Macromolecules of bromobutyl rubber consist of polyisobutylene blocks, statistically separated by saturated and unsaturated bromine-containing isoprenyl units. The structure of bromobutyl rubber macromolecules is not clear yet, but, obviously, the distribution of fragments in them is more complex, than in chlorobutyl rubbers. The reason is that 2–3 halogen atoms enter the reaction of butyl rubber bromination at one double C=C-bond, and only 25–30 mass% of HBr is observed in the reaction products. This proves that the substitution reaction is accompanied by addition. Incidentally, bromobutyl rubber Polycar Bromobutyl X-2 contains 90% of allyl bromine. The degree of unsaturation of the polymer after bromination hardly decreases. No gel is formed. The reaction scheme of bromination suggested is analogous to chlorination of butyl rubber:

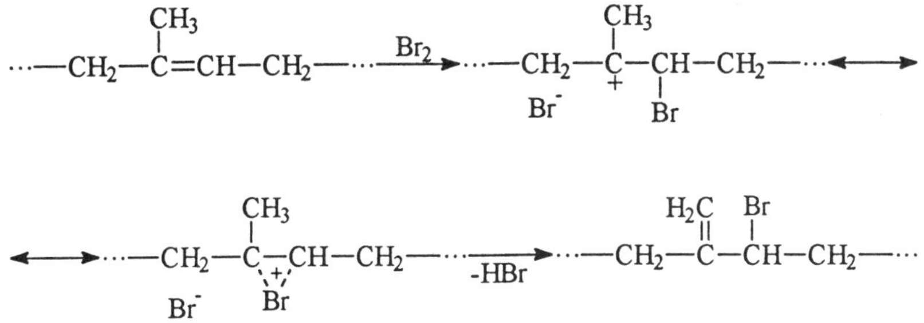

Bromination of the model compound (298 K, 5 min) *cis-:trans*-2,2,4,8,8-pentamethylnonene-4 (48:52) has displayed the presence of four structures in the products. These structures are formed by the scheme of butyl rubber chlorination. Quantitatively, the process shows poor reproduction and is quite sensitive to temperature, concentration of the reagent, reaction duration, working conditions of bromination, and light [21].

Analysis has displayed that in the product containing 1.33 of bromine equivalent, 70–90% of substitution products and 30–10% of addition products of bromine atoms are present. In this case, the relation of structures:

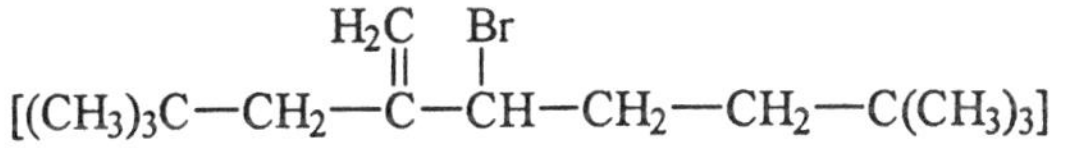

and

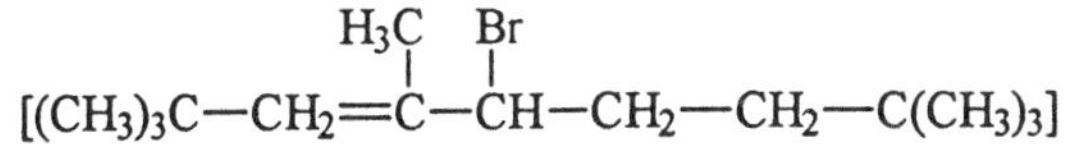

is 6:1, approximately. Contrary to chlorobutyl rubbers, endochains

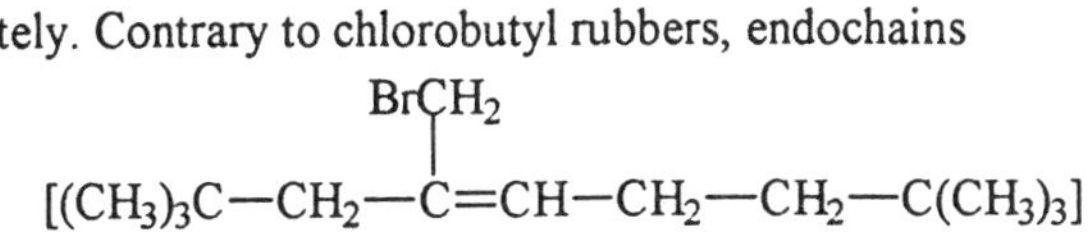

also participate in the composition of macromolecules. These endochains are formed at regrouping of the exo-structure. Exo-endo-regrouping in bromobutyl rubber proceeds more easily, than in chlorobutyl rubbers. The following structure is absent in the composition of macromolecules of bromobutyl rubber:

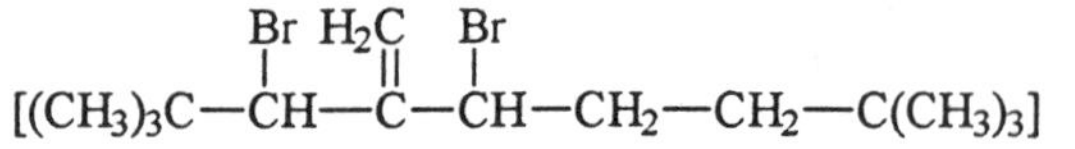

Experimental data indicate that there are probably two mechanisms of butyl rubber bromination – ionic and free-radical ones.

Investigation of industrial bromobutyl rubber Polycar Bromobutyl X-2 has confirmed results of analysis of model compounds. The following structures were identified

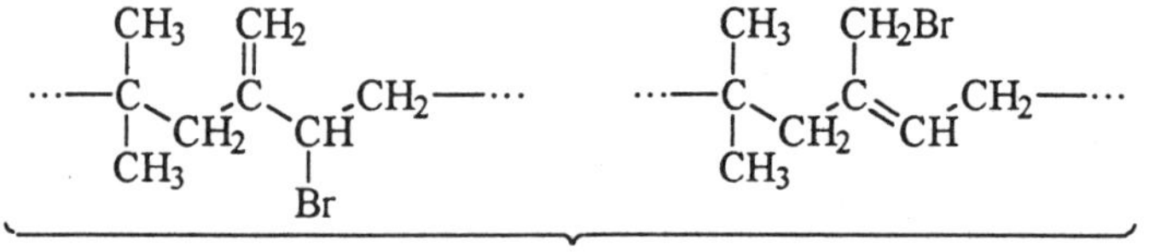

30-82%   18-70%

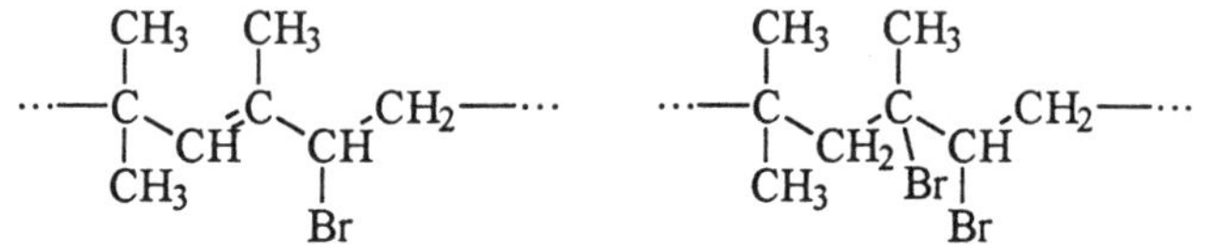

5%   10%

This polymer differs by high stability. Consequently, bromobutyl rubber is low heat aged: after exposure of the rubber vulcanizate at 450 K for 22 hours, strength at elongation is approximately 20-fold greater than the analogous index for butyl rubber.

Bromobutyl rubber is readily soluble in aliphatic and aromatic hydrocarbons. Introduction of bromine into butyl rubber sharply improves its adhesive properties, much better than at chlorination of butyl rubber (see Table 6.8).

6.2.2.2. Chemical properties.

The main chemical properties of bromobutyl rubbers are analogous to those of chlorobutyl rubbers.

The bromine atom in the rubber is more mobile than the chlorine atom. This produces a high rate of vulcanization of bromobutyl rubber, which grows with the content of bonded bromine (Figure 6.4), in the presence of the same vulcanizing systems. Double bonds in bromobutyl rubber are more active than in butyl rubber and chlorobutyl rubber, which is confirmed by its ability to be vulcanized by just sulfur in the absence of metal oxides and accelerators.

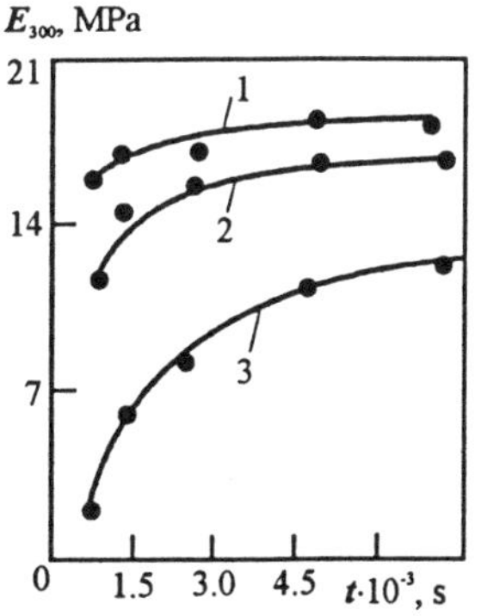

Figure 6.4. Influence of halogenation on the rate of butyl rubber vulcanization (415 K):
1 – bromobutyl rubber ([η] = 1.46, iodine number is 4.8, bonded halogen – 2.13 mass%); 2 – chlorobutyl rubber ([η] = 1.40, iodine number is 4.7, bonded halogen – 1.27 mass%); 3 – butyl rubber of Enjey Butyl 268 trademark ([η] = 1.52, iodine number is 11.0).

6.2.2.3. Application of bromobutyl rubber.

Practically, bromobutyl rubber is applied in the same fields, as chlorobutyl rubber, but unlike the latter, the former possesses increased adhesion to other rubbers, plastics and metals, atmosphere- and thermal resistance, and improved amortization properties [37–39]. Thanks to good compatibility with unsaturated elastomers, it covulcanizes with natural rubber, butadiene–styrene, chloroprene, triple ethylene–propylene, nitrile and other rubbers in any relations, which enables properties of rubbers to be regulated over a wide ranges.

The main customer of bromobutyl rubber is the tire industry. It is used for preparation of the internal layer of tubeless tires, heavy vehicle tubes, as sidewalls of tires. Optimal properties for production of tire sidewalls are displayed by the mixture of 60% of bromobutyl rubber, 20% of triple ethylene-propylene copolymer, and 20% of natural rubber (combination of adhesion, resistance to ozone, elasticity and other properties). For internal sealing layer of tires, a mixture of 75% bromobutyl rubber and 25% natural rubber is recommended (thermal resistance below 425 K).

Bromobutyl rubber is also used in production of various rubber technical articles, for example, heat resistant transportation belts, insulation for electric cables, shock absorbers, linings, structuring thickeners, in pharmacology, in production of soles of shoes and boots basements, in the food industry, medicine, etc..

Bromobutyl rubber is widely applied in production of intermediate or gluing interlayers in multilayer rubber articles, promoting growth of the bond strength between butyl rubber layers from butyl rubber and unsaturated rubbers, in production of glues, dedicated to rubber mounting on metal, as well as additives for activation of butyl rubber vulcanization by alkylphenolformaldehyde resins. It is also possible to apply bromobutyl rubber to obtain self-extinguishing articles, used in construction, productions of detail for automobiles, etc. These conditions are fulfilled, for example, by a composition consisting of 80% bromobutyl rubber and 20% neoprene with additives of chlorinated wax and antimony trioxide. Main technological properties of bromobutyl rubber are shown in Table 6.8.

6.2.3. Other modified butyl rubbers.

Other modified butyl rubbers are not as well developed yet as halogenated butyl rubbers. Among possible rubbers, the following butyl rubbers should be mentioned: bromine chlorinated, perhalogenated, highly chlorinated, and other butyl rubbers [18].

Halogenation with introduction of Cl and Br is performed by bromine monochloride (BrCl). The relation of bonded bromine and chlorine in the polymer is 4:1 instead of expected 1:1. Decrease of polymer viscosity at halogenation is 5-15% and depends on the molecular mass of the initial butyl rubber. At 127°C, the Muni viscosity of chlorobromorubber is 40 or higher. Chlorobromobutyl rubber, containing 1.6–1.8 mass% of Br_2 and 0.18–0.20 mass% of Cl_2, exceeds chlorobutyl rubber in its ability to be covulcanized with highly unsaturated rubbers and bromobutyl rubber and resistance to detachment of hydrogen halide.

Perhalogenation is performed in three stages: halogenation of butyl rubber, dehydrohalogenation of chlorinated and bromobutyl rubbers, and repeated halogenation of diene butyl rubber. Unlike the usual halogenated butyl rubber, the perhalogenated one contains two halogen atoms per unit of isoprene and is of the following structure:

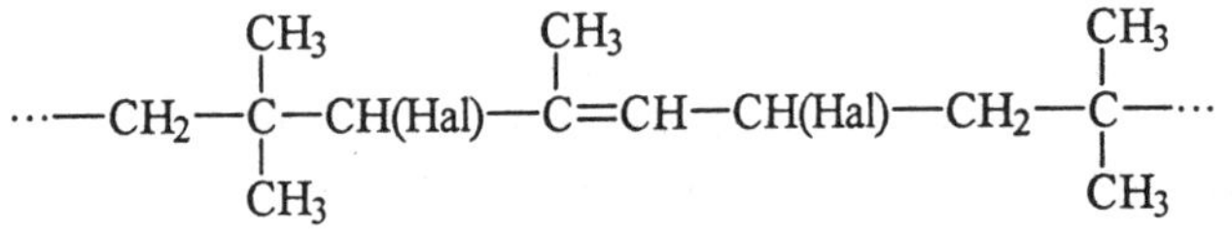

Chlorination of butyl rubber in order to obtain the highly chlorinated product (1.5–1.9 Cl atoms per isoprene unit) is performed in the presence of aliphatic diamines or tertiary diarylamines. The product outpreforms chlorobutyl rubber in vulcanization rate, adhesion and ability to covulcanization with highly unsaturated polymers.

Production of modified types of butyl rubber can be also based on other chemical reactions.

Regulation of butyl rubber cross-linking during polymerization is caused by strictly dosed addition to stock of dienes with nonconjugated double bonds, for example, divinylbenzene, dimethallyl, etc. In accordance with the amount of diene, soluble, structured butyl rubbers, as well as cross-linked products with various gel content (up to 80%) are obtained. Cross-linked butyl rubber possesses lower creep, higher ability for reduction, and somewhat improved physicomechanical indexes of vulcanizates. Decrease of the extrusion rate of stocks from cross-linked butyl rubber can be prevented by increasing the diene dosage (above 4%).

Rubbers with high heat resistance are produced at vulcanization of butyl rubber by peroxides. Application of cross-linked butyl rubber allows improvement of properties of digestion chambers, automobile articles, in particular, thickeners for glass, sheet coverings with increased resistance to weather and ozone.

Sulfonation of butyl rubber leads to formation of sulfobutyl rubber [40]. The reaction is conducted under mild conditions (293 K, 20 min) in a hydrocarbonic solvent in the presence of SO_3 complex – triethyl phosphate, acetyl sulfates and other sulfonating agents. In analogue to halogenation, isoprenyl units are modified; SO_3H-groups dislocate in the allyl position to double bonds. The polymeric product contains 0.5–1.5 mol% of SO_3H-groups. As they are neutralized by NaOH, rigidity, carcass structure and cohesive strength of the rubber are improved, and at content of SO_3Na-groups over 0.6 mol%, the rubber acquires properties of an ionomer.

Sulfobutyl rubber displays lower creeping and cold flow, higher stability in non-filled mixtures, adhesion in metals, and improved bulk modulus. Water dispersions of neutralized rubber are used as linkages and coatings. Combined with butyl rubber and chlorobutyl rubber, sulfobutyl rubber increases the basic skeletal structure of articles.

Sulfochlorination of butyl rubber proceeds according to the mixed mechanism of chlorination and sulfochlorination, and is accompanied by a significant degradation of the main chain of the polymer.

The synthesis of polymeric products containing carboxy groups is based on the Diels–Alder reaction. Preliminarily, chlorobutyl rubber is dehydrochlorinated down to diene butyl rubber. The resulting system of conjugated C=C-bonds in macromolecules easily interacts with maleic anhydride. At neutralization of anhydride groups by solutions of metal salts, amines or their derivatives, an ionostructured polymer (ionomer) is formed, characterized by increased resistance to ozone [18].

Aromatic mononitrosocompounds interact with butyl rubber forming polymeric products, which form the basis for obtaining vulcanizates with increased strength at elongation, improved bulk modulus, and deformation at elongation after filling and vulcanization, as well as with resistance to ozone, electric resistance and compatibility with other rubbers, resins, adhesion to tire cord, metals, fabric, paper, etc. [18]. These elastomers are recommended for application as impregnating compounds for fibrous natural and synthetic materials — tire cord, for example.

In the presence of aromatic dinitrosocompounds, for example, N-methyl-N-4-dinitrosoaniline, the modulus, blow viscosity, resistance to wear and electric resistance of modified butyl rubber increase, and the hysteresis decreases. Moreover, the polymeric product is easier to process [18].

Isocyanates and polyisocyanates promote formation of a three-dimensional network structure, which prevents fluidity of butyl rubber. Modification by isocyanates promotes compatibility of butyl rubber in any relations with natural or synthetic rubbers, in particular, with butadiene–styrene, chloroprene and acrylonitrile ones.

Nitrogen-, phosphorus- or oxygen-containing organic compounds, for example, acrylonitrile, methacrilonitrile, vinylpyridine and its derivatives, acrylic and methacrylic esters, vinylisobutylic ester, vinylacetate, mercaptobenzoic acid, form grafted copolymers at the interaction with butyl rubber, initiated by organic peroxides. These copolymers can be used as adhesives and glues to fix butyl rubber to natural and synthetic fibers, metals, and various elastomers. It is stated that butyl rubber is modified at the interaction with anhydrides of organic acids and aldehydes, as well as by carboxylation, oxidation and epoxidation [18]. Practical application of these polymeric products is still limited. Of special interest are mixtures of butyl rubber and its halogen-derivatives with other elastomers.

6.2.4. Mixtures of halogenated butyl rubbers with other elastomers.

Butyl rubber does not bond with highly unsaturated compounds without application of specific methods. Usually, spongy vulcanizates are obtained already in the presence of 1–5 mass% of highly unsaturated additives.

Halogenated butyl rubbers differ by increased compatibility at vulcanization with other elastomers of equal or higher unsaturation, which enables its useful properties to be widely regulated at combination of chlorinated and bromobutyl rubbers with other elastomers during vulcanization by zinc oxide [17, p. 280]. It is hardly probable that in this case the real covulcanization of polymeric mixtures with formation of chemical bonds between phases of polymers takes place. In most consisting cases, heterogeneous systems, consisting of dispersion areas sized within 0.5–6 μm, are formed at mixing of elastomers with halogenated butyl rubber (on rolls, in a mixer, via solutions, etc.). Heterogeneity of the systems increases at mixing of elastomers with different viscosity.

In the general case, compatibility of pairs of polymers correlates with parameters of solubility and polymer–solvent interaction or with surface tension of polymers (Table 6.12) [41].

Table 6.12

Parameters of solubility, polymer–solvent interaction and surface tension of some rubbers.

Index	Butyl rubber	Ethylene–propylene rubber	Butadiene –styrene rubber	Natural rubber	1,4-*cis*-polybuta-diene rubber	Poly-chloro-prene, neoprene
Surface tension, N/m	0	0.02	0.06	0.04	0.05	0.11
Polymer–solvent interaction parameter	0	0.02	0.180	0.157	0.264	0.315
Solubility parameter, $(J/cm^3)^{1/2}$	0	0.23	1.0	0.53	1.23	2.07

Microscopic investigations have shown that chlorobutyl rubber and bromobutyl rubber are quite compatible with butyl rubber only. Phases of polymers with ethylene–propylene rubber are of the undefined form, and the phase boundary is blurred. For all other elastomers, a clear boundary is observed, the dispersion area usually being spherical.

Halogenated rubbers may be mixed with natural rubber in any relation. Introduction of butyl rubber into natural rubber improves resistance to ozone of rubbers, skeletal structuring of rubber mixtures, their confection stickiness, adhesion to highly unsaturated rubbers after vulcanization, exploitation properties of vulcanizates at low (below zero) temperatures, resistance of articles to weather conditions, improve resistance to crack growing at multiple deformations of rubbers, air permeability of vulcanizates (Figure 6.5) [18, 42]. For example, substitution of

25% of bromobutyl rubber by natural rubber decreases friability temperature of vulcanizates by 8–10°.

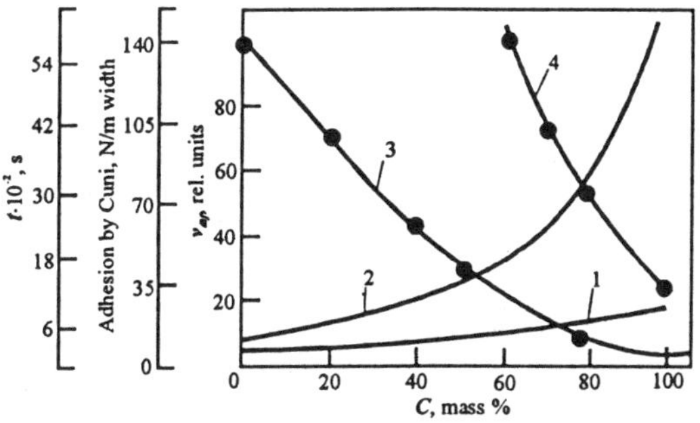

Figure 6.5. Influence of halogenated butyl rubber content in the mixture with natural rubber on indexes of ozone resistance of rubbers (concentration 0.2 vol.%, elongation 50%) according to formation of visible cracks (1) and destruction of the sample (2) and on the relative rate of air diffusion (3), as well as on adhesion (373 K) to rubber, based on the mixture of 50% of natural rubber and 50% butadiene–styrene rubber (4): 1, 2 — for chlorobutyl rubber; 3, 4 — for bromobutyl rubber.

Mixtures containing up to 10 mass% of natural rubber are used for preparation of digesting chambers, and the sealing layer with increased heat resistance for tubeless tires for vehicles. Radiation of vulcanizates of bromobutyl rubber and natural rubber mixture by accelerated electrons increases their cohesive stability with some decrease of fatigue endurance and enables mass of the sealing layer to be decreased while preserving its thickness and cord shift at mounting and vulcanization of tires [43]. Mixtures with 30–40 mass% of natural rubber are used for preparation of the internal layer of tubeless tires.

Butadiene–styrene rubbers mix with a halogenated butyl rubber in any proportion. They increase adhesion after vulcanization more than natural rubber, and increase resistance to ozone and heat resistance [44].

Butadiene–nitrile rubbers are less compatible with halogenated rubbers than others discussed above. The mixture contains up to 40% of butadiene–nitrile rubber. If 20 mass% of rubber with high concentration of acrylonitrile (SKN-40M) is introduced, solidity and resistance to swelling of vulcanizates in hydrocarbonic media increase, and 25 mass% gives rise of oil resistance of rubbers [45]. Combined with neoprene, butadiene–nitrile rubber is used in rubber mixtures with chlorobutyl rubber, applied in production of filler plates, collars, joint rings and other articles. Properties of mixtures of halogenated butyl rubbers and butadiene–nitrile rubbers depend on the degree of homogenization.

Chloroprene rubbers mixed with halogenated butyl rubbers improve oil and fire resistance of the articles. The mixture of halogenated butyl rubbers with polychloroprene in the relation 80:20, containing standard antiprenes (chlorinated waxes and antimony trioxide), possesses self-extinguishing properties; halogenated rubbers are not fireproof. Resistance to oil grows in proportion to chloroprene concentration in the mixture and vulcanization degree (Figure 6.6). However, highly cross-linked vulcanizates possess decreased relative elongation at break [18].

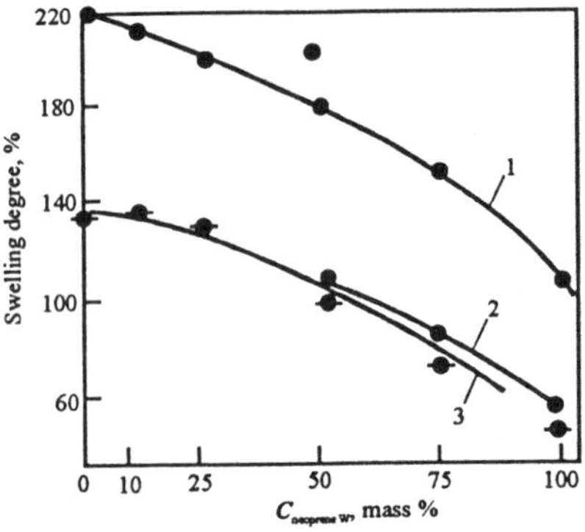

Figure 6.6. Influence of content of neoprene *W* in the mixture with chlorobutyl rubber on resistance to oil of vulcanizates, obtained with various vulcanizing systems: 1 — TMTDS/DBTDS[3]; 2 — permalux; 3 — mercaptoimidasoline (oil No. 3 by ASTM 704, 373 K).

Ethylene–propylene triple elastomers with halogenated butyl rubbers differ by high homogeneity of mixtures, although in this case diffusion boundaries between separate phases are also observed [39]. Vulcanizates possess low air and humidity permeability, good adhesion, and excellent resistance to weather conditions, ozone and cold. Rubbers may be mixed in any proportion.

Of special interest are addition of halogenated butyl rubbers to mixtures of natural rubber with ethylene–propylene copolymer, which are poorly covulcanized with each other. Introduction of a halogenated butyl rubber creates vulcanizates with good dynamic properties (durable at multiple deformations) combined with high resistance to weather conditions and ozone, and good adhesion to highly unsaturated rubbers (Table 6.13) [18].

Butyl rubber mixes well with its halogen-derivative products in any proportion. Vulcanizates are characterized by decreased residual elongation. Heat aging coefficient of mixtures grows with concentration of halogenated butyl rubber. Introduction of halogenated rubbers increases the rate of vulcanization of butyl rubber with low nonsaturation causing no decrease of the resistance to ozone, and improves the latter of rubbers based on butyl rubber with high unsaturation. Substitution of 25 mass% of low unsaturated butyl rubber by bromobutyl rubber significantly increases heat resistance of vulcanizates under harsh conditions of aging, decreases creeping of rubbers as a result of thermal degradation decrease, but deteriorates resistance to humidity, because vulcanizates of halogenated butyl rubbers absorb H_2O much better than vulcanizates of the usual butyl rubbers.

Mixtures of halogenated butyl rubbers with dienebutyl rubbers, obtained from them, are easily processed and are used for preparation of a sealing layer of tubeless tires with high strength at break. The composite from partially dehydrochlorinated chlorobutyl rubber and crystal polyisobutylene, vulcanized by high-energy particles, possesses higher strength indexes comparing with irradiated individual rubbers.

Chlorosulfopolyethylene combined with chlorobutyl rubber displays improved resistance to oil and ozone and is applied as adhesive to fix chlorosulfopolyethylene

[3] Tetramethylthiuramdisulfide/dibenzylthiazolyldisulfide.

and various rubbers to metals and fabrics, to cover fabrics and non-fabric materials [35, 46].

Table 6.13

Properties of vulcanizates, based on mixtures of chlorobutyl rubber (butyl NT-10-68) with ethylene–propylene copolymer and natural rubber.

Index	Composition of mixture, mass parts				
Chlorobutyl rubber	—	15	20	25	30
STEPR[4]	25	20	20	20	20
Natural rubber	75	65	60	55	50
Vulcanization system	A	A	B	B	B
Properties of vulcanizates (vulcanization regime: 30 min, 425 K)					
Hardness by Shore	63	53	54	54	57
Modulus at elongation by 300%, kPa	—	0.65	0.65	0.66	0.69
Strength at break, kPa	1.2	1.2	1.1	1.05	1.1
Relative elongation at break, %	430	490	490	450	470
Statistic resistance to ozone (elongation by 20%, ozone concentration $0.5 \cdot 10^{-4}$%, 311 K, 24 hours) or time before formation of cracks, h	≥ 500	148	≥ 500	≥ 500	≥ 50
Dynamic resistance to ozone (elongation by 25–30%, ozone concentration $0.5 \cdot 10^{-4}$%, 311 K, 30 cycle/min) or time before formation of cracks, h	0.5	1	1	16	30
Dynamic endurance (bending angle 180–60°, 343 K, 300 cycle/min) or time before formation of cracks, h	1	≥ 46	24	≥ 138	≥ 138

Note. Prescription contains (in mass parts): rubbers — 100, technical carbon HAF-LS — 20, paraffin oil — 10, wax — 3, stearic acid — 1. The vulcanization system contains (in mass parts): system A - ZnO — 5.0; sulfur — 1.1; TMTDS — 0.2; DBTDS — 1.5; system B – ZnO — 5.0; sulfur — 0.5; DBTDS — 0.75; alkylphenol disulfide — 1.25.

Beside selective vulcanization, unsaturated elastomers (SKI-3, SKD, and others) combined with halogenbutyl rubbers are capable of covulcanization [47]. Resulting modified rubbers, which contain 10–30 mass% of halogenated rubber, exceed standard samples by resistance to tearing, ozone, heat and atmospheric aging, gas permeability, strength of connection to cord, and dynamic endurance [48]. Combination of these rubbers in different proportions (chlorobutyl rubber:SKD = 70:30) improves resistance to cold and relative elongation at break for vulcanizates of low-saturated elastomer. It was shown [49] that the maximal strength of polyisoprene rubber vulcanizate and chlorobutyl rubber is reached in microheterogeneous systems, when vulcanizing group provides covulcanization of rubbers on the phase boundary.

[4] Synthetic triple ethylene–propylene rubber.

References to Chapter 6

1. Encyclopedia of Polymers, 1972, v. 1, 351–364, Moscow, Sov. Encyclopedia. (Rus)
2. Prokofiev, Ya.N., Shcherbakova, N.V., and Bugrov, V.P., In: Synthetic Rubber, 1976, pp. 342–355, Leningrad, Khimia. (Rus)
3. Study and Development of Production Technology of Monomers and Synthetic Rubbers, Collection, 1981, p. 169, Moscow, CNIITENeftekhim. (Rus)
4. Guizard, C. and Cheradame, H., Europ. Polymer J., 1979, v. 15(7), pp. 689–693.
5. Beresnew, W.W., Sapirov, I.N., and Kirpitschnikov, P.A., Plaste und Kautschuk, 1975, Bd. 22(5), S. 397–399.
6. Kirpichnikov, P.A., Highmolec. Comp., 1979, Ser. A, v. 21(11), pp. 2457–2468. (Rus)
7. Beresnev, V.V., Zapirov, I.N., and Kirpichnikov, P.A., Kouchuk i Resina, 1975, No. 6, pp. 6–8. (Rus)
8. Kennedy, J.P. and Phillips, R.R., J. Macromol. Sci., 1970, Part A, v. 4(8), pp. 1759–1784.
9. Ivanova, S.R., Romanko, T.V., Shaekhova, V.G., Sangalov, Yu.A., Berlin, Al.Al., and Minsker, K.S., Highmolec. Comp., 1985, Ser. A, v. 27(2), pp. 244–250. (Rus)
10. Edwards, D.C., Hopkins, W., and Kumbhani, K.J., Polym. Mater. Sci. And Eng., Proc. ACS Div. Polym. Mater.: Sci. And Eng., v. 57, Spring Meeting, Denver, CO, 1987; Washington, D.C., 1987, pp. 69–76.
11. Edwards, D.C., Hopkins, W., and Kumbhani, K.J., Abstr. Pap., 194th ACS Nat. Meeting (Amer. Chem. Soc.), New Orleans, LA, Aug. 30 – Sept. 4, 1987; Washington, D.C., p. 1062.
12. Labutin, A.L., Anticorrosion and Sealing Materials Based on Synthetic Rubbers, 1982, 213 p., Leningrad, Khimia. (Rus)
13. Ritchie, K., Hule Mex. y. Plast., 1980, v. 36(4), pp. 5–18.
14. Beresnev, V.V., Serobyan, A.K., and Kirpichnikov, P.A., Chemistry and Technology of Petroleum and Oils, 1978, No. 8, pp. 23–26. (Rus)
15. Beresnev, V.V., Kauchuk i Resina, 1981, No. 3, pp. 10–12. (Rus)
16. Beresnev, V.V., Kauchuk i Resina, 1981, No. 5, pp. 26–28. (Rus)
17. Polymeric Mixtures, (Transl. From English), Ed. by Pole D., Newman S., 1981, v. 1, 550 p., Moscow, Mir. (Rus)
18. Shmarlin, V.S., Synthesis, Properties and Application of Modified Butyl Rubbers, Ser.: Industry of Synthetic Rubbers, 1973, p. 80, Moscow, CNIITENeftekhim. (Rus)
19. Mikhailovski, G.A., In: Industry of Synthetic Rubbers, 1974, No. 9, pp. 12–15, Moscow, CNIITENeftekhim. (Rus)
20. Ind. Della Gomma, 1972, v. 16(7), pp. 36, 37, 56, 57; Industry of Synthetic Rubbers, Expr.–Inform., 1973, No. 3, Ref. 3, Moscow, CNIITENeftekhim. (Rus)
21. Van Tongerloo, A. and Vukov, R., In: Proc. Intern. Rubb. Conf., Milan, 1979, pp. 70–79.
22. Yablonski, O.P., In: Chemistry of High–molecular Compounds and Oil Chemistry, 1973, p. 127, Ufa. (Rus)
23. Knaepen, M., Plastica, 1980, v. 33(9), pp. 286–291.
24. Ivan, B. *et al.*, Polymer Bull., 1980, Bd. 2(7), S. 461–467.
25. Minsker, K.S., Gataulin, R.F., Lisitski, V.V., Berseneva, I.S., Salnikov, S.B., and Shmarlin, V.S., Highmolec. Comp., 1983, Ser. B, v. 25(8), pp. 1686–1690. (Rus)

26. Pebalk, A.V., Kauchuk i Resina, 1972, No. 8, pp. 51–52. (Rus)
27. Minsker, K.S., Berlin, Al.Al., and Lisitski, V.V., Highmolec. Comp., 1976, Ser. B, v. 18(1), pp. 54–58. (Rus)
28. Lisitski, V.V., Yanborisov, V.M., Berlin, Al.Al., and Minsker, K.S., Highmolec. Comp., 1982, Ser. A, v. 24(5), pp. 931–938. (Rus)
29. Grechanovski, V.A. and Poddubny, I.Ya., 1968, Ser. B., v. 10, pp. 626–629. (Rus)
30. Minsker, K.S., Gataullin, R.F., Yanborisov, V.M., Krasnova, T.A., Salnikov, S.B., and Shmarlin, V.C., Highmolec. Comp., 1984, Ser. A, v. 26(4), pp. 781–784. (Rus)
31. Shmarlin, V.S., Coll. Sci. W. of SRI of Monomers for Synthetic Rubber, 1981, No. 4, pp. 149–154. (Rus)
32. Dodson, B. and McNeill, I.C., J. Polymer Chem. Ed., 1974, v. 12(10), pp. 2305–2310.
33. Minsker, K.S., Zaikov, G.E., and Kolesov, S.V., Aging and Stabilization of Polymers Based on Vinyl Chloride, 1982, 272 p., Moscow, Nauka. (Rus)
34. Eizner, Yu.E. and Erusalimski, B.L., Electronic Aspect of Polymerization Reactions, 1976, 180 p., Leningrad, Nauka. (Rus)
35. Chem. Eng. News, 1972, v. 50(34), p. 5.
36. Europ. Chem. News, 1972, v. 21(524), p. 13.
37. Law, C., Rubb. World, 1974, v. 170(5), pp. 48–49, 52–53.
38. Veen, W.J., Plastica, 1975, v. 28(6), pp. 236–238.
39. Walker, J., Jones, R.H., and Feniak, G., Rubb. Age, 1976, v. 108(8), pp. 33–39.
40. Kabina, T.S., Kauchuk i Rezina, 1978, No. 1, pp. 6–9. (Rus)
41. Gardiner, J.B., Rubb. Chem. Technol., 1970, v. 43(2), pp. 370–379.
42. Gardiner, J.B., Rubb. Chem. Technol., 1968, v. 41(5), pp. 1312–1323.
43. Mohammed, S.A.H., Timar, J., and Walker, J., Radiat. Phys. and Chem., 1981, v. 18(3–4), pp. 433–444.
44. Vulcanization of Elastomers, Ed. by G. Alliger, I. Suetun, (Transl. From English), 1967, 428 p., Moscow, Khimia. (Rus)
45. Production of Tires, Technical Rubber and Technical Asbestos Articles, Collection, 1978, No. 6, pp. 1–4, Moscow, CNIITENeftekhim. (Rus)
46. Ronkin, G.M., Chlorosulfonated Polyethylene, Them. Review, 1977, 101 p., Moscow, CNIITENeftekhim. (Rus)
47. Levit, E.Z., Ognevskaya, T.E., and Dedusenko, V.N., Kauchuk i Resina, 1982, No. 8, pp. 11–13. (Rus)
48. Turina, V.E. and Shwartz, A.G., Kauchuk i Resina, 1975, No. 3, pp. 27–29. (Rus)
49. Tager, A.A., Adamova, L.V., Shershnev, V.A., Yurlovski, V.D., Lifshits, S.S., and Shigaikina, N.I., Highmolec. Comp., 1989, Ser. A., v. 31(11), pp. 2327–2332. (Rus)

Chapter 7. Industrial production of isobutylene polymers.

7.1. Standard technologies for production of oligo- and polyisobutylene.

Oligo- and polyisobutylene with molecular mass of 300–225,000 are synthesized at temperatures from 273 to 173 K.

In modern production, polymer and oligomers of butylenes are obtained by continuous technology with application of volumetric polymerization reactors of mixing, rather complex in construction, with a working volume of 1.5–30 m^3 [1, 2].

The presence of intensively developed thermally controlling surface (several internal heat exchangers combined with external heat elimination with liquid ethylene and ammonia) is thought obligatory, as well as of powerful mixer, which provides linear rates of the reaction mass motion of 1–10 m/s.

Reactors of Amoco Company (USA) and Lenkhimmash are typical (Figure 7.1).

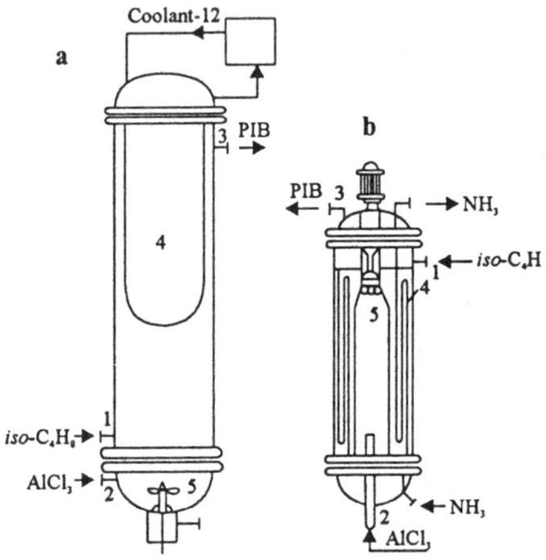

Figure 7.1. Volumetric reactors–polymerizers of the ideal mixing of Amoco company (a) and Lenkhimmash (b); branch pipes of feeding of the monomer (1), catalyst (2) and polymerizate release (3); 4 — thermally controlling elements; 5 — mixing device.

As was imagined in theory, to exclude intensive temperature overloads during polymerization, the reactor provides multiple recirculation of polymerizate in the volume under relatively low (less than 15–20 mass%) of the monomer conversion in one operation. In some cases, monomer content in the initial raw material is limited (up to 21 mass%), which must provide smoother proceeding of the reaction, but requires high expenses of energy for pumping, elimination, cooling down and return of the solvent (often isobutylene) [39].

In various productions raw materials can be used, the composition of which changes from highly purified isobutylene to isobutane–isobutylene mixtures and fractions of C_4 hydrocarbons. Preferably, applied catalysts are solutions of $AlCl_3$ in haloid-hydrocarbons or arenes combined with H_2O, C_2H_5Cl, CH_3Cl and BF_3 with H_2O or R_2O, etc., as well as some modifications of them. The process is conducted preferably under increased and low pressures with application of internal (boiling) or external (cooling agent) heat elimination, various methods of deactivation and regeneration of both catalyst and non-reacted monomer.

A specific feature of fast chemical reactions is significant difficulties in providing isothermal regime of the process conduction under industrial conditions. Usually, in this case, local heating up of the reaction mass in the reaction zone and heterogeneous spatial-temporal distributions of temperature, monomer conversion, and concentrations of active centers and the monomer take place.

Existing technologies include the following general stages [2]: 1) Preparation of the initial raw material (purification, elimination of humidity and inhibiting admixtures, raising the monomer concentration in the blend up to the required content, etc.); 2) Preparation and production of the catalyst; 3) Polymerization of the monomer up to oligomers or polymeric products; 4) Deactivation, elimination and regeneration of the catalyst; 5) Degasification and purification of the polymer from catalyst; 6) Regeneration of non-reacted components of the reaction (monomer, solvent, etc.); Separation (with utilization) of sub-standard oligomeric and/or polymeric products.

Technologically, the most difficult stage in industrial production is polymerization of isobutylene. Therewith it should be taken into account that volumetric reactors of the ideal mixing used operate in a non-optimal and uneconomic "flare" regime (see Figure 3.10, c–f). The process significantly depends on even low variations of the composition and temperature of the initial raw material and the catalyst, the method of catalyst addition into the reaction zone, efficiency of mixing, etc. This, most of all, is what which determines instability of the reactor operation, difference in properties of resulting polymer with time, frequent noncorrespondence of properties to technical requirements, and so on, which leads to formation of relatively large amounts of sub-standard products and technological wastes, which are usually burnt off.

Kinetic features of the process in volumetric reactors are such that the temperature in microareas of the reaction zone is always higher than that of the reaction medium. This predetermines formation of a polymer with lower yields than those from calculations based on kinetic data, lower molecular mass and wider MMD. Relatively wide MMD (up to 5-6-fold and more) usually causes rise of viscosity of the resulting polymer, which is undesirable, as a rule.

7.1.1. Production of high-molecular polyisobutylene.

In industry, high-molecular polyisobutylene is produced by two main methods. Polymers of Oppanol type are obtained by polymerization of isobutylene in the presence of BF_3 in the medium of boiling ethylene. In production of polymer of Vistanex type, the process is conducted in the presence of $AlCl_3$ in ethyl chloride or methyl chloride solution.

Producing polyisobutylene in the medium of liquid ethylene, heat released during polymerization is eliminated at the sacrifice of solvent evaporation, which enables low temperature to be preserved in the reaction zone (the boiling point of ethylene under atmospheric pressure is 169 K), required for obtaining of quite high-molecular polymer. Polyisobutylene with a molecular mass over 100,000 can be obtained at temperatures below 188 K (Figure 7.2).

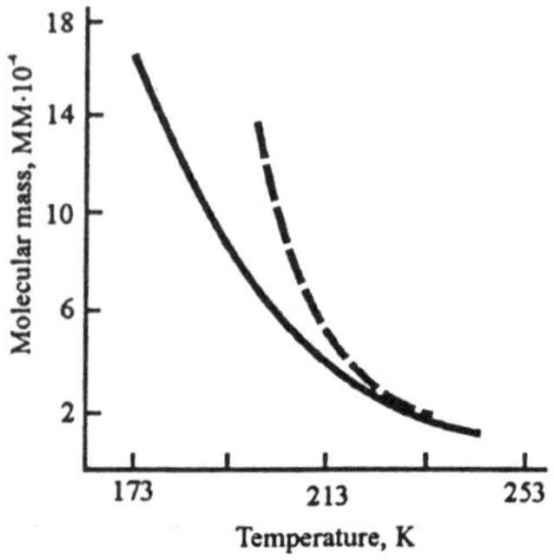

Figure 7.2. Influence of temperature on the molecular mass of polyisobutylene during polymerization in the medium of ethylene (—) and ethyl chloride (——).

High-molecular polyisobutylene of Oppanol type (Badische Anilin und Soda-Fabrik A.G., FRG) with molecular mass of 80,000–225,000 is produced on a continuous moving belt 35–45 cm wide and 800–1000 cm long (Figure 7.3) in the medium of boiling ethylene in the presence of BF_3 at temperatures below 190 K. Catalyst BF_3 (T_{boil} = 172 K) is easily separated from admixtures, in particular, from accompanying SO_2 (T_{cr} = 200.6 K) and is eliminated from the polymer at mixing and heating up [1, p. 85].

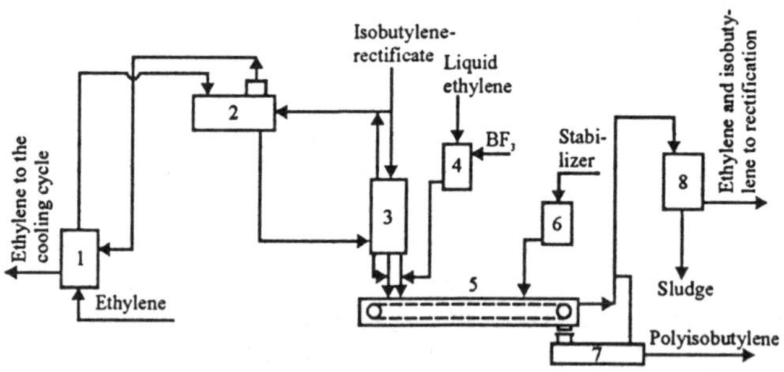

Figure 7.3. Scheme of isobutylene polymerization process in the medium of ethylene: 1, 3 — coolers; 2 — separator; 4, 6 — batchers; 5 — polymerizer; 7 — degasifier; 8 — adsorber.

Isobutylene rectificate, purified from admixtures (99.99 mass%) and cooled down to 233 K, is added into the pipe space of the cooler (3), where it is cooled down to 188 K by liquid ethylene, added from the separator (2). Vapors of ethylene from the separator (2) are directed through interpipe space of recuperative heat exchanger (1) to compression and return to the process through the pipe space of the cooler (1), separator (2) and interpipe space of the cooler (3).

Liquid ethylene releases from the interpipe space of the cooler (3) by two flows: the first flow is mixed, in the assigned relation, with cooled down isobutylene in the pipeline and is directed to the moving belt of the polymerizer (5); the second flow is mixed with the catalyst, BF_3, which is prepared in the batcher (4). Diluted catalyst (0.3% BF_3) is delivered onto the belt of the polymerizer (5) – a cylindrical metal hermetically sealed box in which the infinite belt of a special stainless steel is moving at the rate from 0.25 to 1 m/s. The polymerization time varies from parts of a second to several seconds. Time of rubber presence on the belt may vary from 9 to 36 seconds. The greater part of the ethylene is evaporated at the sacrifice of heat released during polymerization, and a layer of the polymer, 2–3 cm thick, remains on the belt. At the end of the horizontal part of the belt pathway, a solution of stabilizer[1] in isobutylene is added from the batcher (6) into the polymerizer. The polymer is removed from the belt by a special knife and is directed for degasification into two warm mixer-presses (7), heated by vapor, in which homogenization of the polymer proceeds simultaneously. After processing (if necessary) on hot rolls (415–425 K) for 5–10 min, the ready product is packed.

Gaseous ethylene and isobutylene, released from the polymerizer (5), are united with the products, released at degasification, and are directed to the adsorber (8) for residual BF_3 (0.2–0.3%) and isobutylene oligomers to be absorbed, after which they are compensated and directed to rectification.

The great problem in production of high-molecular polyisobutylene is elimination of "overgrowing" of polymerizer walls by a layer of the polymer, which complicates continuous conduction of the process and makes it necessary to stop the process in order to clean the polymerizer. Hence, practically, a periodic process with a cycle of, at best, several days is realized. To perform continuous technology of production of polyisobutylene, it is appropriate to have at least two or even several polymerizers, operating in parallel.

It has been found that the polymerization rate and the yield of polyisobutylene grow with dosage of BF_3 (Figure 7.4).

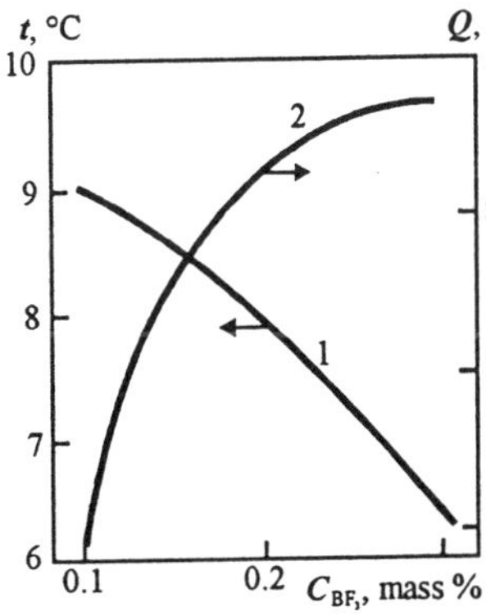

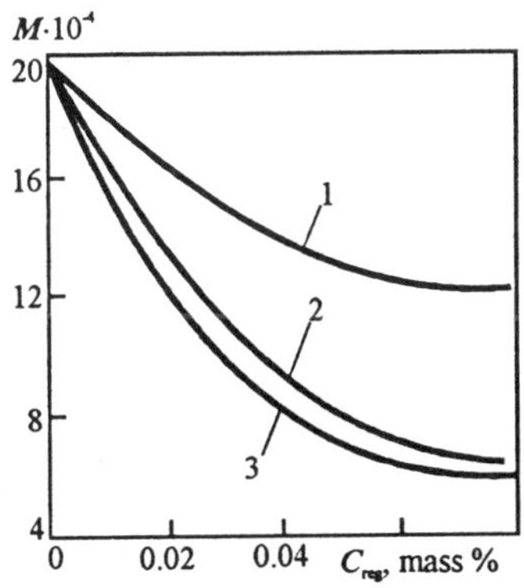

Figure 7.4. Dependence of duration of isobutylene polymerization (1) and polymer yield (2) on the amount of catalyst BF_3.

Figure 7.5. Dependence of the molecular mass of polyisobutylene on the regulator amount:
1 – piperylene; 2 – diisobutylene; 3 – isoprene.

[1] Usually, bis-(3-tert-butyl-6-oxyphenyl)monosulfide.

Admixtures, accompanying the monomer, solvent and catalyst, can be subdivided into three groups: 1) inert substances causing no significant influence on the polymerization process, for example, up to 10 mass% of propylene in the monomer, up to 1.5 mass% of sulfur dioxide to BF_3, which leads just to some deceleration of the process; 2) polymerization inhibitors, in the presence of which the rate of isobutylene polymerization and molecular mass of polyisobutylene decrease (diene hydrocarbons, ammonia, oxygen-containing compounds possessing no mobile atoms of hydrogen, etc.); 3) cocatalysts, which, in small amounts, accelerate isobutylene polymerization, increase molecular mass of resulting polymeric products, and in large amounts usually have a negative influence on the process (alcohols, mercaptans, organic acids, water, etc.). Influence of cocatalytic additives on the cationic polymerization imposes strict demands on the industrial processes of the synthesis of polyisobutylene. That is why fine purification and drying of initial substances are necessary.

Of great importance for production of high-molecular polyisobutylene is the ethylene:isobutylene relation. The rate of polymerization, temperature in the reaction zone and molecular mass of the polymer depend on the monomer content (Table 7.1). A part of the ethylene is introduced in mixture with isobutylene, and the rest of it is mixed with the catalyst solvent.

Table 7.1

Influence of the monomer:solvent relation on polymerization of isobutylene [3, p. 308].

Ethylene:isobutylene relation (by mass)	Reaction duration, s	Polymer yield, mass%	Molecular mass $M \cdot 10^{-3}$
4:1	9	91.6	215
2.5:1	6	89.4	180
2:1	4	84.5	145
1.25:1	Immediately	65.4	100*
1.0:1	Immediately	55.4	72**

Notes. * Intensive boiling.
** Very intensive boiling.

Decrease of the yield and molecular mass of polyisobutylene with content of the solvent (ethylene) is caused by evaporation of a significant part of the monomer (with the solvent) and growth of local temperatures in the reaction zone.

The molecular mass of polyisobutylene can be regulated by introduction of diene and some olefin hydrocarbons (Figure 7.5), as well as (which is less suitable) by change of concentration of the monomer in the batch and the catalyst in the solvent. The molecular mass of polyisobutylene is significantly affected by temperature of polymerization (Figure 7.2).

Polyisobutylene of brands from P-50 to P-118 of the Vistanex type (Enjay Company, USA) are produced in industry in the presence of $AlCl_3$ in solution of ethyl or methyl chloride by polymerization of highly purified isobutylene according to schemes, displayed in Figures 7.6 and 7.7. These are the most suitable schemes, compared with the isobutylene polymerization in the medium of boiling ethylene: higher productivity of devices and homogeneity of the polymer with respect to molecular mass, possibility to obtain not only polyisobutylene of various brands, but also butyl rubber, as well as isobutylene oliomers, on the same equipment (in dependence on the composition and temperature of the initial batch) are some of the advantages.

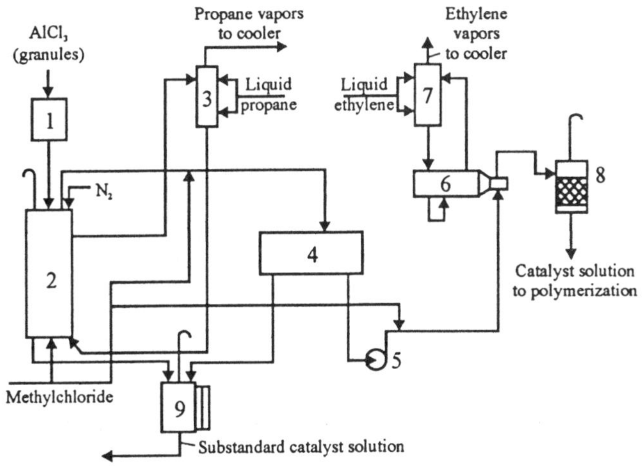

Figure 7.6. Technological scheme of preparation of the catalyst solution in production of polyisobutylene:
1 – hopper; 2 – device for preparation of concentrated solution of the catalyst; 3, 7 – separators; 4 – volume; 5 – pump; 6 – cooler; 8 – filter; 9 – collector of substandard catalyst solution.

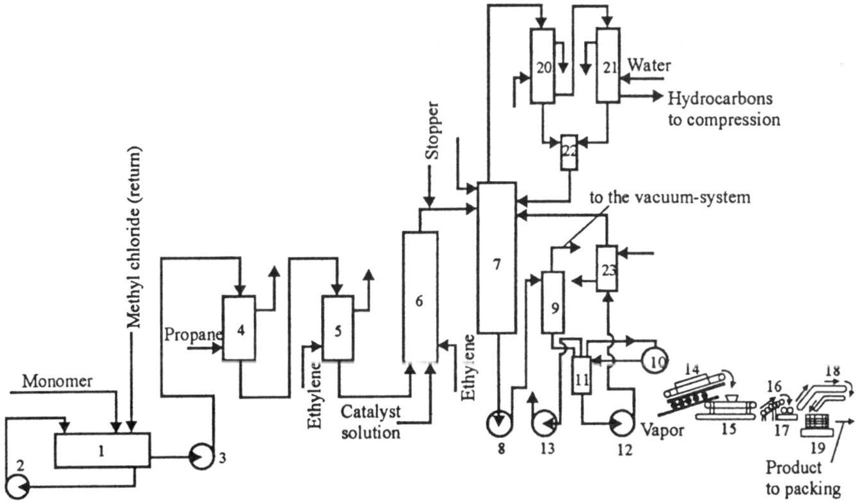

Figure 7.7. Technological scheme of production of polyisobutylene in the solution of methyl or ethyl chloride:
1 — tank for batch preparation; 2, 3, 8, 12 — pumps; 4, 5 — coolers; 6 — polymerizer; 7 — water degasifier; 9 — vacuum degasifier; 10 — vacuum-filter; 11 — vacuum-receiver; 13 — vacuum-pump; 14 — dryer; 15 — injector-machine; 16 — conveyor; 17 — rolls; 18 — cooling conveyor; 19 — brick-forming machine; 20, 21 — capacitors; 22 — separator; 23 — heater.

The catalyst solution is prepared by streaming purified methyl chloride through the device, filled by granulated anhydrous $AlCl_3$ at 243 K. The catalyst solution in methyl chloride of constant concentration (1 mass%) is obtained due to its limited solubility. Solution of $AlCl_3$ is diluted down to the specified working concentration (about 0.1 mass%) directly in the pipeline, then cooled in the ethylene cooler down to 180 K and delivered to polymerization. All operations are performed in an atmosphere of dry inert gas.

The batch is prepared by mixing dry and pure (99.99 mass%) isobutylene with returned isobutylene–methyl chloride fraction. Initial components are dosed according to the results of continuous chromatographic analysis. The prepared batch is delivered from the tank 1 by the pump 3 into propane 4 and ethylene 5 coolers for cooling down to 177–175 K, and then is delivered to the polymerizer 6. Reactors of isobutylene polymerization are unitypical and differ only in some construction details and volume. Usually, they represent cylinder shaped autoclaves, equipped by a central sucking pipe, in the lower part of which a circulation pump is disposed. Around the sucking pipe a great number of peripheral pipes of smaller diameter or blind hollow rings are disposed. The central pipe, peripheral pipes, hollow rings, as well as the bottom, cover plate and walls of the polymerizer are surrounded by a cooling agent (evaporating ethylene in the present case). There is an overflow pipe in the upper part of the reactor for polymerizate release.

Components of the reaction mixture are permanently delivered to the lower part of the reactor and enter sucking branch pipes of the circulation pump. As a result of intensive circulation, flows are turbulized, which is supposed to promote better heat exchange. The reaction mass circulates upward from below. A definite amount of solution or polymer dispersion is permanently released from the upper part of the polymerizer and is delivered to the vacuum degasifier 9, where the isobutylene–methyl chloride fraction is eliminated from the polymeric product. Returned isobutylene–methyl chloride mixture is delivered to appropriate collectors after condensation. Customary polyisobutylene is packed and delivered to customers.

Although polymerization of isobutylene proceeds almost immediately, the mean time of the raw material presence in the reactor is 30–120 min, producing a transformation degree of isobutylene of 75–90%. Changing of batch temperature and introduction of the catalytic solution control temperature in the reaction zone. This method is suitable for controlling the molecular mass of polyisobutylene. Controllers of the molecular mass (diene hydrocarbons, diisobutylene) can also be used (see Figure 7.5).

Kinetic features of the process are such that temperature inside the polymeric-monomeric particles (in microareas of propagating polymeric molecules) is always higher than that of the reaction medium, which, in its turn, is characterized by heterogeneity in the zone of reaction proceeding (non-isothermity). This predetermines formation of the polymer with the molecular mass, lower compared with that expected from calculations based on kinetic data $P = V_p/(\Sigma V_{rev} + \Sigma V_{ini})$, and wider MMD.

7.1.2. Production of moderate- and low-molecular polyisobutylenes.

Polyisobutylene with molecular mass M_n = 300–20,000 is produced by the following scheme (Figure 7.8).

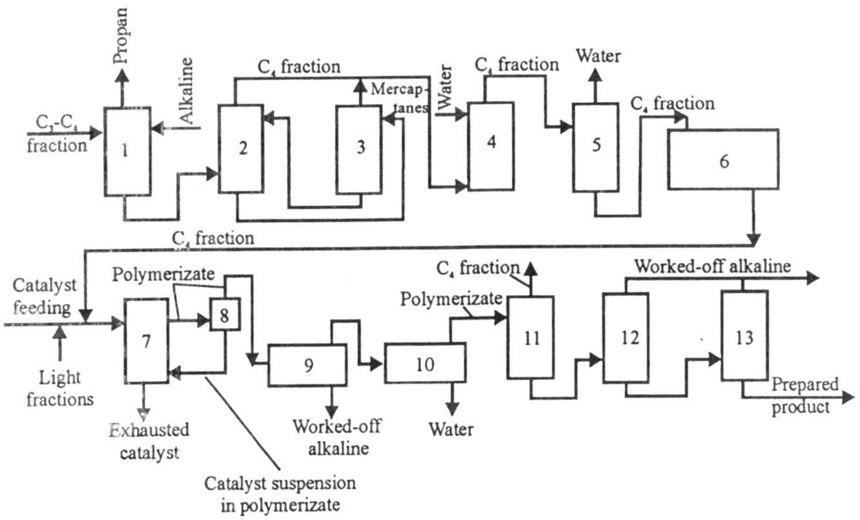

Figure 7.8. Technological scheme of production of low-molecular polyisobutylenes: 1 — column of alkaline washing; 2, 3 — columns of regeneration of alkaline solution; 4 — column of water washing; 5 — column of water elimination and drying; 6 — volume for prepared raw material; 7 — polymerization reactor; 8 — catalyst precipitator; 9 — volume of alkaline deactivation of polymerizate; 10 — volume for water washing of the catalyst; 11 — stripping column; 12 — atmospheric column; 13 — vacuum column.

The following fractions are used as raw materials: C_3–C_4, isobutane–isobutylene mixtures (90:10 - 50:50), butane–butene. Solutions of $AlCl_3$ in chloroalkanes or arenes are catalysts. Molecular mass and yield of the products significantly depend on the content of *n*-butylenes, dienes, sulfur and ammonia compounds in the raw material [4, 5]. However, there is little sense in using regulators in production of low-molecular polyisobutylene. Molecular mass is easily controlled by temperature of the reaction mass (Figure 7.9). That is why, in the case of application of polyisobutylene for viscous additives and series of octols, polymerization is conducted at relatively high (including positive) temperatures. Molecular mass of polyisobutylene and conversion of isobutylene are bound to concentration of the catalyst, introduced into the reaction zone. Molecular mass of the polymer decreases with growth of the catalyst content (up to a definite level), but the monomer conversion increases in this case.

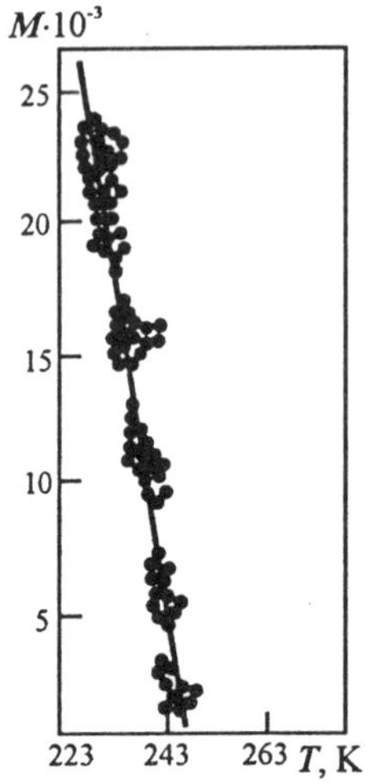

Figure 7.9. Dependence of the molecular mass of polyisobutylene on polymerization temperature (isobutylene — 27.5 mass%, $AlCl_3$ — $7.6 \cdot 10^{-3}$ mol/l, solvent — ethyl chloride).

Due to specific kinetic features of the cationic polymerization of isobutylene, synthesis of low-molecular polyisobutylene usually gives a product with wide MMD ($\overline{M}_w/\overline{M}_n$ = 5–6), which, as a rule, causes a growth of kinematic viscosity of the polymeric product at the regulated average molecular mass. As viscosity determines the exploitation properties of oligoisobutylenes, in the modern industrial production the prepared product after the stage of drying is delivered to the column of clear rectification, in which fractions of oligoisobutylenes of appropriate brands are released along the column height. Therewith quality of the product by width of MMD is improved, and a smaller amount of technological wastes is formed. MMD of the polymer is also narrowed ($\overline{M}_w/\overline{M}_n \cong 2$) at statistical degradation during thermal influence on polyisobutylene (550–650 K).

According to the method of the Cosden Company (USA), low-molecular products (M_n = 300–2,700) are produced at 243-253 K. The raw material is C_3–C_4 fraction containing 15–60% of isobutylene, which is produced, for example, by thermal or catalytic cracking. It is purified from sulfur, dried and then delivered to the raw material volume (Figure 7.8). Then the flow of the raw material, preliminarily mixed with the catalyst ($AlCl_3$), is delivered to the polymerization reactor, consisting of three vertical columns of the total volume of 5 m^3 and the volume by liquid phase of 2.3 m^3, connected by the condenser-cooler in the upper part (Figure 7.10). Mixing of the liquid phase and heat elimination is performed by rising flow of boiling gases, which, condensing in the cooler, return into the reaction zone. The polymerizate, yielding the reactor, is delivered to deactivation after separation from the catalyst in the precipitator. The catalyst with a part of the polymerizate as a suspension is returned into the reactor from the precipitator. Exhausted catalyst is permanently released from the reactor. Deactivation of residues of the catalyst in the polymerizate is performed by water solution of an alkali or ammonia at 300–330 K. The polymerizate, purified from the catalyst, is washed by water and delivered to the steaming column (510–533 K, 700–900 kPa) to eliminate unreacted C_4-hydrocarbons, returned to the system for the secondary use. After atmospheric (510–533 K, 90–290 kPa) and vacuum (0.5 kPa) columns, in which light polymers are separated, the final product is obtained. To obtain thermally stable polymer, the final product may be hydrogenated before elimination of light fractions [6].

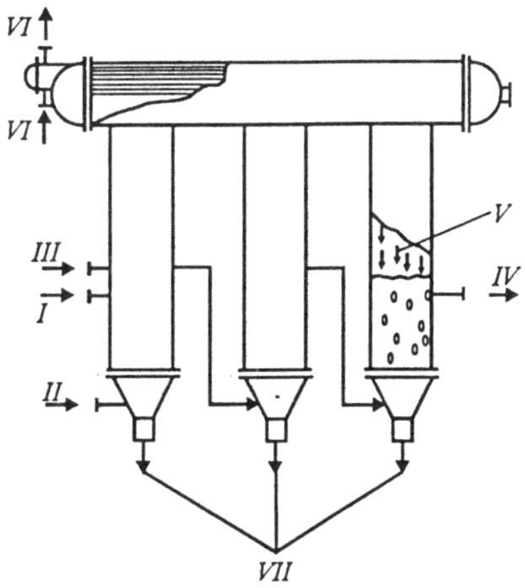

Figure 7.10. The polymerization reactor for obtaining low-molecular polyisobutylene:
I – raw material; II – recycle and catalyst; III – slime from sedimentation tank; IV – ammonia; V – condensate; VI – polymerizate; VII – worked out catalyst.

Generally, the processing for Amoko Company octols reproduces that of the Cosden Company. It differs by the fact that the catalyst, $AlCl_3$, is used dissolved in butane, is prepared at increased temperature and pressure (200-260 kPa), and forms a highly active dispersion, when introduced into the reaction zone. A significant amount (up to 10 mass%) of light polymers, separated at 543 K, is obtained on account of the process conduction at higher temperature (273–278 K).

Technological schemes of production processes of low-molecular oligo- and polyisobutylenes, applied in Russia, display some specific features. In particular, they differ by construction of polymerization reactors and elimination of the reaction heat, as well as by type of catalytic system ($AlCl_3$ solution in chloroethane or in xylene fraction of hydrocarbons). If reactors are usually used, in which heat elimination is performed preferably at the sacrifice of the internal heat elimination by evaporation of components of the raw material mixture (boiling), in Russia, as a rule, thermostatic control is performed at the sacrifice of intensive heat transmission through the walls or cooling surfaces to the circulating agent (ammonia, ethylene), which, of course, is less effective. Provision is made for possibility of variation in definite limits of the technological regime of the process procedure (pressure, temperature, catalyst expenditure, etc.), which enable products with specific required properties to be obtained [6].

Industrial synthesis of polyisobutylene with M_n = 10,000–20,000 provides for production of the concentrate of polymers in a mineral oil (viscous additives P-10 and P-20) (Figure 7.11). After averaging, the raw material is delivered to the rectification column, where separation of heavy C_5 hydrocarbons and higher is performed at 550–580 kPa, temperatures of 328–333 K (cube) and 318–328 K (the upper part of the column). The rectificate, containing up to 45 mass% of isobutylene, is cooled down and dried in the column, filled by $CaCl_2$ by 2/3 of the volume and the remaining 1/3 by NaOH. Final drying of the raw material is conducted in the cooler with a filter for separation of water crystals (258–263 K).

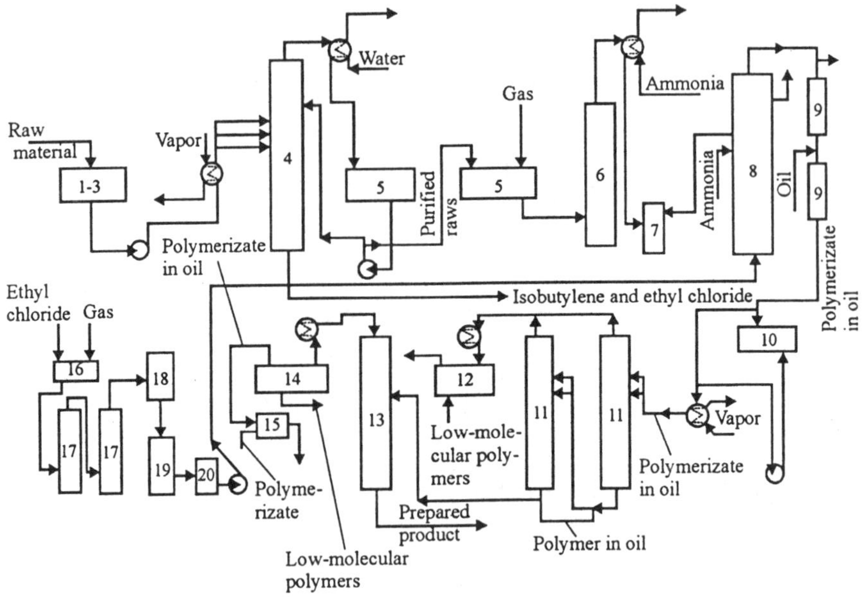

Figure 7.11. Technological scheme of production of polyisobutylene with the molecular mass of 8,000–30,000:
1, 2, 3 – volumes for raw material; 4 – rectification column; 5, 12 – volumes; 6 – drying column; 7 – filter; 8 – polymerization reactor; 9 – mixing tank; 10 – volumes for polymerizate; 11 – degasification columns; 13 – vacuum column; 14 – volume for low-molecular polymers; 15 – vapor-ejection column; 16 – catalyst volume; 17 – adsorbers; 18 – mixer; 19 – tank for catalyst; 20 – filter.

Polymerization is conducted in vertical reactors for complete mixing, and working in parallel, with the mixing device of the axis pump type (1500 rot/min). The reaction mass is cooled down both with the help of jacket and two built-in cylinder cooling chambers, disposed along the vertical axis of the reactor. Cooling down is performed at the sacrifice of liquid ammonia circulation through the system of reactor cooling and ammonia evaporation. The reactor volume is 1.5–15 m^3, the lowest possible temperature is 238 K, and pressure is up to 0.5 MPa.

The reaction mixture is mixed up from below at the center of the reactor and down by periphery (between cooling chambers). The raw material is delivered from above into descending flow; the catalyst (2% solution of $AlCl_3$ in chloroethyl) is delivered from below into the rising flow up to the mixing device. At the reactor exit the polymerizate is treated by water solution of an alkali in order to deactivate the catalyst in the system of diaphragm mixers. After elimination of products of catalyst deactivation, the polymerizate is delivered to drying, and then is mixed with hydrocarbonic oil in the required relation and, after heating in the heat exchanger up to 383–403 K, is delivered to degasification columns (393–433 K, 150–300 kPa) and further on to the vacuum column (433–443 K, 1.3–4 kPa) with built-in film evaporator. After cooling, unreacted gases and highly boiling fractions of di-, tri- and tetramers of isobutylene are accumulated in the volume and are directed to separation and utilization. The prepared product, polyisobutylene concentrate in oil, is released from the bottom of the vacuum column into collectors of the final product, and then delivered to customers.

In production of octols with M_n = 800–1,000, the polymerizate is mixed up with alcohol or with water solution of an alkali in a special vessel with further water washing. Furthermore, it is delivered through sedimentation tanks to the degasifier according to the scheme of separation of P-10 and P-20 additives.

Production of polyisobutylene with M_n = 600–3,500 from butane–butylene fractions is based on the two-stage polymerization of the raw material in reactors of complete mixing. The principal technological scheme of synthesis of octols is shown in Figure 7.12 [5].

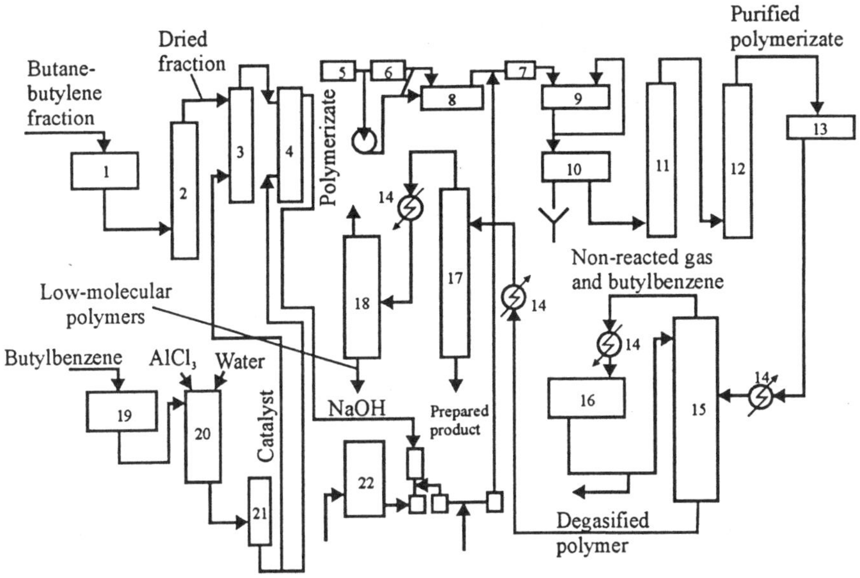

Figure 7.12. Technological scheme of production of octols:
1 – volume for raw material; 2 – column-dryer; 3 – the first step reactor; 4 – the second step reactor; 5, 6, 7 – mixing tanks; 8, 9, 10 – sedimentation tanks; 11 – polymerizate dryer; 12 – percolator; 13 – filter; 14 – heat exchanger; 15 – degasification column; 16 – collector of exhausted gases; 17 – vacuum column; 18 – collector of low-molecular polymers; 19 – volume for butylbenzene; 20 – mixer; 21 – dosing volume; 22 – NaOH preparation block.

Prepared raw material is delivered to the isothermal reactor of the first stage of polymerization. Simultaneously, the catalyst — $AlCl_3$ complex with butylbenzene — is also dosed there. Generally, isobutylene is polymerized at 297–300 K. The residual isobutylene and *n*-butenes enter the reaction in the reactor of the second stage (289–292 K) in the presence of additionally introduced catalyst.

Operations of separation and purification of the products are analogous to those in production of octols 800–1,000. The final vacuum degasification is conducted at 573 K and pressure of 0.13–0.40 kPa.

7.2. New generation of resource and energy preserving technologies for production of higher ecologically friendly oligo- and polyisobutylenes.

7.2.1. Main prerequisites.

The scientific elucidation of the mechanism of fast polymerization, discussed in Chapter 3, formed the basis for creation of principally new technology of production of isobutylene polymers in the flux. Most basic is realization or maximally possible approximation to the main requirement: the reaction time must be greater than the mixing time

$$\tau_{react} = L/k > \tau_{mix} = R^2/D_t,$$

where L is the reaction zone length, D_t is the turbulent diffusion coefficient, R is the reaction zone radius. If dependence of the turbulent diffusion coefficient on the linear rate of the reaction mixture motion is known, the upper border of the apparatus radius, required for the optimal conditions of isobutylene polymerization under a quasi-isothermal regime (Table 7.2), can be estimated.

Table 7.2

Parameters of the reaction zone in oligomerization of isobutylene depending on $V(D_t)$ ([isobutylene] = 20 mass %, $AlCl_3$ in chloroethane, 5 mass %, 243 K, accompanying single injection of the catalyst into the monomer solution, $k_{prop} = 10^6$ l/(mol·s), k_{term} = 17.5 s^{-1}) [1].

V, m/s	D_t, m^2/s	R_{cr}, m	L, m
1	0.004	0.03	0.05
2	0.010	0.07	0.12
3	0.015	0.10	0.17
4	0.020	0.13	0.23
5	0.025	0.17	0.29
6	0.032	0.22	0.35
10	0.054	0.36	0.56

Growth of the linear flux rate does not change the character of the process, but intensifies mass- and heat exchange, which also significantly improves the process proceeding with simultaneous improvement of molecular-mass characteristics of the resulting polymeric product. Therewith, the admissible range of change of the reactor diameter is broadened (R_{cr} rises). Controlling factors of conduction of fast polymerization in the liquid phase in turbulent flows at the application of pipe turbulent reactor, operating in the regime of quasi-ideal displacement (macroscopic type A), are also established.

The scheme displays the interconnection of various input and output parameters of the processes, proceeding in the turbulent pipe reactor, which demonstrates possibilities of the control of the technological process.

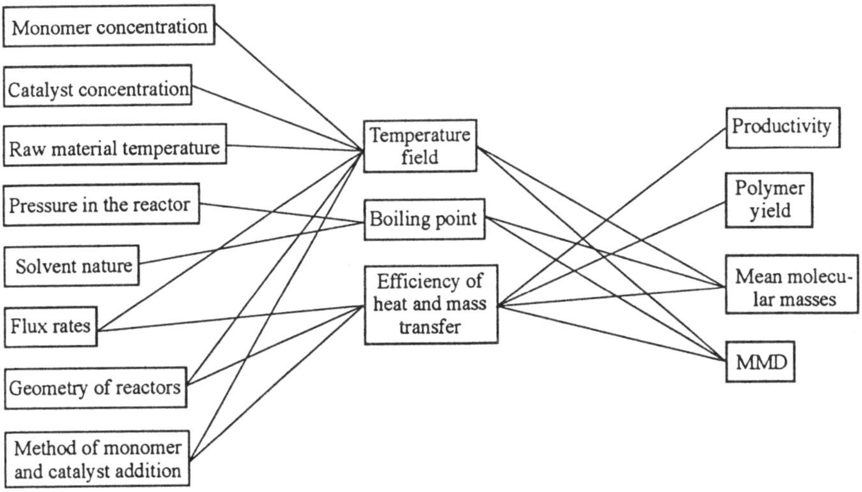

As is obvious, the determining factors for the isobutylene polymerization process in the flux are size of the reaction zone, initial concentration of the monomer, initial temperature of the monomer solution (the amount of catalyst introduced with a solvent is small, and its temperature causes no significant influence on temperature of the reacting system at the reactor input), the rate of reacting flux motion and its turbulization at the point of mixing of monomer and catalyst solutions, total catalyst concentration or gross depth of the monomer transformation, the boiling point depending on the chemical nature of the solvent and pressure in the reactor, relation of mass- and heat transmission, presence of devices which enable an anisotropic mechanism of heat transmission to be created, application of the zonal multistage delivery of the catalyst and/or the monomer, etc. [2, 7–28].

Hence there exists a complex of conditions, which actively affect the whole process, and the main exploitation indexes of the resulting polymeric products as well, its molecular-mass characteristics, in particular. Neglecting these features, no stable or even reproducible and reliable production of polymeric products with assigned properties can be obtained. Experimental and calculated results have shown that the regime of quasi-ideal displacement is optimal in turbulent fluxes with its own specific conditions of the process conduction and overall size of the reaction zone.

The universal catalyst for isobutylene polymerization, especially up to low- and moderate-molecular polymers of isobutylene, is $AlCl_3$. However, its application does not always solve some urgent problems, for example, guaranteed production of standardized brands of polymeric products, especially from industrial fractions of C_4 hydrocarbons. After purification from catalytic poisons (butadiene, sulfur compounds, etc.), the fraction C_4 contains butane and butylene isomers, small amounts of C_2, C_3 and C_5 hydrocarbons, the relation between which changes depending on conditions of the fraction production. Polyisobutylene of decreased molecular mass or products of mixed polymerization of unsaturated hydrocarbons are obtained by isobutylene polymerization from C_4 fractions in the presence of $AlCl_3$.

As has been mentioned already, initiating properties of electrophilic catalysts are determined by the measure of acidic properties. Catalytic systems, based on Lewis acids and weak proton-donors (H_2O, ROH), differing by decreased acidic strength, are quite active and selective. Their application usually solves the problem of synthesis of industrial brands of polyisobutylene from practically every raw material, and from mixtures of C_4 hydrocarbons, in particular. These systems solve the problem of economical complex application of all components of C_4 hydrocarbons fraction [2, 5, 29]. For example, just isobutylene enters the reaction in the presence of $R_nAlCl_{3-n}\cdot H_2O$ in the temperature range of 210–260 K. Such selectivity is not observed with $AlCl_3$ and individual alkyl-aluminum chlorides (Table 7.3).

Table 7.3

Polymerization of isobutylene from the fraction of C_4 hydrocarbons (C_4 fraction – 18 g, catalyst – 0.14 mol/g of hexane, 600 s; tests 1, 2, 4, 5 – 248 K; 3, 6 – 258 K).

Catalyst	Initial mixture / Mixture after polymerization, mass%				Isobutylene conversion, %	Selectivity, rel. un.
	C_3H_6	C_4H_{10}	*iso*-C_4H_8	α- and β-C_4H_8		
$AlCl_3\cdot H_2O\cdot 3CH_3C_6H_5$*	3.0/0	55.5/100	37.9/0	3.5/0	100	0.33±0.02
$C_2H_5AlCl_2\cdot H_2O$*	4.9/3.8	63.2/88.1	8.5/0	23.3/8.1	100	0.48±0.03
$C_2H_5AlCl_2$**	3.9/4.4	64.8/74.7	11.3/0.3	20/20.5	97.5	0.89±0.05
$C_2H_5AlCl_2$	3.0/3.7	55.5/90.7	37.9/3.4	3.5/2.1	95.0	0.51±0.03
$C_2H_5AlCl_2\cdot H_2O$	3.9/4.5	67.5/75.6	11.3/0.3	17.3/19.5	97.5	1.0±0.05
$C_2H_5AlCl_2\cdot H_2O$	0.9/1.3	45.1/65.3	31.1/0.4	22.9/33.0	99.1	1.0±0.05

* In methyl chloride or ethyl chloride.

** Contains 0.1 mole of H_2O per mole of $C_2H_5AlCl_2$.

Besides high conversion of isobutylene (residual content of the monomer is 0.5-1.5 mass%), polymeric products with quite high molecular mass are formed in the presence of $R_nAlCl_{3-n}\cdot H_2O$ hydrated complexes. The exhausted fraction represents practically pure butane–butylene mixture (with concentration up to 98 mass%), suitable for application in organic synthesis, in production of methylethylketone, in particular. Unlike $AlCl_3$ combined with a cocatalyst (H_2O, RX, etc.), $R_nAlCl_{3-n}\cdot H_2O$ hydrated complexes are less sensitive to inhibiting admixtures, cause no corrosion of equipment, and eliminate all poisoning waste gases ejected into the atmosphere. As a consequence of high selectivity of $C_2H_5AlCl_2\cdot H_2O$ aquacomplexes, the molecular mass of polyisobutylene is 2–4-fold increased (from $(5–8)\cdot 10^3$ to $(15–30)\cdot 10^3$ at 243 K), the amount of the catalyst required to achieve 100% transformation of isobutylene decreases: $2\cdot 10^3$ for $3CH_3C_6H_5\cdot HCl\cdot Al_2Cl_6$, $1\cdot 10^{-3}$ for $C_2H_5AlCl_2\cdot 0.1H_2O$ and $4\cdot 10^{-4}$ mole for $C_2H_5AlCl_2\cdot H_2O$ (see Table 7.3).

Constants of catalysts poisoning (k_{pois}) in isobutylene polymerization in the presence of butene-1 are also found in accordance with these data: 1 ± 1 for $C_2H_5AlCl_2\cdot H_2O$ and 15.6 ± 2 for $C_2H_5AlCl_2$ (Figure 7.13).

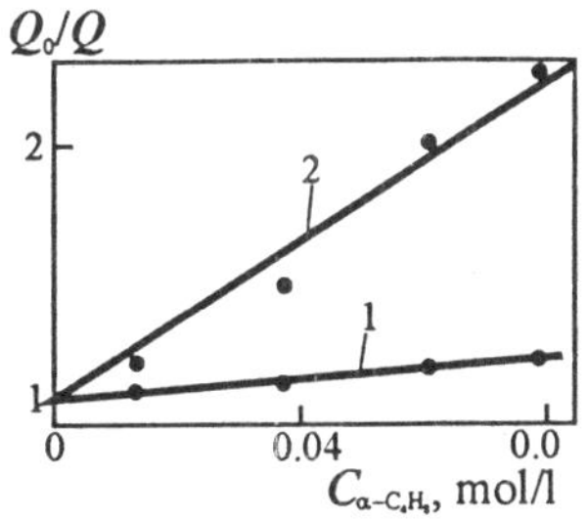

Figure 7.13. Dependence of the relative yield of polymeric products on α–butene concentration during isobutylene polymerization (*iso*-C_4H_8 — 1.5 mol/l, catalyst — $6.3 \cdot 10^{-3}$ mol/l, solvent — *n*-heptane, 253 K, 900 s): 1 — $C_2H_5AlCl_2 \cdot H_2O$; 2 — $C_2H_5AlCl_2$.

Noteworthy is the selectivity of the isobutylene polymerization from the C_4 fraction (0.9 rel. units) under the influence of the $CuSO_4 \cdot HCl$ complex, which displays low acidic strength ($H_0 \approx -8$) and with a specific influence on the solid counterion leveling the role of the medium [30–32]. In this case, nature of the catalyst does not affect the selectivity of the process. Heterogeneous catalysts based on sulfocationites of KU-2 and KU-2(8), modified by aluminum chlorides, may also be effective. This type of catalytic systems is attractive, because it provides the possibility of producing isobutylene polymers in one stage with no washing of polymeric products and deactivation of the catalyst [32, 33].

7.2.2. Production of high-molecular polyisobutylene.

As has been mentioned above, the problem of production of high-molecular polyisobutylene is elimination of spray of the prepared product and overgrowing of the polymerization reactor walls by a layer of the polymer, which complicates permanent conduction of the process and makes necessary interruption of the production process to clean the equipment.

To eliminate spraying of the polymer and its adhering to the walls of the reactor, solution of the catalyst and the monomer was suggested, preliminarily mixed for 0.06–0.62 s at the sacrifice of the catalyst solution delivery into the turbulent moving accompanying flux of the monomer. For this purpose, the device is equipped with a cylinder mixer, one face of which is made as a cone, the top of which is directed against the monomer and catalyst fluxes, and the other face is turned to the belt-polymerizer; as a result, the branch pipe for the monomer addition lies in parallel to the mixer axis at a distance equal to half the radius of the mixer cylinder (Figure 7.14).

Modernization has reduced overgrowing of the polymerization reactor walls and time between repairs of the technological set-up to be increased.

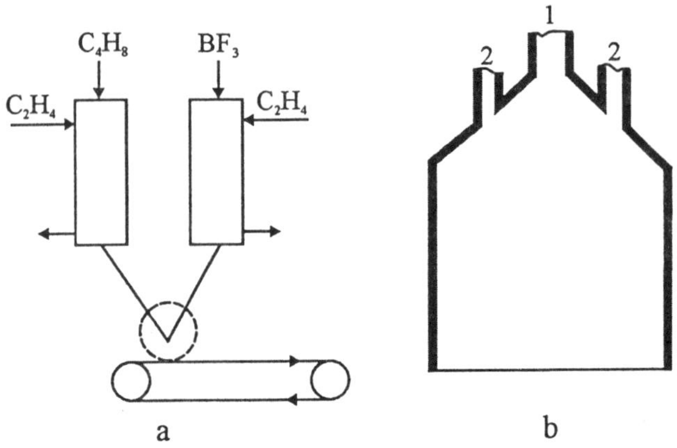

Figure 7.14. The principal scheme of production of polyisobutylene P-200 (a) and the mixer (b): 1 – branch pipe for introduction of the monomer, 2 – branch pipes for introduction of the catalyst (length/diameter = 1:1 – 1:2).

7.2.3. Production of moderate- and low-molecular polyisobutylenes.

As shown above, electrophilic polymerization of isobutylene proceeds at an extremely high rate ($k_p \sim 10^6$ l/mol·s) in the limited reactionary volume of the flare type with various zones temperature, concentrations of reagents and reaction rates steps of the reaction. Because isobutylene polymerization is such a fast reaction, this predetermines complexity (and essentially impossibility) of thermostatic control of the process and its regulation, if standard volumetric mixing reactors are used. It is also difficult to decrease molecular mass and broaden MMD of the polymeric products formed compared with the calculated ones, as well as to decrease productivity of the main device. This is the reason that volumetric mixing reactors, used in the modern industry, are ineffective and not optimal in construction or volume, or productivity, or consumption of the raw material and electric energy. Development of other, more efficient methods of isobutylene polymerization in industry are clearly urgently needed.

In particular, polymerization of isobutylene from fractions of C_4 hydrocarbons is performed in a steel, well heat-insulated pipe of volume 0.5 l (0.7 m long, and 0.02–0.03 m in diameter) without mixing under adiabatic conditions, i.e. the polymerization heat is eliminated at the sacrifice of heating the preliminary cooled mass [34]. The small diameter of the pipe and low rates of the flux motion in it (0.01–0.02 m/s) promote the laminar type of flux, preserved owing to increase of the medium viscosity during the reaction. Relative constancy of the monomer concentration by the reactor coordinate and distribution of the catalyst, $C_2H_5AlCl_2$, in the volume are predetermined by preliminary effective mixing of the raw material, preliminarily cooled down in the cooler, and the catalyst in the mixing tank, and short length of the reactor. The mean time of the raw material presence in the reactor is 35–70 s, as a result of which nearly complete transformation of isobutylene into high-molecular products is achieved. The adiabatic conditions of the process promote no isothermity – the temperature difference at input and output of the reactor varies from 40 to 100 K.

This method of production of isobutylene polymers is not attractive, because it may proceed under the conditions described at low productivity only (low concentrations of the monomer and the catalyst).

More favorable conditions for thermostatic control are found in polymerization of isobutylene in a pipe reactor 0.001–0.1 m in diameter and 0.5–800 m long, made as a spiral (or in any other form) and equipped with a cooling envelope with liquid ammonia [35].

Temperature and pressure are selected in such a way as to conduct the process in the liquid phase although, due to high heat release boiling up of the reaction mass is inevitable at the moment of catalyst injection. Thanks to high rates of turbulent flux of the raw material (0.5–20 m/s), productivity of the reactor-pipe is increased (about 100 kg/h of polymer with molecular mass of 500–5,000 instead of 1–3 kg/h, according to [12]). Full conversion of the monomer during 1-40 s of exposure in the reactor confirms high rates of the polymerization reaction.

Despite their relative simplicity, the methods of production of isobutylene polymers discussed are not optimal from the point of view of realization in industrial production. In particular, productivity of devices is low. So the considered fundamental regularities of fast polymerization of isobutylene display clearly (existence of R_{cr}, in particular) that the scale transfer, widely used in chemical technology for usual volumetric devices such as polymerizers, cannot be applied to creation of new pipe reactors for realization of fast cationic reactions, because of the process localization in small reactor volumes.

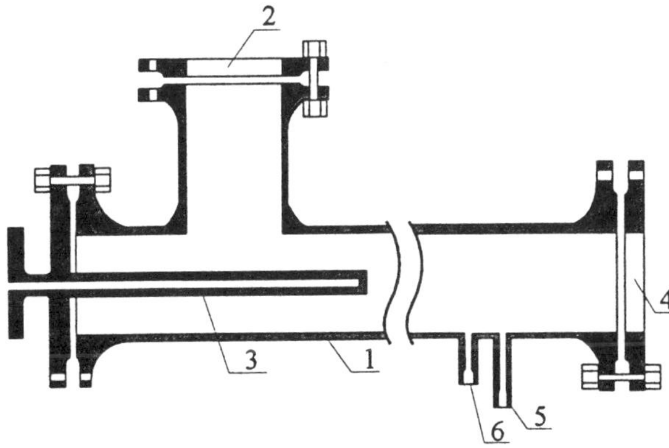

Figure 7.15. Small-scale pipe turbulent reactor:
1 — reactor pipe; 2 — branch pipe for the monomer addition; 3 — branch pipe for the catalyst addition; 4 — exit for polymerizate; 5 — sample selector; 6 — pocket for thermocouple.

A fundamentally new method of production of butylene polymers with molecular mass from 200 to 50,000 was created in Russia. This method enables the technology and equipment support of the production process of production of oligo- and polyisobutylenes to be greatly simplified [2, 7, 10, 15, 36–39].

Quite simple (compared with current designs) energy and material saving technology of production of butylene polymers differs by application of a new pipe turbulent reactor (see, for example, Figure 7.15) of significantly small bulk instead of volumetric (V = 2–3 m^3) mixing polymerization reactors of complex construction. These new reactors possess no special mixing devices or cooling system, but operate in intensive regimes of quasi-ideal displacement in turbulent flows and provide quite high productivity of the process, high quality of the resulting polymeric products, and in many cases, increases ecological friendliness of the production [2, 7, 10, 15, 36–45].

Table 7.4 compares the main (average) indexes of new pipe reactor, working in the regime of highly developed turbulence, with those of volumetric reactors of ideal mixing, used at the present time.

Table 7.4

Technical and economic indexes of pipe turbulent reactor, working in the regime of quasi-ideal displacement in turbulent fluxes, compared with volumetric reactors of ideal mixing.

Index	Pipe turbulent reactor	Volumetric reactor Lenkhimmash, Russia	Volumetric reactor of Stratco Company, USA	Volumetric reactor of Amoco Company, USA
Reactor volume, m^3	0.04±0.02	1.5–4.0	Up to 29	20–30
Specific quantity of metal, tons	0.05	7.5	Up to 40	50
Power capacity, rel. un.	0.08-0.085	1	1	1
Time of reaching the regime, h	0.01	5.5±0.5	4±1	4±1
Exposure time, h	0.003	1.5±0.5	1	1
Productivity, ton/h	Above 10	2±1	Up to 5	Up to 10
Specific productivity, rel. un.	Above 1000	1	1	1.5±0.5
Specific expenses in, rel. un.				
Water	0.8	1	1	1
Raw material	0.75	1	0.9	1
Catalyst	0.55	1	1	1

The new method of production of isobutylene polymers is widely used in industry and guarantees [6, 10, 14, 15, 36–38, 46]:

— Retention of the main principles of the valuable technological scheme of production of isobutylene polymers;
— Decreasing of production squares, including global construction of the polymerization department;
— Significant increase (2–4-fold and greater) in the total productivity of the production of isobutylene polymers;
— Sharp (1000-fold and greater) increase of the specific productivity of the polymerization reactor;

— 100–1000-fold decrease of the specific quantity of metal and volume of a single polymerization reactor;
— Possibility to use raw materials of a broad composition representing up to 80% of the isobutylene concentration, as well as substandard raw materials without additional purification;
— Decrease of specific expenses in water and electric power by 15–20%;
— Decrease in consumption of raw materials (10–15%) and catalyst (1.5–2-fold);
— Abrupt shortening of exposure of the reaction mass in the polymerization reactor — from 1 to $3 \cdot 10^{-3}$ h;
— Decrease in the masses formed from side reactions and secondary processes;
— Significantly lower sensitivity of the reaction to poisoning admixtures;
— Decrease of operator personnel;
— Shortening of duration of some ancillary operations, in particular, activation of the technological regime of the reactor from 1.5-2 hours down to 10-20 minutes;
— Full automatic control of the isobutylene polymerization;
— Versatility of the process, which allows obtaining of a broad assortment of isobutylene polymers with molecular masses from 200 to 50,000 according to the single technological scheme.

Generally, the process differs by its increased ecological friendliness and higher safety. The experience of production of isobutylene polymers in industry using small-scale pipe turbulent reactors and observed advantages are described in [39-45]. Therewith polyisobutylenes, meeting the standards, are produced from practically any sorts of raw materials (isobutane–isobutylene, butane–butylene, butene–isobutylene fractions, etc.) (Table 7.5) and $AlCl_3$ and R_nAlCl_{3-n} in various solvents (in particular, in chloroethyl, methylene chloride, butane, alkylaromatic hydrocarbons, etc.).

Table 7.5

Raw materials used for production of isobutylene polymers in pipe turbulent reactors.

Hydrocarbon	Fraction of hydrocarbons		
	Isobuane–isobutylene (IIF)	Butane–butylene (BBF)	Butene–isobutylene (BIF)
n-Butane	1–2	10–31	—
Isobutane	50–68	42–59	—
Isobutylene	30–50	5–11	40–45
α-Butylene	—	2.8–10	27–30
trans-Butylene-2	—	2–7	8–10
cis-Butylene-2	—	4–10	3.5–4.5
C_3+C_5 hydrocarbons	1–2	2.0–9.5	5–15
Butadiene-1,3	—	—	0.1–0.2

It should be noted that the butene–isobutylene fraction must not be used in volumetric mixing reactors. The process will not proceed.

Oligoisobutylenes with molecular mass from 600 to 1,800 may easily be produced in the pipe turbulent reactor under the displacement regime in intensive turbulent whirls. The required temperature is supported at the sacrifice of the solvent boiling (isobutane, etc.). Any brands of low-molecular polymers of isobutylene (octols), in particular, with molecular mass from 300 to 2,000–5,000, can be obtained in the pipe reactor by pressure regulation (Figure 7.15).

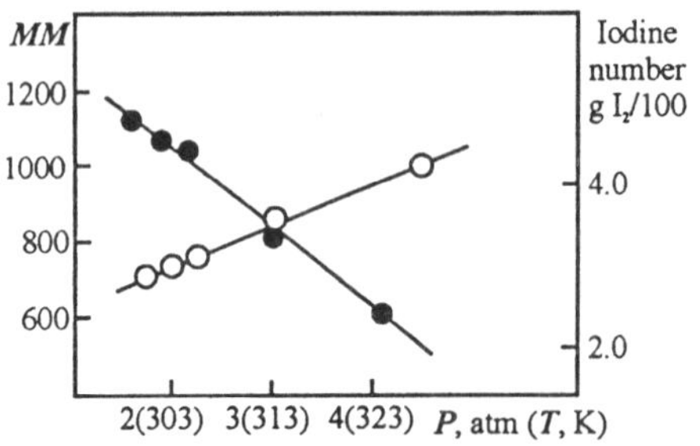

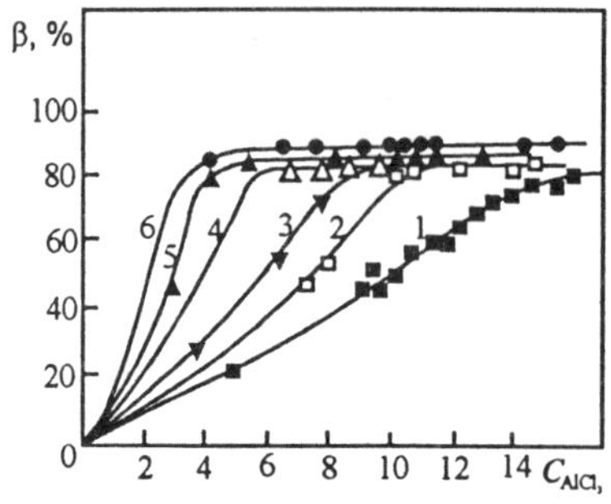

Figure 7.16. Dependence of molecular mass and degree of unsaturation of the polymeric product (as assesses by its iodine number) on pressure in the system and the reaction temperature (conditions of Figure 7.17).

Figure 7.17. Changing of the isobutylene conversion due to the catalyst concentration at T_0, K: 1 — 278; 2 — 273; 3 — 263; 4 — 253; 5 — 248; 6 — 238.

Polyisobutylenes with molecular mass of 8,000–15,000 are produced under conditions of heating up of the reaction mixture, close to the adiabatic one. The required store of cold is imparted to the initial raw material (cooling down to 238 ± 10 K).

It is impossible to use low temperatures (180 ± 10 K) in production of polyisobutylenes with molecular mass of 17,000–25,000, because the resulting polymer does not dissolve completely, owing to which the efficiency of heat elimination is decreased, and the polymer adheres to the armor and reactor walls.

The process must be conducted at temperature above 213 K. The zonal model of the pipe turbulent reactor would also be appropriate for use.

Figure 7.16 displays dependence of isobutylene conversion of $AlCl_3$ content in ethyl chloride under various temperatures of the raw material on the entry of the pipe turbulent reactor. Areas of temperatures and concentrations of reagents are selected so that oligoisobutylene with M_n = 700–1,200 is obtained. Influence of the amount of catalyst added to the reaction mixture grows with temperature of the raw material entering the pipe turbulent reactor.

Width of MMD of the resulting polymeric product strictly depends on ΔT — the difference between temperature of the reaction mixture in the reaction zone (T_r) and the initial temperature of the raw material entering the reactor (T_0) (Figure 7.18). This enables the width of the MMD to be controlled in broad ranges.

Typical curves of MMD of isobutylene polymers, resulting under industrial conditions, compared with volumetric reactors of the ideal mixing are displayed in Figure 7.19. Decrease of T_0 provides growth of content of high-molecular fraction in the resulting polymeric product and consequent broadening of MMD. It is noteworthy that oligo- and polymers of isobutylene, obtained in volumetric reactors of "ideal" mixing generally represent a mixture of low-molecular and high-molecular products, often in equal quantities, although they meet technical requirements by value of the average molecular mass. However, a part of the target product, for example, of macromolecules with M_n = 800–1,000 (octol-1000), in common production of isobutylene oligomers does not exceed 30–35 mass%. As a plane reaction front P (macroscopic type A) can be easily formulated in the pipe turbulent reactor [46], then it is practically possible to conduct the process in the absence of the temperature gradient over the reactor radius; temperature grows over the reactor length only. This allows quite accurate experimental determination of the real temperature of the reacting mixture flux in the assigned element of the volume and easy control of the quality of polymeric product, resulting in the current volume of the reaction zone, molecular mass and MMD, most of all. This is impossible with fast polymerization in volumetric mixing reactors. The dependence of P_w/P_n on $\Delta T = T_r - T_0$ (see Figure 7.18) enables values of temperature gradients really occurring in the volumetric reactor of the ideal mixing to be estimated. As shown by experiment, they can reach 60-70° or higher, despite the application of intensive thermostatic control and mixing up of the reaction mixture.

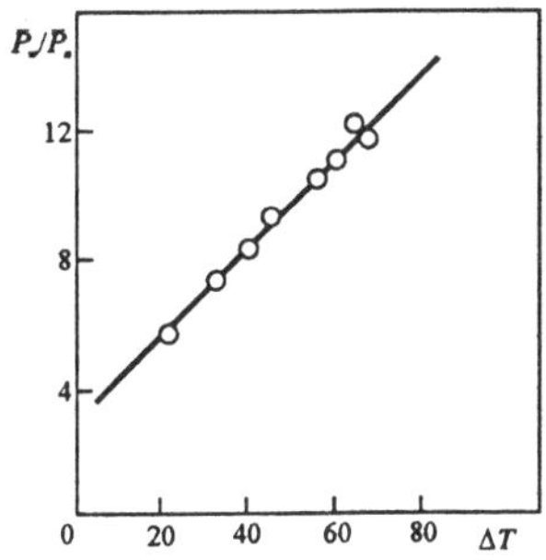

Figure 7.18. Dependence of the polydispersion index of resulting polymeric product (P_w/P_n) on the temperature difference in the reaction zone and at the reactor entry (T_0).

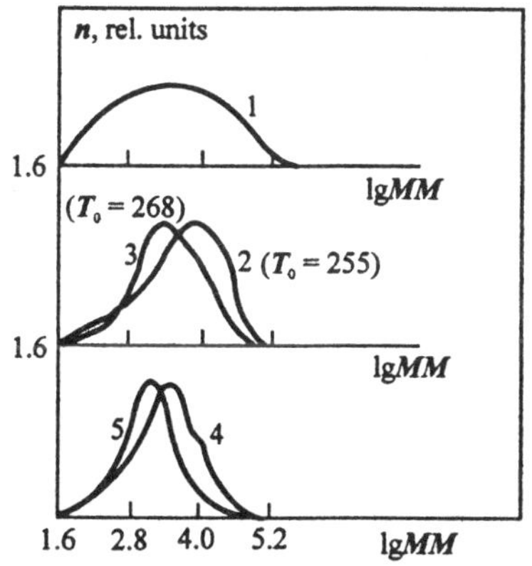

Figure 7.19. Typical curves of MMD change for polyisobutylenes, obtained in the volumetric (with ideal mixing) (1) and turbulent pipe (2-5) reactors.

It must be noted that if the process of fast isobutylene polymerization is conducted according to the new technology, applying pipe turbulent reactors, under conditions far from the macroscopic regime of quasi-ideal displacement in turbulent fluxes P [46], then additional polymerization of the residual monomer under permanently growing temperature with removal from the point of the catalyst and the monomer injection into the reactor will cause a rise in unsaturation of the resulting polymer and, consequently, decrease of M_n with simultaneous widening of MMD. Decrease of the molecular mass and widening of MMD of polyisobutylene will proceed, generally, at the sacrifice of accumulation of a low-molecular fraction (up to 2–3 mass%).

The important feature of application of pipe turbulent reactors for realization of fast polymerization processes is limiting from below of the quantity of added raw material. The pipe reactor operates ineffectively under low loads of raw material: total conversion of the monomer decreases, MMD widens, molecular mass of the resulting polymeric product decreases, productivity of the reactor is sufficiently decreased, and so on (Figure 7.20). The reason is that D_T decreases at low linear rates of the raw material motion in the pipe reactor, and therefore radius R of the reactor becomes greater than R_{cr}. In this case, the process shifts from the quasi-isotermal regime (the regime of quasi-ideal displacement in turbulent flows) into the flare one with all the negative consequences that follow. As productivity of the pipe turbulent reactor is limited from below, the process initiation at conduction of fast polymerization under arbitrary conditions must be performed at working consumption of the monomer and catalyst solution, determined by real productivity of the device.

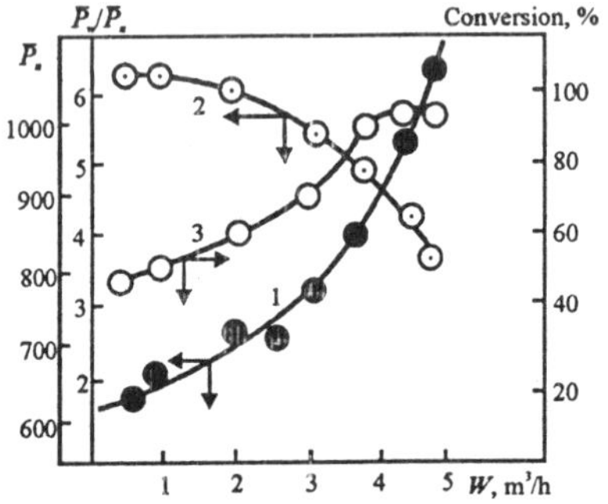

Figure 7.20. Changing of the molecular mass (1), width of molecular-mass distribution (2) and conversion of isobutylene (3) with quantity of the raw material added to the pipe turbulent reactor ($[M]_0$ = 43 mass%, R = 0.025 m; T_0 = 263 K).

Evident advantages are displayed by application of a small-scale pipe turbulent reactor, operating in the regime of quasi-ideal displacement in turbulent fluxes, in production of low-molecular polymers of isobutylene from fractions of C_4 hydrocarbons, depleted of isobutylene (content of iso-C_4H_8 from 1.5 to 15%), as well as from the butene–isobutylene one, possessing no qualified application. In the latter case, maximum implication into the process of not only isobutylene, but also butylenes, the content of which may reach 30 mass% or higher, is appropriate [39–41, 44, 47].

Two-stage technology is usually used for product of polybutenes based on the butane–butylene fraction (BBF) with volumetric mixing reactors applied (isobutylene is polymerized in the first volumetric reactor, and butylenes in the second one), because transformation of various monomers requires different conditions.

Table 7.6

Technological parameters of the device of production of polybutenes from butane–butylene fraction (the fraction composition: C_3-hydrocarbons 2-8; isobutane 42-58; *n*-butane 10-31; α–butylene 2.8-10; isobutylene 5-11; *trans*-butylene-2 2-7; *cis*-butylene-2 4-10; C_5-hydrocarbons 0.1-1.3; sulfur-organic compounds 0.003-0.004.

Index	Polybutene for LCL (TU 38 101743-81)			Electroinsulating liquid – octol (GOST 12869-77)			Polybutene for dispersion medium (TU 38 101679-74)		
	In the mixing reactor		In pipe reactor	In the mixing reactor		In pipe reactor	In the mixing reactor		In pipe reactor
	I stage	II stage		I stage	II stage		I stage	II stage	
Temperature of, °C									
raw material at the entry	—	—	From –8 to –10	—	—	From –10 to –15	—	—	From –10 to –15
in the reactor	13-18	18-23	40-50	From –25 to –28	From –10 to –18	0-20	From –18 to –28	From –14 to –20	0-20
Pressure, kPa	405-490	405-490	490-686	405-490	686	490-686	405-490	405-490	490-686
Raw material consumption, m^3/h	0.6-0.9	0.6-0.9	1-2.5	1.1-1.2	1.1-1.2	1-2.5	0.6-0.9	0.6-0.9	1-2.5
Exposure in the reactor, min	100-150	100-150	0.5-1.5	70-80	70-80	0.5-1.5	100-150	100-150	0.5-1.5
Catalyst consumption, kg per totality of C_4	148	110.2	157.5-189	51.2-55.1	59.1-63	82.7-86.6	51.2-55.1	59.1-63	82.7-86.6
Conversion degree of, mass %									
isobutylene	90-100	100	95-100	85-95	100	95-100	85-95	100	95-100
butylenes	34-72	83-96.5	85-95	23-35	85-95	75-100	25-35	85-95	75-100

Table 7.7

Some properties of polyisobutylenes.

Indexes	Polybutene for LCL (TU 38 101743-81)			Octol (GOST 12869-77)			Polybutene for dispersion medium (TU 38 101679-74)		
	Standard by TU	In mixing reactor	In pipe reactor	Standard by TU	In mixing reactor	In pipe reactor	Standard by TU	In mixing reactor	In pipe reactor
Viscosity at 100°C, mm^2/s	3-10	5.4-9	4-9.5	75-115	93-95	75-108	≤ 115	45	49-79
Molecular mass	340-550	380-390	350-390	600-700	660	648-702	—	—	—
Flash temperature, °C	≥ 130	134-160	140-154	≥ 165	173-194	168-176	≥ 140	171	140-178
Iodine number, g of I_2/100 g of product	60-110	74-91	83-101	—	—	—	—	—	—

Note. Water is absent in all products.

Application of a small-scale pipe turbulent reactor operated under optimal parameters allowed combination of two processes in one device and enables both isobutylene and butylenes to be used in polymerization.

Table 7.6 displays the technological parameters of operation of the polymerization equipment in production of three types of products: polybutene, the raw component for production of lubricating–cooling liquids (LCL), electroinsulating liquid and polybutene, the component of the dispersion medium. Properties of resulting products are shown in Table 7.7.

Comparison of operation of pipe turbulent reactors with volumetric polymerization ones of mixing indicates that productivity of the former from a given amount of raw material is 2–2.5-fold higher than of the latter, at significantly lower volume of the reaction space (by 75 times) and metal consumption (by 150 times). Under application of the pipe reactor, specific catalyst consumption is decreased by 1.5–1.6 times. The conversion degree of butylenes reaches 95–100%. Moreover, as a consequence of the absence of mixing devices electric power consumption is decreased by 20–25% per 1 ton of processed raw material.

Decrease of light polymer yield, which is the industrial waste, is typical for the pipe reactor. The resulting polybutenes are characterized by narrower molecular-mass distribution, than polybutenes of the same brand, produced by the traditional technology. Consequently, consumption of the raw material (butane–butylene fraction) is decreased by 14–16% (see Table 7.6). Economy of electric power, raw material, catalyst, as well as shortening of exploitation expenses allow a decrease in the cost price of polybutene. In physicochemical properties, polybutenes produced in the pipe reactor completely fulfill requirements of appropriate technical conditions (Table 7.7).

Hence, high efficiency of application of the pipe reactor is typical for production of low-viscous polybutenes from industrial butane–butylene fraction. Resulting products comply with all requirements, and in width of MMD significantly exceed analogous products obtained by the traditional technology. Application of the pipe reactor simplifies technology of the production and decreases cost price of polybutene due to significant shortening of expense coefficients for the raw material and catalyst.

Table 7.8

Dependence of polyisobutylene yield on composition of the raw material and the type of device (isobutylene content in the raw material is 40 mass%, butylenes – 30%; catalyst – $AlCl_3$ in C_2H_5Cl, the rate of the raw material introduction into the reactor is 4 m^3/h, 241 K).

Raw material, mass%	*R*, m	Conversion, %	M_n
Butadiene – 0.2	0.5	47	7,000
	0.05	65	8,300
Butadiene – 0.4	0.5	13	5,400
	0.05	36	7,000
Butadiene – 0.2	0.5	39	6,000
Copper-ammonia	0.05	79	6,500
complex – 0.1		67	7,000

Concerning synthesis of polybutenes from butene–isobutylene fraction (BIF), Table 7.8 displays the experimental data, obtained under industrial conditions at the application of volumetric reactors of mixing (R = 0.5 m) and small-scale pipe reactors operating in the regime of quasi-ideal displacement in turbulent fluxes.

Applying the butene–isobutylene fraction, not only isobutylene oligomers with $\overline{M}_n$ = 700–1,500 (octol-K, and other) [44], but higher-molecular products can be produced in one stage, in particular, industrially important polymers of KP-5 ($\overline{M}_n$ = 4,000–6,000) and KP-10 ($\overline{M}_n$ = 9,000–15,000) brands [42].

As the rate of the main reaction of isobutylene polymerization is significantly higher than that of catalyst poisoning ($AlCl_3$) by butadiene, copper-ammonia complex, etc., isobutylene polymers are formed preferentially. Low rates of polymerization of *n*-, *trans*- and *cis*-butylenes (Figure 7.21), as well as the presence of great amounts of catalytic poisons (butadiene, residues of the copper-ammonia complex) make it necessary to introduce a somewhat greater amount of the catalyst into the reaction zone, but still at least 1.5–2 times less than in the same process, conducted in volumetric reactors with ideal mixing.

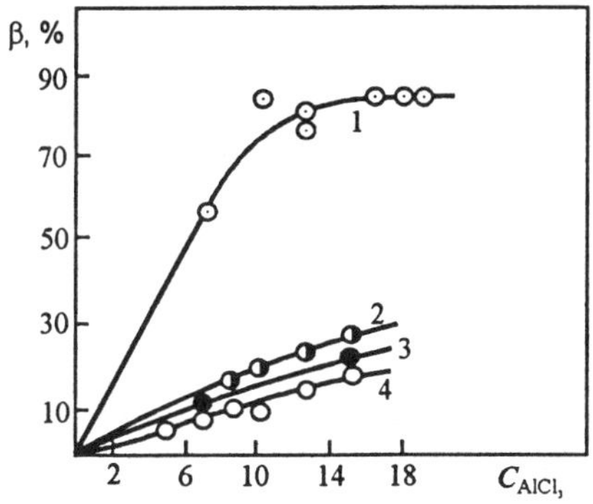

Figure 7.21. Influence of the catalyst concentration ($AlCl_3$) on conversion of butene isomers – the component of butene–isobutylene fraction: 1 – isobutylene; 2 – butene-1; 3 – *trans*-butene-2; 4 – *cis*-butene-2.

Let us note in conclusion that as pressure in the system (determining the boiling point of the catalyst, i.e. temperature in the reaction zone) is the main factor regulating molecular-mass parameters of the polymers, the resulting isobutylene polymerization from IIF, BIF and BBF and other fractions of C_4 hydrocarbons in small-scale pipe reactors, operating in the regime of quasi-ideal displacement in turbulent fluxes with internal heat elimination, enables isobutylene oligomers of any industrial brands to be produced without changing the technological scheme, reconstruction stoppage of the main technological equipment. The process is universal.

$AlCl_3$ dissolved in C_2H_5Cl is appropriate as catalyst for production of isobutylene polymers (238 ± 5 K) with M_n = 9,000–22,000 (KP-10 and KP-20 brands). However, for ecological safety, it is better to use $AlCl_3$ solutions in aromatic hydrocarbons. However, it must be taken into account that the $AlCl_3$·arene complex is insoluble in isobutane and must not be diluted by low-molecular paraffin hydrocarbons. In this case, $AlCl_3$ in arenes displays high viscosity, and at 253 K and higher the catalytic process loses fluidity; also, it is more than two-fold more dense than IIF. Estimation of kinetic constants at the application of catalysts of this type gave values $k_p \cong 10^3$ l/(mol·s), and $k_t \cong$ s^{-1}, i.e. significantly lower than for isobutylene polymerization in the presence of $AlCl_3$ dissolved in ethyl chloride ($k_p \cong 10^6$ l/(mol·s), $k_{ter} \cong$ s^{-1}) [18–20].

The above-discussed makes catalyst mixing with the monomer a critical factor in conducting the process under real (industrial) conditions (the reaction temperature is usually below 253 K).

To improve working of the pipe turbulent reactor industrial polymer production from isobutylene using complexes of $AlCl_3$ in arenes as catalysts, the zonal model of the reactor with internal heat elimination is better [16, 17]. In this case, in accordance with values of kinetic constants (k_p, k_{ter}) and the flux rate v = 1–2 m/s, the length of the reactor L in isobutylene polymerization in the presence of $AlCl_3$, dissolved in arenes, must not be less than 10 m.

To obtain polyisobutylene with molecular mass of 9,000–15,000 (NGNPZ), a pipe turbulent reactor, of the type shown in Figure 7.22, was used. This reactor provided catalyst delivery into three zones and its division in the volume at the sacrifice of turbulent whirls at the catalyst injection into the monomer flux. IIF was added at 256 ± 1 K. External heat elimination provided a controlled temperature field of the reaction, and low (normal) pressure provided boiling of the solvent (isobutane) in the second and the third zones of the reactor. Results of operation of the zonal model of the reactor for obtaining polyisobutylene of P-10 brand are displayed in Table 7.9.

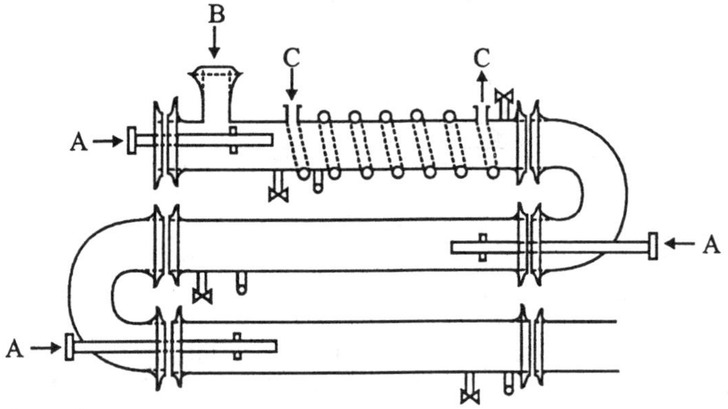

Figure 7.22. Principal scheme of the pipe reactor with zonal introduction of the catalyst: A – catalyst; B – monomer; C – cooling agent.

Hence application of multistage catalyst feeding (the zonal model of the pipe turbulent polymerization reactor) to production of isobutylene polymers from isobutene–isobutylene fraction in the presence of $AlCl_3$ dissolved in arenes provide quantitative polymerization of isobutylene with controlled quality of the final product.

To produce polyisobutylenes of P-20 brand with M_n = 16,000–22,000 (by Staudinger), raw materials of very high qualification are used in industrial production, in particular, isobutylene rectificates. Therewith, $AlCl_3$ solution in ethyl chloride is appropriate as catalyst. In this case, the main problem of conduction of isobutylene polymerization up to polymeric products of P-20 brand in volumetric reactors of mixing is heat elimination.

Table 7.9

Isobutylene polymerization in the presence of $AlCl_3 \cdot CH_3C_6H_5$ as a catalyst (35 mass%) in the pipe turbulent reactor (R = 0.036 m) with zonal introduction of the catalyst (255 K) (Novogroznensky Oil-Processing Plant).

Rate of introduction		Molecular mass of the polymer, M_n				Conversion, mass%
Raw material, m^3/h	Catalyst, l/h	I	II	III	Final product	
2.0	0.8	8,000	8,600	8,800	9,300	97.1
2.5	0.8	8,000	9,200	9,400	9,600	93.8
3.0	0.8	9,000	9,100	9,200	9,800	94.1
3.0*	0.6	5,000	—	—	5,300	93.2
4.0	0.8	9,000	8,700	9,300	10,000	94.7
5.0	0.8	9,400	9,500	10,000	11,000	76.3
5.0	1.2	9,200	9,400	9,500	10,200	91.3

Note. Composition of the raw material (IIF), vol.%: C_3 = 0.4; *iso*-C_4H_8 = 35.9; *iso*-C_4H_{10} = 42.8; *n*-C_4H_{10} = 19.8; C_4H_6 = 0.2; totality of *cis*- and *trans*-C_4H_8 = 0.6.

* The catalyst was introduced into one zone.

Insufficient external heat elimination during the process, combined with permanent recirculation of the reaction mixture due to low conversions of the monomer in one cycle and long exposure of the reaction mixture in the reactor (up to 2 hours) deteriorate quality of the resulting product, promote its clotting, etc. The process needs to be conducted at rather low temperature (190–200 K), for the purpose of which the raw material is cooled down to 183–193 K (the cooling agent is liquid ethylene). The process is characterized by high cost price, metal and power consumption, and difficulties in control. In this connection, it is appropriate to use the zonal model of the pipe turbulent reactor for production of polyisobutylenes of this type. This reactor enables the molecular mass of the resulting polymer to be increased without changing temperature of the reaction (Figure 7.22), which, in its turn, allows conduction of the process at temperatures some 30–40° higher. Operation of the pipe turbulent reactor with multistage addition of the catalyst (n = 10) without heat elimination for polyisobutylene of P-20 brand are displayed in Table 7.10.

Table 7.10

Isobutylene polymerization (20% in batch) in the presence of the catalyst $AlCl_3$ in ethyl chloride (2 g/l) in the pipe turbulent reactor (R = 0,036 m) with zonal addition of the catalyst (10 non-overlapping reaction zones) in Efremovsky Synthetic Rubber Plant.

Addition rate		Temperature, K			Monomer conversion, %	Molecular mass (by Staudinger)
Raw material, m^3/h	Catalyst, l/h	Raw material	Catalyst	In reaction zone (mean)		
3.5	600	216	182	282±1	99.9	19,600
4.0	600	210	180	276±1	99.9	20,700
5.0	700	210	180	282±1	93.1	21,100
5.0	700	223	183	287±2	98.0	17,000
5.0	700	223	183	287±2	96.2	18,400

Application of the pipe turbulent reactor allows regulation in a wide range of not only molecular mass of the resulting polymer (at the sacrifice of change of technological parameters of the process), but also of its MMD, which changes by both length of the reactor (in accord to temperature change in the reaction zone along the flux of the reaction mixture) and in dependence on the method of the catalyst addition (application of the zonal model of the reactor). Therewith, the greater is the number of zones and the longer is the distance between points of the catalyst addition into the reaction zone, the wider is the MMD of polyisobutylene (Figure 7.23).

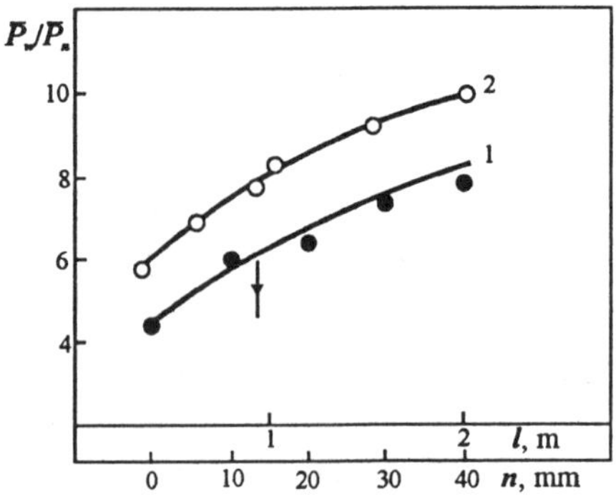

Figure 7.23. Change of MMD width of polyisobutylene with reaction zone length (1) in dependence on the system of the catalyst addition into the monomer solution (2); number of holes – 10, *n* – the distance between holes by length of the branch pipe of the catalyst addition ([M] = 20 mass%, v = 1 m/s, R = 0.025 m, T_0 = 213 m).

7.3. Specific features of butyl rubber production.

Production of butyl rubber is a well-assimilated process. However, there are still some problems that require special attention.

As in the case of polyisobutylene, one of them is bound to kinetic features of isobutylene copolymerization with isoprene. The reaction in the presence of BF_3 and $AlCl_3$ and other similar catalysts proceeds very fast. The reaction proceeds almost immediately on mixing up the reaction mixture with the catalyst solution as fluxes enter the reactor. Each drop is covered by a thin film of the polymer, and the chain propagation is limited by diffusion of monomers into the resulting polymeric–monomeric particle. As the heat conductivity coefficient of the polymer is low, and the front of spreading of temperatures and rates of the process is of flare or close to flare type (similar to polyisobutylene), temperature inside polymer–monomeric particles is always significantly higher than the average temperature of the reaction mixture in the polymerization reactor, as recorded by devices. Clearly, this is the reason for decrease of the molecular mass of butyl rubber and deviation of the average unsaturation degree from the expected one. Although the reaction mixture stays in the reactor not less than 30–40 minutes, the transformation degree of isobutylene is about 75 mass%. That is why, fast and fine dispersion of the catalyst solution in the volume of the reaction zone at its mixing with the monomer solution is of critical importance.

As in the case of polyisobutylene production, kinetic parameters and topochemical features of butyl rubber synthesis must be known. In this case, some change in construction of the polymerization reactor is also possible.

The second problem is bound to polymer precipitation on cooling surfaces under conditions of polymerization. Both butyl rubber, and a polymer of different properties, are partially precipitated on the internal surface of the polymerization reactor during synthesis of butyl rubbers. This reduces heat elimination and creates conditions for piling peripheral pipes and other reaction volume of the reactor by polymer. Consequently, the regime of the reactor operation is broken, the heat transmission coefficient decreases abruptly, and, finally, duration of the polymerization cycle is shortened.

An important role is also played by the resulting thin layer of frozen polymer (batch ice), formed at the reactor filling, which protects the reactor walls from polymer adhering and prevents temperature rise at sharp changes of the reactor productivity [48]. To preserve the batch ice during the process, the temperature of the reaction mass is kept 0.5–1° below the freezing point of methyl chloride. Nevertheless, a layer of polymer gradually adheres on the reactor walls during the reaction. When it becomes 5 mm thick, addition of the catalyst solution and cooling agent must be stopped.

Hence operation of the polymerization reactor in a permanent regime is conceptually problematic. Usually, the reactor is operated in the cyclic regime: the system operates in the working cycle for 20–100 h, and then is emptied for 10–16 h, washed by heated solvent (hexane fraction, benzene, petroleum ether, etc.) and prepared for work. Besides washing of the reactor by hydrocarbonic solvent, additional washing out by hydrocarbonic 0.1–1 mass% solution of catalyst (hydrated alkylaluminum chloride complexes) is appropriate at 270–300 K. Under these conditions, a gel of the polymeric product degrades and moves into solution, and the internal surface of the reactor becomes completely clean. Clearly, in this case, the productivity of reactors decreases; a noticeable amount of substandard product is formed, which is separated from solutions in hydrocarbon in a special apparatus, designed by usual schemes of precipitation of dissolved rubbers, is dried in a warm-drying press and delivered as a second quality product. It is technologically better not to separate rubber from the solution, but to use it as a glue, applied in construction and some other spheres. Rectification of the hexane fraction (or other hydrocarbonic solvent), used for washing out polymerization reactors, is performed in a special three-column set.

The third problem is bound to cyclic recurrence of the polymerization reactor operation on account of not only precipitation of the polymer on heat exchange surfaces, but also of obtaining substandard rubber at the reactor start up until the optimum operating regime is established. This creates formation of technological wastes. Polymeric products with wide MMD, different degrees of unsaturation and branching are formed. In many cases, substandard brands are appropriate for catalytic depolymerization with regeneration of isobutylene of polymerization purity.

The fourth problem is bound to prescription-technological factors. Application of chlorinated hydrocarbons as solvent-cocatalysts to formation of the catalytic system, as well as the dispersion medium, first of all, sharply increases the poisoning hazard of the production, and secondly, significantly complicates the technological scheme of the production – separation of recurrent products consumes high power, and a part of the products is not regenerated and is burnt off. In this connection, advantages of butyl rubber production in hydrocarbon solvent are obvious.

Moreover, drawbacks of the polymeric product itself must be noted, in particular, increased creeping, relatively low rate of vulcanization, non-covulanization with rubbers of general purpose, insufficient adhesion, bad compatibility with some ingredients, low elasticity at room temperature, and high heat formation at multiple deformations. Only some of the mentioned drawbacks could be eliminated by changing the compounding of rubber mixtures and conditions of their processing. However, radical changing of properties of butyl rubbers and, most of all, increase of the vulcanization rate, could be achieved only chemically.

One direction is production of new polymeric products, for example, isobutylene copolymers with β–pinene [49], differing by high resistance to ozone(Figure 7.24) and good ability for vulcanization by sulfur (Figure 7.25).

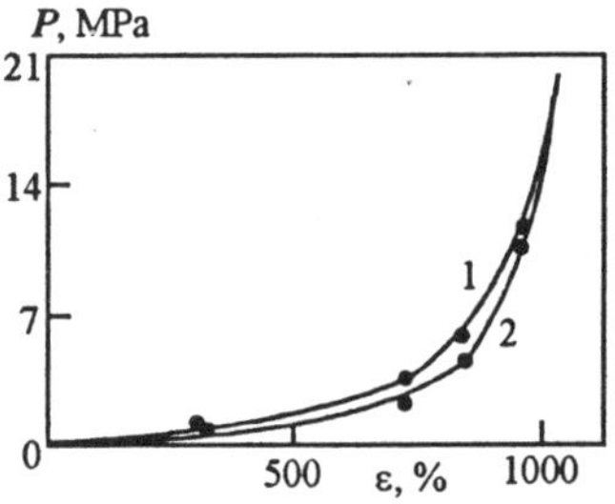

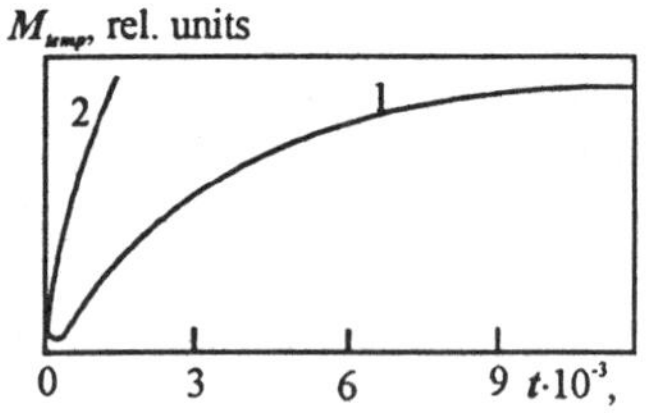

Figure 7.24. Load-elongation curves for isobutylene copolymer with β–pinene (unsaturation of 3%) before (1) and after (2) exposure in ozone (elongation rate is 10cm/min, ozone concentration is 0.00278 kg/m^3, 100% elongation, duration 4.54 s).

Figure 7.25. Vulcanization rate of isobutylene copolymers with 3% (1) and 10% (2) of β–pinene (Monsanto rheometer, 433 K).

A copolymeric product with the following chain structure

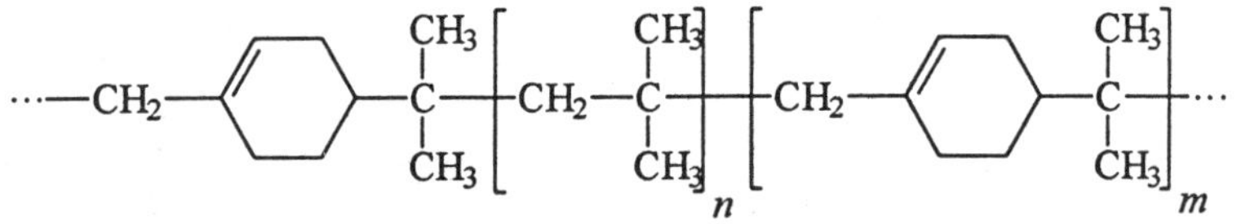

can be considered as a statistic copolymer of isobutylene with 1,4-disubstituted cyclohexene (β–pinene homopolymer is formally the ideally alternating copolymer of isobutylene and cyclohexene).

Change of the relation of polyisobutylene blocks and cyclohexene chains in the copolymer gives a possibility to vary properties of products in wide ranges (from rubber to plastics).

Industrial butyl rubber is aged by ozone with chain break at the isoprene fragment. Introduction of cyclohexene chains into polyisobutylene chains promotes protected "unsaturation" of polymeric products, i.e. the main polymeric chain is preserved even after splitting ozonolysis of macromolecules:

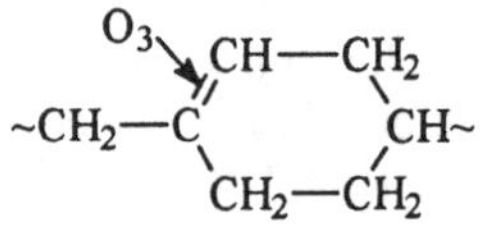

Under standard conditions of synthesis (catalyst $C_2H_5AlCl_2$, solvent–cocatalyst C_2H_5Cl, 250–265 K) an analog of butyl rubber with $\overline{M}_w$ of about 600,000–700,000, MMD ($\overline{M}_w/\overline{M}_n$) of 3–5 and T_g below 220 K is formed.

Butyl rubber with side reactive cyclopentadienyl groups, which can be added, for example, after treatment of partially polymerized chlorobutyl rubber with $(CH_3)_2C_5H_5Al$, possesses an interesting complex of properties, in particular, ability to form thermally reversible nets [50]. Depending on temperature and existence of compounds, capable for reaction with cyclopentadienyl groups, parameters of the vulcanization network can be varied up to obtaining cross-linked or completely soluble samples:

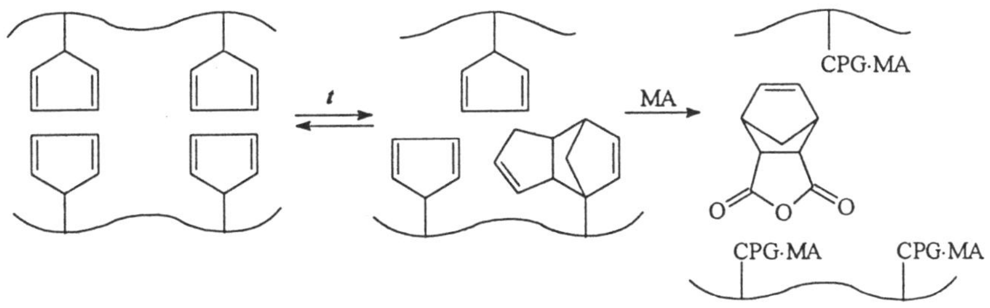

where CPD is the cyclopentadienyl group; MA is maleic anhydride.

Possibilities to modify butyl rubbers are not completely clear yet. At the present time, the main pathway to improve the complex of exploitation characteristics of the polymer is its halogenation.

7.4. Standard technologies of butyl rubber production.

7.4.1. Main premises.

Butyl rubber is the product of isobutylene copolymerization with a small amount of isoprene in the presence of aluminum trichloride in ethyl chloride or a complex catalyst based on an aluminum-organic compound in isopentane at low temperature:

$$CH_2{=}C(CH_3)_2 + CH_2{=}C(CH_3){-}CH{=}CH_2 \longrightarrow \left[\sim CH_2{-}C(CH_3)_2{\sim}\right]_n \left[\sim CH_2{-}C(CH_3){=}CH{-}CH_2{\sim}\right]_m$$

97-99% 3-1%

Various diene monomers can be used for copolymerization with isobutylene, but only isoprene enables butyl rubbers with stable properties to be obtained.

High-molecular polymers and copolymers based on isobutylene can be obtained only by cationic polymerization under conditions of extremely low temperatures (Figure 7.26). That is why, in spite of technological difficulties, connected with application of temperatures of about 173 K, the process was widely used in practice.

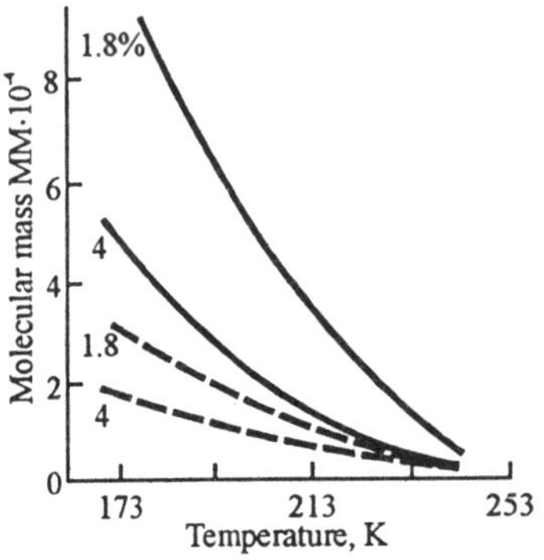

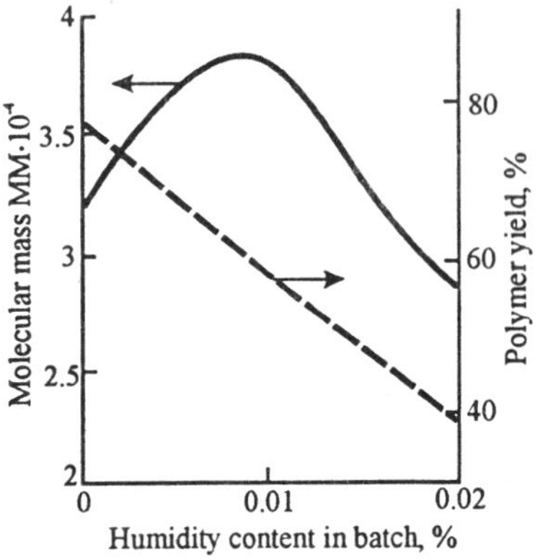

Figure 7.26. Dependence of the molecular mass of butyl rubber on temperature at polymerization in ethyl chloride (—) and isopentane (- - -). Numbers on curves mark content of isoprene.

Figure 7.27. Dependence of the molecular mass (—) and yield (- - -) of butyl rubber on humidity content in the batch.

The lowest alkyl chlorides are used as solvents for their ability to play the role of a cocatalyst in the process, as well as by the ability to obtain a polymerizate as dispersion. As dispersion, comparing with the solvent, possesses lower viscosity at equal concentration of the polymer, this allows application of quite high concentrations of monomers in the reaction mixture. As content of monomers in the batch increases, the molecular mass of butyl rubber hardly changes, which indicates the absence of kinetic chain transfer via the solvent. That is why content of monomers in the reacting mixture can reach 22–35 mass%.

Processes of cationic polymerization, forming the grounds for production of butyl rubber, are extremely sensitive to microadmixtures, especially those playing the role of cocatalysts, if they exist in significant quantities (humidity, alcohols, hydrogen chloride, halogenalkyls):

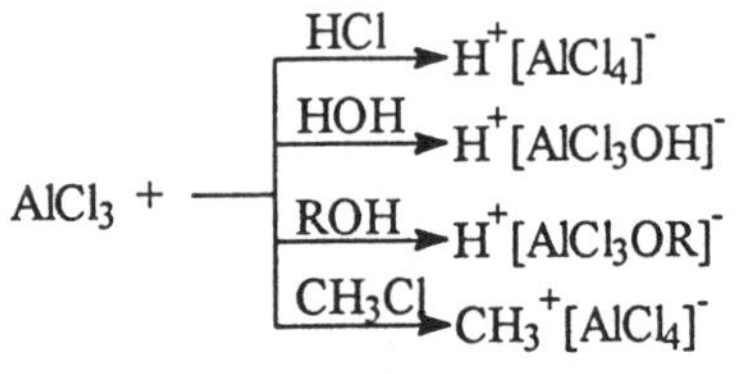

Usually, the highest initiating ability is possessed by complexes, formed at the molar relation catalyst:cocatalyst = 1:1. Hence, in the absolute absence of humidity, copolymerization of isobutylene with isoprene in $AlCl_3$ in hydrocarbon solvent does not proceed. The presence of humidity in the system causes copolymerization; however, the process is strongly inhibited if it is present in significant excess. A copolymer with the maximal molecular mass is produced at humidity content in the batch of about 0.01 mass% (Figure 7.27). This quantity of humidity corresponds to equimolar relation $AlCl_3$:H_2O. A similar effect on the process is caused by hydrogen chloride. Oxygen and sulfur-containing admixtures cause a decrease of the molecular mass of rubber.

The main admixtures accompanying isobutylene are normal olefins, among which β–butylene is the most harmful. It causes decrease of the polymer yield. Usually, isobutylene containing not less than 99.7% of the main substance and not greater than 0.2% of *n*-butylenes is used. The higher the concentration of isobutylene, the more stable and efficient is the copolymerization process.

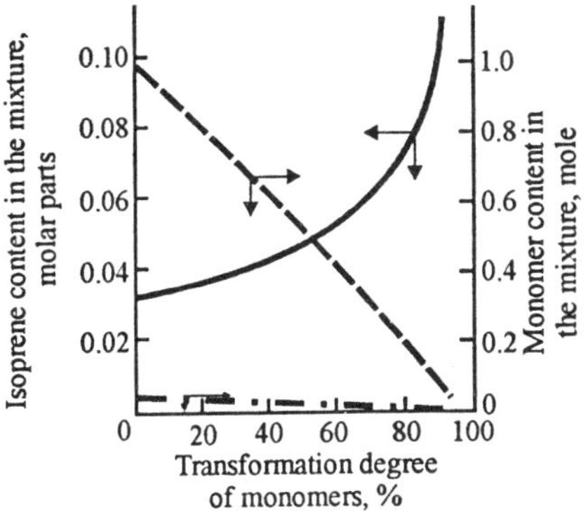

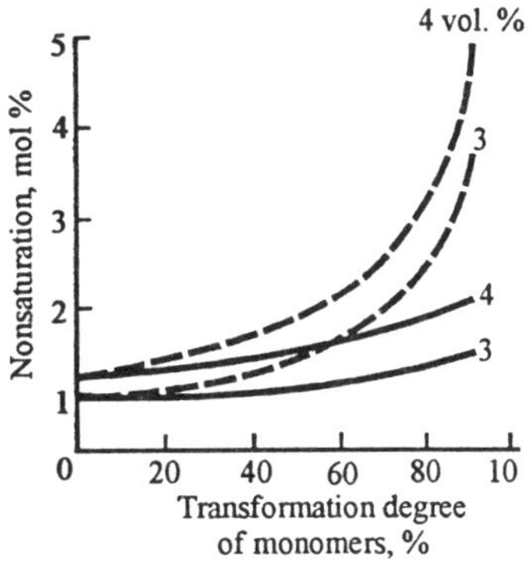

Figure 7.28. Dependence of isobutylene content (- - -) and isoprene (-·-·-) in the reaction mixture, as well as the molar part of isoprene in the mixture, (—) on the transformation degree of monomers.

Figure 7.29. Dependence of integral (—) and differential (- - -) composition of butyl rubber on the transformation degree of monomers at different content of isoprene in the initial mixture.

Isoprene, containing no more than 1.0% of piperylene and isoamylenes, is purified by passing its vapors over dry alkali with further condensation.

To purify methyl chloride, the method of its treatment by aluminum chloride is suitable; therewith all admixtures, able to interact with the catalyst, are bound. Concentration of the main substance in a solvent must be no less than 99.5%; the presence of dimethyl ester and hydrogen chloride must be avoided.

Among other possible admixtures, the following are allowed in the composition of monomers and the solvent, up to the values given (in mass%):

Alcohols (in isobutylene)	0.002
Carbonyl compounds (in isoprene)	0.0009
Unsaturated compounds (in methyl chloride)	0.007
Humidity (in all reagents)	0.002

Activity of isoprene at cationic copolymerization is significantly lower than that of isobutylene. When the process is conducted on $AlCl_3$ in methyl chloride at 170±1 K, r_1(isobutylene) = 2.25, r_2(isoprene) = 0.40. Such a great difference in constants of copolymerization causes the concentration of isobutylene in the solution of monomers to be decreased faster than the concentration of isoprene, and the proportion of isoprene in the mixture of monomers rises with the transformation degree (Figure 7.28). This causes the total unsaturation of butyl rubber to be increased with the transformation degree, and the difference in the composition of macromolecules, formed during different periods of time. If total nonsaturation, appropriate to the integral composition of the copolymer, changes comparatively slowly during copolymerization, then differential unsaturation changes significantly faster (Figure 7.29).

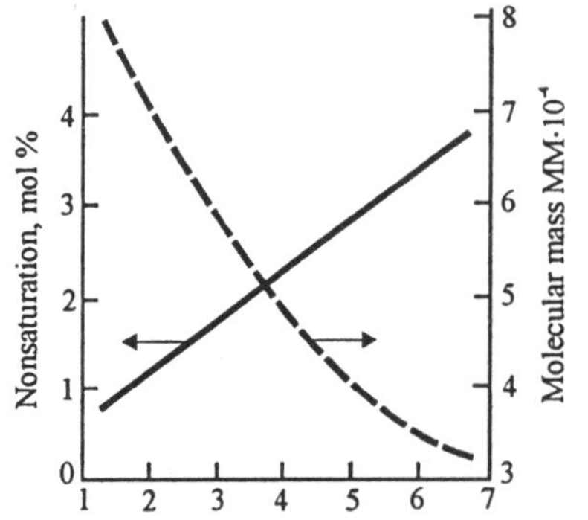

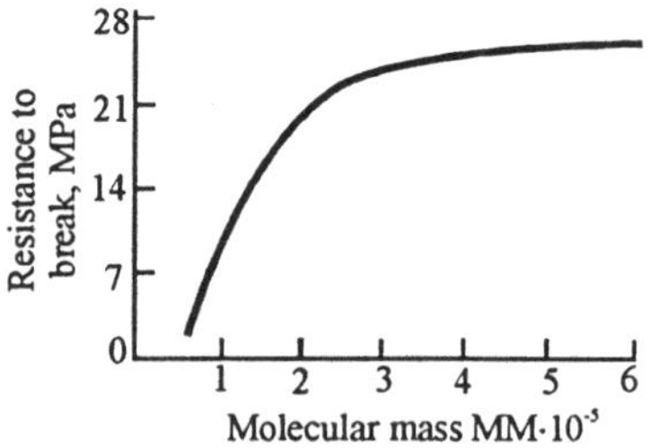

Figure 7.30. Dependence of unsaturation (—) and molecular mass (- - -) of butyl rubber on the isoprene amount in the mixture.

Figure 7.31. Dependence of strength of unsaturated rubbers on the molecular mass of butyl rubber.

Unsaturation of the resulting copolymer grows with isoprene content in the initial batch, but its molecular mass decreases (Figure 7.31). This is explained by isoprene participation in reactions of chain termination. This is the reason that the concentration of isoprene does not exceed 4% of the mass of isobutylene in obtaining industrial brands of butyl rubber. Investigation of the structure of butyl rubber has shown that isoprene chains in the macromolecule are mostly attached by the 1,4-*trans*-position; about 1.0% of isoprene chains are attached in 1,2- and 3,4-positions. Isoprene chains in butyl rubber macromolecules are not chemically bound with each other and are statistically distributed in all macromolecules.

Butyl rubber, obtained with $AlCl_3$ in the medium of methyl chloride, displays high linearity of macromolecules and narrow molecular-mass distribution.

Butyl rubbers with molecular mass by Staudinger[2] over 30,000 are widely applied in practice. At lower values of the molecular mass, vulcanizates of butyl rubber possess insufficient strength (Figure 7.31) and increased stickiness, which make difficulties in obtaining the polymer according to the permanent scheme due to polymer adhesion on the cooling surfaces. To obtain butyl rubber with the required molecular mass, the process is conducted at a temperature no higher than 177 K, because the molecular mass of the resulting copolymer abruptly decreases at even a small rise of temperature (Figure 7.26). At such low temperatures, the resulting

[2] Calculating molecular mass by Staudinger, it is suggested that $[\eta]=K\overline{M}_\eta$, the constant K being determined for the lowest homologues of the sequence. Real values of the molecular mass of polymers are usually greater than those, calculated by the Staudinger method by an order of magnitude.

polymer is insoluble in methyl chloride and exists in the glassy state (glass transition temperature for butyl rubber is 204 K), which makes dispersion of the polymer quite stable and transportable.

Required dosage of the catalyst solution is determined by purity of applied monomers and the solvent. Usually, aluminum trichloride in amount of 0.025–0.035% from the mass of monomers is consumed for obtaining butyl rubber. Therewith, 2.5–3.5 tons of the polymer per each kilogram of the catalyst are formed.

7.4.2. Technology of production of butyl rubber.

The catalytic solution is prepared by passing purified methyl chloride through a device, filled with granulated anhydrous aluminum trichloride, at 243 K. Therewith, due to limited solubility of $AlCl_3$ in CH_3Cl, a solution is obtained, possessing constant concentration of the catalyst (1 mass%). The saturated catalytic solution is diluted by methyl chloride in a pipeline down to the working concentration (0.1 mass%), is cooled down to 180 K in an ethylene cooler and is delivered to the polymerization reactor. All operations involved in preparation of the catalytic solution are performed in an atmosphere of dried nitrogen.

The batch is prepared by mixing of isobutylene, isoprene and recurrent isobutylene–methyl chloride fraction, dried and purified from harmful admixtures and pumped from corresponding tanks. To support the assigned composition of the batch, its chromatographic analysis is performed, in accordance with results of which the dosage of initial components is corrected. The batch composition is determined by the brand of produced butyl rubber. Prepared batch is delivered from the tank 1 by the pump 3 (Figure 7.32) for cooling into propane 4 and ethylene 5 coolers, and is delivered to polymerizer 6 at a temperature 176 ± 1 K.

All reactors, applied in industry to production of butyl rubber, are standardized and differ only in trivial construction details. They are of cylindrical form, are equipped with central sucking pipe with the circulation pump in the lower part of it. A great number of peripheral pipes of smaller diameter or a one-piece hollow ring are disposed around the sucking pipe. Both the central pipe and peripheral pipes, as well as the bottom and cover plate of the polymerization reactor are surrounded by the cooling agent (evaporating ethylene). There is a run-over pipe for the polymerizate output in the upper part of the reactor.

The batch and catalytic solution are permanently delivered to the lower part of the reactor and enter sucking branch pipes of the circulation pump manifold. As it rotates intensively, fluxes are turbulized, which promotes better heat exchange, and directed motion of the reaction mass by the circulation pipe up from below is performed. Back to the lower part of the polymerization reactor the reaction mass is delivered by the circle tube still. A definite amount of the polymer dispersion, equal to the amount of delivered batch, is permanently released from the upper part of the polymerization reactor.

Copolymerization of isobutylene with isoprene proceeds at a very high rate. Already at batch mixing with the catalyst solution, each drop of it is covered by a thin film of the polymer, and the chain propagation is performed at the sacrifice of diffusion of monomers into the polymeric–monomeric particle formed. As the heat conductivity coefficient of the polymer is small, the temperature inside polymeric–monomeric particles may be much higher than in the reaction medium,

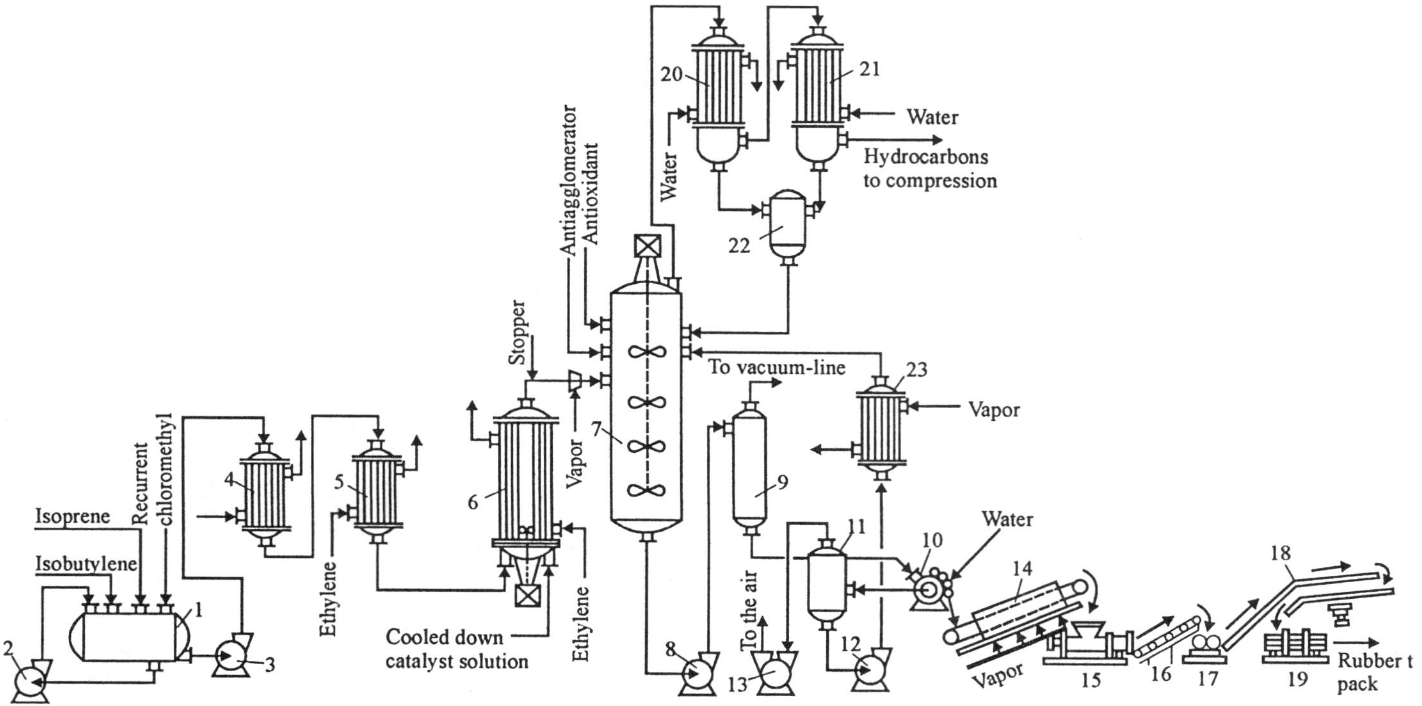

Figure 7.32. Scheme of butyl rubber formed in methyl chloride:

1 – tank for batch preparation; 2, 3, 8, 12 – pumps; 4, 5, 20, 21 – coolers; 6 – polymerizer; 7 – water degasifier; 9 – vacuum degasifier; 10 – vacuum-filter; 11 – vacuum-receiver; 14 – dryer; 15 – extruder; 16 – conveyor; 17 – rolls; 18 – cooling conveyor; 19 – brick-making machine /brick former/; 22 – separator; 23 – heater.

which leads to a decrease of the molecular mass of copolymer. That is why, fast and fine dispersion of the catalyst solution at its mixing with solution of monomers is of critical importance.

Although copolymerization of isobutylene with isoprene proceeds almost immediately, the resulting dispersion of polymer in methyl chloride, formed at mixing of solutions, is exposed for 30–40 minutes in the reaction zone. Therewith, the transformation degree of isobutylene becomes 75%. Temperature in the reaction zone is regulated by pressure of evaporating ethylene or by addition of the catalytic solution. During polymerization, polymer precipitation on cooling surfaces is possible, which reduces heat exchange, and cleaning of the reactor becomes necessary. Washing over of the polymerization reactor is performed by heated up solvent (gasoline, petroleum ether, hexane fraction, etc.).

The product containing 8–12% of the polymer, non-polymerized monomers and methyl chloride, is mixed with a stopper (methyl or isopropyl alcohol) in the run-over pipe in order to deactivate the catalyst, and then is delivered through the granule-former into water degasifier (Figure 7.31). Elimination of the main mass of methyl chloride and unsaturated hydrocarbons is performed at 345 ± 3 K in the degasifier of the first stage. Heat necessary for elimination of volatile products is delivered at the sacrifice of heating up of circulation water and sharp high-pressure vapor, delivered into the granule-former. An anti-agglomerator (freshly precipitated zinc or potassium stearate) and antioxidant dispersion in water (for example, neosone D or product 2246) are added to the degasifier. Vapors of the solvent and monomers pass through coolers 20 and 21, in which water vapors are condensed, and are delivered for compression, separation and processing of recurrent products.

Passing through the degasifier of the first stage, the pulp containing 3% of the polymer, is delivered to the vacuum degasifier 9 by the pump 8. Here, residues of the solvent and other volatile products are eliminated from the rubber at residual pressure of 19.6–39.2 kPa (0.2–0.4 kGs/cm^2), created by the vacuum-pump. After a vacuum the degasifier, a dispersion of the rubber in water is delivered to the drum-type vacuum-filter 10, where the rubber is additionally washed by softened water and pressed off from humidity on pressing rolls. Water, separated on the vacuum-filter, after vacuum-receiver 11 is returned to the degasifier of the first stage through the heater 23 by the pump 12. Drying of the rubber is performed in the air drier 14 by hot air at 388 ± 5 K.

After processing in the drier butyl rubber contains 0.5% humidity. To eliminate sticking of the rubber, the belt transporter of the drier is sprayed by an emulsion of silicon liquid. Heated butyl rubber is homogenated in the extruder 15, and is delivered to rolls 17 as a belt through the conveyor 16. As the temperature of the rolls is 388 ± 5 K, all traces of humidity are eliminated. Passing the rolls, the belt is delivered to the brick-former 19 through the cooling conveyor 18. Bricks of the rubber with mass 30 ± 2 kg are coated by polyethylene film, packed into containers and transported to the finished products storage area.

Worm aggregates are also used for drying butyl rubber, which have certain advantages in processing of dissolving rubbers for general purposes.

Due to development of various processes for obtaining modified butyl rubbers and latexes, the process of polymerizate degasification can be performed by applying organic solvents instead of water (hexane, cyclohexane, etc.). In this case, a homogeneous solution of the butyl rubber is obtained as the final product.

7.5. New technologies in production of ecologically friendly butyl rubber.

Two unique methods of production of butyl rubber, possessing increased ecological friendliness, have been developed in Russia. These methods are also energy and resource saving.

7.5.1. Synthesis of high-molecular butyl rubber in hydrocarbon solution.

The process of butyl rubber production in a hydrocarbon solvent (isopentane) at 188 ± 5 K was developed and assimilated. As a catalyst, this process applies hydrated complexes of haloid aluminum-organic compounds $R_nAlCl_{3-n}{\cdot}H_2O$ in 1:1 relation.

The efficiency of catalytic systems grows in the sequence: $RAlCl_2{\cdot}H_2O$ < $(RAlCl)_2O{\cdot}H_2O$ < $[RAlCl_2{\cdot}H_2O + (RAlCl)_2O{\cdot}H_2O]$.

The process of isobutylene copolymerization with isoprene in a hydrocarbon solvent is characterized by some specific features, compared with the known methods of obtaining butyl rubber in chlorohydrocarbons under the effect of $AlCl_3$.

In particular, constants of isobutylene copolymerization with isoprene are significantly changed. Composition of the catalyst affects the yield of the polymer and molecular characteristics of the resulting butyl rubber (Table 7.11).

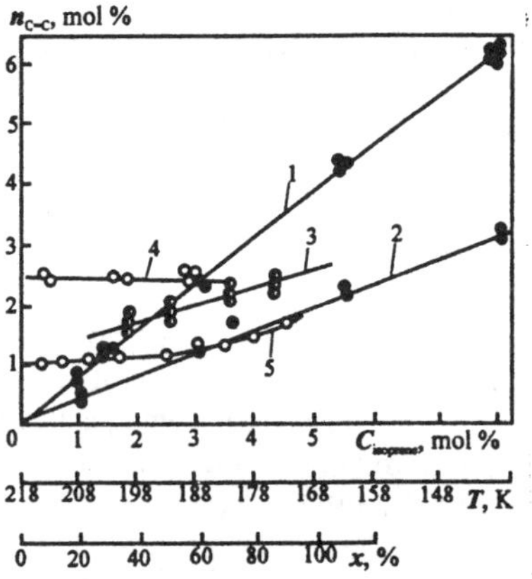

Figure 7.33. Dependence of C=C-bonds content in butyl rubber on the concentration of isoprene (1, 2), temperature (3) and conversion of monomers (4, 5): 1, 3, 4 — catalyst $(C_2H_5)_{1.5}AlCl_{1.5}{\cdot}H_2O$, solvent — isopentane; 2, 5 — catalyst $AlCl_3$, solvent — chloromethane (173 K, 900 s); Concentration of monomers, mol/l: 1, 2 — 3.1; 3 — 4.88; 4, 5 — 2.44.

As concentration of monomers in the reaction mixture grows and polymerization temperature decreases, the polymerization degree of butyl rubber increases. The most high-molecular fractions (10–20 mass% of the polymer) with

$\overline{M}_w$ more than $1.5 \cdot 10^6$ represent branched polymers with the mean mass number of branching tangles of 10–12.

A typical feature of the process is the almost complete absence of the reaction of chain transmission via the monomer, increase of the relative reactivity of isoprene (Figure 7.33) and independence of C=C-bonds content in butyl rubber on conversion. Content of C=C-bonds in the copolymer decreases with temperature rise at the same composition of the monomer mixture (Figure 7.33, curve 2).

Consequently, molecular mass (chain transmission to the monomer is absent during the copolymerization of isobutylene with isoprene), unsaturation degree of macromolecules and polymer yield grow in the absence of $R_nAlCl_{3-n} \cdot H_2O$ (depending on its content in the mixture with R_nAlCl_{3-n}). In the presence of $R_nAlCl_{3-n} \cdot H_2O$, a copolymer is formed, isobutylene sequences of which are separated by a single macromolecule of isoprene ($r_1 \cdot r_2 \to 0$), whereas application of $AlCl_3$ in CH_3Cl leads to statistic copolymerization of isobutylene with isoprene ($r_1 \cdot r_2 \to 1$). As a consequence, properties of rubber mixtures based on butyl rubber are also improved: they display affinity to plasticization, possess increased basic skeletal structure, higher elastic regeneration and swelling in a bridle at extrusion.

Table 7.11

Influence of the catalyst composition on copolymerization constants, and also on properties and yield of butyl rubber (isopentane, isoprene concentration 2.34%, catalyst content $6.4 \cdot 10^{-4}$ mol/l, 188 K, 15 min).

Indexes	Content of aquaomplex $RAlCl_2 \cdot H_2O$ calculated for $RAlCl_2$, mass%								
	20	30	40	50	60	70	80	90	100
Copolymerization constants ± 0.003									
r_1	1.39	1.21	1.14	—	1.04	—	0.98	0.98	0.98
r_2	0	0	0		0		0	0	0
Polymer yield (recalculated for $RAlCl_2 \cdot H_2O$ complex)	110	200	300	400	450	550	650	810	1000
Polymerization degree of macromolecules ($P_n \times 10^{-3}$)	6.5	11.6	14.2	16.5	19.0	20.0	23.0	24.1	26.2
Content of unsaturated C=C-bonds in the composition of butyl rubber macromolecules (mass %)	1.63	1.75	1.90	2.10	2.20	2.28	2.32	2.30	2.34

Note. For $AlCl_3$ in CH_3Cl, $r_1 = 2.25 \pm 0.003$; $r_2 = 0.4 \pm 0.001$.

The new technology allows regulation of the molecular mass and molecular-mass distribution of butyl rubber in wide ranges and guaranteed obtaining of polymeric products of higher exploitation chatacteristics compared with the butyl rubber, obtained according to the standard technology in chlorohydrocarbons in the presence of $AlCl_3$.

The scheme of butyl rubber production in hydrocarbon solution is displayed in Figure 7.34. The role of solvent for preparation of the catalytic complex in obtaining soluble butyl rubber is mostly played by isopentane. The reaction mixture is prepared by mixing up isopentane, isobutylene and isoprene in the device 1.

Relation of components is determined by the brand of resulting butyl rubber. The circulation pump 2 conducts mixing up and delivery of the batch to polymerization. Control and regulation of the batch composition is performed by

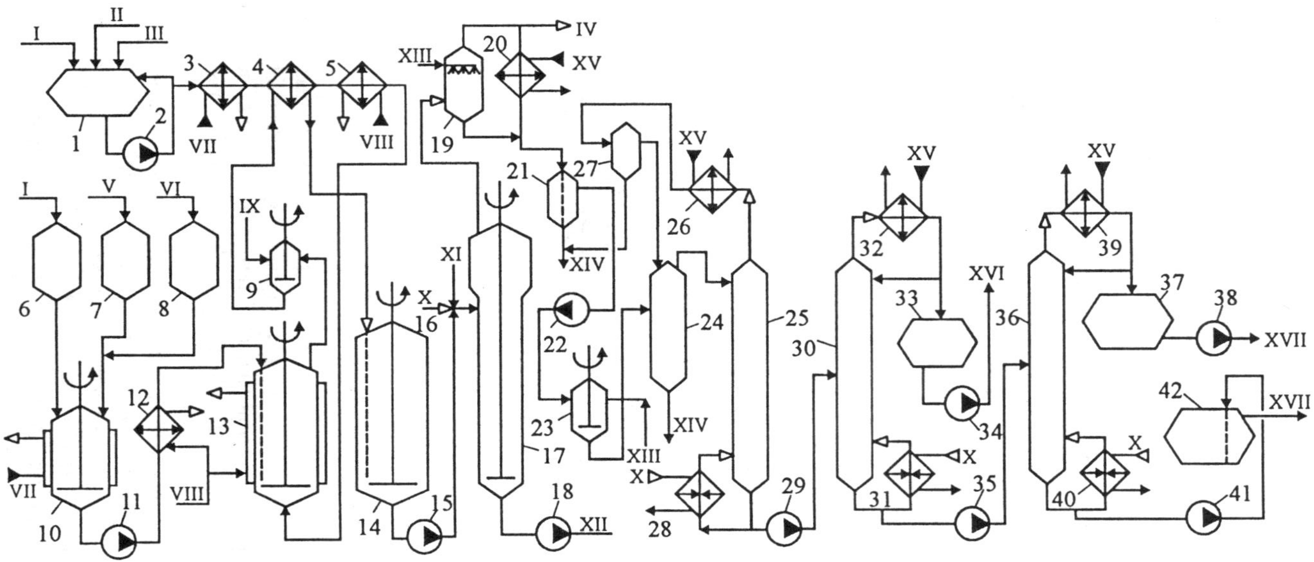

Figure 7.34. The scheme of production of butyl rubber in isopentane:

1 — tank for batch preparation; 2, 11, 15, 18, 22, 29, 34, 35, 38, 41 — pumps; 3, 4, 5, 12 — coolers; 6, 7, 8 — batchers; 9, 23 — intensive mixers; 10 — device for catalyst preparation; 13 — polymerization reactor; 14 — averaging machine; 16 — injector; 17 — degasifier; 19 — separator; 20, 26, 32, 39 — condensers; 21 — hydraulic lock; 24, 27 — presettling tanks; 25, 30, 36 — rectification columns; 28, 31, 40 — boilers; 33, 37, 42 — collectors.

I — isopentane; II — isoprene; III — isobutylene; IV — to compressor; V — catalyst; VI — modifier; VII — liquid propane; VIII — liquid ethylene; IX — stopper; X — vapor; XI — stabilizer, anti-agglomerator, circulating water; XII — pulp to concentrator; XIII — water; XIV — water for stripping of organic compounds; XV — brine solution; XVI — condensate for batch preparation; XVII — isobutylene to the store area; XVIII — isopentane to the store area.

automatic chromatograph. The prepared mixture is cooled down to 183 K in propane 3, recuperative 4 and ethylene 5 coolers and is delivered to copolymerization.

The catalytic complex is prepared in the device 10. Isopentane, ethylaluminum sesquichloride and modifier are dosed in the assigned relation from batchers 6, 7 and 8, respectively. Heat released during preparation of the complex is eliminated by liquid propane, delivered into the coat of the complex preparing device 10. Prepared complex is delivered to copolymerization by the pump 11 through ethylene cooler 12.

Copolymerization of isobutylene with isoprene is conducted in a usual mixing reactor 13, equipped by a mixer and a coat (external heat elimination), fed by liquid ethylene in order to eliminate the reaction heat.

A temperature of 188 ± 5 K is preserved in the reactor. Cooled down solution of the catalytic complex is dosed in the assigned relation into the lower part of the polymerization reactor. The polymerizate yielding from the reactor, containing 10–15 mass% of polymer in solution, is mixed in the intensive mixing tank 9 with a stopper (ethyl alcohol) in order to deactivate the catalytic complex. Then it is delivered to the averaging machine 14 through the recuperative heat exchanger 4.

Precipitation of butyl rubber from the solution is conducted in accord with the usual scheme, assumed for precipitation of dissolved rubbers. From the averaging machine 14, the polymerizate is delivered to the injector 16 by the pump 15, where it is mixed with water vapor, anti-agglomerator and suspension of a stabilizer. The mixture is delivered to the upper plate of the degasifier 17. From the degasifier cube 17, degasified granulate of the rubber is delivered by the pump 18 for concentration, purification from water and drying, which are performed according to usual schemes applying, for example, aggregates LK-4, LK-8 or the one by the Andersen Company (USA). Recurrent products, separated in the separation part of the degasifier 17, are delivered to separation and utilization.

Vapors of hydrocarbons and water are delivered to the separator 19, irrigated by water, in which recurrent products are partially condensed, granules of rubber are captured, and volatile products, directed to compensation, are separated. Condensation of recurrent products is performed in the condenser 20, cooled by the brine solution. The condensate, united with the one from the separator 19, is poured off into the hydraulic seal 21, in which stratification is made. The lower water layer is delivered for steaming of organic products, and the upper hydrocarbonic layer is delivered by the pump 22 to the intensive mixer 23 for mixing up with water and washing off from the stopper. Hydrocarbons, washed off from water-soluble admixtures, are delivered from the upper part of the presettling tank 24 into the aerotropic drying column 25, equipped by the boiler 28, and the water layer is delivered to stopper separation.

The azeotrope, eliminated from the upper part of the column 25, is condensed in the condenser 26, cooled by the brine solution, and is delivered to the presettling tank 27. The lower water layer from the presettling tank 27 is delivered for steaming into the presettling tank 24. Dried hydrocarbons from the cube of the column 25 are delivered by the pump 29 to the column 30, equipped by the boiler 31 and the fractionating column 32. The condensate is partially delivered for irrigation of the column 30, partially through the collector 33 by the pump 34 – for preparation of the batch. Liquid from the cube of the column 30 is delivered by the pump 35 for stratification of hydrocarbons into the column 36, equipped by the boiler 40 and fractionating column 39. Isobutylene, condensed in the fractionating column 39, is pumped off by the pump 38 through the collector 37 and is delivered to the store area. A part of it is returned to irrigate the column 36. C_5 hydrocarbons are pumped off by the pump 41 into the tank 42, wherefrom they are delivered to processing or to the store area.

The new process possesses some significant advantages: 1) application of highly toxic and volatile CH_3Cl is excluded; 2) polymerization temperature is some 20–40° higher; 3) working time of the reactor before washing out is increased by 8–12 times, most of all by less adhesion of the polymer to the reactor walls; 4) elimination of catalyst residues from the final product is much more simple; 5) as the rate of the process is significantly lower than that of the reaction of butyl rubber production in the presence of $AlCl_3$ in CH_3Cl, the external heat elimination becomes quite effective; 6) the process is easy to control; 7) molecular mass of the resulting polymer is strictly controlled because there is less temperature change in the reaction zone (M_n from 500 to 100,000); 8) translation of the polymerization system on production of another brand is performed without stopping the equipment and changes of the raw material quality; 9) usage of substandard raw materials is possible because of specificity and increased selectivity of the catalytic system applied; 10) usage of worked off gases in the production cycle is possible; 11) low corrosion effect of the reaction mixture; 12) sanitation of working conditions.

Application of catalysts of various modifications allows control of technological properties of butyl rubber, obtaining of a polymer with different polydispersion and creep, with the lowest scatter of Muni viscosity. Application of polymerizate cold for cooling the polymerization mixture, exclusion of the adsorption drying assembly, special compressors and highly efficient columns, and other technological advantages enable energy consumption of the process, total price and cost price of butyl rubber to be decreased.

Consequently, compared with standard processes ($AlCl_3$ in CH_3Cl), consumption of capital of the new process for the set with productivity of 40 thousand tons per year is some (20 ± 5)% lower, and the cost price of butyl rubber with improved exploitation characteristics is about (15 ± 5)% lower.

7.5.2. Synthesis of low-molecular butyl rubber in turbulent pipes.

In 1981, a new method of production of butyl rubber with MM = 20,000–40,000 (by Staudinger) was agreed for development, in which the role of the main polymerization reactor was played by a small-scale pipe turbulent reactor, 10 cm in diameter and 600 cm long, instead of the volumetric mixing reactor of 8 m^3 by bulk (engine 75 kW/h power; consumption of liquid ethylene for heat elimination, 1.8 ton/h). A typical feature of the pipe turbulent reactor is its construction as a pipe without cooling cover, with a branch pipe for accompanying introduction of the catalyst ($AlCl_3$ in ethylene chloride solution) and a branch pipe for radial introduction of comonomers dissolved in ethylene chloride. Besides low metal consumption (900–1000 times lower than for volumetric mixing reactor in the standard process), the pipe turbulent polymerization reactor differs by simplicity of construction, servicing and control of the process absence of electric energy consumption for mixing devices and cooling agent, delivered to the reactor. Decrease of electric power consumption at permanent operation amounts to more than 650,000 kW/h. Repair and servicing is much easier. It also differs by the absence of unproductive losses at preservation of the main technological scheme.

The process can be made automatic.

The scheme of the process control provides for regulation of delivery of the reacting mass and solvent (ethyl chloride). Control of the reaction temperature is performed by cyclic delivery of catalyst using a generator set with alternating duty factor of precedence block and regulation of the influence by rate of switching off the controlled value. Introduction of differentiating chain generally stabilizes operation of the control contour (deviations in the range of 3–5%).

The operation regime of the pipe reactor for synthesis of butyl rubber is the following: consumption of the reaction medium (relation of isobutylene is 20 ± 5% to the isoprene volume – 1–8 vol.%) is 3–4 ton/h at the catalyst consumption (0.2% solution of $AlCl_3$ in chloroethyl) of 0.1–0.2 m^3/h. The linear rate of the reaction mixture motion in the reactor is 0.2–1 m/s. Butyl rubber without gel-fraction is obtained.

7.6. Standard production technologies of modified butyl rubbers (bromobutyl rubber, chlorobutyl rubber).

7.6.1. Main premises.

The most used industrial method of butyl rubber modification to improve and widen its exploitation properties is the halogenation reaction. In different countries, from ¼ to ¾ of the total production of butyl rubber is accounted for by chlorinated and/or brominated butyl rubber. A series of methods of butyl rubber halogenation can be used: in solution, melt, dispersion, in the solid phase. However, only the first one is used in industry. The solvent must not chemically interact with halogenating agent. Most often, the process is performed in carbon tetrachloride or hydrocarbons.

Chlorination in water dispersion requires preliminary preparation of 15–25 mass% solution of butyl rubber in saturated hydrocarbon (*n*-hexane) and its dispersion in 150–800 mass parts of the water phase containing 1–10% of ionogenic emulsifier of Na nonylphenoxypolyethylenesulfate type and 1–5 mass% of emulsion stabilizer of Na dihydroorthophosphate type. Then the solvent is removed down to a content less than 1–2 g per 100 g of the polymer. Prepared latex with the average diameter of particles of 0.5–2 μm (dry residue 30–60 mass%) is delivered for chlorination (290–355 K). Chlorobutyl rubber latex is used directly, or the polymer is separated by usual methods (coagulation, concentration).

Chlorination of butyl rubber in melt is performed on rolls, using dichloroamine T as chlorinating agent.

At the present time, chlorination of butyl rubber is the only method widely applied in industry.

To obtain chlorobutyl rubber, a mixture of chlorine with nitrogen in ratio from 1:5 to 1:10 is pumped through the polymer solution. IR- and NMR-spectroscopy methods have proved the structure of polymers, resulting in this case. Generally, reactions of hydrogen atom substitution in isoprene chains proceed; therewith, 75% of the unsaturation of the initial butyl rubber is preserved, and chains of the following types are formed:

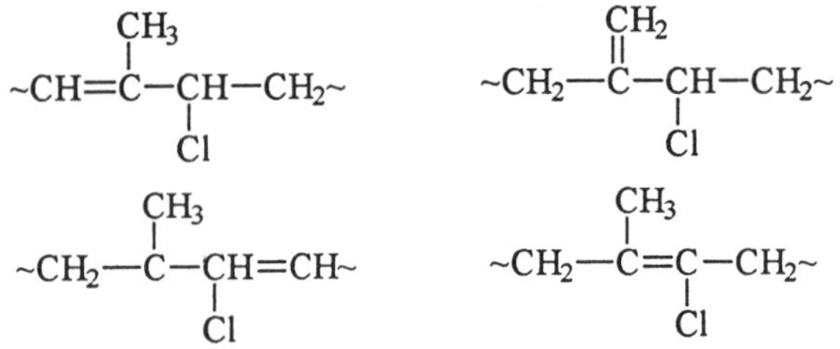

The chlorine content in the modified butyl rubber is 1.1–1.3 mass%. Chlorine atoms in the allyl position differ by high mobility, and halogenated chains are able to participate in vulcanization reactions.

Initial solutions are prepared by two different methods: dissolution of dry granules of the rubber or substitution of the solvent. Ethyl chloride, chloroform, toluene, hexane, heptane, cyclohexane, the fraction of C_4–C_7 hydrocarbons, purified from olefins (bromine number less than 0.1)[3] can be used as solvents.

The concentration of butyl rubber in solution is 5–20% for high-molecular product and 30–60% for the low-molecular one (molecular mass of 5,000–10,000). Chlorination of butyl rubber is performed by Cl_2, and as well by monochloroiodite, hydrogen halides in the presence of oxidants, *tert*-butyl hypochlorite or hypochlorites of alkali metals, sulfurylchloride, chlorosulfonic acid, chloroacetamide, or aromatic sulfenylchlorides [51, 52]. Chlorine must be diluted (by 5–10 times) by an inert gas: N_2, CO_2, CH_4, C_2H_6, etc.

The amount of chlorine, necessary to be introduced into the elastomer, depends on the degree of unsaturation of the initial butyl rubber and must equal, approximately, to 1 atom of Cl per double bond – the "critical" concentration, over which degradation of the polymer becomes deeper, and useful characteristics of butyl rubber deteriorate. The critical concentration is estimated from the relation:

$$\overline{K} = \frac{aD}{100M_1 + D(M_2 - M_1 - a)} \cdot 100 = \frac{D}{1.58 - 0.0134D},$$

where $\overline{K}$ is the chlorine content in chlorobutyl rubber, mass%; D is the content of C=C-bonds (unsaturation) in butyl rubber, mol%; M_1 and M_2 are molecular masses of isobutylene (56.11) and isoprene (68.12), respectively; a is the atomic weight of chlorine (35.46).

Reaction of substitution of H atoms by Cl proceeds quite fast. The reaction time is shorter than 60 s. The temperature range is quite wide (from 220 to 475 K, preferably 290-325 K). During chlorination of butyl rubber, the content of internal C=C-bonds decreases and content of exomethylene groupings increases in the final product (Table 7.12).

For rubbers containing no more than 1 mass% of chlorine, approximately 90% of initial unsaturation is preserved. At deeper chlorination, unsaturation and molecular mass of the polymer decrease. However, to estimate accurately the molecular mass, it is necessary to determine values of constants K and α in the Kuhn–Mark–Hauvink

---

[3] For example, hexane fraction of the following approximate composition, mass%: 2,2-dimethylbutane – 0.1; 2,3-dimethylbutane - 2.4; 2-methylpentane – 10.75; 3-methylpentane – 12.45; *n*-hexane – 43.5; methylcyclopentane – 20.5; 2,2-dimethylpentane – 0.4; benzene – 7.7; cyclohexane – 0.85; C_6 olefins – 1.35; hydrogen sulfide, etc.

equation with regard to chlorination degree and total nonsaturation of butyl rubber. Application of the correlation $[\eta] = KM_\eta^{\alpha}$, suggested to estimate $\overline{M}_\eta$ of butyl rubber, is incorrect for estimation of its chlorinated derivatives.

Table 7.12

Molecular characteristics of chlorobutyl rubber (10 mass% solution of butyl rubber in *n*-hexane, Cl_2:N_2 = 1:6, 293 K).

Sample*	$\overline{M}_\eta \cdot 10^{-3}$	Calculated content $G \cdot 10^2$ of, mol/mol of polymer			Content $G' \cdot 10^2$ of, mol/mol of polymer	
		Unsaturated C=C-bonds	Isoprene chains	Chloroprene chains	Internal unsaturated C=C-bonds	Exomethylene groupings
Butyl rubber	835	2.1	2.1	0	2.1	0
CBR-0.27	391	1.8	1.4	0.4	1.2	0.6
CBR-0.62	333	1.8	0.8	1.0	0.7	1.1
CBR-0.81	412	1.8	0.5	1.3	0.4	1.4
CBR-1.05	478	1.8	0.2	1.6	0.3	1.5
CBR-1.55	578	1.5	0	2.1	0.1	1.4
CBR-1.90	617	1.5	0	2.1	0.2	1.3

* Index marks chlorine content in mass%.

As additives, preventing decrease of the molecular mass at butyl rubber chlorination, water, alcohol or ester is used, which, however, somewhat decelerates the process of chlorination. Aliphatic diamines and tertiary diarylamines (0.05–1 mass%) or their mixtures were found to be effective at introduction of 1.5–1.9 Cl atoms per double bond in butyl rubber.

High degradation of the polymer down to obtaining liquid products (chlorinated oligoisobutylenes) is observed at intensive chlorination by large amounts of chlorine [53]. Decrease of the molecular mass at chlorination is strictly determined by the Cl/C=C relation. Evidently, to obtain chlorobutyl rubber with optimal physicomechanical properties, it is necessary to use butyl rubber with the maximally high molecular mass and relatively high unsaturation (1.9–2.2 mol %).

Contrary to chlorine, bromine interacting with butyl rubber is inclined to attachment reactions. However, if the process is conducted under controlled conditions, introduction up to 90% of bromine into the allyl position has been achieved:

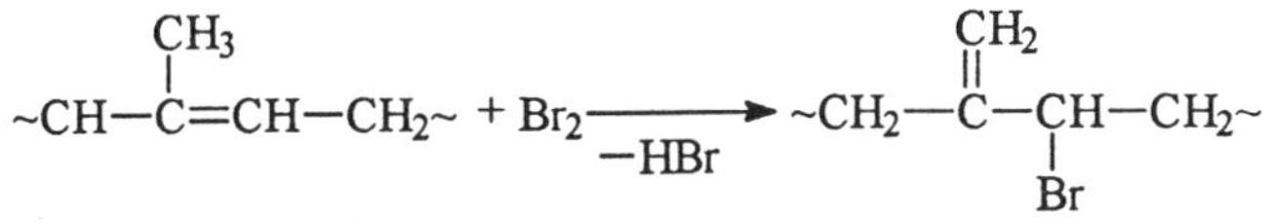

Bromination of butyl rubber is performed by mixing rubber and bromine dissolved in the same solvent. In the range from 213 to 323 K, temperature causes no significant influence on the process, which is explained by the specificity of fast halogenation.

To prevent corrosion of technological equipment during halogenation and to exclude appearance of metals in halogenated butyl rubber, the main devices and connective lines must be enameled or made from non-metallic materials.

7.6.2. Productions of halogenated butyl rubbers.

7.6.2.1. Chlorobutyl rubber.

The scheme of production of chlorobutyl rubber is displayed in Figure 7.35. Rubber is delivered into the worm press, equipped with a granulating device. Granules of rubber are loaded through the manhole into the device 3, equipped with mixer and heating coat. Simultaneously, accurately dried gasoline, containing no unsaturated compounds, is delivered from the batcher 1 by the pump 2 to obtain 13% solution of the rubber. Dissolution is performed at mixing and heating by hot water, delivered into the coat of the device 3 at 300–315 K for 10 ± 2 hours. A prepared solution is delivered by the pump 4 into chloinator 5, into which gaseous chlorine, diluted by dried nitrogen in the ratio 1:6 (by volume), is dosed simultaneously. Chlorination proceeds at high rate and ends in the device with mixer 6, wherefrom a solution of chlorobutyl rubber is delivered for neutralization.

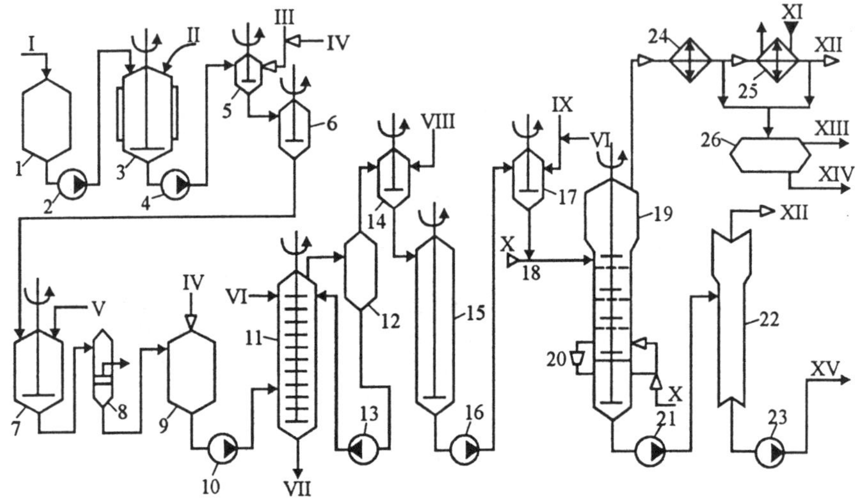

Figure 7.35. Production scheme for chlorobutyl rubber:
1 – batcher; 2, 4, 10, 13, 16, 21, 23 – pumps; 3 – device for preparation of butyl rubber solution; 5, 14, 17 – intensive mixing tanks; 6 – device with mixer; 7 – neutralizer; 8 – filter; 9 – collector; 11 – washing column; 12, 26 – presettling tanks; 15 – averaging machine; 18 – injector; 19 – degasifier; 20 – throttling device; 22 – vacuum degasifier; 24, 25 – condensers.
I – gasoline; II – butyl rubber granules; III – chlorine; IV – nitrogen; V – alkaline solution; VI – water; VII – washing water to purification; VIII – stabilizer; IX – anti-agglomerator; X – steam; XI – brine solution; XII – to the vacuum line; XIII – gasoline to drying; XIV – water to steaming of organic compounds; XV – pulp to concentrator.

The neutralizing agent is 10% alkaline solution, delivered simultaneously with chlorobutyl rubber solution into the neutralizer 7. A neutralized solution of chlorobutyl rubber is delivered through the filter 8 into the collector 9, where nitrogen is separated, returned to recycle after drying. A solution of chlorobutyl rubber is delivered by the pump 10 to the washing column 11 for washing from neutralization products. Water in 1:1 (by volume) ratio is delivered into the upper part of the washing column 11, equipped by a mixer. Washed solvent is delivered for stratification into the presettling tank 12. The lower water layer from the presettling tank 12 is delivered by the pump 13 into the washing column, and a volume of liquid from the column 11 is delivered for purification. Washed solution of butyl rubber from the upper part of the presettling tank 12 is delivered into the intensive mixing tank 14 for filling by stabilizer, and is averaged in the device with mixer 15. Then, after mixing with circulating water and anti-agglomerator in the intensive mixing tank 17, it is delivered for degasification.

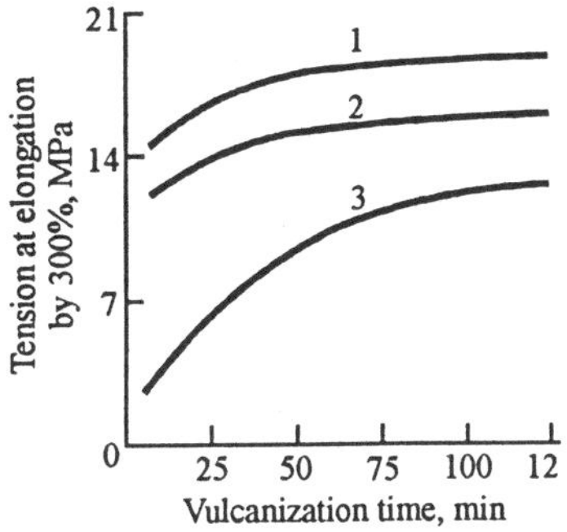

Figure 7.36. Influence of halogenation on the rate of butyl rubber vulcanization at 142°C: 1 – bromobutyl rubber; 2 – chlorobutyl rubber; 3 – butyl rubber.

Degasification of chlorobutyl rubber is performed on the two-stage degasifying set at a temperature of 358–363 K and pressure of 0.11–0.12 MPa on the first stage and 368–371 K under vacuum on the second stage. The solution of chlorobutyl rubber is delivered through the injector 18 into the degasifier of the first stage 19, equipped by the blind plate, mixer and the throttling device 20. The heat necessary for solvent elimination is added with water steam and secondary vapors, delivered from the upper part of degasifier of the second stage 22. Partially degasified pulp, containing 3 mass% of chlorobutyl rubber, is delivered by the pump 21 for the final degasification into the device 22, working under vacuum, and is pumped out from the cubic part by the pump 23 for concentration. Hydrocarbons with water, eliminated from the degasifier 19, are condensed in the condensers 24 and 25, cooled down by water and brine solution, respectively. The condensate is stratified in the presettling tank 26. The upper layer – gasoline – is delivered for drying and returned into recycle, the lower layer – water – is delivered for elimination of hydrocarbons.

Separation of chlorobutyl rubber from the pulp, drying and packing are performed by the usual schemes utilized in the synthetic rubber industry for dissolved rubbers.

To stabilize chlorobutyl rubber, naphtham-2 or di-*tert*-butyl-*n*-cresol (0.3% of rubber mass) and potassium stearate (0.8-1.2% of rubber mass) are used.

To prepare the initial solution at combined production of butyl rubber and chlorobutyl rubber, butyl rubber granules are used, taken from the vibroreceiver of separation and drying apparatus and delivered by the belt conveyor.

In the USA, production of chlorobutyl rubber uses a semiproduct, obtained by degasification of butyl rubber, synthesized in methyl chloride, and heated up to 422 K by hexane steam. The polymer solution, formed at degasification, is concentrated up to rubber content of 15–16% and is delivered for chlorination, performed at 55°C during 5–10 s under chlorine dosage of 3–3.5% of the rubber mass.

Usually, halogenation of butyl rubber is accompanied by decrease of molecular mass of the initial polymer.

Halogenated butyl rubber is more active at vulcanization compared with unmodified polymer (Figure 7.36); to vulcanize it, metal oxides, amines, dioles, polyoxyaromatic compounds, phenolformaldehyde resins, etc. can be used.

Vulcanizates of halogenated butyl rubber, obtained with application of zinc oxide, display extremely high heat resistance, keeping all the unique properties of vulcanizates of usual butyl rubber.

7.6.2.2. Bromobutyl rubber.

Bromobutyl rubber can be obtained by halogenation in the solid phase, in solution or in water dispersion.

One of the first industrial products (Haicar-2202) was obtained by reaction of butyl rubber bromination in the solid phase (350–405 K; 75 ± 45 min). Of great importance was the regime of preparation of the master mixture, temperature of brominating agent decomposition, type of solvent chosen, etc. The temperature was kept below that of halogenating agent degradation while the components were mixed. However, in this case, obtaining a homogeneous stable product of constant composition was rather difficult. Moreover, Haicar-2202 was very expensive. Bromination of butyl rubber in solution or water dispersion are more thorough and efficient processes, analogous to those considered in description of chlorobutyl rubber.

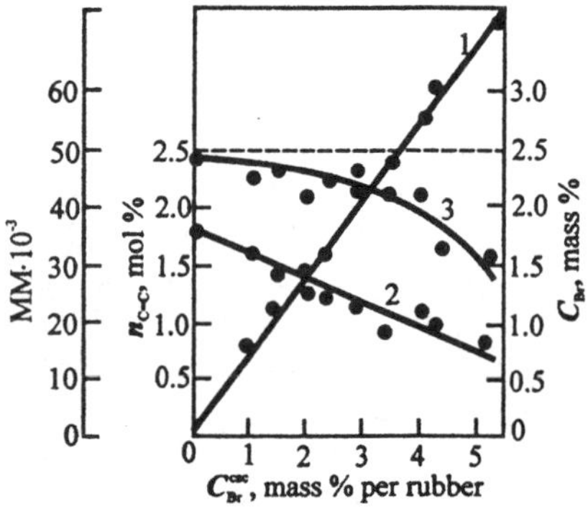

Figure 7.37. Bromination of butyl rubber with molecular mass of 49,000 by Staudinger and nonsaturation of 1.8 mol % by bromine solution in CCl_4 (298 K, 1200 s): 1 – bromine content in the polymer; 2 – nonsaturation; 3 – molecular mass.

The steps involved in butyl rubber bromination (Figure 7.37) are mostly analogous to those of chlorination [52]. If the concentration of bromine is increased, its content in the polymer grows linearly, and unsaturation of the rubber decreases linearly. However, some differences do also exist. For example, bromination of butyl rubber in dry solvent causes deeper degradation of macromolecules than with chlorination. Introduction of small quantities of water or alcohol almost eliminates these unwanted events.

7.7. *New technology of chlorobutyl rubber production in small-scale pipe turbulent reactors.*

In 1997–1998, a suitable new and quite efficient method of butyl rubber production applying pipe turbulent reactor-chlorinator with productivity of about 150–200 kg/h of chlorobutyl rubber (diameter 0.07 m, length 3 m) was developed in Russia.

The process is conducted in the device for butyl rubber halogenation, consisted of a cylindrical body, a nozzle, connecting pipes for adding elastomer solution, gas or gas mixture and release of the gas-liquid flux.

Construction of the system guarantees constant turbulence along the flux in the absence of the shell regime of gas mixture passing (nitrogen:chlorine = 2:1 to 10:1) through (13 ± 2)% solution of butyl rubber in nephrace.

Experiment–industrial tests have shown that the rate constant of butyl rubber chlorination is quite high ($K = 10^2$ to 10^4 at 300 K), and time of reaction mass exposure in the reaction zone may not exceed 5–10 s.

Obtaining of chlorobutyl rubber fulfilling the demands of Russian standards is guaranteed.

Bromobutyl rubber can also be obtained by the same technology. Corresponding solutions of halogens in the solvent can be used instead of gaseous halogens.

It must be noted that the stages of catalyst decomposition, neutralization of acidic products in the reaction mixture and in the solvent, as well as at solvent washing from ionogenic admixtures, are fast reactions. That is why, at these stages, it is appropriate to use small-scale pipe turbulent apparatuses of jet type in both new technologies and for butyl rubber halogenation in standard volumetric mixing reactors.

The new process is extremely simple, cheap, power and material saving, etc. In the stages of degradation of Ziegler-Natta catalytic systems and electrophilic catalysts ($AlCl_3$) pipe turbulent reactors are used with high efficiency in Russian industry, in production of synthetic *cis*-1,4-isoprene and ethylene–propylene rubbers, etc.

7.8. *Ecological aspects of technology of production of isobutylene polymers.*

Traditionally, the problem in production of isobutylene polymers in industry is in decomposition of the catalyst and washing off the resulting degradation products. Usually, water, water–alkaline or water–alcohol decomposition of the catalyst is performed, which represents rather long and power-consuming process with high consumption of water [54], if volumetric mixing reactors or diaphragm mixers are used. The resulting chlorine in wastewater makes its industrial use difficult ecologically.

In this stage, it is also appropriate to use small-scale pipe turbulent aggregates, working according to the principle of conservation of high turbulence level along the flux, which was mentioned above.

Another prospective direction is development and application of non-water methods of polymerizate purification under normal or increased temperatures, in particular, of inert filtering materials, aprotic organic solvents or chemically active complex formers [5, 54]. These methods are quite economic, and have considerable potential because the chemical–additives are used in quantities comparable with the quantity of the catalyst applied. In this connection, of particular interest are solid sorbents of the type of aluminum oxide, alumosilicate and, especially, ionites [5, 54, 55]. It seems optimal to use accessible cationites, for example, sulfonated styrene copolymers with divinylbenzene, sulfonated gossipole of gossipolic resin in the *n*-form in the variant with preliminary treatment of the polymerizate with small amounts of C_2–C_5-alcohols (for $AlCl_3$) [54]. In this case, ionite is capable of ionic exchange and physical sorption, which is illustrated by the following reactions:

$$K{-}SO_3H + ROAlCl_2 \rightarrow K{-}\underset{\underset{O}{\parallel}}{\overset{\overset{OH}{|}}{S}}{=}O\cdot Al(RO)Cl_2$$

$$K{-}SO_3H + ROAlCl_2 \rightarrow K{-}SO_3\underset{\underset{Cl}{|}}{Al}{-}OR + HCl$$

$$K{-}SO_3H + (RO)_2AlCl_2 \rightarrow K{-}SO_3Al(OR)_2 + HCl$$

$$2K{-}SO_3H + ROAlCl_2 \rightarrow (K{-}SO_3)_2AlOR + 2HCl$$

$$K{-}SO_3H + HCl \rightarrow K{-}\underset{\underset{O}{\parallel}}{\overset{\overset{OH}{|}}{S}}{=}O\cdot HCl$$

Duration of the sorbent work (which purifies up to 40 volumes of the polymerizate per volume of ionite) at temperatures of 283–373 K, little decrease of activity after ~300 sorption-regeneration cycles, ability to use concentrated acidic effluents in sorbent regeneration as a coagulant instead of $Al_2(SO_4)_3$ at floatation purification of wastewater from oil-processing enterprises under satisfactory (sometimes high) degree of product purification — all this allows consideration of this method as an important component of polyisobutylene production by low-waste or non-waste technology. Besides high ecological safety, methods of non-water elimination of the catalyst from polymeric products are technically simple (filtration of products of catalyst interaction with additives, polymerizate pressing though solid sorbent) enable corrosion of the equipment to be eliminated and cause no rise in the final product cost price. There is no doubt that the presence of water in production of polyisobutylene is undesirable in all stages of the process from polymerization to creating the finished product.

A specific feature of polyisobutylene production is formation of polymeric technological wastes, brought about by the high rates of the process, alternations in temperature regime, influence of admixtures, etc. Formation of substandard polymers takes place during starting up reactors, forcing the system to move from the mode of one polyisobutylene brand synthesis to another.

Some substandard oligomeric products can be used directly, for example, as lubricants (in axle boxes of railway wheel pairs), sealing composites (in construction), etc. But in the general case, simple and economic uses must be found for the technological rejects of isobutylene oligomers.

One of the main methods of reject utilization is pyrolysis (depolymerization) of polymeric products in order to obtain isobutylene [56–58]. The low heat of isobutylene polymerization (72 kJ/mol) is the thermodynamic basis of such processes. Gasification and burning off methods are less economic, although they are quite widespread.

Secondary utilization of polyisobutylene, as well as of many other polymers, by burning off (gasification) is performed for the purpose of recuperation of energy costs [57, 58]. Various devices are used for burning off, the working principle of which is based on spraying burning-off polymer in fuel chambers in the presence of an oxidizing agent (oxygen). The heat energy obtained is used for steam production, warming up compartments and production areas, hothouses, greenhouses, etc.

Of particular interest are methods of thermal degradation of high-molecular polyisobutylenes down to low-molecular products of the type of oligomers, oils, etc., completely excluding formation of gaseous substances. Temperature control of cracking separately by zones in the reactor gives practically 100% conversion of the raw material, which comprisesend-product rejects, up to final products with controlled molecular mass and composition.

One of the methods of polyisobutylene rejects degradation is photolysis of polymeric products down to mixture of low-molecular products of isobutylene, diisobutylene and saturated hydrocarbons [59].

Depolymerization of the lowest isobutylene oligomers during pyrolysis takes place at 560–1070 K [2]. The main product (with yield of about 50 mass%) is isobutylene. When diluted by an inert gas to 3–25 volumes per specific volume of the olefin, the monomer yield can be increased up to 80 mass% by decreasing the contribution of secondary reactions (isomerization, coke formation, etc.).

The technological scheme of the process includes degradation of isobutylene oligomers down to gas products and extraction of the monomer from the mixture with cold H_2SO_4. Isobutylene oligomers are delivered by pump from the tank into the steam heater, and further into a heated pipe reactor with a nozzle. The resulting gas mixture is delivered through the cooler into the absorber with 65% H_2SO_4. The resulting isobutylsulfuric acid is decomposed in the device by overheated water steam. The isobutylene obtained (purity no less than 95 mass%) is collected in tanks.

For high-molecular products, thermal depolymerization of polyisobutylene is only one of several possible directions of reject utilization. However, contrary to low-molecular products, a complex mixture of hydrocarbons with a wide selection of molecular masses (up to 400) at isobutylene content no more than 20–25 mass% is formed [60, 61]. This is achieved from parallel reactions of isomerization, hydrogenation, intramolecular disproportioning and intramolecular chain transmission, which lead to formation of a great number of compounds, multiples and non-multiples of isobutylene: H_2, CH_4, C_2H_4, C_2H_6, C_3H_6, n-mers of isobutylene. Such selection of products decreases their value as a source of secondary raw material in production of isobutylene polymers, although pyrolytic gases can be delivered to separation by fractionation in respect to standard industrial schemes.

A greater degree of chemical sophistication is found in the process of thermocatalytic degradation of oligo- and polyisobutylenes, which is conducted in the presence of natural and synthetic alumosilicates, phosphorus-containing compounds, activated aluminum oxide and other catalysts (Table 7.13).

Table 7.13

Catalytic depolymerization of isobutylene oligomers.

Catalyst	Tempera ture, K	Yield, mass%	Isobutylene content, mol%	Note
Fuller's earth	583–699	80–100	60–90	For di- and trimers of isobutylene only
Diatomaceous earth with phosphoric acid	477–644	80–100	64–92	For di- and trimers of isobutylene only
Aluminum oxide	477–671	80–93	70–95	For isobutylene dimers at 1 vol. of olefin diluted by 1–3.5 vol. of inert gas
Synthetic zeolites	485	93.9	76.0	Isobutylene contains up to 20% of paraffin hydrocarbons as admixtures
	529	86.9	64.5	
	633	86.6	85.0	
Phosphoroacidic salts	490	99.3	91.3	Isobutulene contains 5–10% of paraffin hydrocarbons as admixtures
	536	97.9	91.2	
	591	96.6	87.0	
Alumosilicate	798	100.0	100.0	For isobutylene dimers only
Bleaching clay 'atapulgis'*	489	100.0	97.0	For di- and trimers of isobutylene only
	592	100.0	100.0	
	701	96.6	100.0	

* Composition: SiO_2 — 66.2%; Al_2O_3 — 12.52%; Fe_2O_3 — 3.78%; MgO — 10.19%; CaO — 3.41%; other admixtures — 4.08%.

The best among them is breaching clay "atapulgis", which generates isobutylene with quite high conversion at low temperature. Application of standard catalysts of hydrocarbon cracking is not appropriate because of the ability to transfer hydrogen atoms, which leads to formation of up to 20% of saturated hydrocarbons in reaction products of polyisobutylene degradation. Industrial alumosilicate catalysts, characterized by a wide selection of acidic centers, display relatively high activity in isomerization (content of α–butylene is up to 2 mass%). Mostly, isomerization also proceeds on activated aluminum oxide. Introduction of alkaline additives leads to decrease of content of α–butylenes in the reaction products of thermocatalytic depolymerization of polyisobutylene, whereas rise of concentration and strength of acidic centers significantly increases the role of isomerization.

Synthetic zeolites [62, 63] are quite suitable catalysts. They increase selectivity of the catalytic degradation of polyisobutylene. Depending on the conditions under which the degradation is conducted, divalent forms of synthetic zeolites (types A, X, Y) lead to formation of 50–90 mass% of isobutylene with reasonably low content of α–butylene as accompanying product (Table 7.14). Introduction of CO_2 during degradation of polyisobutylene on zeolites increases the conversion. Simultaneously, the role of coke formation processes is reduced.

Synthetic zeolites as catalysts of selective degradation of polyisobutylene rejects are appropriate only for depolymerization of di- and trimers of isobutylenes.

Table 7.14

Catalytic degradation of isobutylene oligomers on synthetic zeolites.

Catalyst	Contact duration, s	*T*, K	Conversion, mass%	Yield, mass% per reacted raw material				Liquid products
				iso-butane	*iso*-C_4H_8	C_5-C_7	coke	
NaX	6	573	36.8	—	85.0	—	2.1	Mixture 75%
	12	523	44.4	0.4	67.8	5.7	1.7	TMP-1 and 25%
	40	523	65.0	3.4	51.0	13.4	3.4	TMP-2
	6	573	33.0	—	86.8	—	2.3	TMP-1
	6	573	24.5	—	90.0	—	3.2	TMP-1
NaY	6	523	25.2	—	90.3	—	0.4	Mixture of iso-
	12	523	83.3	2.0	44.2	15.3	2.5	butylene dimers
	40	523	84.0	12.7	31.8	29.7	5.3	
	8	573	100.0	—	41.2	7.1	1.2	Isobutylene trimers
0.53 CaNaX	6	493	69.2	5.8	33.7	22.1	4.0	Mixture of isobutylene dimers
	6	493	66.8	6.1	37.8	6.1	5.7	TMP-1
	6	493	63.3	9.4	40.5	18.1	7.5	TMP-2
0.9 CaNaX	6	493	43.7	10.2	31.0	23.3	7.0	Mixture of isobutylene dimers
	6	493	50.0	21.1	46.8	4.8	3.7	TMP-1
	6	493	44.1	20.3	45.4	3.7	6.3	TMP-2

* TMP-1 is 2,4,4-trimethylpentene-1; TMP-2 is 2,4,4-trimethypentene-2.

A suitable universal method of utilizing technological rejects of non-vulcanized butyl rubber of practically every molecular mass (from 112 to 20,000) is selective degradation (550–600 K) applying salt complexes of aluminum chlorides of $Me[C_2H_5AlCl_3]$ or $Me[AlCl_4]$ structure [2, 64, 65] by the scheme of utilization of polyisobutylene rejects. Increased, compared with polyisobutylene, concentration of internal unsaturated groups of >C=C<, which are centers of polymer degradation, in macromolecules of butyl rubber promotes the isobutylene elimination reaction at lower temperatures (520–600 K; 570–675 K for polyisobutylene). The yield of isobutylene exceeds 90 mass%. Simultaneously, less acidic and, consequently, technologically more suitable catalytic systems, for example, those based on alkaline-earth metal chlorides can be used. As viscosity of butyl rubber is high, preliminary homogenization of the rubber–catalyst system (application of the polymer as melts, solutions in saturated or chlorinated hydrocarbons) is favorable for the efficient degradation of butyl rubber [66]. The catalyst as powder, solution in organic solvents, melt or, in the "unsaturated" view, on a mineral support, is injected into the reaction zone. They display high thermal stability (up to 770–875 K), decreased sensitivity to admixtures, and low acidity, which determines the absence of the coordinating action. Any polyisobutylenes without preliminary purification, differing in fractional composition, are easily processed according to this method, and high (80–95%) conversion at the average isobutylene concentration in products of 75–95% and α–, β–butylenes no more than 2.2% is reached. The coke content is small, and even in the worst case gives only 0.02–0.03 mass% from the total quantity of processed raw material.

Some data characterizing activity and selectivity of salt complex catalysts in the form of Brönsted acids in thermocatalytic degradation of polyisobutylene and butyl rubber are summarized in Table 7.15.

Table 7.15

Thermocatalytic degradation of polyolefins.

Catalyst	Yield of gaseous products [a], mass%	Composition of C_4 hydrocarbons, mass%			Catalyst selectivity, rel. units
		i-C_4H_{10}	i-C_4H_8	ΣC_4	
Polyisobutylene (653 K, 1 h, $\overline{M}_n$ = 1100, [Cat] = 0.001 mol/l)					
Without catalyst	76.0	2.1	16.0	19.1	0.2
$H[AlCl_3OH]$	63.0	2.1	19.1	22.1	0.2
$H[NaAlCl_4OH]$	38.2	7.0	88.8	98.1	1.0
$H[Li(C_2H_5)AlCl_3OH]$	27.0	7.0	87.5	97.5	1.0
$H[Na(C_2H_5)AlCl_3OH]$	24.2	6.9	88.3	98.3.	1.0
$H[K(C_2H_5)AlCl_3OH]$	19.8	7.2	87.3	97.8	1.0
$MgCl_2{\cdot}AlCl_3{\cdot}H_2O$	93.8	24.2	73.4	98.0	1.0
$H[Mg(C_2H_5)AlCl_4OH]$	59.6	7.0	91.3	99.1	1.0
$H[Ca(C_2H_5)AlCl_4OH]$	54.0	7.2	90.5	98.3	1.0
$H[Ba(C_2H_5)AlCl_4OH]$	45.2	8.1	86.5	98.6	1.0
Polyisobutylene (643 K, 1 h, $\overline{M}_n$ = 1100, [Cat] = 0.001 mol/l)					
Without catalyst	68.1	2.2	18.7	22.0	0.2
$H[NaAlCl_4OH]$	100.0	1.0	90.1	91.1	1.0
$H[NaAlCl_4OH]$	96.0	1.7	65.1	67.1	0.7
Butyl rubber (643 K, 1 h, $\overline{M}_n$ = 25,700, [Cat] = 0.001 mol/l)					
Without catalyst	50.5 [b]	Traces	3.0	36.7	0.6
$H[AlCl_3OH]$	75.0	58.3	0.5	58.8	0.6
$H[LiAlCl_4OH]$	39.8	23.6	39.1	76.1	0.75
$H[KAlCl_4OH]$	86.0	30.3	24.8	62.3	0.62
$MgCl_2{\cdot}AlCl_3{\cdot}H_2O$	75.4	2.1	93.8	97.2	1.0

Notes. a) Molecular hydrogen is identified in the composition of gaseous products.

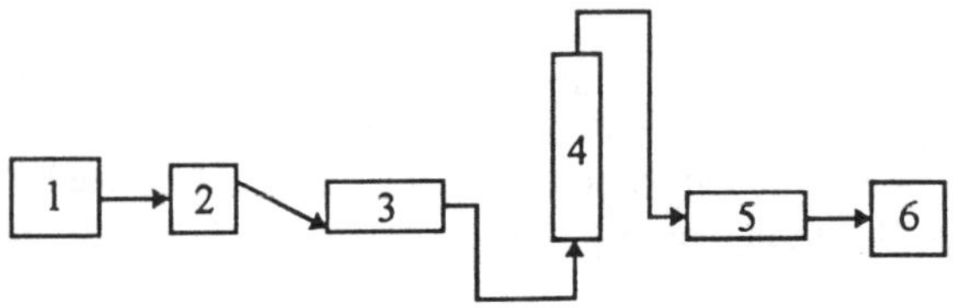

Figure 7.38. Scheme for thermocatalytic degradation of polyisobutylene reject products.

The method of thermocatalytic degradation of substandard polyisobutylenes enables efficiency of production of isobutylene oligomers to be increased at the sacrifice of more complete exhaustion of the olefin raw materials, and allows creation of practically reject-less technology of production.

The main features of the process scheme (Figure 7.38) include the pipe contact aggregate with the catalyst (4), into which the raw material is delivered from the raw material storing tank (1) and the heating device (3) services by the pump (2). After cooling down (5), isobutylene is accumulated in special tanks (6).

Productivity of the process is no less than 5 tons of the polymer per kg of the complex catalyst.

Modification and development of the method of selective thermocatalytic degradation of isobutylene polymers in the presence of salt complexes of aluminum chlorides are: possibility to activate metallocomplex catalysts by dry hydrogen chloride or alkylchloride flow, as well as application of catalytic systems in melt under permanent addition of the polymer. Under definite conditions of activation of metallocomplexes, the temperature of selective degradation of isobutylene polymers may be decreased to 448 ± 25 K.

Hence, based on the above, it follows that, in the realization of technological processes of isobutylene polymer synthesis, the process must be conducted in hydrocarbons, soluble catalytic systems of the type of chloroaluminum-organic hydrated complexes ($R_nAlCl_{3-n} \cdot H_2O$) must be used, and polymerization in small-scale pipe turbulent reactors in the displacement mode in turbulent fluxes must be performed as well. This allows fast chemical processes to be conducted under quasi-isothermal conditions, applying methods at the stage of catalyst elimination, which exclude usage of water or water solutions, regenerating the C_4 fraction of hydrocarbons by selective thermocatalytic degradation of substandard polymeric products and production rejects.

References to Chapter 7.

1. Kirpichnikov, P.A., Bersenev, V.V., and Popova, L.M., Album of Technological Schemes of Main Productions in Synthetic Rubber Industry, 1976, 112 p., Leningrad, Khimia. (Rus)
2. Minsker, K.S. and Sangalov, Yu.A., Isobutylene and Its Polymers, 1986, 224 p., Moscow, Khimia. (Rus)
3. Vaculic, P., Chemistry of monomers, Translation from Czech, 1960, v. 2, 738 p., Moscow, Izdatinlit. (Rus)
4. Minsker, K.S., Kirillov, A.P., Sangalov, Yu.A., Yudaev, A.I., and Yanu, A.Yu., Chem. And Technol. of Petroleum and Oils, 1977, No. 1, pp. 27–31. (Rus)
5. Prokofiev, K.V. and Sangalov, Yu.A., Chem. and Technol. of Petroleum and Oils, 1986, No. 3, pp. 2–4. (Rus)
6. Prokofiev, K.V., Verbitski, B.G., and Rogov, S.A., Low–molecular Polybutenes, Thematic review, CNIITENeftekhim, 1982, 51 p. (Rus)
7. Minsker, K.S. and Berlin, Al.Al., Fast Polymerization Processes, Gordon and Breach Publ., Inc.: Switzerl., Austr., Japan, USA, 1996, p. 146.
8. Berlin, Al.Al., Minsker, K.S., Prochukhan, Yu.A., and Enicolopyan, N.S., Highmolec. Comp., 1989, Ser. A, v. 31(9), pp. 1779–1798. (Rus)
9. Erlin, Al.Al., Minsker, K.S., Prochukhan, Yu.A., and Enicolopyan, N.S., Polym. Plast. Technolog. Eng., 1991, v. 30(2–3), pp. 253–297.
10. Berlin, Al.Al., Dyumaev, K.M., Minsker, K.S., Khalafov, F.R., and Kolesov, S.V., Chemical Industry, 1995, No. 9, pp. 550–556. (Rus)
11. Berlin, Al.Al., Minsker, K.S., Sangalov, Yu.A., Oshmyan, V.G., Svinukhov, A.G., Kirillov, A.P., and Enicolopyan, N.S., Highmolec. Comp., 1980, Ser. A, v. 23(3), pp. 566–574. (Rus)
12. Minsker, K.S., Berlin, Al.Al., Svinukhov, A.G., Prochukhan, Yu.A., and Enikolopyan, N.S., Doklady AN SSSR, 1986, v. 286(5), pp. 1171–1177. (Rus)
13. Berlin, Al.Al., Minsker, K.S., Prochukhan, Yu.A., Karpasas, M.M., and Enicolopyan, N.S., Highmolec. Comp., 1986, Ser. B, v. 28(6), pp. 461–465. (Rus)
14. Berlin, Al.Al., Minsker, K.S., Prochukhan, Yu.A., Karpasas, M.M., and Enicolopyan, N.S., Internat. Polym. Sci. And Technology, 1986, v. 13(11), pp. 95–97. (Rus)
15. Berlin, Al.Al., Minsker, K.S., Prochukhan, Yu.A., Karpasas, M.M., and Enicolopyan, N.S., Doklady AN SSSR, 1986, v. 287(1), pp. 145–148. (Rus)
16. Berlin, Al.Al., Prochukhan, Yu.A., Minsker, K.S., Alexanyan, G.G., Grobov, S.V., and Enicolopyan, N.S., Highmolec. Comp., 1988, Ser. A, v. 30(11), pp. 2441–2446. (Rus)
17. Berlin, Al.Al., Prochukhan, Yu.A., Minsker, K.S., and Enicolopyan, N.S., Highmolec. Comp., 1991, Ser. A, v. 33(2), pp. 243–269. (Rus)
18. Minsker, K.S., Berlin, Al.Al., Prochukhan, Yu.A., and Enicolopyan, N.S., Highmolec. Comp., 1986, Ser. B, v.28(6), pp. 466–469. (Rus)
19. Minsker, K.S., Berlin, Al.Al., Prochukhan, Yu.A., and Enicolopyan, N.S., Internat. Polym. Sci. And Technolog., 1986, v. 13(12), pp. 90–93.
20. Prochukhan, Yu.A., Minsker, K.S., Berlin, Al.Al., Karpasas, M.M., Kompaniets, V.Z., and Enicolopyan, N.S., Doklady AN SSSR, 1988, v. 28(6), pp. 1428–1430. (Rus)
21. Prochukhan, Yu.A., Minsker, K.S., Berlin, Al.Al., Tumanyan, E.S., and Enicolopyan, N.S., Doklady AN SSSR, 1986, v. 291(6), pp. 1425–1427. (Rus)

22. Companiets, V.Z., Konoplev, A.A., Berlin, Al.Al., Prochukhan, Yu.A., Minsker, K.S., Karpasas, M.M., and Enicolopyan, N.S., Doklady AN SSSR, 1987, v. 287(5), pp. 1129–1132. (Rus)
23. Prochukhan, Yu.A., Minsker, K.S., Karpasas, M.M., Berlin, Al.Al., Bakhitova, R.Kh., and Enicolopyan, N.S., Highmolec. Comp., 1988, Ser. A, v. 30(6), pp. 1259–1262. (Rus)
24. Berlin, Al.Al., Karpasas, M.M., Companiets, V.Z., Minsker, K.S., Prochukhan, Yu.A., and Enicolopyan, N.S., Doklady AN SSSR, 1989, v. 305(2), pp. 365–368. (Rus)
25. Berlin, Al.Al., Companiets, V.Z., Konoplev, A.A., Minsker, K.S., Prochukhan, Yu.A., Ryabenko, E.A., and Enicolopyan, N.S., Doklady AN SSSR, 1989, v. 305(5), pp. 1143–1146. (Rus)
26. Berlin, Al.Al., Prochukhan, Yu.A., Minsker, K.S., Tumanyan, E.A., Alexanyan, G.G., and Enicolopyan, N.S., Highmolec. Comp., 1988, Ser. A, v. 30(11), pp. 2436–2440. (Rus)
27. Minsker, K.S., Berlin, Al.Al., Prochukhan, Yu.A., Tumanyan, E.A., Karpasas, M.M., and Enicolopyan, N.S., Doklady AN SSSR, 1986, v. 291(1), pp. 114–116. (Rus)
28. Berlin, Al.Al., Prochukhan, Yu.A., Minsker, K.S., ., Alexanyan, G.G., Grobov, S.V., and Enicolopyan, N.S., Highmolc. Comp., 1989, Ser. A, v. 31(3), pp. 612–616. (Rus)
29. Sangalov, Yu.A., Minsker, K.S., Prochukhan, Yu.A., and Kirillov, A.P., In: Achievements of Science To Oil–Chemical Production, Ufa, 1975, p. 74. (Rus)
30. Razuvaev, G.A., Sangalov, Yu.A., Yasman, Yu.B., and Minsker, K.S., Doklady AN SSSR, 1978, v. 238(2), pp. 338–341. (Rus)
31. Sangalov, Yu.A., Yasman, Yu.B., Khudaiberdina, Z.I., and Ilkaeva, E.M., Highmolec. Comp., 1978, Ser. A, v. 20(1), pp. 180–190. (Rus)
32. Sangalov, Yu.A., Yasman, Yu.B., Odinokov, V.N., Svinukhov, A.G., and Minsker, K.S., Neftekhimia, 1977, v. 17(6), pp. 865–872. (Rus)
33. Sangalov, Yu.A., Gladkikh, I.F., and Minsker, K.S., Highmolec. Comp., 1982, Ser. B, v. 24(5), pp. 356–358. (Rus)
34. Pat. No. 1396193 (France), 1965.
35. Pat. No. 2904314 (France), 1980.
36. Prokofiev, K.V., Kotov, S.V., Berlin, Al.Al., Minsker, K.S., Prochukhan, Yu.A., Sangalov, Yu.A., and Alexanyan, G.G., Bulletin of MChT Interoilproduct, Sofia, PRB, 1989, No. 1, pp. 19–22.
37. Berlin, Al.Al., Minsker, K.S., Wirtschaft Technic, 1990, No. 2, S. 54–56.
38. Minsker, K.S., Prochukhan, Yu.A., In Coll.: Modern Power and Resource Saving Ecologically Pure Technologies, Ufa, Izd. AN RB i VkhO im. D.I. Mendeleev, 1992, pp. 45–53. (Rus)
39. Kotov, S.V., Prokofiev, K.V., Minsker, K.S., Sangalov, Yu.A., and Berlin, Al.Al., Chem. and Technol. of Petroleum and Oils, 1990, No. 4, pp. 14–15. (Rus)
40. Berlin, Al.Al., Prokofiev, K.V., Minsker, K.S., Prochukhan, Yu.A., Kotov, S.V., Bulankov, B.F., Yasinenko, V.A., and Naumova, T.N., Chem. and Technol. of Petroleum and Oils, 1988, No. 7, pp. 8–9. (Rus)
41. Berlin, Al.Al., Minsker, K.S., Prokofiev, K.V., Prochukhan, Yu.A., Sangalov, Yu.A., Kotov, S.V., Yasinenko, V.A., and Alexanyan, G.G., Neftepererabotka i Neftekhimia, 1988, No. 2, pp. 25–28. (Rus)

42. Kotov, S.V., Prokofiev, K.V., Prochukhan, Yu.A., Minsker, K.S., Adilov, N.A., Velishov, D.A., Yasinenko, V.A., and Alexanyan, G.G., Neftepererabotka i Neftekhimia, 1989, No. 1, pp. 17–18. (Rus)
43. Minsker, K.S., Prochukhan, Yu.A., Kolesov, S.V., Berlin, Al.Al., Kotov, S.V., Kutluguzhina, I.Kh., Prokofiev, K.V., Kozlov, V.G., and Alexanyan, G.G., Neftepererabotka i Neftekhimia, 1989, No. 11, pp. 40–42. (Rus)
44. Kotov, S.V., Berlin, Al.Al., Prokofiev, K.V., Minsker, K.S., Sangalov, Yu.A., Prochukhan, Yu.A., Adilov, N.A., and Alexanyan, G.G., Chem. and Technol. of Petroleum and Oils, 1990, No. 6, pp. 10–11. (Rus)
45. Kotov, S.V., Atmadzhaev, V.E., Minsker, K.S., Yasinenko, V.A., Prokofiev, K.V., and Berlin, Al.Al., Neftepererabotka i Neftekhimia, 1992, No. 1, pp. 38–41. (Rus)
46. Berlin, Al.Al., Minsker, K.S., Dyumaev, K.M., Novel Unified Power and Resource Saving of Highly Productive Technologies of Increased Ecological Purity Based on Pipe Reactors, OSC "NIITEKHIM", Moscow, 1996, 188 p. (Rus)
47. Kotov, S.V., Atmadzhev, V.E., Berlin, Al.Al., Filin, V.N., Yasinenko, V.A., and Prokofiev, K.V., Neftepererabotka i Neftekhimia, 1996, No. 5, pp. 31–33. (Rus)
48. Mikhailovski, G.A., In: Industry of Synthetic Rubbers, Moscow, CNIITENeftekhim, 1974, No. 9, pp. 12–15. (Rus)
49. Kennedy, J.P., Macromol. Chem., 1974, Bd. 175(4), pp. 1101–1115.
50. Kennedy, J.P., J. Macromol. Sci., 1980, Part A, v. 14(1), pp. 1–10.
51. Encyclopedia of Polymers, Moscow, Sov. Encyclopedia, 1972, v. 1, pp. 351–364.
52. Shmarlin, V.S., Synthesis, Properties and Application of Modified Butyl Rubbers, Ser.: Industry of Synthetic Rubbers, Moscow, CNIITENfeftekhim, 1973, p. 80. (Rus)
53. Ibragimov, A.D., Scientific Notes of M. Azizbekov Oil and Gas Institute, 1972, v. 9(??), pp. 82–84. (Rus)
54. Vlasov, A.V., Prokofiev, K.V., Sangalov, Yu.A., and Kotov, S.V., Chem. and Technol. of Petroleum and Oils, 1988, No. 2, pp. 6–7. (Rus)
55. Karpanova, V.E., Pinsker, A.E., and Korol'kov, N.M., Zh. Prikl. Khimii, 1980, v. 53(8), pp. 1759–1762. (Rus)
56. Abbas, K.B., Polym. Eng. Sci., 1980, v. 2(10), pp. 653–707.
57. Poller, R.C., J. Chem. Technol. Biotechnol., 1980, v. 30(4), pp. 152–160.
58. Sangalov, Yu.A. and Minsker, K.S., Secondary Application of Isobutylene Polymers, In Coll.: Secondary Application of Polymeric Materials, Ed. by E.G. Lyubeshkin, Moscow, 1985, pp. 124–134. (Rus)
59. Chandra, R., Progr. Polym. Sci., 1985, v. 11(1–2), pp. 1–22.
60. Kiran, E. and Gilham, J.K., J. Appl. Polymer, Sci., 1976, v. 20(8), pp. 2045–2048.
61. Khmelnitsky, R.A., Lukashenko, I.M., and Brodsky, E.S., Pyrolytic Mass-spectroscopy of High-molecular Compounds, Moscow, Khimia, 1980, 279 p. (Rus)
62. Lapidus, A.L., Neftekhimia, 1976, v. 16(5), pp. 723–728. (Rus)
63. Lapidus, A.L., In: Catalytic Properties of Zeolites, Novosibirsk, Nauka, 1976, pp. 53–60. (Rus)
64. Sangalov, Yu.A., Prochukhan, Yu.A., and Minsker, K.S., Highmolec. Comp., 1978, Ser. B, v. 20(3), pp. 686–688. (Rus)
65. Sangalov, Yu.A., Minsker, K.S., Berlin, Al.Al., and Prochukhan, Yu.A., Highmolec. Comp., 1983, Ser. A, v. 25(7), pp. 1451–1457. (Rus)
66. Ivanova, S.R. and Minsker, K.S., Highmolec. Comp., 1984, Ser. A, v. 26(8), pp. 1686–1691. (Rus)

Chapter 8. Composites and fields of application of polyisobutylenes.

8.1. Main areas of application of oligo- and polyisobutylenes.

Industrial production of polyisobutylenes was already established in 1938–1939. At the present time, industry produces a great number of brands of oligo- and polyisobutylenes, which differ mainly in molecular mass (Table 8.1), and which are widely applied in many branches of the national economy.

The main technical characteristics of oligo- and polyisobutylenes are given in Table 8.2.

The re are many fields of application of polyisobutylenes [1-11]. Di-, tri- and tetramers of isobutylene are used as high-octane motor petroleum (polymer-gasoline). Oligoisobutylenes with MM = 200-500 are applied to obtain highly efficient lubricating-cooling fluids. Usually, these composites contain an antioxidant injected. For articles in the electrotechnical industry, products with MM = 600-700 are used, which possess high dielectric characteristics, for example, electroinsulating synthetic oil (capacitor octol). Octol-600, polyisobutylenes of P-5, P-10 and P-20 brands are mainly applied as viscous additives to lubricating oils, thickeners of consistent lubricants, etc. Octol-600 of the brand A displays high mechanical and thermal stability in synthetic oils, designed for use in highly loaded equipment, operating in zones of increased temperatures. Octol-600 of the brand B is applied to synthesize additives, resistant to wear and friction. The additive P-5 (TU 38 10-12-09-72) represents the concentrated (no less than 65%) solution of polymer in transformer oil. Thickening additive P-10 is a 30% solution of polyisobutylene with MM = 9,000-15,000 in light industrial (I-12A) or transformer oil (TU 38 10-12-09-72). The improved variant of the additive P-10 is a thickening electroinsulating additive (TU 38 10-16-88-72), which is a 15-20% solution of polyisobutylene of the same molecular mass in industrial oil I-20A. It is applied to cable oils and promotes complete substitution or reduction to minimum application of pine rosin in impregnating compositions for power cables. The additive P-20 (TU 38 10-12-09-72 is the concentrated (no less than 25%) solution of polyisobutylenes in industrial oil I-12A.

Octol-1000 (TU 38 10-11-34-76) is used as a raw material in production of succinimide and other additives, for preparation of sealing composites, as a thickening component and a basis for producing synthetic and mixed lubricating oils of various application. Succinimide additives possess extremely high washing-dispersant properties and relative resistance to water, low and high temperatures. The total concentration of additives decreases from 13-18 to 8-9% when washing additives of the succinimide type are used.

Native industry also produces such modifications of polyisobutylenes as octol-200 (TU 38 00-11-79-74), octol-K (TU 38 10-18-08-80) and octol-component of the sealing (TU 38 10-18-36-80). Octol-200 is used as a component of synthetic and mineral oils, and octol-K as a raw material for production of a lubricant of

Table 8.1

Industrial brands of oligo- and polyisobutylnes.

Country of origin	Molecular mass										
	300-500	600-800	800-1200	2000-3500	4000-6000	9000-15000	16000-25000	40000-55000	70000-99000	100000-134000	135000-174000	175000-225000
	Low-molecular							High-molecular				
USSR	Octol-300	Capacitor octol	Octol-1000	Octol-600	P-5	P-10	P-20	P-50	P-85	P-118	P-155	P-200
USA	Polybutenes Indopol									Vistanex	Paratac	
	L-4; LV-10	H-25	H-100	H-1500								
	L-50	H-35	H-300	H-1900								
	L-100	H-50										
Great Britain	Polybutenes Hivis									Vistanex		
FRG	Oppanol, Dinagen											
Japan	Polybutenes by Furukawa Chemical Co.											
	LV-50	HV-15	HV-100	HV-1900								
	LV-100	HV-15E	HV-300									
		HV-35										

Note. Main producers of oligo- and polyisobutylenes are: Amoco, Lubrisol, Standard Oil Co., Chevron, Consed, Exxon (USA); Edvin Cooper, British Petroleum, Brayton (Great Britain); Badische Anilin- und Soda-Fabrik A.G. (FRG); Sumitomo, Nichiman (Japan).

Table 8.2

Physicochemical characteristics of native and foreign oligo- and polyisobutylenes.[1]

Brand	Molecular mass	Density at 293 K, kg/m^3	Kinematic viscosity at 373 K, mm^2/s	Flash temperature, K	Iodine number, g I_2/100 g	Content of mechanical admixtures, mass %	Content of Cl ions, %	Color, CNT units	Acidic number, mg KOH/g	Additionally regulated indexes: Solidification temperature, K		
Polybutene for SCL (TU 38 10-17-43-81)	340-350	830-850	3-10	≥403	60-100			1.5	≤ 0.1	≤ 243		
Polybutene by Amoco Co., USA	370	832	5.2	424	100			0.5	0.02	243		
Polybutene by Sumitomo Co., Japan	370	837	7.2	411	99.2			0.5	0.15			
Polybutene by Nichiman Co., Japan	428	842	10.6	411	73			0.5	0.01			
										ρ_V at 373 K, Ohm·m	ε at 1000 Hz	Tanδ at 50 Hz and 373 K
Capacitor octol of A brand, GOST 12869-77			75-115	≥438			Absent	1		$2.5\cdot10^{11}$	2.0-2.2	$1\cdot10^{-4}$ ($\leq1\cdot10^{-3}$)
Polybutene Indopol H-50 E by Amoco Co., USA	720		103	441			Absent	0.5		$6.1\cdot10^{11}$	2.12 (298 K)	$5\cdot10^{-4}$
Polybutene HV-15E by Furukawa Chemical Co., Japan	700		87	451			Absent	0.5		$3.1\cdot10^{11}$	2.16 (353 K)	$7\cdot10^{-4}$
Polybutene Hivis-5 by British Petroleum Co., Great Britain	780		97-120	440			Absent	0.5		$2\cdot10^{11}$		$3\cdot10^{-4}$
Octol-200	700-1000		190-250	≥453	—	—	—					
Octol-K	700-1500		200-600	≥413	—	≤0.2	≤0.05					
Octol-component for sealing	700-1000	870-890	190-350	≥413	—	≤0.2						
Octol-100 succinimide (TU 38 10-1134-76)	800-1000	870-890	200-300	≥413	25-45	0.06		1.5	0.1			
Indopol H-1000 by Standard Oil Co., USA	915	889	210	467	37.4	0.02		0.5	0.01			
Oppanol BZ by BASF A.G., FRG	720	880	230	447	46.2	0.08		1	0.12			

[1] All brands are marked in accordance with standards of the country-producer.

Polybutene HV-100 by "Furukawa Chemical" Co., Japan	870	886	240	466	37.7	Absent		0.5	0.05			
Polybutene Hivis-10 by "British Petroleum" Co., Great Britain	930	882	240	471	29.7	0.03		0.5	0.13			
Octol-600 of brand A (TU 38 00-11-79-74)			≥550	≥493		≤0.05			≤0.05	Thickening ability in relation to oily basis recalculated per 5% solution, m^2/s	Resistance to depolymerization at 373 K (viscosity decrease), m^2/s	
Octol-600 of brand B (TU 38 00-11-79-74)			≥550	≥493		≤0.05			≤0.05		Without additive	With additive VNIINP –360
Viscous additive P-5 (TU 38 10-12-09-72)	4000-6000		≥1000	≥423		≤0.30						
P-10 (TU 38 10-12-09-72)	9000-15000		≥1000	≥438		≤0.1				3.5-8.5	$4\cdot10^{-6}$	$2\cdot10^{-6}$
Low-molecular polyisobutylene, P-20 (TU 38 10-32-57-80)	15000-25000					≤0.35		Light tones		13-29	$6.6\cdot10^{-3}$	
Thickening additive P-10 (electroinsulating)	9000-15000		≥500	≥438	10-16	≤0.02	Absent		≤0.07	ρ_V at 373 K and voltage ≥ 100 V, Ohm·m	tanδ at 50 Hz and 373 K	E at 373 K and 50 Hz, mV/m
										≥$5\cdot10^{10}$	$5\cdot10^{-3}$	≥120
High-molecular polyisobutylene (P-85, P-118, P-155, P-200), TU 13-303-67	For physicomechanical and electrical properties see Table 5.1									Heat resistance by Martance 338-353 K		
										Resistance to cold 213 K		
										Shrinkage 2-3%		

Note. On account of mixtures of C_4-olefins applied as the raw material, industrial brands of polyisobutylenes can represent copolymers of isobutylene with small amounts of other olefins.

traction gears in diesel engines. The octol-component of a sealing material participates as a plasticizer in prescription of the sealing belt.

P-50 brand is widely used in preparation of various glues, in particular, in production of adhesive belts.

Based on isobutylene and butylene oligomers, prescriptions of synthetic lubricating oils have been developed in the Soviet Union: TMP-200 oil for lubricating of fluid friction bearing of high-speed cold rolling mill of aluminum and alloys (it is able to undergo increased loads without carbonization at the metal annealing); Symbol-80 oil for traction of pipe from aluminum and alloys; high-temperature industrial oils VIP-130, MKM-110[2] for lubricating metalloceramic sliding bearings for rolling equipment, for example, calender, rolls, etc., operating at 463–483 K; Epol-200 oil for face thickenings of mixers in production of high pressure polyethylene, etc.

Table 8.3

Composition and characteristics of filled mixtures based on high-molecular isobutylene.

Index	Mixture					
	Facing				Laying	
	PSG	PS-2	GI	PT	PG	PTA
Composition of mixtures, mass parts						
PIB P-200	100	100	100	100	100	100
Technical carbon						
DG-100	100	150	100	—	—	—
PM-15	—	50	—	—	—	—
Graphite	100	—	100	—	100	—
Talc	—	—	—	300	—	100
Asbestos	—	—	—	—	—	150
Paraffin	—	5	20	—	—	—
Petrolatum oil	—	—	—	5	—	—
Physicomechanical properties						
Density, kg/m^3	1320	1370	1360	1530	1180	2430
Conditional strength at elongation, MPa	4.5-5.0	3.5-4.5	3.0-4.0	2.5-4.0	3.0-3.5	7.0-9.0
Relative elongation at break, %	475-550	150-200	275-600	150-300	430-480	18-20
Residual elongation, %	150-200	135-150	100-300	50-100	20-30	0
Shore hardness, A	67	88	55	50	30	93
Rigidity, N	65	90	40	21	33	33
Elastic restoration, mm	4	2	6	5	30	Non-elastic
Friability temperature, K	249	249	249	237	—	—

Note. Delivery is performed in plates 2.5-4 mm thick, no less than 800 mm wide and 1800–3000 mm long. Sheets are powdered by talc, chalk or layered with a polymeric film.

[2] TMP, VIP, MKM, etc. are Russian brands.

Super high-molecular polymer (MM over 1 million) is used in the shoe industry and for special purposes.

Polyisobutylene is combined with natural and synthetic rubbers, some elastomers (polyisoprene, butadiene copolymer with styrene, etc.), thermoplasts (polyethylene and polypropylene), waxes, mineral oils, bitumen, asphalt, etc. It is also combined with various mineral fillers and pigments (technical carbon, graphite, talc, magnesium oxide, zinc and titanium white, chalk). Introduction of fillers decreases cold fluidity, increases stability and hardness, and improves resistance to light.

Industrial plate materials designated "Plates from polyisobutylene of PSG brand" were produced in the Soviet Union, obtained by mixing of polyisobutylene, technical carbon and graphite. The same material in FRG is produced under the brand of Oppanol ORG, and Ecelestic Z in Great Britain. Modifications of this material differ in content of technical carbon, as well as stability (soft and hard PSG, TU 38-105203-70).

Typical composition and physicomechanical indexes of filled mixtures based on high-molecular polymer are displayed in Table 8.3.

The material PSG is used to protect equipment and building constructions from corrosion. Protective facings from PSG material are widely applied, in particular, in chemical industry (in production of acids, alkali, salts, fertilizers, chemical poisons, in aniline paint productions, etc.), which saves on envirnomentally scarce or potentially polluting materials such as lead, chromonickel steel, etc. Table 8.4 gives information on chemical stability of polyisobutylene of PSG brand [6].

High resistance to aging combined with water and impermeability makes PSG high-quality waterproofing material, which is appropriate to be used in construction of swimming pools, bridge constructions, dams, mines, tunnels, and other constructions, designed for durable operation without repair.

Non-hardening and non-drying sealing materials based on polyisobutylene of various brands, used in construction, include butyl rubber, ethylene–propylene copolymer, nitrile rubber and other elastomers as additives, which decrease fluidity of the composition at increased temperature, increase resistance and deformation properties, resistance to effects of aggressive media stabilizing stickiness and other properties. Mineral fillers, bitumen, tar, asphalt, highly boiling oils are also used in prescriptions. The latter promote homogenization and required working viscosity. In cases when expensive drying oils are used, it is necessary to provide for protection of the main mass of the sealing material from contact with air and to prevent creeping. Quite high adhesion of the sealing to concrete, bricks and other materials is increased with introduction into compositions of epoxy, phenoloformaldehyde, coumarone-indene resins, colophone, chlorinated or brominated butyl rubber, which, however, increase the cost price of the sealing. Tables 8.5–8.8 display typical compositions and characteristics of some simple and complex sealing materials, as well as comparative data of polyisobutylene-sealings of another type (Tables 8.9, 8.10) [6, 11].

Sealing materials based on oligo- and polyisobutylenes require a high degree of filling which does not preclude their durable exploitation even at low temperatures (down to 223 K). For example, 15–35 mass part of PIB (low-molecular), 25–35 of PIB (high-molecular), 100 of bitumen, 60 of mineral oil, 500–600 of milled limestone, 80–90 mass parts of asbestos are introduced into the sealing composition designed for thickening welds in large-panel concrete constructions. Another widely used non-drying sealing-mastic UMS-50 (5 mass parts of PIB P-118, 20 of neutral oil,

Table 8.4

Resistance of PSG material to aggressive media.

Medium	Concentration, mass%	Temperature, K	Medium	Concentration, mass%	Temperature, K
Inorganic acids			Salts and bases		
Nitric	30-508	323	Al, Fe, Sn, Mg, Ni sulfates, chlorides	Any	353
Boric	5-10	333	ammonia	Saturated water solution 100 (dry gas)	353 333
hydrobromide	48	353	ammonia nitrate, chloride, fluoride, sulfate	Saturated	333
silicic	Any	333	alumochromopotassium alum	Saturated	373
Hydroflurosilicic	Up to 32	333	lead acetate	Saturated	373
Arsenic	80	353	zinc chloride	Saturated	373
Sulfuric	Up to 80	333	Other inorganic substances		
Sulfurous	Any	353	water	Any	373
Hydrochloric	Over 30	353	hydrogen, oxygen, ozone, chlorine	Any	293
Chloric	10 and 20	333	fluorine	100	313
Phosphoric	80	373	hydrogen peroxide	Up to 30	293
Hydrofluoric	40 and 60	293-333	Organic substances		
Organic acids			aniline chloride	Saturated	293
Adipic	Technical	293	acetaldehyde	40 (water solution)	313
Benzoic	Any	333	glycerol	100	343
Formic	50 and 100	333	methanol, ethanol	96	333
Tartaric	Saturated	333	formaldehyde	40 (water solution)	333
Acetic	80	313			
Chloroacetic	80	363			

75 mass parts of chalk) is more effective than its competitors competition among sealing rubber composites. The most typical drawbacks of sealing materials based on PIB are low cohesive stability, thermoplasticity, fluidity under load, low resistance to oils, fats, lubricants and many solvents.

Table 8.5

Composition of polyisobutylene profile sealing materials.

Ingredients	Sealing composition, mass%	
	Solid	Less solid
PIB and vegetable oil (7:3)	9	10
Petrolatum oil	17	15
Asbestos fiber	16	—
Asbestos mill	58	44
Quartz mill	—	31

Table 8.6

Characteristics of non-solidifying sealing materials of various types.

Sealing	Penetration, mm^{-1}	Density, kg/m^3	Dry residue, mass%	Bond strength with metal at shear and break, MPa
Plastic or plastoelastic mass	70-120	1170-1700	≥ 90	≥ 0.01
Molded belts, braids, etc.	200-300	1170-1700	≥ 90	≥ 0.01

Table 8.7

Prescriptions of complex polyisobutylene sealing materials.

Components	Content of examples, mass%		
Polyisobutylene	5	15	20
Chalk or limestone	85	78	76
Nitrile rubber SKN-18	1.67	0.83	0.17
Petrolatum	8.33	4.17	0.83
Concrete	1	2	3

Table 8.8

Properties of complex sealing materials from polyisobutylene.

Properties	Indexes of properties of examples, Table 8.7		
Strength at elongation, MPa	0.016	0.028	0.037
Relative elongation, %	19	29	38
Resistance to 5% HCl after 24 h			
Strength at elongation, MPa	0.016	0.018	0.016
Relative elongation, %	32	30	29

Table 8.9

Characteristics of non-drying sealing materials.

Brand	Basis	Range of exploitation temperatures, °C	Penetration, 10^{-1} mm	Storage duration, months	Application field
Thiocol paste	Thiocol DA	-35…+50	—	12	Aviation
Thiocol puttying	Thiocol DA	-35…+50	40-110	12	Aviation
51-G-3	Ethylene-propylene rubber	-60…+100 (up to +140 for short time)	45-70	12	Instrument making and radio engineering
51-G-4	The same	The same	160-200	36	The same
51-G-6	Polyisobutylene P-20 (low-molecular)	-50…+90	220-230	6	Car industry
51-G-7	The same	-50…+90	80-120	6	The same
U-20A	Polyisobutylene P-118 or P-80 (high-molecular)	-50…+70	60-90	12	Car industry and aviation
U-22	The same	-50…+70	80-100	12	The same
Mastic UMS-50	The same	-50…+70	125-150	Unlimited	Construction
Buteprol	Ethylene-propylene rubber + butyl rubber	-50…+70	—	—	The same
PI-KN	Degraded waste of SKI-3	-50…+100	60-120	12	Construction, car industry, agriculture

Table 8.10

Comparative properties non-drying sealing materials.

Indexes	Brand, company, country			
	Bestic, Prestik, Great Britain	Sigma, Sigma Rubb. Sealant 8480, Netherlands	51-G-7 (USSR)	PI-KN (USSR)
Appearance	Homogeneous plastic mass			
Density, kg/m^3	1080	1240-1280	1600	1700
Penetration (with the help of cone), 10^{-1} mm	40-110	—	80-120	60-120
Resistance to tearing off from steel, no less than, MPa	0.018	—	0.018	0.018
Adhesion to usual construction materials	Good	Good	Good	Good
Temperature limit of workability in air, °C	-30…+90	—	-50…+90	-50…+90
UV-radiation effect	Stickiness preserved	—	Stickiness preserved	Surface structuring with stickiness loss
Guaranteed storage time, months	—	6	6	12

High-molecular PIB mixed with polyethylene and polystyrene (displaying no cold fluidity) is used as an insulating material for high-frequency pulse cables of high voltage. Typical composition (in mass%) is: PIB – 20–60, polyethylene – 10–40, technical carbon – 25–50, microcrystalline paraffin – up to 5.

In the rubber industry, PIB is used in combination with natural and synthetic rubbers and fillers. Rubbers based on polyisobutylene display quite high physicomechanical indexes, possess increased heat resistance, resistance to ozone, water and gas impermeability and resistance to acids. Such rubbers are applied to prepare water-impermeable fabrics, coats, tents, acid resistant hoses, transporter belts, as well as to provide protection from aggressive media.

Glues based on high-molecular PIB are applied to join together wood, metal, glass, fibers, fabric, paper, films, and leather.

Due to the absence of double bonds in the polymeric chain (excluding the end ones), PIB is not susceptible to the usual (sulfur) vulcanization. However, under definite conditions, it can be structured forming vulcanizates with high physicomechanical characteristics. One f possible prescription is displayed below:

Composition, mass parts	
PIB P-200	100
Technical carbon (gaseous)	60
Neosone D	1
Manganese peroxide	4
n-Quinone dioxime	1
Physicomechanical properties of vulcanizates	
Conditional strength at elongation, MPa	23
Tension at elongation by 300%, MPa	4.2
Relative elongation at break, %	642
Residual deformation, %	52
Elasticity by recoiling, %	9
Shore hardness	60

Existence of end C=C-bonds in low-molecular polymeric products enables their functionalization to be performed according to various schemes, which is used in industrial synthesis for obtaining various additives to oils and fuels [1, 2, 8]. In particular, native and foreign additives are known: a) a washing-dispersing ashless compound (succinimides C-5A, Lubrisol 890, 894 by Lubrisol Company, OZOA 1200 by Orobis Company, etc.) – the product of the reaction between polyisobutylene amber anhydride and tetraethylene pentaamine or other polyamines; b) a metal-containing washing product (Ba phosphonate, Amoco 121 by Amoco Company, etc.) – the product of interaction between low-molecular polyisobutylene and P_2S_5 and further treatment by Ba oxide or hydroxide; c) anti-wear and anti-friction (sulfured octol, Hitech E 312 by E. Cooper Company, etc. Usually, due to decreased reactivity of double bonds in PIB (compared with low-molecular analogs) and necessity to use increased temperatures, processes for the synthesis of additives differ by insufficiently high yields of target products and require development of more modern methods.

8.2. *Specific and new fields of application of polyisobutylenes and composites derived from them.*

Introduction of functional groups and fragments into the PIB chain (modification by end double bonds, fragmentary, block- and grafted copolymers) and, in some cases, defunctionalization of PIB (elimination of double bonds), sufficiently broadens the technical possibilities and fields of application of polymeric products derived from isobutylene. This is bound to the fact that functionalization, beside conservation of basic useful properties of PIB, assigns new characteristics to them. For example, compared with unsaturated polymers, low-molecular PIB, containing heteroatomic fragments at the chain end, are characterized by increased resistance to oxidation (Table 8.11) [12].

Table 8.11

Oxidation of functional derivatives of PIB by oxygen (473 K, P_{O_2} = 250 mm Hg).

End groups in PIB	Induction period of oxidation, min
tert-C_4H_9~, ~$C(CH_3)=CH_2$, ~$CH=C(CH_3)_2$	0
~COOH, ~$C(O)CH_3$	10
~OH	35
~NH_2	45
~$C(O)SP(O)(OR)_2$	55
~C_6H_4OH	135

High resistance to oxidation, mechano- and thermodegradation characterize polyisobutylenearomatic compounds [13–17] and products of hydrogenation of oligoisobutylenes (white oils) [18–20], applied in various branches of industry, medicine, etc. In particular, products of catalytic hydrogenation of PIB by hydrogen or ionic (electrophilic) hydrogenation by aluminum chloride–isopropylbenzene system are suitable for insulation of capacitors with increased stability, showing 3–4 times higher durability than those involving unsaturated PIB [15, 18]. They are also used as lubricants in various branches of industry, medicine, cosmetics [19].

Oils based on polyisobutylenearomatic compounds differ by low temperature of solidification [21]. As additives to synthetic and mineral oils, these compounds are characterized by increased thickening properties [22].

The presence of phenolic fragments and phosphite derivatives, obtained from them, in the composition of functionalized PIB imparts properties of antioxidants and stabilizers to the polymeric products for the sequence of chlorine-containing and diene polymers, as well as monomers (Tables 8.12–8.14) [17, 22–25]. As seen from the Tables, polyisobutylenephenols and polyisobutyleneaminophenols show significantly greater effects, compared with low-molecular 2,6-di-*tert*-butyl-4-methylphenol. Phosphorylated polyisobutylenephenols and aminophenols display high inhibiting activity in processes of butadiene (piperylene) thermopolymerization and thermooxidation of rubbers.

PIBs with end carbonyl and carboxy groups, that come from ozonolytic degradation of butyl rubber, possess properties of efficient viscous additives to oils, moreover, with higher thickening ability than that of its unsaturated analog.

Increasing the viscosity index, they do not affect the freezing temperature of the oil, but improve its thermal and mechanical stability [26, 27].

PIB derivatives with end N, S and P-containing groups:

$$\sim HN-C_6H_4-NH_2,\ \sim S-C\overset{N}{\underset{S}{\langle\rangle}}C_6H_5,\ -NH-(CH_2)_6NH_2,\ -S-\underset{\overset{\|}{S}}{P}(iso\text{-}C_3H_7O)_2,\ \text{etc.}$$

synthesized by chemical transformations of the above- mentioned compounds, display multifunctional action in lubricating oils, possessing not only thickening, but also antioxidant anti-frictional properties [27]. Creation of all-season oils on their basis is in prospect.

Viscous additives may be low-molecular copolymers of isobutylene with α–methylstyrene, *para*-fluorostyrene and other vinylaromatic and carbocyclic compounds [28, 29]. PIB derivatives with end OH- and COOH-groups show potential for creation of liquid rubbers of a new type [30]. Such liquid rubbers, obtained by the method of ozonolytic degradation of isobutylene copolymer with piperylene, are produced by Enjay Company (USA) on the industrial scale. Combination of unsaturation of isobutylene chain, imparting high chemical resistance to articles, gas impermeability, compatibility with several polymers, with advantages of processing of liquid rubbers, containing highly reactive groups, in functional oligomers must provide various fields of application for these compounds.

Solidification of oligoisobutylene-dicarbonic acid by epoxy compounds in the presence of tertiary amines as the catalyst is accompanied by formation of rubbers with a regular 3D-network [31]. Using isobutylene polymers and copolymers with amine and amidoamine groups, elastic polyureas, polyuretanes and polyepoxides were obtained [32, 33].

From the point of view of regulating properties of spatial and filled polymeric systems (strength, permeability, etc.), block- and grafted copolymers (thermoplasts) are characterized by wide possibilities. Separately, Si-containing products must be noted. PIB with reactive (hydrolyzing) silicon-containing end groups forms weather-resistant, waterproof and heat resistant gas impermeable electroinsulating materials [34]. Hydrolysis (condensation) of tri- and tetraetoxilanes in the matrix of preliminarily cross-linked PIB enables structures to be strengthened, which is bound to higher ability to deformation of filler particles and increase in its protective qualities [35, 36]. Taking into account low values of diffusion coefficients of various gases (H_2, O_2, N_2, Cl_2) for PIB compared with natural, neoprene and butyl rubbers [37], the mentioned examples significantly broaden abilities of polyisobutylene in creation of protective polymeric coatings.

As in several application fields, such as protection from corrosion, sealing, etc. (production of thickening and film materials, compounds, sealing materials, unsaturated mastics, sticky belts, consistent lubricants), PIB is used mixed with fillers. Existence of functional groups in the polymer is of great importance for improving compatibility of components in mixture. Such traditional drawbacks of PIB as low cohesion stability, instability in oils, fats and may solvents, and low adhesion may be eliminated, apparently, by changing polarity of macromolecules at functionalization. For example, films from isobutylene copolymer with low content of chloronorbornene display 10-fold higher adhesion strength to aluminum, than PIB [38]. Adhesive properties of PIB can be improved by its liquid-phase oxidation [39].

Table 8.12

Influence of high-molecular phenols and aminophenols on thermal and color stability of halogen-containing polymers (175°C, in air), c_0 = 1 mass part.

Compound	Thermal and color stability of halogen-containing polymers*			
	PVC	CPE	SCEPR	CSPE
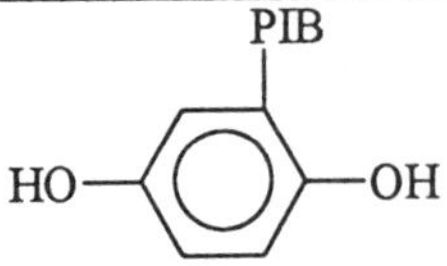	53/2.0	—	44/1.0	—
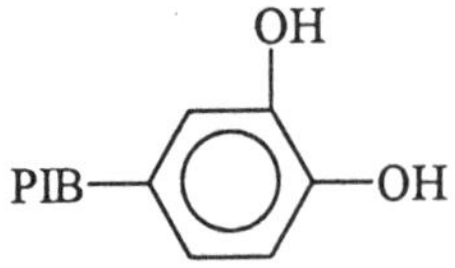	56/2.0	48/2.0	45/1.0	—
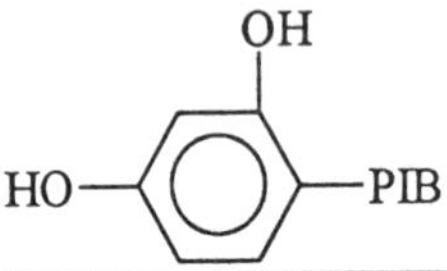	—	—	45/1.5	34/2.5
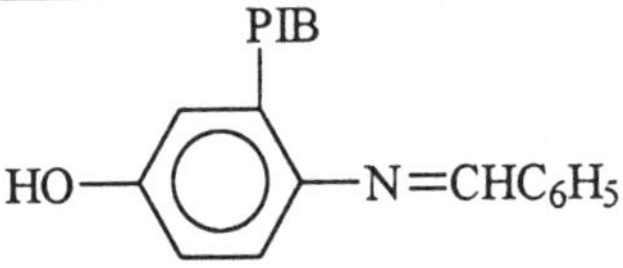	41/2.0	40/2.0	31/2.5	27/2.5
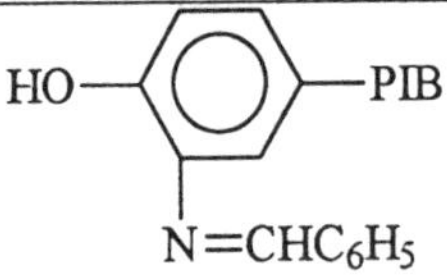	42/2.0	39/2.0	33/2.0	28/2.5
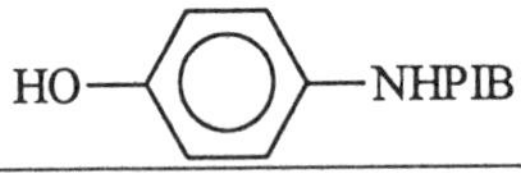	40/2.0	42/2.0	35/2.0	28/3.0
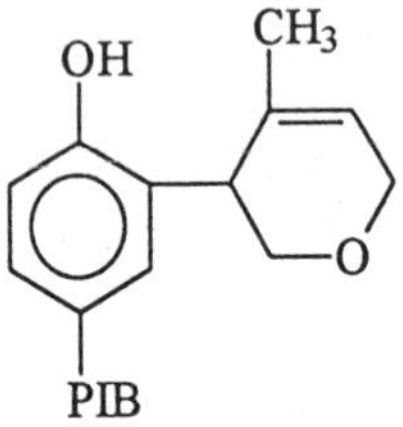	54/1.5	58/1.5	—	47/1.5
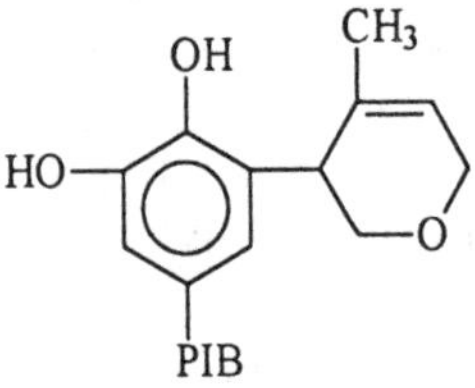	79/1.5	61/1.5	—	51/1.5

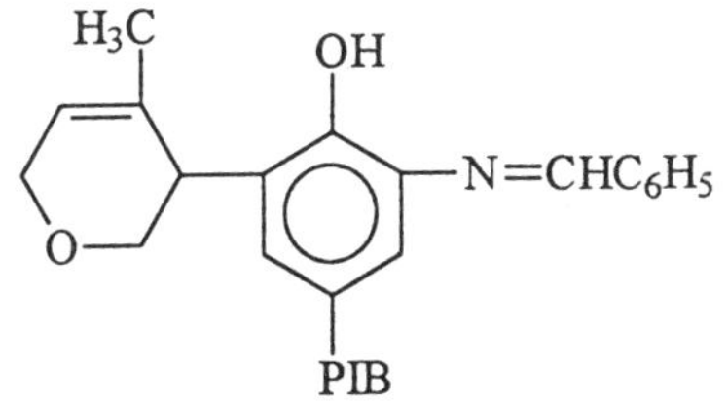	43/2.0	—	33/2.5	29/2.5
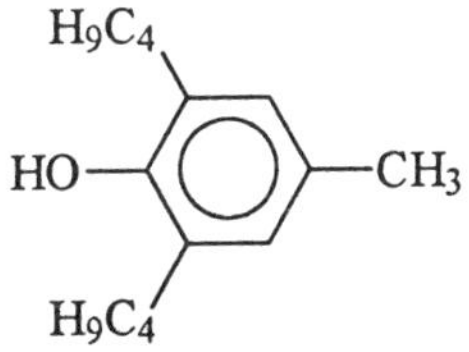	20/4.5	24/4.5	20/4.5	17/5.0

* Numerator marks time of thermal stability τ, min; denominator marks color stability is points of Synmero scale.

Note. PVC – polyvinyl chloride; CPE – chlorinated polyethylene (33 mass% of Cl); SCEPR – synthetic chlorinated ethylene-propylene rubber (25.6 mass% of Cl); CSPE – chlorosulfopolyethylene (29.5 mass% of Cl, 1.4 mass% of S).

Table 8.13

Comparative inhibiting activity of high-molecular phosphorylated phenols and aminophenols during thermal polymerization of dienes.

Monomer	Inhibitor	Inhibitor concentration, mass%	Conditions of inhibiting		Inhibiting effect, %
			τ, h	I, °C	
Butadiene	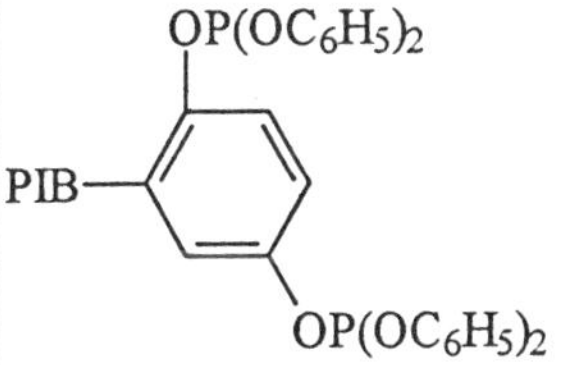	0.005	24	70	99.4
		0.030	24	70	99.5
		1.000	24	70	99.5
Piperylene	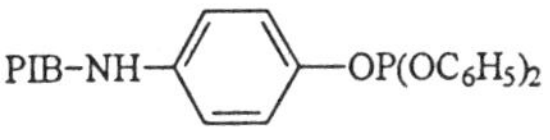	0.005	24	80	99.3
		0.030	24	80	99.5
		1.000	24	80	99.7

New possibilities to apply PIB are presented by compositional modification of its properties. Resistant and thermophysical properties of PIB can be regulated by introducing powders of metal carbonyls (Fe, Cu), as well as copper ferrites [40, 41]. Highly filled (up to 90 mass%) composites were likely obtained.

Valuable complex of properties (combination of high viscosity index, low temperatures of solidification with heat resistance), realized in composites of lubricating-cooling fluids and synthetic oils, is stated for mixtures of low-molecular PIB with polyethylsiloxanes (PES) [42–44]. Of interest is the fact of unlimited compatibility of PIB and PES up to MM values of polyethylsiloxanes of ~1,500 [45]. In the general case, compatible practically useful mixtures of PIB with polysiloxanes, especially polydimethylsiloxanes, are obtained under application of combining additives — block-copolymers of mentioned monomers [42, 44].

Table 8.14

Efficiency characteristics of high-molecular phosphites at oxidation by SKI-3 and SKMS-30 (140°C, 30 min).

Stabilizer	Stabilizer concentration, mass%	PPI, %	
		SKI-3	SKMS-30
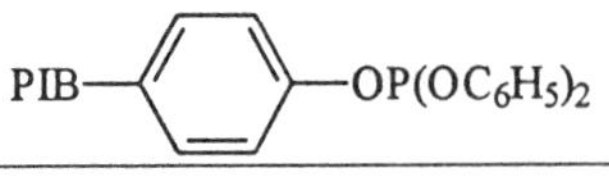	0.2	75	87
	0.4	80	90
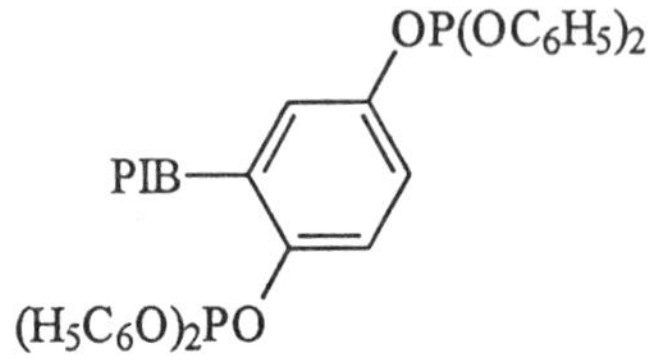	0.2	78	90
	0.4	82	93
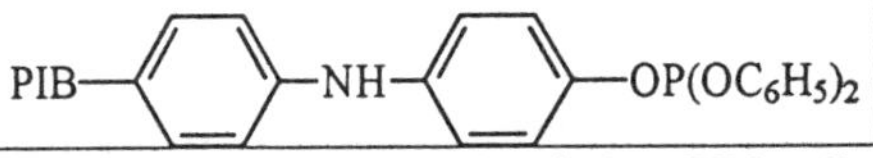	0.2	81	91
	0.4	83	94

<u>Note.</u> Plasticity preservation index $PPI = P_{ag}/P_0 \cdot 100\%$, where P_0 and P_{ag} are Walles indices of plasticity of rubber samples before and after aging, respectively; SKI-3 – polyisobutylene rubber, SKMS-30 – butadiene copolymer with 30 mass% of styrene.

Application fields of PIB are broadening, bound to solution of new technical, including ecological tasks. Oligoisobutylnes with various end groups (hydroxy, sulfo-, epoxy, etc.) are used as resistant to aging linkages for missile propellants, instead of previously used polybutadienes [46]. In this case, workability of the linkage at ~350 K is 4 months.

Modified low-molecular PIB of P-20M brand is suitable for restoration of exploitation properties of worked-off industrial lubricating oils [47]. High-molecular PIB as a powder is a non-toxic agent, Elastol, for collecting oil [48]. It imparts viscoelastic properties to oil, increased adhesion and makes collecting of oil with the help of a disk foam remover simpler.

Non-toxic water emulsions of PIB for medicinal purposes were developed using complex saccharose and fatty acid esters, complex esters of polyoxyethylene sorbite and fatty acid, ammonia salt of branched dodecylbenzeneculfoacid, etc. as additives [49]. The emulsion preserves humidity and weakly adheres to skin, i.e. it is suitable for application.

A new dynamically developing field of PIB application is connected with its specific stickigenic properties, which has been used for a long time in preparation of sticky belts and other materials with gluing properties. Stickiness of materials is partially bound to adhesion; therewith, a transition from adhesive break to cohesive one takes place in the sphere of the maximal stickiness effect [50]. In this connection, values of so-called low-temperature caking of polymers are demonstrative, i.e. decrease of the surface area of polymer powders after finish grinding in a vibromill and further temperature increase [51].

For PIB, this index significantly exceeds these of other polymers (polymethylmethacrylates, polystyrene, polyethylene, polypropylene, etc.). As low-temperature caking proceeds at overcoming the barrier for rotation around carbon-carbon bonds and at obtaining a definite mobility by macrochains, it is connected with T_g and cohesive strength of the polymer, which are some of the lowest for PIB. Considering the above-said, as well as high water repellence and resistance to atmospheric factors of the polymer, difficulty of oxidation, etc., PIB is suited to the requirements of entomological (non-solidifying) glues [52, 53]. These glues represent a new variety of glues, the main requirement to which is high and desirably positive with time stickiness, not complicated by secondary factors [53]. As the main application field of entomological glues is gluing (fixing) support in modern ecologically friendly biotechnological means for protection of cultural plants from pests, they must also be characterized by low dissolving ability, absence of coloring and repellent properties, i.e. biocompatibility providing registration and identification of insects. The properties mentioned are completely inherent in PIB, which, combined with the ability to preserve surface stickiness over a long period (months and years), make it one of the most attractive substances as the basis of entomological glues. Evidently, not by accident the glue of Bird Tanglefoot type (Zoecon Co., USA) is one of the best. It represents a composition based on polyisobutylene.

Although prescriptions of entomological glues based on PIB depend on the method of usage or their application on the object (trap): manual, mechanized, spraying, i.e. on the required level of technological properties of the glue, it is necessary, as a rule, to introduce structure forming additives providing creation of a spatial net in the system and appearance of the fluidity limit under exploitation temperatures of 293-333 K [54, 55]. The best among them are waxes and wax-like materials with limited compatibility to PIB, especially, natural wax, low-molecular polyethylene. They are efficient at concentrations over 4 mass%. Although PIB-wax systems are characterized by thixotropic properties (non-Newtonian fluids) and hysteresis phenomena, finally, their properties at heating and cooling change strictly reversibly, the basic parameter in this case, entomological stickiness (0.5–0.8 kGs/cm^2), being firmly reproduced.

It must be noted that properties of PIB-wax compositions can be directly regulated by various additives for functional purposes. For example, butyl rubber increases gluing properties and provides fixation of big insects by glue [53]. Fat-soluble paints, in particular, 3-methyl-5-hydroxy-1-phenyl-4-(*o*-diiso-metaxylyl) pyrasol, impart additional biological activity owing to yellow-green range of the spectrum, attractive for some insect-pests [56, 57]. Application of the propellent of C_4 hydrocarbons fraction as a solvent enables the aerosol variant of the gluing composition to be realized.

Entomological glues based on PIB participate in the composition of gluing, pheromone-gluing traps, insect-trapping belts and other biotechnological protection means for agricultural, green-house plants and forests [58–60]. Combined with classic methods of plant protection (chemical and biological), they provide high efficiency of plant growing for producing ecologically pure products, especially under conditions of green-housing farms.

The application fields of PIB and its derivatives mentioned are far from exhausting their abilities. Numerous examples of thermoplasts and elastomers of various purposes are known, in which PIB is used as a technological additive improving processing and separate properties of materials. Varying of molecular masses and the type of functionality provides technological or higher level of PIB compatibility with components of composites, which determines the high quality of the articles produced.

References to Chapter 8

1. Guterbock, G., Polyisobutylene and isobutylene copolymers, 1962, 363 p., Leningrad, Gostoptekhizdat. (Rus)
2. Minsker, K.S. and Sanglov, Yu.A., Isobutylene and Its Polymers, 1986, 224 p., Moscow, Khimia. (Rus)
3. Kaplan, S.Z. and Radzevenchuk, I.F., Viscous Additives and Thickened Oils, 1982, 136 p., Leningrad, Khimia. (Rus)
4. Prokofiev, I.K., Verbitsky, B.G., Rogov, S.A., and Kirichenko, L.N., Low-molecular Polybutenes, Them. Review, Ser. "Oil Chemistry and Flange Processing", 1982, 51 p., Moscow, CNIITENeftekhim. (Rus)
5. Anosov, V.I., In: Synthetic Rubber, 1976, pp. 328–341, Leningrad, Khimia. (Rus)
6. Labutin, A.L., Anti-corrosion and Sealing Materials Based on Synthetic Rubbers, 1982, 213 p., Leningrad, Khimia. (Rus)
7. Polymeric Mixtures, Ed. by D. Pole and S. Newman (Transl. From English), 1981, v. 1, 550 p., Moscow, Mir. (Rus)
8. Wipper, A.B., Vilenkin, F.V., and Gaisner, D.A., Foreign Oils and Additives, 1981, 192 p., Moscow, Khimia. (Rus)
9. Bekker, R., Coating, 1978, v. 11(9), p. 306.
10. Jordan, R., Adhasion, 1980, Bd. 24(5), S. 120, 122–124; Bd. 24(9), S. 256, 258–262.
11. Shitov, V.S., Ryazanova, M.P., Khodikova, L.V., Anti-corrosion Ebonite Covers and Non-drying Sealing Materials, Ser. Synthetic Rubber Industry, 1990, iss. 7, 54 p., Moscow, CNIITENeftekhim. (Rus)
12. Kirpichnikov, P.A., Highmolec. Comp., 1979, v. A21(11), pp. 2457–2468. (Rus)
13. Yasman, Yu.B., Sangalov, Yu.A., Prokofiev, K.V., and Nelkenbaum, E.M., Chem. and Technol. of Petroleum and Oils, 1983, No. 8, pp. 5–7. (Rus)
14. Sangalov, Yu.A., Yasman, Yu.B., and Minsker, K.S., In: Thes. Rep. V Intern. Microsymp. on Ionic Polymerization, 1982, p. 29, Prague. (Rus)
15. Grigoriev, V.V., Suvorov, B.V., Sangalov, Yu.A., Prokofiev, K.V., and Uzunkoyan, P.N., In: Thes. Rep. 7th All-Union Conf. on Catalytic Reactions in the Liquid Phase, 1988, p. 53, Alma-Ata. (Rus)
16. Akhmedov, A.I., Levshina, A.N., and Sadyikov, Z.I., Heftekhimia, 1979, pp. 134–139. (Rus)
17. Roney, J.M., J. Appl. Sci., 1980, v. 25, pp. 1365–1372.
18. Yasman, Yu.B., Sangalov, Yu.A., Prokofiev, K.V., Khudaiberdina, Z.I., Gladkikh, I.F., and Nelkenbaum, E.M., Neftekhimia, 1983, v. 23(4), pp. 500–507. (Rus)
19. Weisser, O., Chem. Prum., 1979, v. 29(7), pp. 360–364.
20. Yasman, Yu.B., Khudaiberdina, Z.I., and Sangalov, Yu.A., In: Thes. Rep. VI All-Union Conf. on Catalytic Reactions in the Liquid Phase, 1983, pp. 37–38, Alma-Ata. (Rus)
21. Sangalov, Yu.A., Yasman, Yu.B., Nelkenbaum, E.M., Kozlov, V.G., and Minsker, K.S., Highmolec. Comp., 1983, v. A25(2), pp. 227–231. (Rus)
22. Prokofiev, K.V., Alekseeva, L.P., Lobanova, G.N., and Sangalov, Yu.A., Chem. and Technol. of Petroleum and Oils, 1985, No. 2, pp. 27–28. (Rus)
23. Biglova, R.Z., Malinskaya, V.P., and Minsker, K.S., Highmolec. Comp., 1994, v. 36(8), pp. 1276–1280. (Rus)
24. Biglova, R.Z., Malinskaya, V.P., Sagitdinova, Kh. F., and Minsker, K.S., Highmolec. Comp., 1995, v. A37(9), pp. 1474–1481. (Rus)

25. Mukmeneva, N.A., Cherezova, E.N., Biglova, R.Z., Malinskaya, V.P., and Minsker, K.S., Highmolec. Comp., 1997, v. A39(6), pp. 953–959. (Rus)
26. Kaplan, S.Z. and Padzevenchuk, I.F., Viscous Additives and Thickening Oils, 1982, 136 p., Leningrad, Khimia. (Rus)
27. Beresnev, V.V., Stepanov, E.A., and Kirpichnikov, P.A., Chem. and Technol. of Petroleum and Oils, 1991, No. 4, pp. 37–39. (Rus)
28. Akhmedov, A.I., Chem. and Technol. of Petroleum and Oils, 1986, No. 12, pp. 36–39. (Rus)
29. Aliev, F.M., Synthesis and Transformations of Elementorganic Compounds, 1987, pp. 11–15, Baku. (Rus)
30. Mogilevitch, M.M., Turov, B.S., Morozov, Yu.L., and Ustavshchikov, B.F., Liquid Hydrocarbonic Rubbers, 1983, 200 p., Moscow, Khimia. (Rus)
31. Zapirov, I.N., Nasyibullin, Sh.A., Lazareva, L.M., and Faizullin, I.N., Highmolec. Comp., 1980, v. A22(7), p. 640. (Rus)
32. Patent No. 3608975 (FRG), MKI C 08 F 8/03, C 08 F 210/10, 1987.
33. Beresnev, V.V., Yunusov, O.A., and Kirpichnikov, P.A., Kauchuk i Resina, 1981, No. 3, pp. 10–12. (Rus)
34. Claim No. 6438407 (Japan), MKI C 08 F 8/42, No. 62–194938, 1989.
35. Sun, C.C. and Mark, J.E., Chem. Soc. Polym. Prepr., 1986, v. 27(2), pp. 230–231.
36. Sun, C.C. and Mark, J.E., J. Polym. Sci., Polym. Phys., 1987, v. 13(7), pp. 1561–1564.
37. Sukhareva, L.A. and Kipnis, Yu.B., Protective Polymeric Coatings in Production of Artificial Leather, 1989, 256 p., Moscow, Khimia. (Rus)
38. Sangalov, Yu.A., Nelkenbaum, Yu.Ya., Khusnutdinov, R.I., Jemilev, U.M., and Minsker, K.S., Highmolec. Comp., 1984, Ser. A, v. 26, pp. 600–605. (Rus)
39. Chernyak, B.I., Kripp, I.M., Miklosh, R.B., and Britan, M.S., Theoretical and Experimental Chemistry of Catalytic Reactions and Polymers, 1987, pp. 84–87, Cheboksary. (Rus)
40. Makhmudbekova, N.L., Pavlov, A.P., and Gumerov, A.G., Thes. Rep. of All-Union Sci.-Tech. Conf., v. 2, 1986, p. 61, Moscow. (Rus)
41. Makhmudbekova, N.L., Kurilkin, N.E., Lyutikova, I.V., and Pavlov, A.P., Plasticheskie Massy, 1988, No. 6, pp. 35–37. (Rus)
42. Prokofiev, K.V., Nelkenbaum, Yu.Ya., and Sangalov, Yu.A., Zh. Prikl. Khimii, 1989, v. 62(2), pp. 436–438. (Rus)
43. Sangalov, Yu.A., Nelkenbaum, Yu.Ya., and Prokofiev, K.V., Interdep. Coll. Phys.-Chem. Bases of Synthesis and Processing of Polymers, 1977, pp. 52–56, Gorky. (Rus)
44. Sangalov, Yu.A., Nelkenbaum, Yu.Ya., and Prokofiev, K.V., Chem. and Technol. of Petroleum and Oils, 1986, No. 5, pp. 31–34. (Rus)
45. Sangalov, Yu.A., Nelkenbaum, Yu.Ya., Prokofiev, K.V., Sadyikov, R.A., and Shishlov, N.M., Highmolec. Comp., 1987, Ser. A, v. 29(3), pp. 583–587. (Rus)
46. Harman, K.O. and Bohrer, E.K., Polym. Bull., 1987, v. 18(5), pp. 403–410.
47. Kuznetsova, M.B., Askarov, R.Z., and Beresnev, V.V., In: Thes. Rep. All-Union Sci.-Techn. Conf. "Chemmotology–90", 1990, p. 159, Dnepropetrovsk. (Rus)
48. Fingas, M. and Tennyson, E., Spill Technol. News Lett., 1988, v. 13(2), pp. 40–46.
49. Claim No. 59-122534 (Japan); MKI C 08 L 23/22, A 61 L 15/00.
50. Tiyama Mitsuo, Technol. Adhes. and Seal., 1990, v. 34(7), pp. 316–324.
51. Khainike, G., Tribochemistry, 1987, 680 p., Moscow, Mir. (Rus)

52. Vakhitov, V.A., Amirkhanov, D.V., and Sangalov, Yu.A., In: Novel Means and Methods for Plant Protection, 1992, 154 p., Ufa. (Rus)
53. Sagalov, Yu.A., Romanko, T.V., and Musin, M.A., Bashk. Khim. Zh., 1996, v. 3(4), pp. 17–22. (Rus)
54. Sangalov, Yu.A., Nelkenbaum, Yu.Ya., Katsyutsevitch, E.V., Romanko, T.V., Ponedel'kina, I.Yu., Kolloid. Zh., 1992, v. 54(4), pp. 153–158. (Rus)
55. Sangalov, Yu.A., Romanko, T.V., Ponedelkina, I.Yu., Kutsyuntsevitch, E.V., Colloid. J., 1995, v. 57(5), pp. 695–699.
56. Sangalov, Yu.A., Katsyutsevitch, E.V., Romanko, T.V., and Musin, M.A., Protection of Plants, 1995, No. 10, p. 26. (Rus).
57. Sangalov, Yu.A., Nelkenbaum, Yu.Ya., and Odinikov, V.N., Protection of Plants, 1993, No. 2, p. 28. (Rus)
58. Sangalov, Yu.A., Romanko, T.V., and Nelkenbaum, Yu.Ya., In: Thes. Rep. XVI Symp. "Rheology-92", 1992, pp. 139–141. (Rus)
59. Sangalov, Yu.A., Romanko, T.V., Katsyutsevitch, E.V., and Musin, V.A., In: Thes. Rep. XVII Intern. Symp. on Rheology, 1994, p. 130, Saratov. (Rus)
60. Sangalov, Yu.A., Musin, M.A., Antonova, N.E., and Turianov, R.A., In: Thes. Rep. VIII Symp. on Rheology, 1996, p. 97, Karacharovo. (Rus)

Subject Index

C

D

E

F

G

H

I

K

L

L

M

N

O

P

Q

R

S

T

U

V

W

Z